International Conference on Mechatronics in Energy and Environment Protection (ICMEEP 2020)

IOP Conference Series: Materials Science and Engineering
Volume 995

Erode, India
16 - 17 October 2020

ISBN: 978-1-7138-3305-5
ISSN: 1757-8981

Printed from e-media with permission by:

Curran Associates, Inc.
57 Morehouse Lane
Red Hook, NY 12571

Some format issues inherent in the e-media version may also appear in this print version.

Institute of Physics
Dirac House, Temple Back
Bristol BS1 6BE UK

Phone: 44 1 17 929 7481
Fax: 44 1 17 920 0979

techtracking@iop.org

Additional copies of this publication are available from:

Curran Associates, Inc.
57 Morehouse Lane
Red Hook, NY 12571 USA
Phone: 845-758-0400
Fax: 845-758-2633
Email: curran@proceedings.com
Web: www.proceedings.com

TABLE OF CONTENTS

IOP Conf. Series: Materials Science and Engineering **995** (2020) 011001 doi:10.1088/1757-899X/995/1/011001

Conference Message

The International Conference on Mechatronics in Energy and Environment Protection - ICMEEP 2020 is held during 16 - 17 October 2020 at Bannari Amman Institute of Technology, Erode through virtual mode. This ICMEEP conference is planned to be organized as an annual series in future and this is the first conference in the series which is organized by the Department of Mechatronics of Bannari Amman Institute of Technology.

The concept of the conference ICMEEP is to bring together leading academicians, scientists, researchers and industrialists and provide them with a forum to exchange and to share their experiences and research findings in Mechatronics, related to energy savings and environmental protection for the growth of the society. Research on topics related to Mechatronics applications in the energy and environmental problems is of significance because they impact the existence of all living organisms.

Many authors from reputed institutions have submitted their papers for this ICMEEP conference. A total of 61 articles have been received and 46 articles are accepted for oral presentation in the virtual conference. All the received papers were undergone peer review by two members of the program committee. Finally program chairs who handle the articles made the final acceptance decisions.

The conference program also includes two keynote sessions from Dr. S. Dhinakaran, R&D, Tata Steel, United Kingdom and Dr. Khalid Rafi, Senior Additive Manufacturing Program Development Lead, ASTM.

Finally, I would like to take this opportunity to thank all members of the organizing committee, the technical program committee, keynote speakers, reviewers and the faculty of the Department of Mechatronics, Bannari Amman Institute of Technology.

Dr. K. L. Senthil Kumar
Organizing secretary

IOP Conf. Series: Materials Science and Engineering **995** (2020) 011001 doi:10.1088/1757-899X/995/1/011001

International Conference on Mechatronics in Energy and Environment Protection | ICMEEP 2020
16 - 17 October 2020

About the Conference

The rapid expansion of energy and environment sectors is necessary for the inclusive growth of the world. At the same time, industries are thriving towards a paradigm shift to ensure sustainability through the revolution of automation. This conference aims to bring together leading academicians, Scientists, researchers and industrialist and provide them with a forum to exchange and to share their experiences and research findings in mechatronics related to energy saving and environmental protection, for the growth of the society.

Topics

- Additive manufacturing
- Artificial intelligence
- Bio-fuels
- Biomass and environmental degradation
- Biomaterials
- Biomimetics
- Composite materials
- Computational intelligence
- Control system
- Energy conservation
- Factory automation
- Green manufacturing
- Health monitoring of E-vehicles
- Human robot interaction
- Industry internet of things
- Machine vision
- MEMS/NEMS
- Micro robots/ nano robots
- Modelling and simulation
- Nanomaterials
- New product development
- Optimization
- Phase change materials
- Precision manufacturing
- Product design and development
- Rehabilitation robots
- Semiconductor materials
- Smart materials
- Soft robotics
- Solar photovoltaic
- Solar thermal system
- Surface treatment
- Tribotronics
- Wind power

IOP Conf. Series: Materials Science and Engineering **995** (2020) 011001 doi:10.1088/1757-899X/995/1/011001

Advisory Committee

- Dr C Palanisamy Principal
- Dr K Sivakumar Dean - PDS
- Dr S Jegadheeswaran Professor - Academics R&D

Keynote Speakers

- Dr H Khalid Rafi, Senior lead ASTM International – AM Singapore
- Dr S Dhinakaran Project Lead R&D Tata Steel Swansea UK

International Committee

- Dr Hal Gurgenci Professor University of Queensland
- Dr Shankar Raman Universidad Nacinoal Autonomade Mexico Mexico
- Dr Adarsh Kumar Pandey Sunway University Malaysia
- Dr Wilfrido Rivera Goez Franco Universidad Nacinoal Autonoma de Mexico
- Dr Senthilarasu Sundaram University of Exeter UK
- Dr Mahendran Samykano Universiti Malaysia Pahang Pahang
- Dr K Kardigarma Universiti Malaysia Pahang Pahang
- Dr Marwan Brouche Saint-Joseph University of Beirut
- Ir Ts Dr Mohamed Thariq Bin Haji Hameed Sultan Universiti Putra Malaysia

National Committee

- Dr S Aravindan Profesor IIT Delhi
- Dr N Arunachalam Associate Professor IIT Madras
- Dr S Dennis Ashok Professor VIT Vellore
- Dr M. Ramji Professor IIT Hyderabad
- Dr M Jayaprakash Assistant Professor IIT Indore
- Dr K Ponappa Assistant Professor IIITDM Jabalpur
- Dr B Premachandran Professor IIT Delhi
- Dr J Ramkumar Professor IIT Kanpur
- Dr M Santhakumar Associate Professor IIT Palakkad
- Dr A K Sharma Associate Professor IIT Roorkee

IOP Conf. Series: Materials Science and Engineering **995** (2020) 011001 doi:10.1088/1757-899X/995/1/011001

Organizing secretary

- Dr K L Senthil Kumar Professor and Head Mechatronics

Convenors

- Dr A Megalingam @ Murugan Associate Professor Mechatronics
- Mr S K Dhinesh AP (Sr Gr) Mechatronics

Organizing Committee

- Dr M C Pravin, AP (Sr Gr) Mechatronics
- Dr P Nagarajan, AP (Sr Gr) Mechatronics
- Mr K S Hanumanth Ramji AP (Sr Gr) Mechatronics
- Dr K Nijandhan, AP Mechatronics
- Mr.S.Vignesh AP Mechatronics
- Ms K K Megavarthini AP Mechatronics
- Mr A Mohamed Ismail AP Mechatronics
- Mr K Yuvaraj AP Mechatronics
- Mr S Ragavendra Prabhu AP Mechatronics
- Mr V Vadivel Vivek AP Mechatronics

Review Committee

- Dr.D.Kesavan IIT Palakad
- Dr.P.Jeyaraj NIT Karnataka
- Dr.K.Hariharan IITMadras
- Dr.K.L.Senthikumar BIT
- Dr.A.Megalingam BIT
- Dr.Ramathirumurugan MCET
- Dr.S.Shankar KEC
- Dr.M.C Lenin Babu VIT
- Dr.G.Rajamurugan VIT
- Dr.Ponappa IIT Jamshedpur
- Dr.M.Govindaraju Amrita University
- Dr.R.Vairavignesh Amrita University
- Dr.P. Senthilkumar BIT
- Dr.P.Nagarajan BIT
- Dr.M C Pravin BIT

Table of Contents

IOP Conf. Series: Materials Science and Engineering **995** (2020) 011001 doi:10.1088/1757-899X/995/1/011001

IOP Conf. Series: Materials Science and Engineering **995** (2020) 011001 doi:10.1088/1757-899X/995/1/011001

IOP Conf. Series: Materials Science and Engineering **995** (2020) 011001 doi:10.1088/1757-899X/995/1/011001

ICMEEP_40	**A Review on Mechanical Behaviour of Aluminium Reinforced Composites** *Raghuvaran P, Suresh M, Aakash S, Balaji M, Dinesh Kumar K and Hari Prasath S*
ICMEEP_41	**A Review on Influencing Welding Parameters of Super Alloys** *Raja S, Saranya R, Sai Yaswanth V, Sachin Adithya A and Rohith KumarM B*
ICMEEP_42	**Predict the Tensile Strength of Friction Welded Steel / ASS304L Dissimilar Joints** *Rajendran C, Surya Prakash E, Surya Prakash P, Sivasankar V and Rohith K*
ICMEEP_43	**Flow conditions analysis of Gas Turbine Combustor** *Abinaya R , Rohini D , Soundarya B ,Balakrishnan S , Dakshinamurthy B and Rajeshwari P*
ICMEEP_44	**Impact of Baffle on Liquid Structure Connection in Fuel Tank Sloshing** *Rohini D, Abinaya R , Lokesharun D, Karthik K, Sovishnuchringth V and Sivaprakash P*
ICMEEP_45	**Experimental Analysis on Water Absorption Behaviour of Natural Composite Materials** *Lakshmanan D , Naveen R, Saravanan P , Nivitha D and Mathi Vathana R R*
ICMEEP_46	**Comparative Analysis of the Conventional Heat Treatment and Electro-Pulsing Treatment for the Microstructural Analysis** *Shubhayan Mukherjee , Rakibul Biswas and Sannik Acharya*

IOP Conf. Series: Materials Science and Engineering **995** (2020) 011002 doi:10.1088/1757-899X/995/1/011002

Peer review declaration

All papers published in this volume of *IOP Conference Series: Materials Science and Engineering* have been peer reviewed through processes administered by the Editors. Reviews were conducted by expert referees to the professional and scientific standards expected of a proceedings journal published by IOP Publishing.

- **Type of peer review: Double-blind**
- **Conference submission management system: EasyChair**
- **Number of submissions received: 63**
- **Number of submissions sent for review: 61**
- **Number of submissions accepted: 46**
- **Acceptance Rate (Number of Submissions Accepted / Number of Submissions Received X 100): 73.02**
- **Average number of reviews per paper: 02**
- **Total number of reviewers involved: 15**
- **Any additional info on review process:**

  ***All the accepted papers are reviewed and checked with novelty and similarity index <15% through Turnitin software.**

- **Contact person for queries:**

Dr.A. Megalingam,
Associate Professor,
Department of Mechatronics,
Bannari Amman Institute of Technology,
Sathyamangalam, Erode,
Tamilnadu,
India-638401.
Mail id: megalingama@bitsathy.ac.in.
Phone no:+91-9884646847.

IOP Conf. Series: Materials Science and Engineering **995** (2020) 012001 doi:10.1088/1757-899X/995/1/012001

Development of Low cost 6 Dot Unicode Braille printer for aiding visually impaired persons

A Mohamed Ismail[1], K Yuvaraj[2], S Vigneswaran[3] and P Nagarajan[4]

1,2,3 & 4 Assistant Professor, Bannari Amman Instititute of Technology, Department of Mechatronics, Sathyamangalam, Tamilnadu, India.

mohamedismail@bitsathy.ac.in , yuvarajk@bitsathy.ac.in, vigneshwaran@bitsathy.ac.in, nagarajanp@bitsathy.ac.in,

Abstract. Numbers of Visually impaired persons are suffering a lot for communicating their thoughts. Braille literacy is not affordable to all since the printers are expensive which costs roughly one Lakhs. This work aims to develop a braille embosser for the affordable price which embosses the A4 size sheet with braille code. Arduino Uno controller is employed to control the angular position of the stepper motor. The input is given as text format to the serial monitor of the arduino IDE. The input texts are converted into Unicode in order to get the corresponding braille code. With the help of stepper motors code is printed on the paper.
Keywords: Braille Printer, Arduino UNO, Braille code, Stepper motors, affordable cost.

1. Introduction

Around 36 million people are visually impaired in this universe more than 80% of people among them are in below the poverty line. The visually impaired people can read and write using a set of 3x2 matrixes of 6 dots which is called as Braille code as shown in Figure 1. Number of commercial Braille printers is available today for aiding the needy persons but they are not affordable to all as it is much expensive. It is decided to make an efficient, cheap and user friendly Braille printer. Generally there are two ways of creating a product which may be either creating from scratch or modifying the already existing system. In this project, the Braille printer is made from the scratch by making use of some spare parts from the old printer. It is developed in such a way that the printer can print what the text given by the user in the sheet. In some existing method the text is given directly to the arduino board using serial communication and then same text is converted [7].

In the proposed system, the GUI is created in which the user can type what he wants to print as Braille code. Advanced Braille printing devices are being developed with object character recognition technology where the text can be recognized with mobile applications [4]. Two step algorithm is implemented where the word is recognized from the continuous speech with the received signal by comparing token passing and baseline algorithms[9] New Braille printer are also developed where the graphical data can also be represented by using edge detection technology to print the images in A4 paper[5].While considering the accuracy on papers researches made trial and error with various papers such as bond paper, A4 sheet, butter paper and thin billing papers and found that billing paper are recommended as the impressions were good comparatively[1].

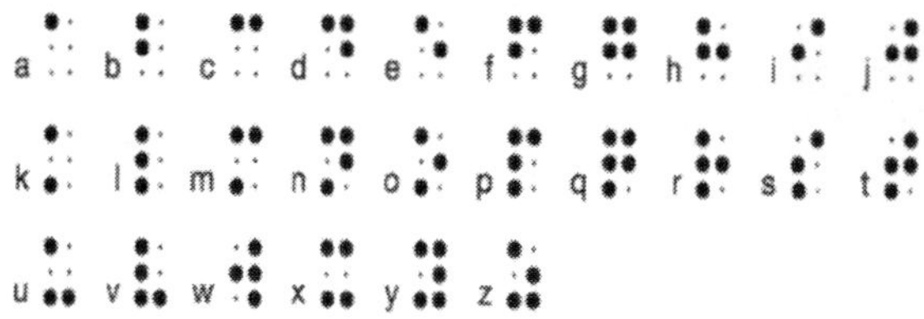

Figure 1. Braille code for alphabets

2. Methodology

The product is developed in such a way, where the entered text is processed and converted the text into Braille format further the Braille code is printed on the sheet. This requires an arduino mega and ramps board for accessing the entered text and convert that into Braille code. Figure 2 shows the block diagram of the motor actuating concept.

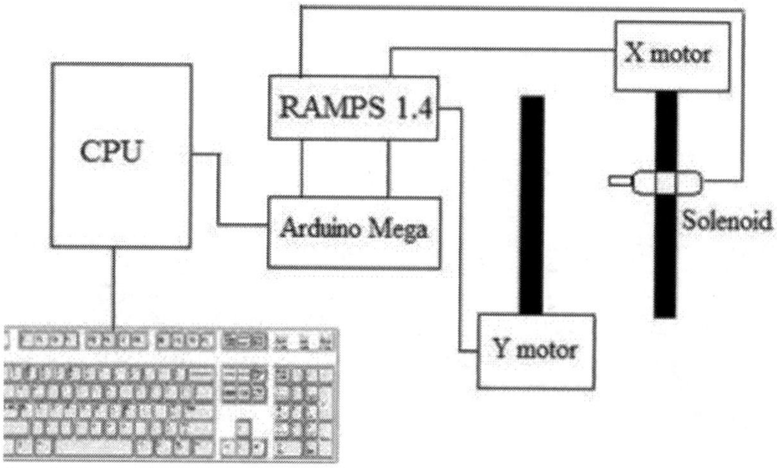

Figure 2. Block diagram of the motors control with Arduino

2.1. Text to Braille conversion

- The GUI which we created based on python script will open a text box where we have to enter the text.
- The entered text will be saved in the form of .txt file.
- Clicking on print button the text will be processed and executed by the python script.
- After the completion of execution the corresponding Gcode will be created with the file name of .gcode.
- The GUI ask for permission to send gcode file to arduino controller, with simply yes or no question
- Proceeding with yes the process of embossing the Braille code will takes place.

The flow chart for the Braille embosser process is shown in Figure 3. The ultimate aim of this work is to provide the common interface to the blind person for the affordable price. Here, the text to Braille code printing system is proposed. In case of voice recognition based Braille printing process, speech is taken as input through microphone. This input speech is processed in the system. Step algorithm is used to convert chopped input into its equivalent binary code [6].

IOP Conf. Series: Materials Science and Engineering **995** (2020) 012001 doi:10.1088/1757-899X/995/1/012001

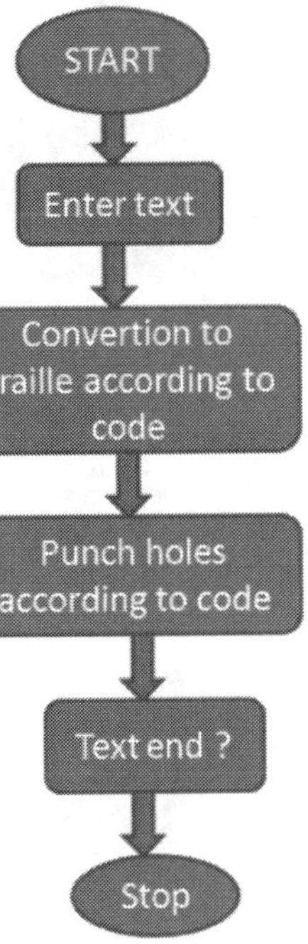

Figure 3. Flow Chart of Braille embosser

3. Experimental Setup

The software and hardware which are used to develop the Braille printer are discussed below.

3.1 *Arduino with marlin firmware*

The Marlin is an open source firmware based on arduino platform used to control the activities of motors. The Marlin firmware supports ramps 1.4 board. This was coded to enable or disable various parameters like LCD display, end stops, servo, stepper motor, cooler fans etc. In Configuration.h file some parameters has to be changed before uploading it on the Arduino Mega controller.
Pronterface software is a Graphical interface for the user with which one can initialize the printer manually. It has more added features of controlling the print speed. It is hosted by Printrun software application. Pronterface GUI page is shown in Figure 4.

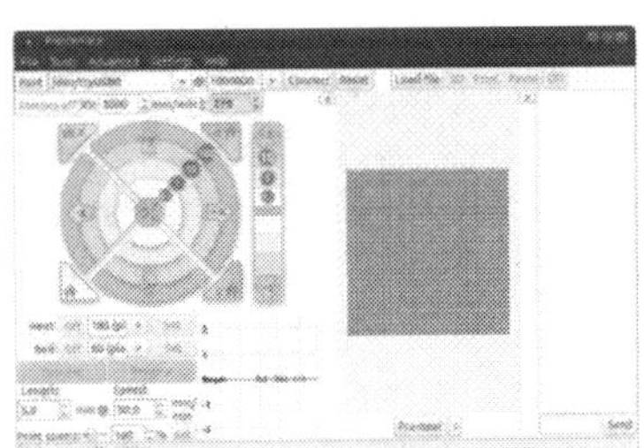

Figure 4. Pronterface GUI page

3.2 Text to braille.py

This script developed to convert text entered to G-codes. After conversion it will be in G-code format. When print button is clicked in GUI, text_to_braille script is executed and G-code file is created, after file is created, GUI send G-code file to the arduino through serial port. The coding is done with python language as shown in Figure 5.

Figure 5. Python script

3.3 Hardware Interface

The Electronic hardware components employed in this project consists of 2x Nema 17 motor, Ramps 1.4,Arduino Mega, 12V SMPS power supply and A4988 motor driver. The Circuit connections of those components are shown in Figure 6.

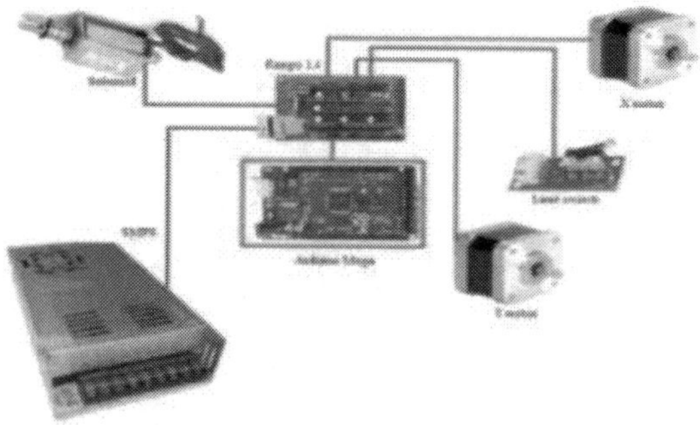

Figure 6. Circuit diagram of electronic components

- The Switch Mode Power Supply (SMPS) unit is connected as power supply unit to the system.
- The positive and negative terminals were connected to the corresponding positive and negative ports in the ramps 1.4 board.
- The two Nema 17 motors were connected to the X and Y ports of the ramps board.
- The limit switch is connected to the end stop port of the ramps 1.4.
- The solenoid push pull actuator is connected to the D9 port in the ramps 1.4.

3.4 Paper feeding mechanism

Among the two stepper motors one is used to feed the paper. A smooth rod is attached to the shaft of the stepper motor using the shaft coupling. The rods itself consist of three rubber rollers aligned with equal space in between. Then another rod of sponge roll is fitted to the base of the system freely. When the Y motor gets rotating the paper which was in between these two rods tends to move forward. That is how the paper feeding occurs. Figure 7 shows the mechanism of rollers conveying paper.

Figure 7. Rollers conveying paper

IOP Conf. Series: Materials Science and Engineering **995** (2020) 012001 doi:10.1088/1757-899X/995/1/012001

3.5 Embossing mechanism

The six Dots Braille Unicode systems is implemented in this system with standard dimension. User can print up to sixty four unique codes [3] with this method. The six dot Unicode pattern with its dimensions is shown in Figure 8.

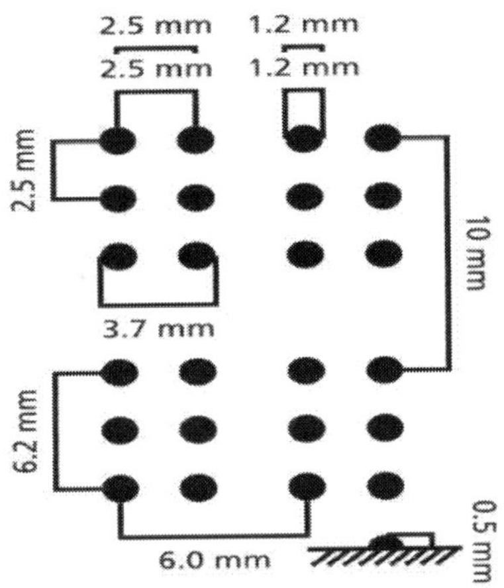

Figure 8. Six Dot pattern with dimension

Solenoid is used to provide linear motion through actuator for embossing on the sheet which is shown in figure 9. At the end of the solenoid plunger an embossing pin has attached. The solenoid is made to move freely by the guide rods. The toothed belt has attached to the solenoid which transforms the rotary motion of the X motor into linear motion. The solenoid is initially in Normally Opened (NO) condition which in turn closed while supply is given once. When the correct position is reached the solenoid will get opened and punch the sheet to make a dot. When one row is done then the motor will be actuated to move the sheet forward of about 2.5mm. The moment when a line of text is printed it will move the sheet for 5mm.

Figure 9. Solenoid to perform Pull, Push tasks

4. Braille Printing

- The input text is given through the GUI named as printergui.py
- Entered text is converted into Braille code and it creates the corresponding G code.
- Generated G code is sent to the Arduino controller.
- Arduino controller is enabled with Marlin firmfare which is used to perform the printing

operation.

- The controller then controls the stepper motors in order to feed the sheet as well as to coordinate the position of embosser.

The solenoid for push pull action is attached with embosser which has actuated to punch the Braille code on papers. Figure 10 shows the Assembly of hardware components with frame.

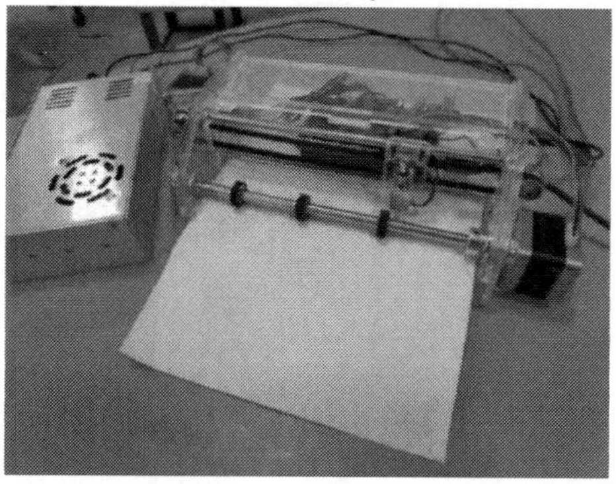

Figure 10. Assembly of hardware components with frame

5. Result and Discussion

The text Welcome is printed in the A4 sheet as shown in Figure11. The same had tested with visually impaired people which were recognized by them appropriately. Printer casing are made with acrylic sheet which in turn the device is made for the affordable cost. The performance of the proposed work is efficient comparing with the existing printers.

Figure 11. Embossed paper with the text: welcome

6. Conclusion

The fabrication of a low cost Braille embosser was done with an estimated cost of around Rs.5000.It is used to print the sheets of books, novels as needed by the blind people. The working with this machine is very easy when compared to other commercial printers.

7. Future work

The speech recognized Braille code printing can be implemented in upcoming days. This will be helpful for the visually impaired people to print by themselves without seeking others assistance.

References

[1] S. Padmavati, Nivedita V, Sankari, Raamprashanth.N.S, Rajat Bohra, "Economic Printing of Braille Documents", International Journal of Emerging Technologies in Computational and Applied Sciences, 8(2), March 2014 - May 2014, pp. 170-173.

[2] Durán Encinas, J. A. Sandoval Bringas, A. I. De Casso Verdugo, C. Sandoval Bringas and J. G. Soto Muñoz, "Low-Cost Braille Printer Prototype Design With OCR Technology," 2019 International Conference on Inclusive Technologies and Education (CONTIE), San Jose del

Cabo, Mexico, 2019, pp. 205-2054, doi: 10.1109/CONTIE49246.2019.00047.

[3] A. K. Garg, "Braille-8 — The unified braille Unicode system: Presenting an ideal unified system around 8-dot Braille Unicode for the braille users world-over," 2016 IEEE International Conference on Advanced Networks and Telecommunications Systems (ANTS), Bangalore, 2016, pp. 1-6, doi: 10.1109/ANTS.2016.7947839.

[4] I. Durán Encinas, J. A. Sandoval Bringas, A. I. De Casso Verdugo, C. Sandoval Bringas and J. G. Soto Muñoz, "Low-Cost Braille Printer Prototype Design With OCR Technology," 2019 International Conference on Inclusive Technologies and Education (CONTIE), San Jose del Cabo, Mexico, 2019, pp. 205-2054, doi: 10.1109/CONTIE49246.2019.00047.

[5] Bo Li, Zhiping Wang, Junbiao Liu and Guangrong Fang, "New blind graphic printing technology based on New Braille Printer," 2011 International Conference on Computer Science and Service System (CSSS), Nanjing, 2011, pp. 1403-1406, doi: 10.1109/CSSS.2011.5973931.

[6] Harshali Gangurde, Sayali Pingale, Dipali Pawar, Jyoti Aher, "Braille Printer" International Journal for Research in Engineering Application & Management (IJREAM) Vol-01, Issue 01, APR 2015.

[7] Shwethashree S, Sowmya S K, Sri Ranjini, Vanaja N, Parameshwara M C," Text Image to Braille Code Converter" International Journal of Engineering Research in Electronics and Communication Engineering (IJERECE) Vol 5, Issue 6, June 2018.

[8] Paul Blenkhorn, "A System for Converting Braille into Print"- 1995, IEEE Transactions on Rehabilitation Engineering, Vol. 3, No. 2.

[9] Cabatabaei,"Methods of reading braille code"- 2015,International Journal for Research in Engineering Application & Management (IJREAM) Vol-01, Issue 01.

[10] Choromanska "Braille coding techniques"-2017,International Journal of Advanced Research in Computer and Communication Engineering Vol. 6, Issue 4.

IOP Conf. Series: Materials Science and Engineering **995** (2020) 012002 doi:10.1088/1757-899X/995/1/012002

Design and Analysis of Sliding Mode Control for Battery Charging Applications

P Sivaraman[1*] , T Logeswaran[2], J S Sakthi Surya Raj[3] and S Boopathimanikandan[4]

[1]Associate Professor, Bannari Amman Institute of Techology, Department of Electrical and Electronics Engineering, Erode, Tamilnadu, India.
[2]Assistnt professor (Sr.G), Kongu Engineering College, Department of Electrical and Electronics Engineering, Erode, Tamilnadu, India.
[3]Research Scholar, Bannari Amman Institute of Techology, Department of Electrical and Electronics Engineering, Erode, Tamilnadu, India.
[4]PG Scholar, Bannari Amman Institute of Techology, Department of Electrical and Electronics Engineering, Erode, Tamilnadu, India.

sivaramanpsr@bitsathy.ac.in , logeskongu@gmail.com, sakthisuriyaraj@bitsathy.ac.in, boopathimanikandan.pe18@bitsathy.ac.in ,

Abstract. In Electric Vehicle (EV) charging stations connected loads are intermittent due to continuous variation of charging level battery. In this situation, the linear controller like P, PI, and PID does not offer a quick and good transient response to the EV charging converter. In some cases, EV charging stations push to instability region for abrupt change in loads. Therefore, this research work focused on investigation of non-linear controllers for EV battery charging station. This paper investigates the performance of Hysteresis Controller (HC) and Sliding Mode Controller (SMC) for EV charging station under dynamic load conditions. The hysteresis controller and sliding mode controller are developed for buck converter fed EV charging stations and its performance is analyzed using pole zero plot and bode plot methods. From the investigation it is found that SMC offer optimum gain margin and phase margin under dynamic load condition for EV charging station.

Keywords: Hysteresis Controller, Sliding Mode Controller, Electric Vehicle, Charging Station and Dynamic Controller.

1. Introduction

Among the various categories of power conversion operations, DC-DC converters are well known for maximum power point tracking [1], DC bus integration, and battery charging application [2]. Due to the transient switching action DC-DC converters are nonlinear in nature [3,4]. Most of the design, controllers are using Linear controller like PID based controllers. In some cases, due to nonlinear change in loads, the transient response of linear controller has peak overshoot, oscillations in the output before settles to final value [5]. There are two categories of control in DC-DC Converters namely, Voltage Mode (VM) and Current Mode (CM) control [6]. The voltage mode control detects the output voltage and compared with a reference voltage to generate error voltage signal. In Current Mode control, both the output voltage and inductor current are sensed. Due to the non-linear and time varying nature of DC-DC converters, the designer controls the output voltage by directly controlling the inductor current (CM) [7]. A fixed-frequency pulse width modulation based Sliding-Mode Controllers (SMC) developed for dc–dc converters that operates in the continuous conduction mode is proposed in [8]. However, the PWM based SMC has transient parameters with high voltage ripples. This type of nonlinear controllers is used to support the converters that needs high dynamic response for a wide range of operating conditions [9]. To

improve the performance of the converter in terms of its transient response an nonlinear hysteresis controller for DC-DC converter is proposed in [13]. However, the attenuation of disturbances in the power supply and sensitivity to unknown loads are need to be analysed [10]. To reduce the overshoot in output voltage a combination of linear and non-linear parts consisting of PID and hysteresis controller is proposed in [11]. Though, the settling time is drastically reduced, there is a negligible amount of overshoot in output voltage during load variation [12]. Hence for Battery Charging Applications, linear controller generates peak overshoot voltages that results increase in charging current will leads to damage the battery. Some nonlinear behaviour of Battery, will results in system instability. The system will operate in unstable condition, which causes unbounded output generated by DC-DC converter [13]. Thus a new controller designed that includes both hysteresis controller and Sliding mode controller to improve the charging of battery.
In this paper the methodology of hysteresis controller and SMC and the design is studied in section 2. The section 3 and 4 gives the control algorithm of PI controller and proposed SMC. In section 5 the results obtained by implementing PI, SMC and hysteresis controller on buck converter for battery charging is analyzed based on the current and voltage wave form. Finally, the comparison is made based on the results obtained.

2. Methodology

A new control design has proposed to improve the design includes both hysteresis controller and SMC. SMC changes the dynamics of the system by applying a discontinuous control signal. This discontinuous signal forces the system to slide along a sector of the system's normal function. The proposed method used Hysteresis based SMC to reduce the transient parameters and improves the stability of the system during nonlinear conditions. Buck Converters are used for low power battery charging applications [14]. Hence, buck converter is used for the proposed methodology. The designed proposed controlled is shown in figure 1.

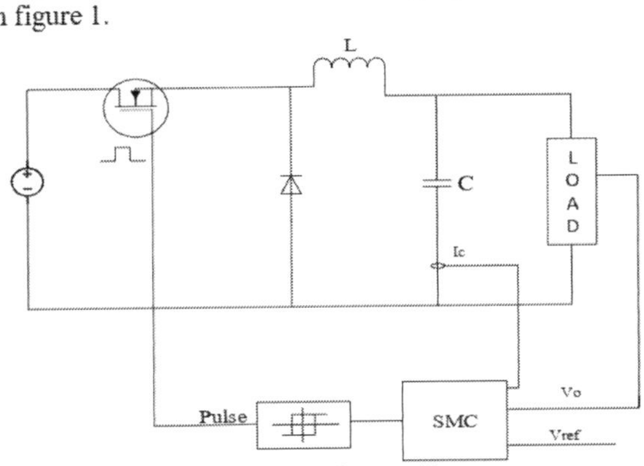

Figure 1. Proposed controller.

2.1. Hysteresis Controller

Hysteresis controller are used to control the converter that has faster load transients [15]. It is also called as ripple control method since it controls by detecting the ripples in the output voltage. Further, the output voltage is monitored directly only with the help of a comparator without an error amplifier. The comparator directly turns on/off the switch by detecting the level of output voltage that exceeds or falls below the threshold value at an instant. Fixed on-time and fixed off-time control scheme are used for detecting the voltage that falls below and above the threshold level respectively [16]. The hysteresis controller to fed buck converter is presented in figure 2.

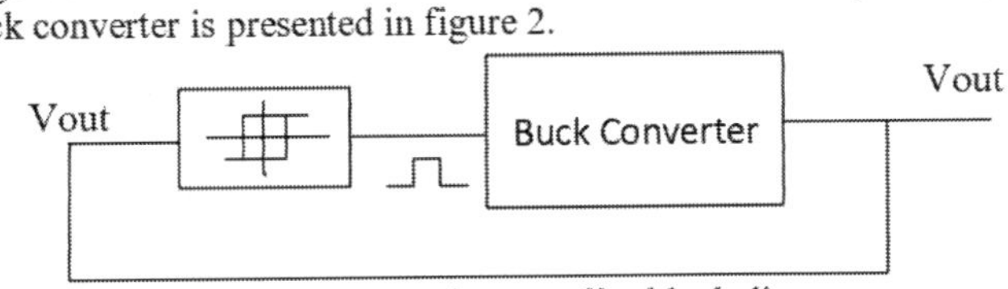

Figure 2. Hysteresis controller block diagram.

It has two types. One is regulating error signal and another one is regulating output voltage directly. It directly output voltage is fed to hysteresis controller. Hysteresis band will regulate the output voltage within threshold limit [17]. When output voltage touches, the upper band HC turns off the switch and output voltage decays and enters to lower band HC turn on the switch. The Peak overshoot and oscillations of output voltages are eliminated. It does not require any Error amplifier and additional PWM

Generator [18]. By applying threshold limits Variable Switching Frequencies (VSF) can be achieved with less ripples on output voltage.

2.2. Sliding Mode Controller

Sliding Mode Controller (SMC) is a nonlinear controller it overcomes the slow response and transient oscillation caused by dynamic change in load. The output of SMC is a discontinuous signal that drives the converter to slide in the cross section of the operating region of converter [19]. The trails always slide towards an adjacent section designed with multiple control structures so that the entire trail will not occur with single control structure. Hence, the trail slides along the borders of the control system [20]. This motion of the system is called sliding mode and the locus consisting of the borders is called the sliding surface. This methodology is used to force the system to reach towards a normal operation or its desired final value by sliding on the surface. This nonlinear dynamic nature of the controller and its robustness against the disturbance makes it an ideal choice for battery charging application.

2.2.1. Design of Sliding Mode Control Equations. To design a SM controller, the first step is to develop a state space model of the converter with desired control variables [21]. The controller designed for the study is a PID sliding mode voltage controller. This controller overcomes the steady state DC error prevailed in the conventional controller by including additional voltage error value in the calculation. The principle of operation of sliding mode is given in figure 3.

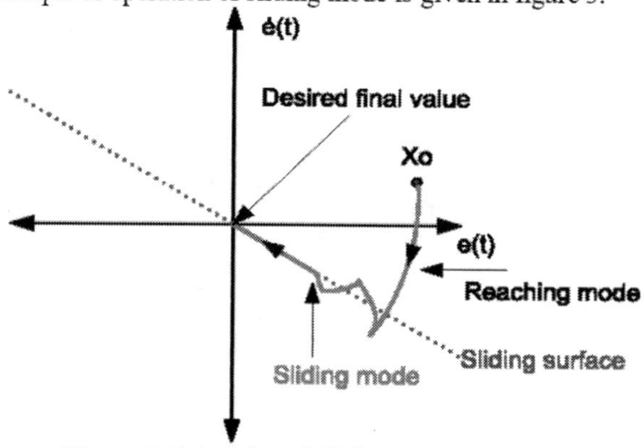

Figure 3. Principles of Sliding mode controller.

After obtaining the state space description, the controller design is developed. The general SM control law that adopts a switching function is suitable for this system, such as

$$U = \begin{cases} 1 \ when\ L > 0 \\ 0 \ when\ L < 0 \end{cases} \tag{1}$$

Where L is the immediate state variable's trail, and is described as

$L = \alpha_1 x_1 + \alpha_2 x_2 + \alpha_3 x_3$ Where $\alpha_1, \alpha_2, \alpha_3$ are sliding coefficients and x_1, x_2, x_3 are voltage error respectively. During SM operation the dynamic response of the converter and the equation relating to the sliding co-efficient are obtained by solving L=0. This effects in second order linear equations and produces over-damped, critically damped and under-damped responses. The SMC block diagram is illustrated in figure 4.

$$\frac{\alpha_1}{\alpha_2} = \frac{10}{T_s}\frac{\alpha_3}{\alpha_2} = \frac{2.5}{\epsilon^2 T_s^2} \tag{2}$$

where T_s *and* ϵ are settling time and damping co-efficient respectively.

The control voltage (Vc) equations are essential for the application of buck or Hysteresis-based SMVC converters is given by

$$V_C = -K_{p1} i_c + K_{p2}(V_{ref} - \beta V_o) + \beta V_o \tag{3}$$

Where $K_{p1} = \frac{1}{R_L C}$ and $K_{p2} = \frac{\beta}{C}\beta = \frac{V_{ref}}{V_{out}}$

K_{p1} and K_{p2} are the constant gain factors for the feedback signals i_c and $V_{ref} - \beta V_o$ respectively.

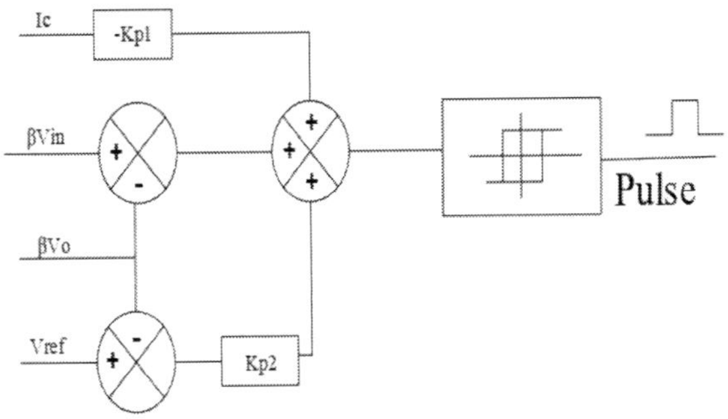

Figure 4. SMC block diagram

3. Control algorithm using PI controller

DC Link Controller is to design to maintain output voltage. Controller measures the output voltage and compares it with the reference voltage given and the error signal generated is fed to PI controller [22]. Low pass filter (PI) is used to regulate error to generate pattern voltage to generate pulses. The DC link controller block diagram is represented in Figure 5.

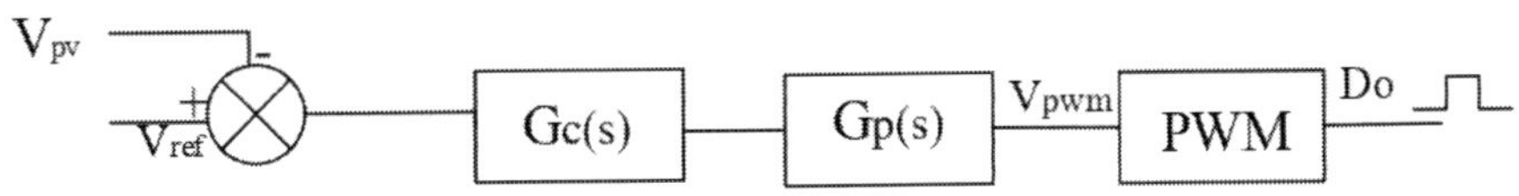

Figure 5. Open loop DC link controller.

The plant transfer function is found by control to input voltage

$$T(s) = \frac{D_o(s)}{V_i(s)}. \quad (4)$$

$$T(s) = \frac{a(Cs+1)}{s^2 + bS + c} \quad (5)$$

Where

$$a = \frac{V_{in}}{LC_1} b = \frac{1}{RC} + \frac{1}{L} c = \frac{1}{LC} \quad (6)$$

The open loop transfer function with PI Compensator is $H(s) = T(s) * G_{PI}$. By proper tuning PI Controller, it can able to operate the controller in stable region. By choosing Kp=0.5, Ki=0.01 controller enter into stable operating region. The Bode and Pole Zero plot of DC converter with compensator is shown in Figure 6.

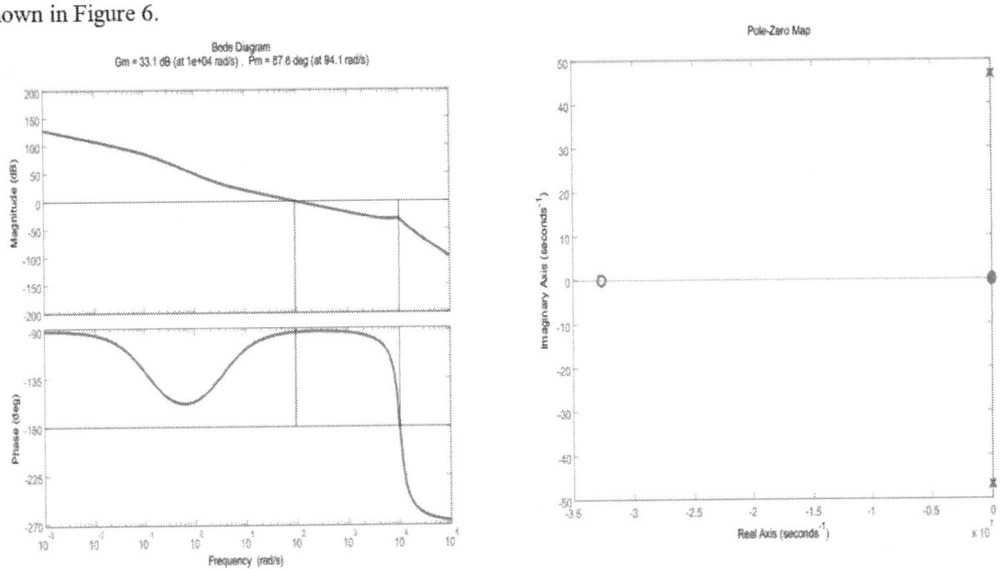

Figure 6. Bode Plot and Pole Zero plot of open loop DC controller with PI.

From the plot, the positive values of Gain margin and Phase margin shows that Controller is in stable region. From the pole zero plot, All the poles are lying in LHS of S-plane and two complex poles are lying in the imaginary axis indicate output voltage have minimum ripples.

4. Control algorithm of proposed SMC

DC Link Controller is to design to maintain output voltage. Controller senses the output voltage is fed to SMC. SMC will regulate the output within the sliding surface which chosen and make the system to operate in desired level. Hysteresis controller is used to regulate output voltage within threshold limit and generate pulses according to output voltage ripples. The DC link controller with SMC is presented in figure 7 and the corresponding bode plot and pole zero plot of open loop DC link controller is shown in figure 8.

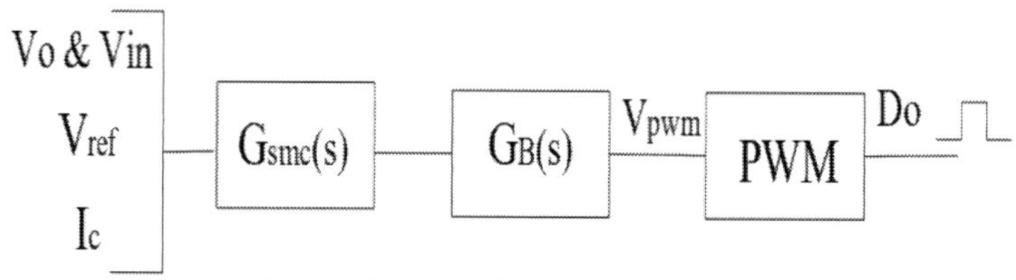

Figure 7. Open loop DC link controller fed SMC.

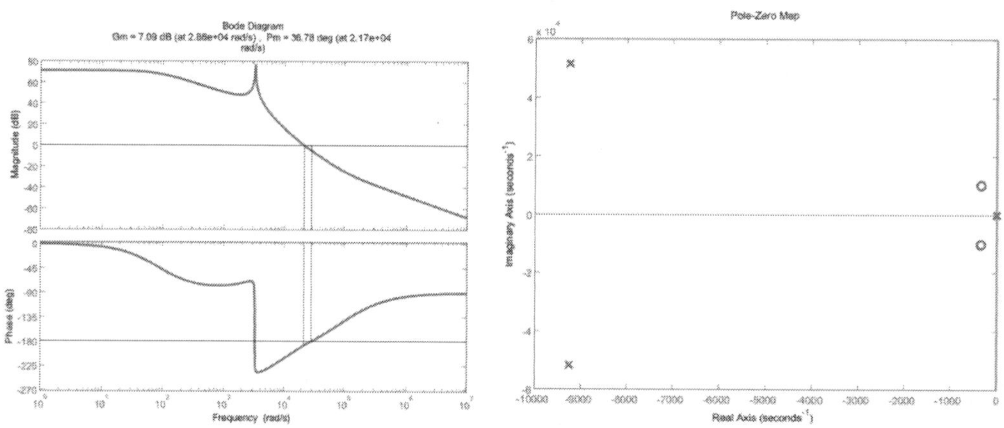

Figure 8. Bode plot and pole zero plot of open loop DC link controller with SMC

All the poles are lying in left of S plane. This shows the controller will be always operates in stable region. From the bode plot, the gain margin and phase margin is 7.09dB and 36.78 degree respectively which is optimsed value.

5. Result and discussion

The Buck converter is designed for 7V(output) with 12V (Input Voltage). The Inductance and Capacitance values of Buck converter is designed by assuming current ripple of 1% and Voltage ripple of 1% and operated in Continuous Conduction mode. The battery chose for this simulation is Lithium-ion battery [23,24]. The voltage and capacity of the battery is 6V and 5Ah respectively. Efficiency of Buck converter in this design is around 97.35%.

5.1. Buck Converter with PI Controller

In this method Voltage mode control method is used. It involves sensing of output voltage, which is compared with the reference voltage to generate error voltage signal. The error signal is fed to PI controller and its output is compared with a saw tooth waveform, which generates a Pulse Width Modulated (PWM) signal that is fed back to the converter as gate pulse. The output waveform of linear controller fed Buck converter is shown in figure 9.

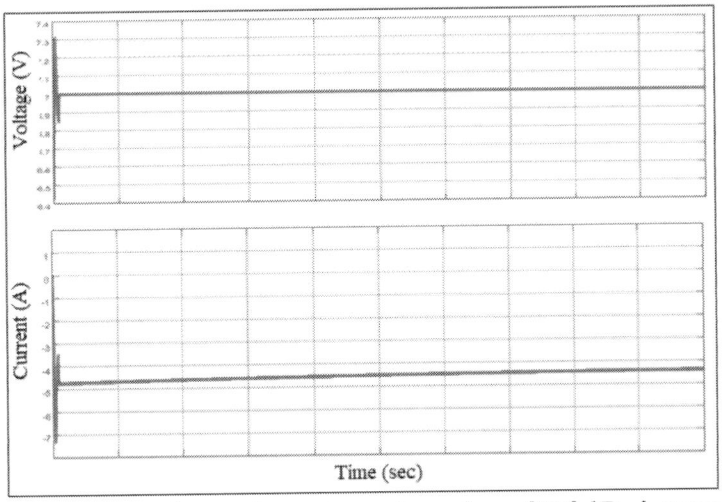

Figure 9. Output voltage and output current waveform of PI fed Buck converter

The output voltage waveform has peak overshoot is around 0.3% and settling time of 0.05sec.

5.2. Buck Converter with Hysteresis Controller

In this method Voltage mode control method is used. The error signal is generated by sensing and comparing the voltage obtained with the reference voltage. The obtained error signal is fed to the Hysteresis Controller. The comparator generates the pulse to turn on/off the switch when the output voltage level has exceeded or fallen below a set threshold level. The hysteresis level set for this design is 0.01. The output voltage is oscillating between 6.99V-7.01V. The output waveform of hysteresis controller fed buck converter is presented in figure 10. The peak overshoot is completely eliminated and settling time obtained is 0.05sec.

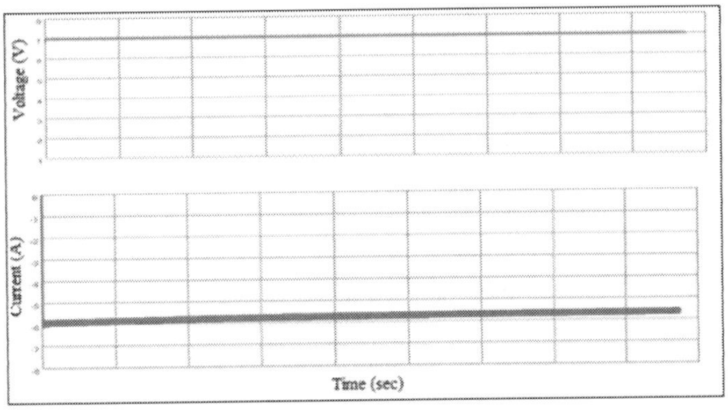

Figure 10. Output Voltage and Current waveform of Hysteresis controller fed Buck converter

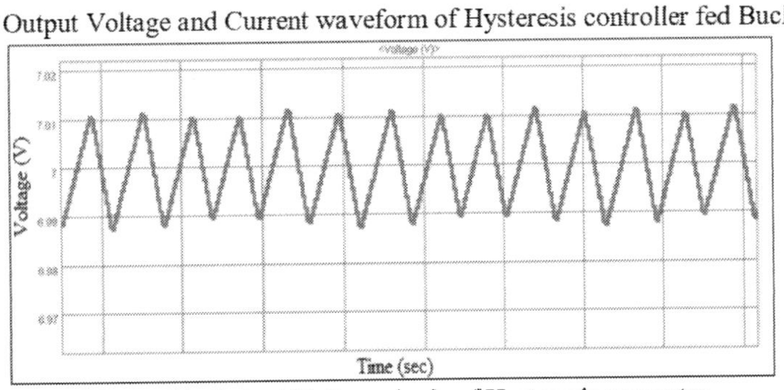

Figure 11. Output Voltage ripple of Hysteresis converter

Figure 11 Displays the output voltage ripple band of hysteresis controller. It shows the output voltage is oscillating between 6.99V to 7.01 V with ripple of 0.02V.

5.3. Buck Converter with proposed Non-Linear Controller

In this method both Voltage control mode method current control mode is used. The nonlinear converter used in this design is Hysteresis based SMC. The simulation model figure of buck converter with SMC is given in figure 12.

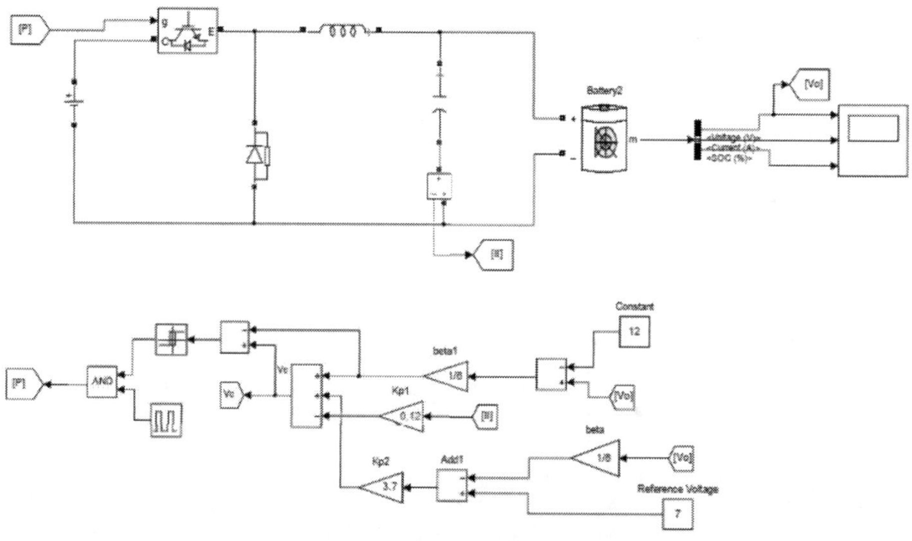

Figure 12. Buck converter fed SMC controller.

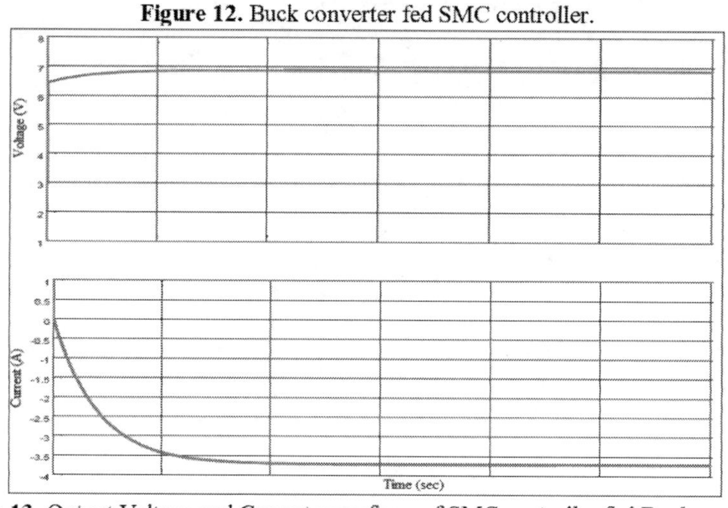

Figure 13. Output Voltage and Current waveform of SMC controller fed Buck converter.

This proposed controller merges the hysteresis band for PWM generation and regulating error using Sliding mode design. The output waveform of non-linear controller fed buck converter is presented in figure 13. The peak overshoot and oscillations is eliminated shown in the output voltage waveform. An Comparative Analysis of PI, Hysteresis and SMC is made and tabulated in table 1.

Table 1. Comparison of various controller parameters

Parameters	**PI Controller**	**Hysteresis Controller**	**Sliding Mode Controller**
Output Voltage	7.1 V	7.007 V	6.88 V
Voltage Ripples	0.2 V	0.02 V	0.004 V
Peak Overshoot	0.03%	0	0
Settling Time	0.1 sec	0.05 sec	0.04 Sec

The output voltage from SMC is slightly less that reference voltage. But voltage ripples are very low compared to PI controller and Hysteresis Controller. PI Controller has peak to peak ripple voltage is 0.2 V with peak overshoot of 0.03% with settling time of 0.1 seconds. Compared to all converter, SMC has low settling time around 0.04 sec with zero peak overshoot, i.e. Peak overshoot is completely eliminated in SMC and hysteresis controller. Hysteresis controller operating at 2.5kHz for the threshold voltage of 0.01 V. The PI Controller and SMC are operating at 50kHz. So switching losses will be more for SMC and PI controller. Hysteresis controller will not able to perform correct control action for nonlinear change in load. For the case of nonlinear control action, SMC will behave smooth operations.

6. Conclusion

Buck Converter for battery charging application was designed. The performance of converter is analyzed with three different control techniques like PI, hysteresis controller and SMC. Hysteresis controller monitors the output voltage and maintained within threshold limits and also eliminates peak overshoot and oscillations generated while transient state. The performance of Hysteresis based SMC is efficient that maintains the converter to be operated in designed sliding surface. Current mode control method is used for design SMC which regulates the output voltage by regulating inductor current. The results confirm that the proposed controller has less ripple, fast settling time and zero peak overshoot.

References

[1] Sivaraman P 2015 A New Method of Maximum Power Point Tracking for Maximizing the Power Generation from a SPV Plant Journal of scientific and Industrial Research Vol 74 No 3 pp 411 – 415.
[2] Prem P Sathik J Sivaraman P Matheswaran A Abdel Aleem S.H.E 2019 A new asymmetric dual source multilevel inverter topology with reduced power switches Journal of Chinese Institute Engineering Vol 42 No 5 pp 460–472.
[3] Kim K Lee H Hong S and Cho G A 2019 Noninverting Buck–Boost Converter with State-Based Current Control for Li-ion Battery Management in Mobile Applications IEEE Transactions on Industrial Electronics Vol 66 No 12 pp 9623-9627.
[4] Ponnusamy P Velliangiri S Ali J.S.M 2020 A Hybrid Switched Capacitor Multi-Level Inverter with High Voltage Gain and Self-Voltage Balancing Ability Electric Power Components System pp 1–14.
[5] Sivaraman P and Prem P 2017 PR Controller Design and Stability Analysis of Single Stage T-Source Inverter Based Solar PV System in Journal of Chinese Institute of Engineers Vol 40 No 3 pp 235 - 245.
[6] Chuang Y and Ke Y 2007 A Novel High-Efficiency Battery Charger with a Buck Zero-Voltage-Switching Resonant Converter in IEEE Transactions on Energy Conversion Vol 22 No 4 pp 848-854.
[7] Chuang Y and Ke Y 2008 High efficiency battery charger with a buck zero-current-switching pulse-width-modulated converter IET Power Electronics Vol 1 No 4 pp 433-444.
[8] Siew-Chong Tan Lai Y M and Chi K Tse 2016 A Unified Approach to the Design of PWM-Base Sliding-Mode Voltage Controllers for Basic DC-DC Converters in Continuous Conduction Mode IEEE Transactions on circuits and systems Vol 8 No 08.
[9] Tan C Lai Y M Cheung M K H and Tse C K 2005 On the practical design of a sliding-mode voltage controlled buck converter IEEE Trans. Power Electronics Vol 20 No 2 pp 425–437.
[10] Escobar G Ortega R Sira-Ramirez H Vilain J and Zein I 1999 An experimental comparison of several nonlinear controllers for power converters IEEE Control Systems Magazine Vol 19 No 1 pp 66-82.
[11] van der Broeck C H De Doncker R W Richter S A and Bloh J V 2015 Unified Control of a Buck Converter for Wide-Load-Range Applications IEEE Transactions on Industry Applications Vol 51 No 5 pp 4061-4071.
[12] Kim T Kim D Shin H Lee S Suh J and Yang B 2019 Low Power Digital PWM Buck Converter With a Clock-Gating Shift-Register International Conference on Electronics Information and Communication (ICEIC) Auckland New Zealand 2019 pp 1-3.
[13] Schild A Lunze J Krupar J and Schwarz W 2009 Design of Generalized Hysteresis Controllers for DC–DC Switching Power Converters IEEE Transactions on Power Electronics Vol 24 No 1 pp 138-146.

[14] Prem P Sivaraman P Sakthi Suriya Raj J S Sathik M J Almakhles D 2020 Fast charging converter and control algorithm for solar PV battery and electrical grid integrated electric vehicle charging station Automatika Vol 61 No 4 pp 614–625.

[15] Zheng Y Chen H and Leung K N 2012 A Fast-Response Pseudo-PWM Buck Converter with PLL-Based Hysteresis Control IEEE Transactions on Very Large Scale Integration (VLSI) Systems Vol 20 No 7 pp 1167-1174.

[16] Chincholkar S H Jiang W and Chan C 2018 A Modified Hysteresis-Modulation-Based Sliding Mode Control for Improved Performance in Hybrid DC–DC Boost Converter IEEE Transactions on Circuits and Systems II: Express Briefs Vol 65 No 11 pp 1683-1687.

[17] Lekić A and Stipanović D M 2016 Hysteresis Switching Control of the Ćuk Converter IEEE Transactions on Circuits and Systems I: Regular Papers Vol 63 No 11 pp 2048-2061.

[18] Deshmukh M 2017 A constant frequency second order sliding mode controller for buck converter 2017 Second International Conference on Electrical Computer and Communication Technologies (ICECCT) Coimbatore 2017 pp 1-5.

[19] Guo S Lin-Shi X Allard B Gao Y and Ruan Y 2010 Digital Sliding-Mode Controller for High-Frequency DC/DC SMPS IEEE Transactions on Power Electronics Vol 25 No 5 pp 1120-1123.

[20] Chuang Y 2010 High-Efficiency ZCS Buck Converter for Rechargeable Batteries in IEEE Transactions on Industrial Electronics Vol 57 No 7 pp 2463-2472.

[21] Sivaraman P 2012 Dynamic modeling and analysis of T-source electronic inverter using state space technique Sci. Res. Essays Vol 7 No 38 pp. 3269–3280.

[22] Tsang K M Chan W L and Wei X L 2008 Robust DC/DC buck converter using conditional integrator compensator in Electronics Letters Vol 44 No 2 pp 152-153 17.

[23] Raviraj V S C and Sen PC Comparative study of proportional-integral sliding mode and fuzzy logic controllers for power converters in IEEE Transactions on Industry Applications Vol 33 No 2 pp 518-524 March-April 1997.

[24] Prem P Sivaraman P Almakhles Dhafer Sanjeevikumar P Leonowicz Zbigniew Matheswaran A and Mohamed Ali Jagabar S 2020 A New Multilevel Inverter Topology with Reduced Power Components for Domestic Solar PV Applications. IEEE Access.

Iterative Learning Control for a Repetitive 2D Process

S Muthumari[1], V Gomathi[1*], R Srimathi[2], Kishore Eswaran[3], V Kamatchi Kannan[4]

1 Tamilnadu Electricity Board, Tamilnadu, India.
1* Assistant Professor-III, SASTRA Deemed University, Thanjavur, Tamilnadu, India.
2 Assistant Professor-SR, Vellore Institute of Technology, Chennai, Tamilnadu, India.
3 Developer, KPIT, Bangalore, Karnataka, India.
4 Associate Professor, Bannari Amman Institute of Technology, Sathyamangalam, Tamilnadu, India.

muthumarikannan@gmail.com, gomathi@eie.sastra.edu, srimathi.r@vit.ac.in, kishoreeswaran23@gmail.com, kamatchikannanv@bitsathy.ac.in

Abstract: To control the systems which operate in a repetitive mode is known to be a difficult task. For example, systems like robotic manipulators are used for doing the repetitive assigned task. This can be overcome by the iterative learning control (ILC) method. In this article, the DC motor model is discussed for controlling the speed of the motor by the design of an ILC algorithm, and its performance is compared with the conventional PID algorithm. The ILC algorithm is implemented for the minimum and non-minimum phase systems. The resulting response, absolute and integral absolute errors are plotted. The Z-N tuning method is used to implement PID control. The performance of the DC motor has been evaluated by comparing the Integral Square Error (ISE), Integral Absolute Error (IAE), and Integral Time Absolute Error (ITAE) with the conventional PID controller.

Keywords: DC motor, Iterative Learning Control (ILC), Z-N tuning, Integral Absolute Error (IAE), Integral Square Error (ISE), Integral Time Absolute Error (ITAE).

IOP Conf. Series: Materials Science and Engineering **995** (2020) 012003 doi:10.1088/1757-899X/995/1/012003

1. Introduction

The performance of a system that executes the same work many times can be improved by learning from the past execution (iteration). The improvement in performance can be achieved by connecting the control for subsequent iterations with error information. This technique is used to control the systems operating in a repetitive or trial-to-trial mode with the requirement that a reference trajectory $y_d(t)$ is defined over a finite interval $0 \leq t \leq \alpha$, where α denotes the trial length, is followed to high precision.

A typical control law is given in (1) the following form:

$$u_{k+1}(p)=u_k(p)+\Delta u_{k+1}(p),\ k\geq 0 \quad (1)$$

The general block diagram for an ILC controller is given in the Figure 1.

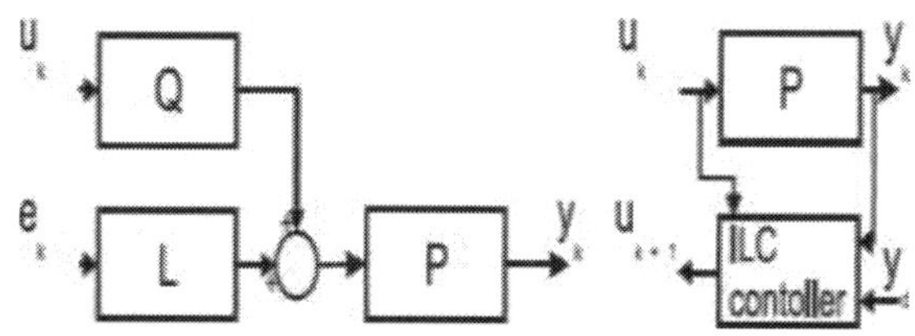

Figure 1. Block Diagram for an ILC controller

Two-dimensional process-based iterative learning control using noncausal finite-time interval data was explained by Błażej.C, Krzysztof.G, and Eric Rogers [1]. M. Rezaei and Kerman [2] have discussed PID parameter selection based on iterative learning control. Iterative learning control law design for an experimentally supported 2D process for error convergence and performance improvement was proposed by Lukasz. Hi and Krzysztof.G [3]. A review of repetitive control, run-to-run control, and ILC was proposed by Y.Wang, Furong.G, and Francis J. Doyle III [4]. Kevin L. Moore, Yang Q.C, and Hyo-Sung A [5] were explained about iterative learning control. Bristow, D.A.Tharayil, and M.Alleyne, A.G. [6] detailed ILC in a survey of iterative learning control paper. Owens D.H. [7],[8] presented optimization in iterative learning control. Jian-Xin Xu; Ying Tan[9] deals with both linear, nonlinear ILC. P. J. Schutyser [10] presented an approach to noncausal iterative learning Control. R.W. Longman [11] has discussed the ILC and repetitive control. Linear iterative learning control schemes were analyzed and explained elaborately by N. Amann, D. H. Owens, E. Rogers [12]. The paper is discussed in the following manner: Section 2 discusses the DC motor model. Section 3 explains the design of the ILC algorithm. Section 4 compares ILC and conventional algorithms. Section 5 includes the simulation results of the DC motor, and the performance criteria were evaluated. Finally, Section 6 concludes how the ILC is better than the conventional PID controller.

2. Modelling of A DC Motor

An electromechanical DC motor system is considered in this work. From the electrical system with field and armature circuit, only the armature is considered for the analysis as the field is excited by a constant voltage. The rotating part of the motor and load connected to the shaft on the mechanical system side. The DC motor speed control system with armature is shown in Figure 2.

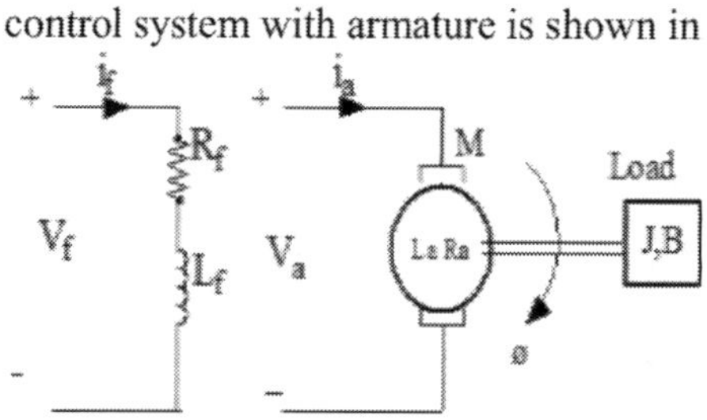

Figure 2. DC Motor Circuit

Where,
Ra is the Armature (Electric) resistance (Ω)
La is the Armature (Electric) inductance (H)
ia is Armature current (A)
Va is Armature voltage (V)
eb is the Back emf (V)
Tq is the torque developed by the motor (kg.cm)
ω is Angular displacement (rad/sec)
J is the Moment of inertia of motor ($kg.m^2$)

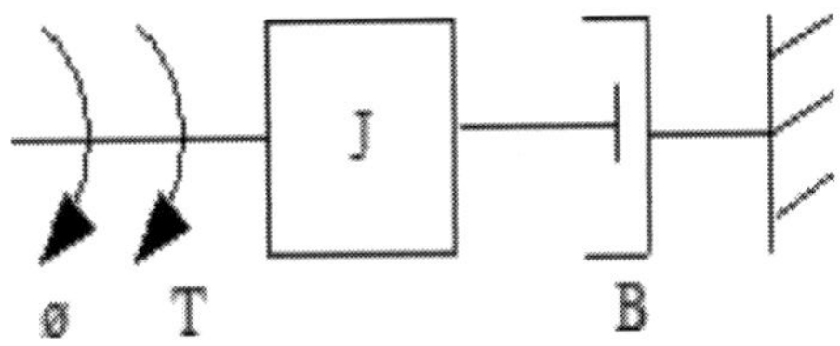

Figure 3. Mechanical Equivalent of the DC Motor

$$J\frac{d^2\theta}{dt^2} + B\frac{d\theta}{dt} = KI \quad (2)$$

$$Js^2\theta(s) + Bs\theta(s) = KI(s) \quad (3)$$

Where,
J is Motor's moment of inertia ($kg.m^2$)
B is Motor and load's frictional coefficient (Nm.s)

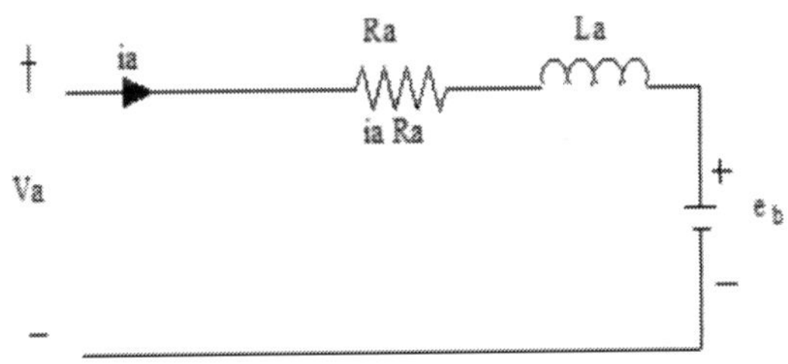

Figure 4. Electrical Equivalent of the DC Motor

$$eb = K\omega = K\frac{d\theta}{dt} \quad (4)$$

eb is the back emf
Applying Kirchhoff's voltage law

$$L\frac{di}{dt} + Ri = V - eb \quad (5)$$

$$LsI(s) + RI(s) = V - Ks\theta(s) \quad (6)$$

$$I(s) = \frac{V(s) - Ks\theta(s)}{R + Ls} \quad (7)$$

Using the equations (2)-(7) obtained from the mechanical equivalent shown in Figure 3 and electrical equivalent shown in Figure 4, we can arrive at the transfer function of the DC motor.

$$G(s) = \frac{\omega(s)}{V(S)} = \frac{K}{[(R+S)(JS+B)+K^2]} \quad (8)$$

3. Design of ILC Algorithm

3.1 Design Guidelines

The closed-loop system transfer function is represented by P(s), and the controller is represented by C(s), and ILC is shown in Figure 5. The tracking error ek is an error at k^{th} iteration and is given as input to the controller. It is found by comparing measured position yk with reference position yd. All iterations are of same finite time length, t ∈ [to tend]. The repetitive part of the error signal is reduced by ILC during the iterations. From one iteration to another, the feedforward signal is calculated by tracking error, and by using the learning filter L, the error is filtered and included in the forward path.

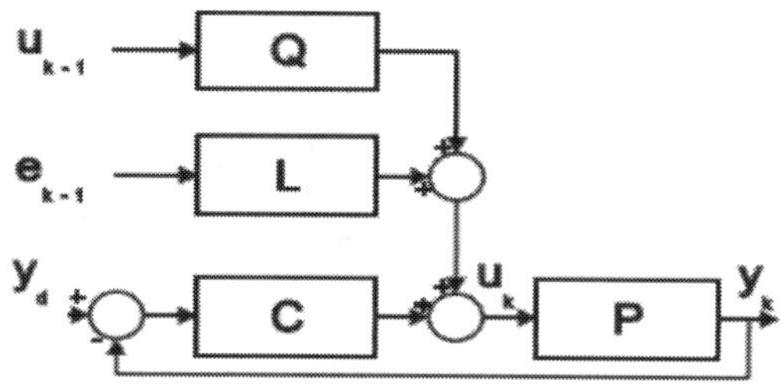

Figure 5. Block diagram of ILC Architecture

L, Q is ILC parameters to be chosen in accordance with the convergence criteria, where P is the process's transfer function, and SP is the sensitivity function of the same.
The control law (9) linear time-invariant (LTI) system yk = P*uk is as

$$uk+1 = Q*uk + L*(yd - yk) \quad (9)$$

A sufficient condition for convergence of the above equation for all yd and all uo is given in (10)-(15) :

$$|Q - SP*L| < 1 \quad (10)$$

Proof

$$ek = yd - yk \text{ , and } yk = P*uk \quad (11)$$
$$uk+1 = Q*uk + L*ek \quad (12)$$
$$ek+1 = yd - P*Q*uk - P*L*ek \quad (13)$$
$$ek+1 = yd - Q*yk - P*L*ek \quad (14)$$
$$ek+1 = (1 - Q)yd + (Q - P*L)ek \quad (15)$$

The error will decrease when Q and L satisfy the criteria equation of convergence. The fixed point error will be reached after one trial when (Q – SP*L) = 0. The error will increase every trial, and finally, the output won't be bounded when Q – SP*L > 1 then. The error will be constant as in (16) after a number of trials because it converges to a fixed point.

$$\lim_{k\to\infty} ek+1 = \lim_{k\to\infty} ek = e^* \quad (16)$$

Then the asymptotic error e* is given by equation (17)

$$e^* = \frac{(1-Q)}{1-Q+P*L} y_d \quad (17)$$

And the asymptotic input is in (18) – (20)

$$u^* = \frac{L}{1-Q+P*L} y_d \quad (18)$$
$$u^* = Qu^* + Le^* \quad (19)$$
$$u^* = \frac{L}{1-Q} e^* \quad (20)$$

3.2 Learning filter Design (Design of L)

The selection of Q and L is necessary for the design of the ILC controller for an LTI process for achieving good performance. From the asymptotic error equation, it can be observed that Q must be chosen close to unity to get a minimum error. From the convergence criteria, the learning filter should

obey the convergence rule and can be obtained as $(SP)^{-1}$ for fast convergence.
Using the closed-loop transfer from the feedforward to the error signal, i.e., the process sensitivity SP = P/(1+P*C). The propagation of the error from iteration to iteration can be written as ek+1 = (Q − SP*L) ek.

3.3 Design of Q

Q should be designed in such a way that L becomes a proper transfer function when multiplied with Q.
Example given in (21)-(23)
For first order Q,

$$Q = 1/(\varepsilon s+1) \quad (21)$$

Second order Q,

$$Q = 1/(\varepsilon s+1)^2 \quad (22)$$

Third order Q,

$$Q = 1/(\varepsilon s+1)^3 \quad (23)$$

The value of ε should be appropriately chosen. As ε value is greater, the settling time for the response is greater. For small values of ε, the settling time is minimum.

4. Comparison of ILC And Conventional Algorithm

The ILC algorithm has been implemented for a minimum phase system, and various values of ε in Q design; the response plots are shown in Figures 6, 7, and 8.

4.1 Simulation Results for Example 1

$$G(s) = \frac{1}{S^3 + 3S^2 + 3S + 1}$$

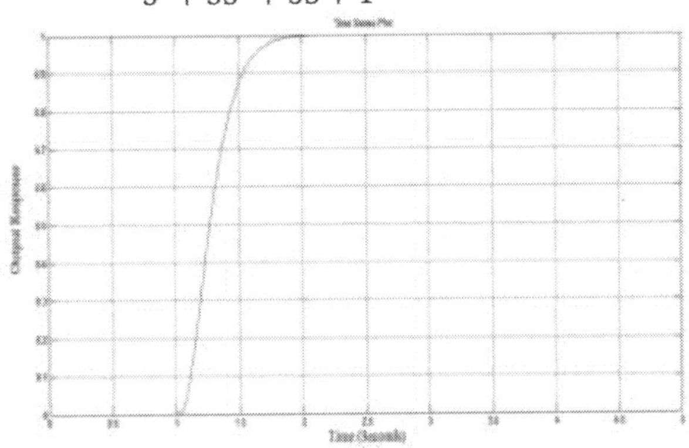

Figure 6. Output response when ε = 0.1

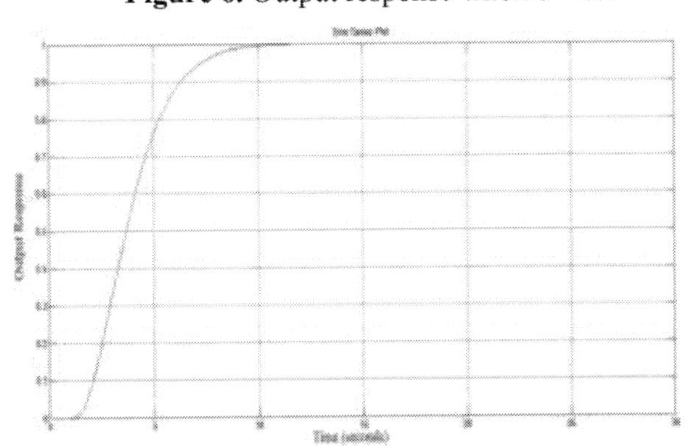

Figure 7. Output response when ε = 1

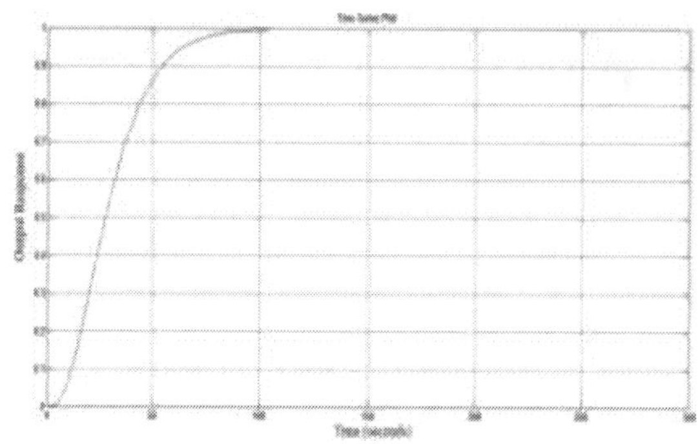

Figure 8. Output response when ε = 1

The PID controller is designed for DC motor velocity control. The mathematical representation of the PID controller is given in (24)

$$C(S)=K_p+\frac{Ki}{S}+K_dS \tag{24}$$

The Z-N tuning method can be used to evaluate K_p, K_i, and K_d.

The Table 1 and 2 show the performance indices ISE, IAE, ITAE for the systems designed under the ILC and Conventional algorithms. The simulations of the algorithms are done for three types of systems, namely: Minimum Phase System, Non-minimum phase system, and Delayed system.

Table 1. Performance Indices of Systems Design Under ILC Algorithm

Systems	Performance Indices	
Minimum Phase System $\frac{1}{S^3+3S^2+3S+1}$	ISE	:0.2062
	IAE	: 0.3
	ITAE	: 0.36
Non-Minimum Phase System $\frac{-S+1}{S^2+S+1}$	ISE	: 0.5
	IAE	: 1.0
	ITAE	: 2
Delayed System $\frac{1}{S^2+3S+1}$	ISE	: 1.25
	IAE	: 2
	ITAE	: 5

Table 2. Performance Indices of Systems Design Under Conventional Algorithm

Systems	Performance Indices	
Minimum Phase System $\frac{1}{S^3+3S^2+3S+1}$	ISE	:0.81
	IAE	: 0.725
	ITAE	:0 .712
Non-Minimum Phase System $\frac{-S+1}{S^2+S+1}$	ISE	: 2.076
	IAE	: 2.22
	ITAE	: 5.602
Delayed System $\frac{1}{S^2+3S+1}$	ISE	:0.541
	IAE	:0.451
	ITAE	:2.41

4.2 Uncertainty Handling

For the minimum phase system discussed above, the ILC algorithm is executed when the system has 10% uncertainty. The output response of a step signal in the presence of 10% uncertainty is shown in Figure 9.

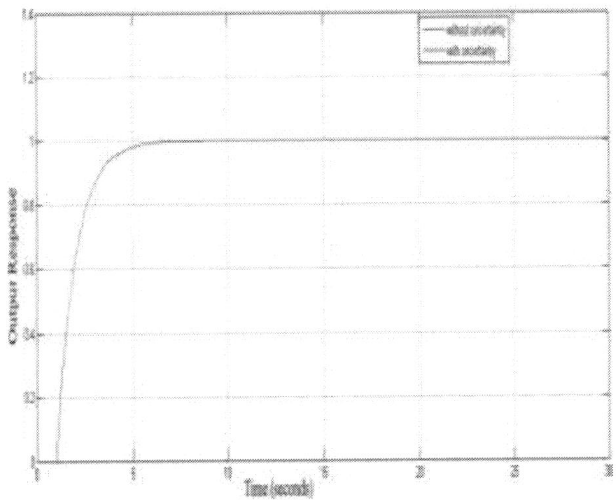

Figure 9. Output Response of ILC under 10% uncertainties

For the same minimum phase system, when handled by a conventional algorithm, the output response of a step signal in the presence of 10% uncertainty is shown in Figure 10.

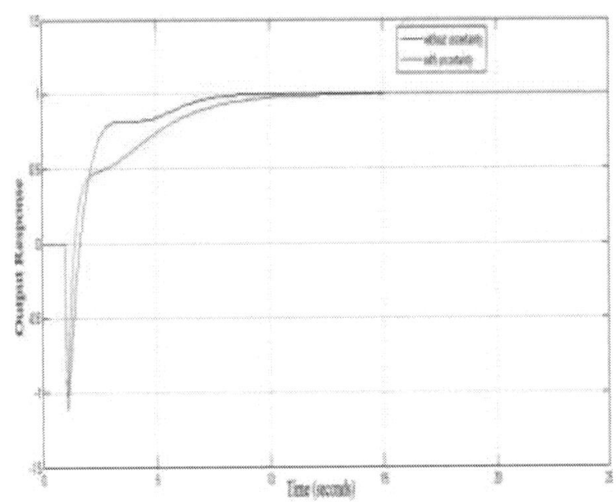

Figure 10. Output Response of conventional algorithm under 10% uncertainties

5. Results and Discussion

The proposed ILC controller is designed for the DC motor through simulation. The same is done using the conventional method. This is done for comparison purposes.

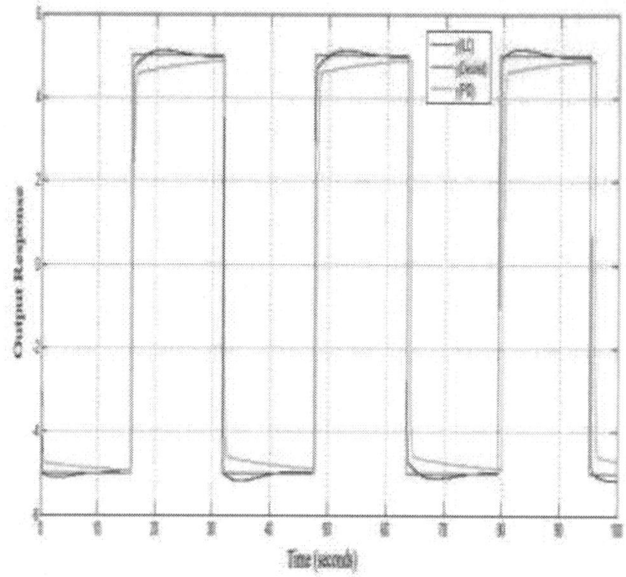

Figure 11. Output responses of ILC & Conventional Algorithms

Figure 11 shows the output response plot of both ILC and conventional algorithms. From the figure, it is clear that the rise time of an ILC algorithm is less when compared to a conventional algorithm. And there is an offset present in the conventional algorithm, and it is eliminated by the ILC algorithm
The Absolute Error versus Time plot is shown in Figure 12, and it reveals that the area under the graph is greater for the conventional algorithm than that of the ILC algorithm.

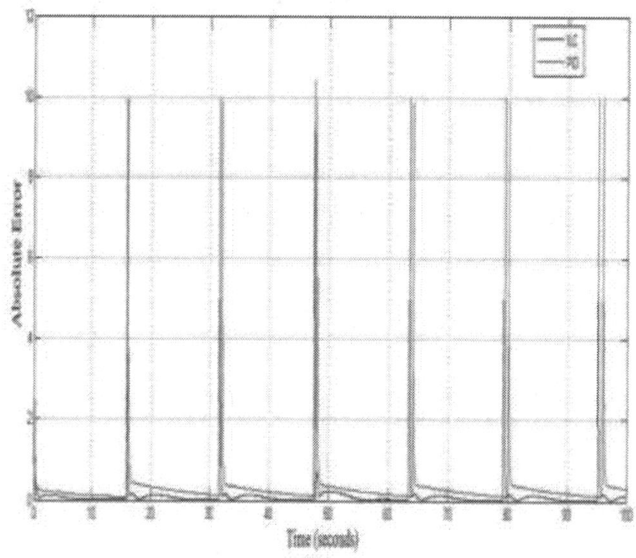

Figure 12. Plot of absolute error using ILC & PID for the DC Motor

The Integral Absolute Error plot is shown in the Figure 13 from that it is observe that the error value for conventional algorithm keeps increasing by large number in comparison with that of the ILC algorithm.

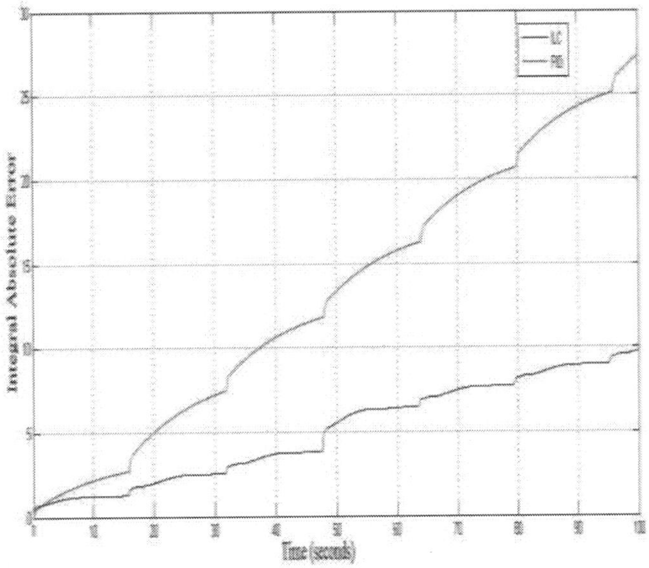

Figure 13. Performance measure of the DC Motor using ILC & PID Algorithms

6. Conclusion

The application-based implementation, as well as the test case systems of the ILC algorithm, are the evidence to validate the superiority of the ILC algorithm over the conventional algorithm. The ILC algorithm offers better speed control than the conventional control algorithm. The rise time is reduced, and the settling time is also lower, thus offering better efficiency. Thus the objective is validated, and hence we can conclude that the ILC algorithm offers better performance.

References

[1] Błażej C Krzysztof G Eric Rogers 2014 2D systems based robust iterative learning control using noncausal finite-time interval data, Systems & Control Letters pp 36–42.
[2] Rezaei M Kerman 2011 PID Parameter Selection Based on Iterative Learning Control Electrical Engineering Department University of Kerman Contemporary Engineering Sciences Vol 4 pp 201 – 220,
[3] Lukasz Hi Krzysztof G Zhonglun C Eric R Chris Freeman T Paul L 2010 Experimentally supported 2D systems based iterative learning control law design for error convergence and performance Control Engineering Practice Vol 18 pp 339–348.
[4] Wang Y Furong G Francis J Doyle III 2009 Survey on iterative learning control repetitive control and run-to-run control Journal of Process Control pp 1589–1600.
[5] Kevin L Moore Yang Q C and Hyo-Sung A 2006 Iterative Learning Control A Tutorial and Big Picture View, Tutorial Sessions - Industry and Applications IEEE CDC.
[6] Bristow DA Tharayil M Alleyne AG 2006 A Survey of Iterative Learning Control A learning based method for high performance tracking control IEEE control systems magazine Vol 26 pp 96-144.
[7] Owens DH Hatonen J 2005 Iterative learning control- An optimization paradigm Annual Review in Control Vol 29 No 1 pp 57-70.
[8] Owens DH Feng K 2003 Parameter optimization in iterative learning control International Journal of Control Vol 76 No 11pp 1059-1069.
[9] Jian-Xin Xu Ying Tan 2003 Linear and Nonlinear Iterative Learning Control Springer-Verla pp 177 ISBN 3-540-40173-3.

[10] Schutyser PJ 2002 Noncausal Iterative Learning Control Control Laboratory Faculty of Electrical Engineering University of Twente.
[11] Longman RW 2000 Iterative Learning Control and Repetitive Control for Engineering Practice International Journal of Control Vol 73 pp 930–954.
[12] Amann N Owens DH Rogers E 2000 Analysis of Linear Iterative Learning Control Schemes - A 2D Systems/Repetitive Processes Approach Multidimensional Systems and Signal Processing Vol 11 pp 125–177.

Implementation of Cardamom Sorting Machine

S P Sriram[1] , M Veera Sivakumar[2] , P V Rithika Shree[3] , K Nagavarshini[4] and K Poomani[5]
[1, 2, 3, 5] Department of Mechatronics, Bannari Amman Institute of Technology, Sathyamangalam, Tamilnadu.
[4] Department of Agriculture, Bannari Amman Institute of Technology, Sathyamangalam, Tamilnadu.
sriram.mc19@bitsathy.ac.in, veerasivakumar.mc19@bitsathy.ac.in, rithikashree.mc19@bitsathy.ac.in, nagavarshini.ag19@bitsathy.ac.in, poomanik@bitsathy.ac.in

Abstract. Cardamom grading is an imperative agricultural invention handing out conscientiousness which continuously seeks for improve methods of mechanization. Categorization the cardamom includes the process of segregating cardamom based on its skin texture such as the husk size, husk identity and peapod colour. Grading includes the procedure of assigning grades and in this way categorizing cardamom to diverse delivery qualities as the cardamom superiority increases with price, therefore it is required to grade the cardamom. Optimization techniques like K means nearest Neighbour algorithm, decision tree are combined with machine vision to find the cardamom colour and can be used for the purpose of grading. The cardamom is compared with the finest elected cardamom for the cataloguing manages and the cardamom is graded based on the colour of the cardamom. The project deals with the design and fabrication of cardamom sorting machine with the help of machine vision and machine learning process.

Keywords: Cardamom grading, K- means algorithm, cardamom colour

1. Introduction

Visual based defect detection is industrial friendly. It is used to detect the different kinds of errors and defects in the industrial product. This error solving visual based defect detector can be used as three thing, which is Quality Control Program, Data Acquisition and Data Analysis. Visual Inspection is uncomplicated and easiest method but the contribution towards industries is ultimately high as this technology recognizes the standard quality of the product. Visual -based defect detection technology enhances the quality and improves the standard of the product and management. therefore, this type of technology is more preferably used among the industries. To bring out the promising standard quality and uniformity the product must be produced without any defect. Therefore, this visual based defect detection technology is used. There are two modes of visual based defect detection technology, they are manual inspection and automated inspection. Both the inspection has their own merits and demerits toward the manufacturing industry. Human based manual inspection causes errors and defects in identifying the quality of the produce, so proper training is essential to improve the skill of the inspector.

2. Literature Survey

Bailey t, et al (2018) proposed the theory of distance-weighted K nearest neighbour algorithm. In this article, Bailey proposed that when the number of training sample is small, then the distance -weighted rule would yield only a smaller probability of error, than the majority rule. They show the admissibility of the distance -weighted k -nearest neighbour rule by demonstrating that the probability of error, for a particular dataset obtained by the distance-weighted rule is strictly lower than that for majority rule. This approach reduces the distance between the required and the present data set [1].

Sheela S, et al. (2016) proposed the theoretical model for object sorting on conveyor belt. They proposed the low cost automation for object sorting in conveyor belt using Raspberry pi 3 and a USB camera for detecting the colour and size of the objects. They intend to discard the objects which are not in desired colour or the suitable size by pushing them out of the conveyor belt .They are using a linear actuator for discarding the undesirable object out of the conveyor. This method is suitable for conducting sorting operation on the conveyor [2].

J. Kim,(2012) proposed a structured light camera based 3D visual perception and tracking system application with robot arm grasping for manufacturing. The light camera is used to track geometry of an object based on the shape and creating 3D point clouds of the object. The tracked 3D points are then converted to DXF or 3DS format for the computer to read the shape of the object. This method is advanced version of the machine vision to further improve the perception and accuracy of the system [3].

Tanishka Singh, et al.(2016) proposed a method for object sorting by Robotic arm using image processing techniques .They proposed to use high resolution camera to capture the image of an object. By doing so the images can be used for reading shape and structure of the object and performing the operations accordingly. This proposed concept is simple. The robot arm can be used to separate and this can be used in quality checking and control [4].

A .Gowen, C.P., et al.(2017) proposed an efficient method for food quality and safety control using hyperspectral imaging of the food particles. They provided an introduction for hyperspectral imaging of food using image processing and image acquisition. This proposed method can be used to analyse the food particle such as its nutrients to estimate the date for the food expiry identification in the food processing industry. A simple approach is followed in agriculture to find the defective product [5].

2.1 Problem Description

In agriculture, immense care and time should be taken for sorting and grading which are the basic key sectors. Sorting is nothing but segregating a particle based on size using sieve of required sieve size. Grading is process of assigning grades and categorizing in different qualities. Grading is based on external appearance. Cardamom which is one of the important spices in Indian food takes much time in sorting and grading. It is sorted based on the size and it is graded with its external appearance.

2.2 Solution of the Problem

To overcome this difficulty various machine learning algorithm combined with image processing is used. The advantage by using this machine learning is precession in answering improvises with each learning procession.

3. Machine Vision

Machine vision is the technology and method used to provide imaging based automatic inspection to learn about the physical characteristic of the process. The process includes converting the real time analog signal to digital format so that the computer can process data from the given object.

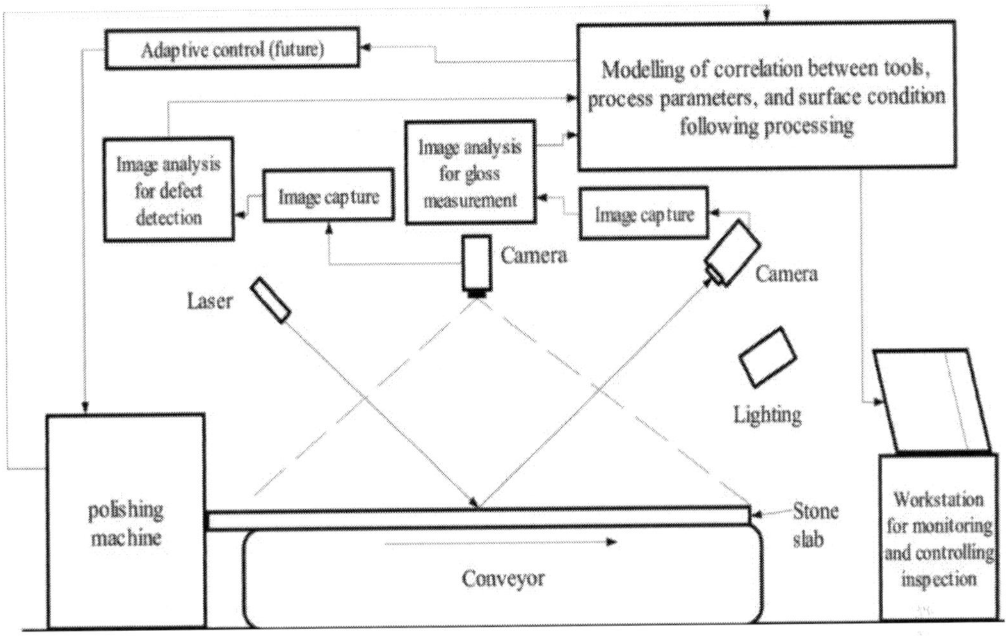

Figure 1. Machine vision in inspection system

The figure 1 shows implementation of the machine vision in the process of inspection in industrial application. The image obtained from the camera acts as input, the captured image is enhanced for processing and the image analysis is done find out whether the product is defective or not.

4. Working Process

The entire working process is shown in the flow chart figure 2. The each and individual steps will be explained in the following pages.

4.1 Initializing

Prior to taking care of the cardamom to the container, the associations are checked and the physical dependability of the surface is checked. Presently the switch is turned on, which thus gives flexibly to the principle engine of the transport. Presently the cardamom is taken care of into the container. If it's not too much trouble guarantee that the container isn't flooding with the cardamom, feed for the most extreme holding ability to lessen sticking of the cardamom.

4.2 Sensor input

When the transport begins moving the cardamom took care of to the container tumbles to the outside of the transport. Presently the transport pushes forward and the cardamom is isolated and sent independently by the plan of the transport. At the point when the cardamom interacts with the light of TCS230 the shade of the cardamom is found. Subsequent to finding the shade of the cardamom the sensor imparts the sign to the microcontroller in the computerized design which is prepared to peruse by the regulator.

4.3Decision making

At the point when the cardamom interacts with the sensor, the choice must be made. In first case if the cardamom is acceptable, the regulator detects it and doesn't do anything. Then again, if the nature of the cardamom is awful, a 12V straight actuator is utilized to jump out the cardamom from the line. This choice is significant as it straightforwardly influences the arranging cycle and thus this cycle is considered as the fundamental factor in arranging the cardamom.

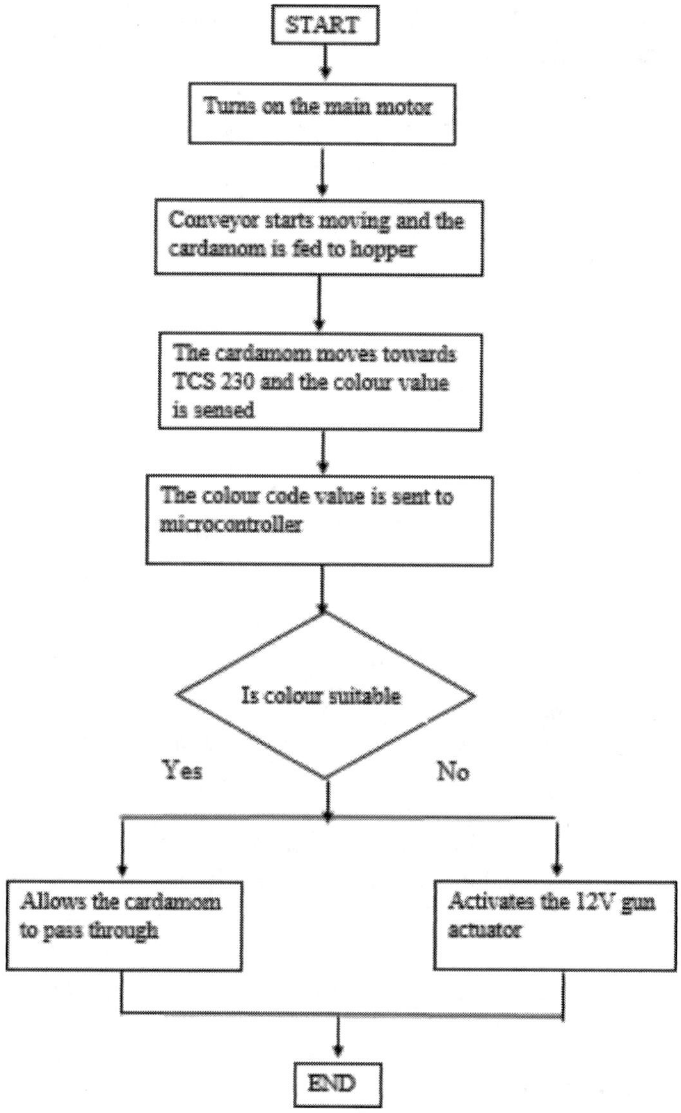

Figure 2. Process flow chart

4.4 Timing

Exactly when the cardamom collaborates with the sensor, the decision must be made. In first case if the cardamom is worthy, the controller identifies it and doesn't do anything. On the other hand, if the idea of the cardamom is dreadful, a 12V direct actuator is used to leap out the cardamom from the line. This decision is noteworthy as it direct impacts the masterminding cycle and consequently this cycle is considered as the essential factor in orchestrating the cardamom.

4.5 Display

Show unit is LCD screen; which is utilized to show the RGB estimations of the current cardamom. By showing we can examine with the hypothetical outcomes whether the test results are the equivalent. The Typical showing of the values are around 20's. However, the showed qualities can be roughly the real qualities along these lines, the genuine qualities will be near + or – 5 of the real qualities.

4.6 Separation of cardamom
After the sign from the sensor arrives at the direct actuator, the actuator pushes ahead which thus pushes the awful cardamom out of the line. The awful cardamom can be gathered from the different zone while the great quality cardamom is permitted to move in the line. the dismissal system relies on the measure of awful quality cardamom in the transport line. In the event that there is all the more terrible cardamom, at that point the proportion of creation to dismissal will be low. Under ordinary conditions the dismissal proportion will be 5:1, where the measure of the terrible cardamom will be 20 percent of the delivered sum. The isolated cardamom can again be sent to the transport for partition however with the diverse dismissal esteems set. Again, the dismissal is legitimately needy upon the pre-set RGB estimations of the cardamom.

4.7 Testing environment
Before testing the right condition must be set to get the strong qualities and the same about qualities ought to be consistent, as the dismissal system is relying onthe steady shading qualities, nature with consistent lighting must be set. Theshading esteems are in a roundabout way identified with the surrounding lighting henceforth the room ought to consistently be kept up with steady light source.

4.8 Steps Involved in KNN Algorithm

K Nearest Neighbour is a supervisor machine learning algorithm used to solve regression problems

1. Find the number of nearest number needed(k)
2. Distance between Labelled and Unlabelled new data should be calculated
3. Based on the shortest distance and K-Neighbour the labelled data should be used.
4. Compare the new data with labelled data
5. The new data should be categorized under any one of the labelled data

5. Product Design

The detailed 2D design of the entire cardamom sorting machine is shown in figure 3. The colour sorting cardamom machine separates the cardamom based on the colour of the cardamom. The product has a conveyor which moves the cardamom for sensing colour and sorting. A colour sensor or a high resolution camera is used to take multiple images of the cardamom, the images are read by the microcontroller computer like raspberry pi. This product uses a machine learning algorithm like KNN algorithm to check the multiple images taken. The colour code values after checking the image will act as a source of input to the controller. After finding out the colour of the cardamom, the controller acts immediately. A 12V dc linear actuator is used to pop out the poor quality cardamom. The separated cardamom can again be sent to the queue separately for second quality sorting. Now the colour values should be adjusted based on the cardamom that have been rejected.

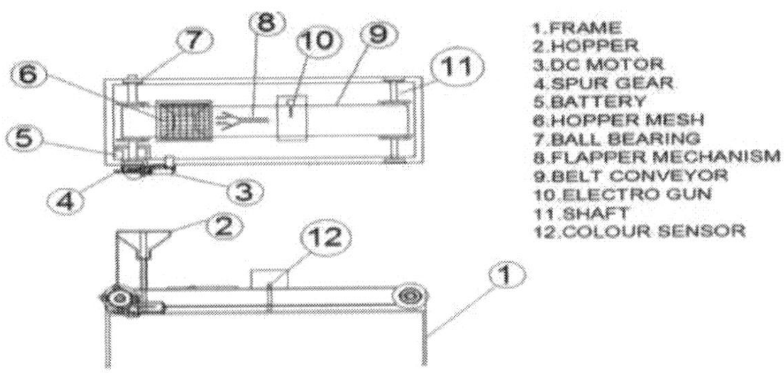

Figure 3. Detailed 2D design

The Design is done using AutoCAD software and is labelled in paint software. The detailed dimension in mm are shown in figure 4 and figure5.

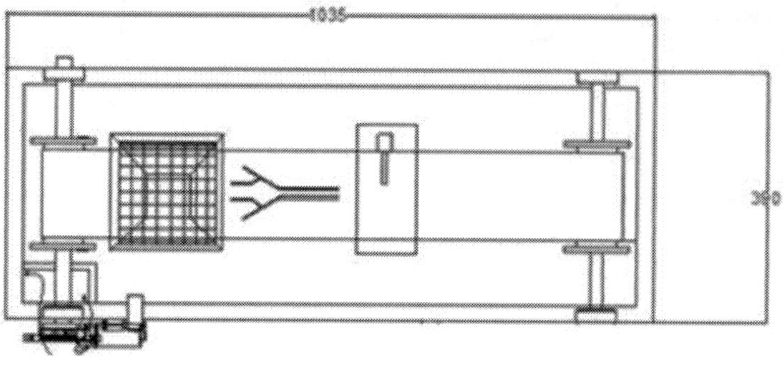

Figure 4. Top view

Figure 5 shows the front view design of the project. Theheight of the frame is 250mm and the hopper is tapered to 40degrees. The hopper has the box pipe installed at the end and the dimension of the pipe is 25x25mm.

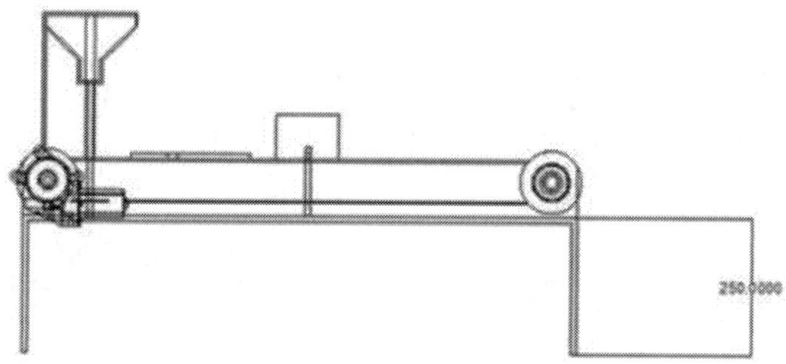

Figure 5. Front view

6. Conclusion

In the areas where cardamom plantation is carried out after the long drying process this cardamom sorting machine can be applicable. Sorting of the cardamom according to color and storing separately helps in fixing high price for the green colored good quality seeds and the brown colored faded cardamom. By using this type of method the owners will be highly profitable comparatively. Visual inspection of the cardamom is done using the KNN Algorithm and simulated. The experimental values may vary depending upon the ambient lighting of the surrounding.

References

[1] Bailey T and Jain A 1978 A note on distance-weighted K-Nearest Neighbor rules IEEE Trans.Systems,Man,Cybernetics. Vol 9 pp 311-313
[2] SheelaS,Shivaram.K.R Meghashree S Monica L Prathima A and Shriya M Kumar 2016 Low Cost Automation for Sorting of Objects on Conveyor Belt International Journal of Innovative Research in Science, Engineering & Technology. Vol 13 pp 195-200
[3] Lee S Kim J and Lee M 2012 3Dvisualperceptionsystemforbinpickingin automotive sub-assembly automation Automation Science and Engineering (CASE) 8th IEEE International Conference on Automation Science and Engineering. South Korea pp 706-713
[4] Taniksha Singh Dnyanesha Dhaytadak PradnyaKadam and Sapkal RJ 2016 Object Sorting by

Robotic Arm Using Image Processing International Research Journal of Engineering and Technology (IRJET). vol 3 1472 - 1476
[5] Gowen A A O'Donnell CP Cullen PJ Downey G and Frias JM 2007 Hyperspectral imaging an emerging process analytical tool for food quality and safety control Trends in FoodScience and Technology. Vol 18 pp 590- 598

Performance and Emission Characterization of Adjustable Cylinder Head Type Variable Compression Ratio in Two-Stroke SI Engine

R Gowtham[1], A Hakeem Meeran Bathusha[2], R Sankara Ganapathy[3], K Mohanraj[1] and N Venkatasubramanian[1]

[1]Assistant Professor, Rajalakshmi Engineering College, Department of Automobile Engineering, Chennai, Tamilnadu, India.
[2]Student , Rajalakshmi Engineering College, Department of Automobile Engineering, Chennai, Tamilnadu, India.
[3]Student , Rajalakshmi Engineering College, Department of Mechanical Engineering, Chennai, Tamilnadu, India.
gowtham.r@rajalakshmi.edu.in,
hakeemmeeranbathusha.a.auto.2017@rajalakshmi.edu.in ,
sankaraganapathy.r.mech.2017@rajalakshmi.edu.in.

Abstract. A lot of technologies have been introduced to retain the petrol engines which exhibit a lot of drawbacks based on emissions and knocking. Variable compression ratio (VCR) technology is applied in the line of SI engines for a long period to give back the standards. This paper is going to analyze and experiment on a commercially available type of VCR technologies. The investigation can provide a great method of varying the compression ratio that is put forth, practically analyzed, and previewed with a change from the previous paper and taken to an account to restrict the compression ratio of +/-1.5. In this analysis, we use the same technology as commercially used two-stroke engines but the clearance volume is changed and introduces a metallic bung on the burning chamber. We have performed the test by varying the compression ratios to more or less than two from the original compression ratio. At lesser weight conditions, a low compression ratio is maintained throughout the experiment to boost up the turbocharger to get a better amount of volumetric efficiency without any trouble. The engine is tested at various compression ratios for the performance study and emissions from the power source at various load conditions. While adopting this method in commercial, we can achieve a potentially higher fuel economy, more power, lower emission compared to other available technologies in ongoing research.

Keywords: Fuel economy, two stroke Engine, Variable compression ratio.

1. Introduction

Based on the working condition of the engine, an engine having a varying compression ratio can be referred to as the variable compression engine .The compression ratio is reduced during higher loading conditions for more power. The compression ratio is kept higher at lesser load conditions for higher output performance. To reduce the knock occurrence, an engine with compression ratio which fits to demands of the load is needed under full load conditions. At lower capacity, the VCR engine works at a better compression ratio to realize the efficiency advantages of fuel, so at high power levels the engine works at a lesser compression ratio to avoid knocking. The ideal compression ratio is the working of vehicle working factors as air inlet factor and temperature of coolant in engine, outlet gas temperature, knocking, fuel octane rating, etc. In a VCR engine, the working temperature is mostly kept at appropriate temperature, where combustion efficiency is higher. Using a two-step compression ratio is to produce a mass engine practically and able to switch to a low compression ratio in 1 second. The variable compression ratio has slight friction in the essential engine [1]. In this study 100% petrol and 20% n-butane alcohol mixture in gasoline is utilized. As a result of which, at low load brake thermal efficiency increases with compression ratio .We can see that the adoption of VCR technology the part load efficiency of gasoline engine is increased and also with fuels having good anti knocking are used[2].Performance and Emissions of pinnae oil-based bio-fuel are analyzed. The fuel mixes up to 1/4th are tested with four variable compression ratios. Fuels which are blended gives great results compared to diesel. At high compression ratio the emissions are low in VCR engine [3]. Effects of Compression ratio and lambda on CNG SI ICE 1st and 2nd law are practically analysed. With its methane decreasing, CNG shows more irreversible in combustion [4]. Tests were done with diesel and its works of a 4- stroke engine for proportional study with an injection force of 220 bar and compression ratio of 19:1[5]. The two-stroke engine was simple in working and in construction and had more advantages such as power, efficiency, and pickup. The engine needs two strokes of the piston for a cycle. The exhaust gases are escaped as the cylinder walls are not closed by the piston and the pressure inside the cylinder is reduced. The fuel-air mixture moves to the combustion chamber and pressurizes in the crankcase to a higher pressure than atmospheric [6-7]. The two stroke engine produces more power than a four stroke engine based on single piston movement. In Two Stroke Spark Ignition engines the thermal efficiency increases with a compression ratio rise. Under half speeding conditions, the fuel consumption shall be improvised with a higher rate of compression ratio. Anyhow, this is frequently accompanied with issues of roughness and knocking. This paper describes a type of new cylinder head design which lets operation at higher compression ratio under half weight circumstances and at low compression ratios under full speed condition. The reviews of all the design approaches and answers used to get VCR with its unique features consider the results of previous research and its corresponding commercial barriers. The gradually more stringent emission norms and to achieve a better economy of fuel have long prolonged difficulties for research in the automobile sector for advancements towards thermal efficiency and less polluting engines. VCR technology has been considered for a long time as a way for increasing the engine performance, mileage and output characteristics as possible for emission reduction. The most important thing about the VCR engine is to work as possible for various loads of compression, by varying the volume of the combustion chamber, together with the performance. The design and results towards achieving VCR with its unique features, consider the results of previous research and its corresponding commercial barriers [8-11]. The comparatively investigated the properties of En19 Steel and En41b Steel employed in Camshaft of diesel engine. The motive of the paper is to review the residual stress of the cylindrical specimen at surface. The engine camshaft is one of the most important parts within the automobile engines. Due to the rotational effect loading on the touching surfaces in the follower and cam it gives a wear in the component and causes a constant working of the gear valve (speed), displacement of valve, and therefore the torque [12]. To review experimentally the residual stresses of cylindrical objects at the surface layer. The camshafts from diesel engines are generally made by the EN19 Steel material. Replacing EN19 steel with EN41B. The outcome was taken from the hardness test rig shows that the hardness of the Nitride steel EN 41B is 31 percent more than EN19 Steel and therefore the hardness of the Nitride steel EN41B is 45 percent greater than EN19 steel after hardening. The impact test shows that the toughness of EN41B is 38 percent greater than EN19. They all developed an idea of a Variable Compression Ratio Engine using a metal plug. Traditionally two stroke petrol engines discover a wide variety of uses in the manufacturing of two-wheelers universally, due to environmental laws these engines are decreasing in numbers. Despite these packed norms, these engines are still used in agriculture, generators, etc. During this present analysing, a great method of adjusting the compression ratio is researched and analysed. By introducing a metal plug on the combustion chamber the clearance volume is changed. This modification acceptable to possess four different values of clearance value has been brought out as two sections. The primary part deals with basic planning and fabricating of engine, and testing at various compression ratios for the study of the performance of the engine. The subsequent part deals with the combustion in the engine by means of FLUENT and the examination of the exhaust system. An increase in the compression ratio improvises fuel efficiency and output. The greatness of this project is to allow the 2 wheel drive to change the compression ratio. They have a new proposed design to increase or decrease the compression ratio by modifying the connecting rod at two stages. By this method, we can get high fuel economy savings in comparison to other systems [13-14]. They introduced the knowledge about the progress in the inquiry and acknowledged his expertise concerning the dynamometer and flow bench applications. They perform an experimental and theoretical investigation on BT efficiency by varying the effect of CR. The effect of compression ratio on brake thermal efficiency, CO & NO_X are analyzed using Greaves MK20 SI Engine [15]. The thermal efficiency is influenced mainly by compression ratio. A Higher compression ratio gives higher thermal efficiency and a lower compression ratio gives lower thermal efficiency in the engine. The operating conditions of a vehicle differ variably, as in traffic, frequent stopping and crowded roads, in freeway highway driving, and constant-speed highway driving, also driver condition depends. In a common SI engine, the maximum compression ratio is set by the state in the cylinder at high load. If the compression ratio is higher than the predetermined limit, the fuel will pre-ignite causing knocking. When the load is increased temperature increases at the end thermal efficiency is dropped which influences the compression ratio and air-fuel mixture. The fuel-air cycle Efficiency decreases so that a high compression ratio could be induced without the risk of knocking in power engines. Increasing the compression ratio from 8 to 14 gives efficiency again from 55 to 70 percent (15%), whereas going from the ratio from 16 to 20 produces a gain from 62 to 65 percent (3%). The figure 1 given below explains the effects of the compression ratio based on thermal efficiency.

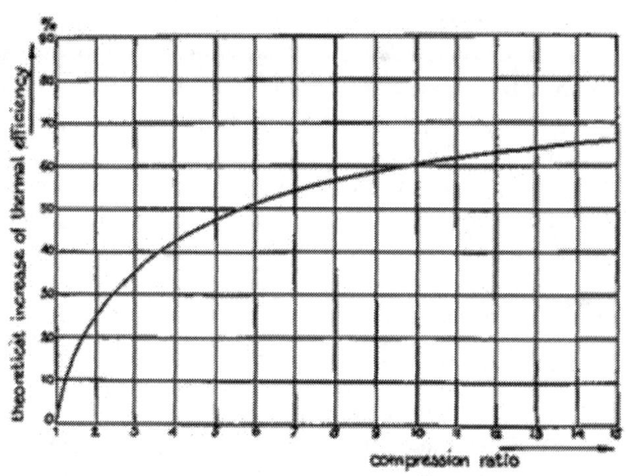

Figure 1: Compression ratio Vs. theoretical increase of thermal efficiency

IOP Conf. Series: Materials Science and Engineering **995** (2020) 012005 doi:10.1088/1757-899X/995/1/012005

2. Materials and Methods

2.1 Experimental Design

The variable compression ratio is gained by modification of volume of the cylinder by dislocating the periphery piston, which is small and meets with the combustion chamber. The compression ratio is altered with the use of a periphery piston. For a particular condition, it is advised to keep the piston at median point near to the best compression ratio, anyhow it needs the cylinder bore at finite length for the piston to move freely. The variation in the piston's compression height gives a correct way to form a variable compression ratio engine. To activate the change of height with high speed reciprocating assembly, the VCR system needs to increase the reciprocating mass. All the pointed concepts above need extra devices to make the engine operation difficult and change the compression ratio. Thus in this work, a change in the later model is made to get a higher difference in the compression ratio and analyze the difference in data.

2.2 Variable Compression ratio

Selection of engine: A center placed spark plug cylinder head is chosen to prevent the contact of the piston and metal plug. With all the above in mind, the YAMAHA RX100 engine with the following specifications was chosen. The scrub capacity of the engine sustained at 96.97cc and the burning ratio changed while changing consent capacity while introducing an additional chamber. By using this only decreased compression ratio, cylinder head drilled with a diameter 9.5mm along with it is threaded with a standard threading diameter of M9 so that the metal plug can be threaded without difficulty. The threaded hole in the cylinder head is drilled at 350 so that the metal plug can be inserted to a Maximum length of 8.5 cm without disturbing the piston and the spark plug. The fins are grinded such that the metal plug is seated perfectly in the cylinder head which prevents leak of gases from the burning chamber. The tabular column 1 is Engine Specification and Tabular column 2 Chemical composition of metal plug.

Table 1: Two-Stroke Engine Specification

MAKE	**YAMAHA RX100**
ENGINE TYPE	Petrol Engine with Centered spark plug
POWER	11kW
SPEED	7500rpm
BORE	50.8 mm
STROKE	50.8mm
DISPLACEMENT	98 cc
COMPRESSION RATIO	6.7:1

Table 2: Chemical Composition of Metal Plug

Material	**COMPOSITION**
Ferrous	97 – 97.79%
Carbon	0.36 – 0.44%
Silicon	0.21%
Manganese	0.7 - 1%
Phosphorus & Sulfur	0.01%
Aluminum & Nickel	4.21% & 0.23%
Chromium	0.9 – 1.2%
molybdenum	0.25 – 0.36%

2.3 Design of Metal Plug

The metal plug is made from 709M40 material, low alloy steel which can withstand higher temperatures up to 1000° C. The chemical composition of 709M40 material has been described below. The selection based on low wear resistance, high quality, high tensile strength and metric round bar. They have a tempered hardness of 201-375HB. It should have a Thermal processing temperature of 850-950°C.

The machined metal plug is shown in figure 2 and Figure 3 is 3D modelling. The machined tool must be precisely made with less tolerance to manufacture the metal plug error-free. Mechanical properties of the metal are: Tensile strength – 700 to 1225 MPa; Yields strength – 445 to 940 MPa; Hardness, Brinell – 201 to 175 HB.

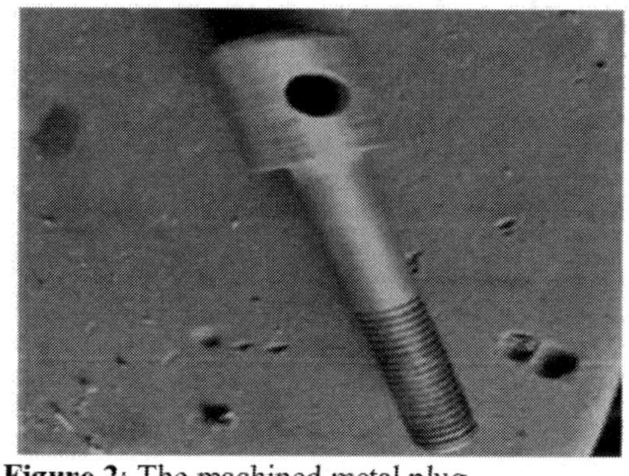

Figure 2: The machined metal plug

Figure 3: 3D Modeling of machined plug

2.4 Experimental Fabrication

The machined metal plug is screwed within the plate and the plate is fitted into the engine block. Acceptable care ought to be taken that the metal plug shouldn't knock with the piston once running. During this paper, the metal plug of length 8.5cm is inserted within the plate that goes to reduce the clearance volume of the cylinder wherever the compression quantitative relation of the engine is altered. The metal plug inserted shouldn't disturb the plug position which could end in incomplete combustion, and fails the novelty during this work. The figure 4 and 5 shows the cylinder head at lower compression ratio and 3D Modeling. The figure 6 and 7 shows the cylinder head at higher compression ratio and 3D Modeling.

Figure 4: Cylinder Head at Lower Compression Ratio

Figure 5: Model Shows Compression Ratio at Low

Figure 6: Cylinder Head at Lower Compression Ratio

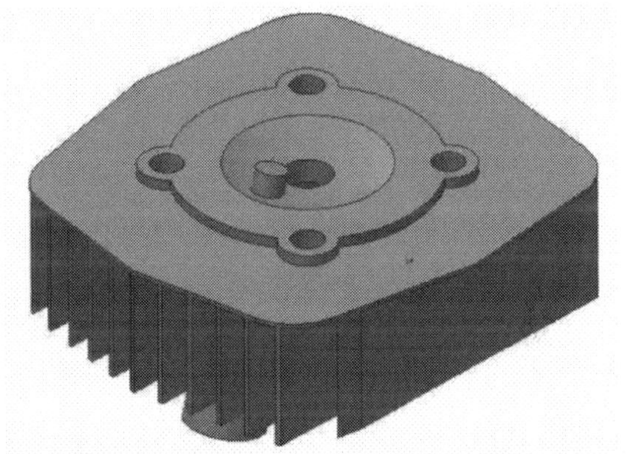

Figure 7: Model Shows Compression Ratio at High

3 Results and Discussion

3.1Perfomance characteristics

3.1.1 Total Fuel Consumption

The below graph represents the difference of complete combustible utilization with brake power at distinct compression ratio .The combustible utilization raised with capacity in least the compression ratios .The complete combustible utilization raised till compression ratio of 6.9. The progress within combustible utilization taken into account to be consequence of decreased heat capacity ratio of functioning gases and raised time loss, mechanical loss and cooling loss.

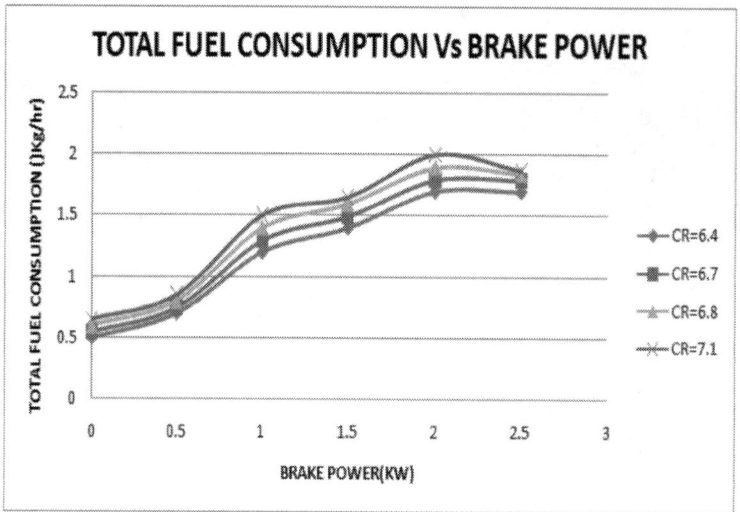

Figure 8: Variation of total fuel consumption with BP at different compression ratio.

3.1.2 Brake Thermal Efficiency

The below graph shows that the change of brake thermal efficiency with brake thermal capacity at various compression ratios. The consequence exhibits that BTE increases with compression ratio. Due to the functioning gas centigrade high-flown by a rise in the compression ratio. Effect of the burning ratio cannot be prevented while changing the functioning gas particular heat ratio

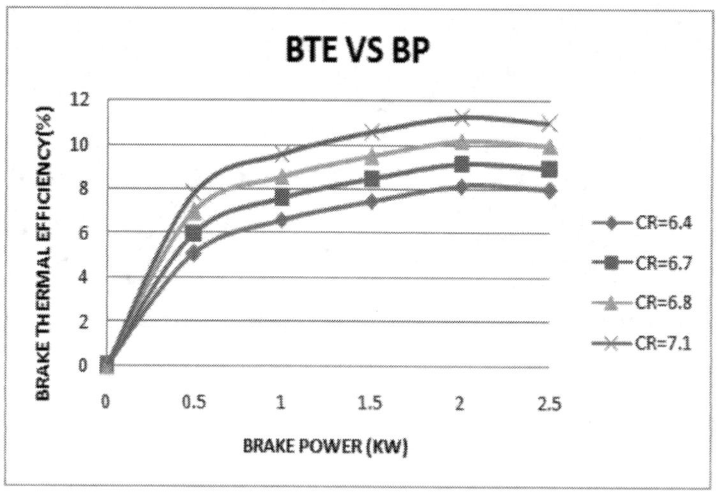

Figure 9: Variation of Brake Thermal Efficiency with compression ratios

3.2 Emissions Characteristics

For internal combustion engines, the challenge is not only to confirm future emissions standards: which is already possible under good conditions as well as to check the CO and NOx emissions along with other products. There is an increase in compression ratio to increase the capacity ratio in the burning chamber. Due to a decreased surface to volume ratio, the hydrocarbon emissions and carbon monoxide are decreasing and will be visible in the graph. The thermal efficiency is lowered and as a result exhaust gas temperature is increased, inside the combustion chamber attaining a complete combustion increases the peak cycle temperature. The use of higher compression ratio results in more blaze gas centigrade and smaller remaining content. It may guide a larger NO_x discharge on a capacity

IOP Conf. Series: Materials Science and Engineering **995** (2020) 012005 doi:10.1088/1757-899X/995/1/012005

base. Although engine regulation rises with rise in compression ratio, brake specific particular oxides of nitrogen discharge reduce. The burning chamber consequence more surface to capacity ratio, a correlatively more crack capacity and smaller fatigue vapor centigrade. Thus the capacity of blaze state regions rises resulting in more HC discharge. Due to this engine CR raises with a rise in hydrocarbon discharge. While raising particular CO discharge is reduced due to lower CR% fuel efficiency. The below graph shows hydrocarbon emission versus brake power at different compression ratios of two stroke engines. The diagram 11 shows the oxides of nitrogen versus brake power for different compression ratios. The figure 12 shows emission of Carbon monoxide and brake power at various burning ratios.

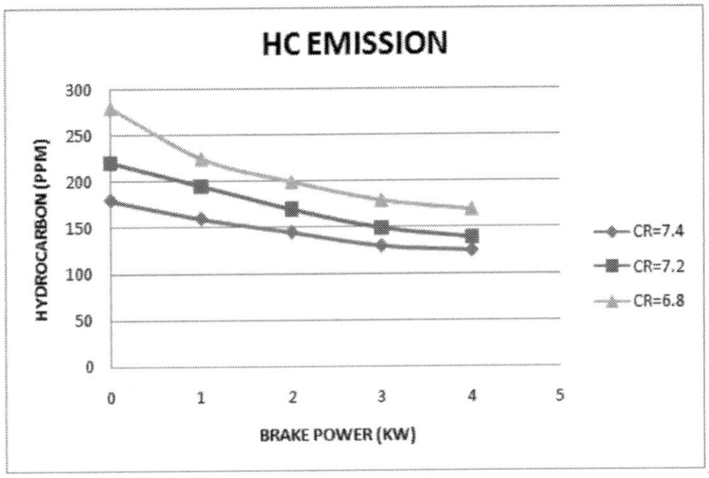

Figure 10: Hydrocarbon VS Brake Power

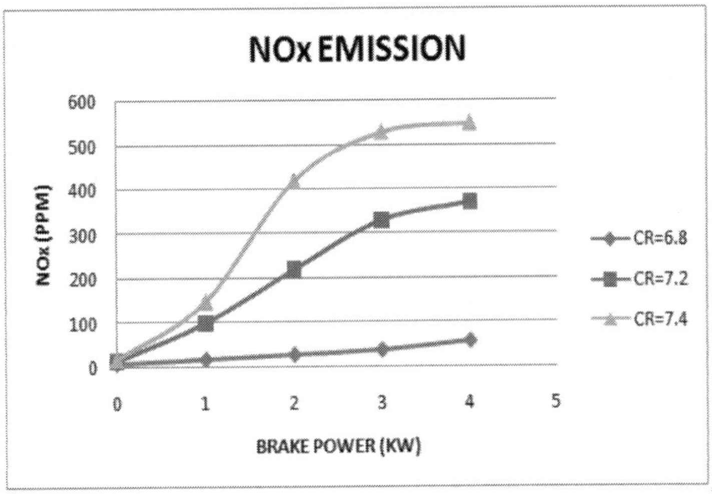

Figure 11: Oxides of nitrogen VS Brake Power

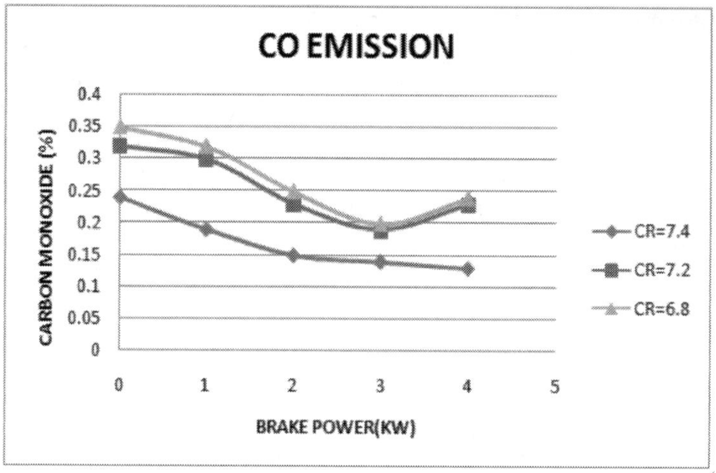

Figure 12: Carbon Monoxide VS Brake Power

4 Conclusions

The experimental study of fabrication and performance of variable ratio engines is being performed using a chassis dynamometer. Lower the compression ratio increases the pulling capacity of the engine due to high torque being achieved. High compression ratio raises the NOx due to high brake thermal efficiency whereas it reduces the CO and HC emission due to complete .The burning ratio is diverse by employing easy metal plugs. The complete combustible utilization was increased with an increase in compression ratio. The particular combustible utilization was decreased with a compression ratio. The brake thermal efficiency raised with the increase in compression ratio for increase in specific CO emission. The mechanical efficiency and torque increased with a compression ratio because of the increase in stroke to bore ratio. No detailed report is needed to vary the compression ratio and new pattern authorize the driving force to work at the burning ratio of his alternative supported land he desires to guide.

References

[1] KarstenWittek Frank Geiger Jakob Andert Mario Martins VitorCogo and Thompson Lanzanova 2019 Experimental investigation of a variable compression ratio system applied to a gasoline passenger car engine Energy conversion and management Vol. 183 Page number 753-763.

[2] Rinu Thomas M Sreesankaran Jeevan Jaidi Dileep M.Paul P.Manjunath 2016 Experimental evaluation of the effect of compression ratio on performance and emission of SI engine fuelled with gasoline and n-butanol blend at different loads Perspectives in Science Vol. 8 743—746.

[3] T Ashok Kumar R Chandramouli T Mohanraj 2015 A study on the performance and emission characteristics of esterified pinnai oil tested in VCR engine Ecotoxicology and Environmental Safety Volume 121 Pages 51-56.

[4] A Javaheri V Esfahanian A Salavati-Zadeh and M Darzi 2014 Energetic and Exegetic analyses of a variable compression ratio spark ignition gas engine Energy Conversion and Management Volume 88:739–748..

[5] Srinivas kommana Balu Naik Banoth Kalyani and Radha Kadavakollu 2016 Performance and Emission of VCR-CI Engine with palm kernel and eucalyptus blends Perspectives in Science Volume 8 Pages 195-197.

[6] M Ayaz Afsar Prafulla V. Pawar and Prathik Dahule S. Papinwar 2001 Experimental investigation of direct air injection scavenged two stroke engines International symposium on computing, communication and control (ISCCC). Proc. Of CSIT vol.1 (2001) PP. 21-24.

IOP Conf. Series: Materials Science and Engineering **995** (2020) 012005 doi:10.1088/1757-899X/995/1/012005

[7] Yuh Tohru G 2010 The effect of higher compression ratio in two-stroke engines Yamaha Motor Co Ltd SAE 931512 Page number 355-362.

[8] Amjad Shaik N Shenbaga Vinayaga Moorthi and R rudramoorthy 2007 Variable compression ratio engine: A future power plant for automobiles- An overview Proceedings of the Institution of Mechanical Engineers Part D: Journal of Automobile Engineering vol. 221 no. 9. pp 1159-1168.

[9] N. Seshaiah 2010 Efficiency and exhaust gas analysis of variable compression ratio spark ignition engine fuelled with alternative fuels International journal of energy and environment, Vol.1 Issue 5 pp 861-870.

[10] Lawankar, S 2013 Influence of Compression Ratio and Ignition Timing on the Performance of LPG Fuelled SI Engine SAE Technical Paper Page number 2013-01-2889.

[11] Mahesh P. Joshi and Aparna V. Kulkarni 2012 Variable Compression Ratio (VCR) Engine- a review of future power plant of automobile at International Journal of Mechanical Engineering Research and Development (IJMERD) Volume 2 Number 1 pp. 09-16.

[12] N.V Diwakar C. Bhagyanathan and J. David Rathnaraj 2014 Analysis of mechanical properties of EN 19 Steel and EN44B Steel used in Diesel Engine Camshaft at International Journal of Current Engineering and Technology Special Issue-2 page no: 162-167.

[13] Dr. A.Srinivas Dr. G.Venkatasubbaiah P.Venkateswar rao and M. Penchal Reddy 2014 Some innovation in design of low cost variable compression ratio two stroke petrol engine International Journal of Mechanical Engineering Volume 2 Issue 5.

[14] Savita Tomar Reena Mishra Sarita Bisht Sanjeev Kumar Aman Balyan and Gaurav Saxena 2013 Optimisation of connecting design to achieve VCR Journal of Engineering Research and Applications ISSN : 2248-9622 Vol. 3 Issue 6 pp.281-286.

[15] Jai Preetham R , 2014 Performance and Emission characteristics of variable Compression ratio S.I engine at International Journal of Innovation and Applied Studies Vol. 6 No. pp. 172-186.

IOP Conf. Series: Materials Science and Engineering **995** (2020) 012006 doi:10.1088/1757-899X/995/1/012006

ROBOTIC ARM BASED INTELLIGENT SYSTEM WITH AUTOMATIC PETROLBUNK AND AMOUNT COLLECTION

K roobini[1], P senthilkumar[2], P kathiresan [3], GO girupaakaran[4]and M saravanan [5]

2 Assistant Professor, Bannari Amman Institute of Technology, Department of Mechatronics Engineering, Erode, Tamilnadu, India.
1 PG Scholar, Bannari Amman Institute of Technology, Department of Mechatronics Engineering, Erode, Tamilnadu, India.
3, 4, 5 UG Scholar, Bannari Amman Institute of Technology, Department of Mechatronics Engineering, Erode, Tamilnadu, India.

roobini.ia19@bitsathy.ac.in, SENTHILKUMARP@bitsathy.ac.in, kathiresan.mc19@bitsathy.ac.in, girupaakaran.mc19@bitsathy.ac.in, Saravanan.me18@bitsathy.ac.in,

Abstract. The world is heading towards automation nowadays, but India is still at the beginning of the automation era. A rise in the number of vehicles in India has led to pollution and traffic in almost all of India's cities in recent years. The supply of fuel to this large number of vehicles at fuel stations has created many problems in India. In our everyday lives, liquid dispensing systems in various places such as offices, bus stands, train stations, gas pumps are very commonly found. Currently, fuel stations are manually operated. Such fuel pumps take time and require more manpower. All these problems are sorted out by the use of unmanned petrol pumps to position fuel stations in distant locations, which take less time to run and are reliable and can be placed wherever the customer goes to make use of the electronic clearing system's services. All these problems are sorted out by the use of unmanned petrol pumps. So, one of the best automation processes in our country is this proposed method. The automation of the gas station is a mandatory activity to be carried out due to the dramatic rise in the number of vehicles. To provide quick access and save time at the gas station, an autonomous petrol station using ATMEGA2560 is recommended. Our task reduces the use of human power often eliminates the possibility of all forms of cheating. For the convenience of the customer, automated billing facilitates swiping of credit / debit cards as well as cash depositor in the fuel counter itself.

Keywords: Image processing, Robotic arm, Automation, ATMEGA2560.

1. Introduction

Approximately all petrol pumps today need a microcontroller to regulate the electrical pumps for automation, driving the electrical pump to turn off accordingly. But it also needs manpower to run and raise revenue. The automation of the gas station is a mandatory activity to be carried out due to the dramatic rise in the number of vehicles. To provide quick access and save time at the petrol station, an automated petrol station using ATMEGA2560 is suggested. Our task reduces time and the use of human strength often decreases the risk of all forms of corruption. For the convenience of the customer, automated billing allows swiping of credit / debit cards as well as cash depositor in the fuel counter itself. This provides real-time details about the density of the fuel. Interfaces such as IR sensors, fire sensor, and c[illegible] / cash unit, pumping unit, camera, robotic arm, keypad and LCD display are regulated by the ATMEGA2560. To detect the vehicle type (4whe[illegible] ensor is used to transmit the signal to the Microcontroller. The cash / card unit allows the [illegible] ash / card (credit / debit card) to fill the required amount of petrol with a voice command [illegible] icrocontroller to connect with the user to

IOP Conf. Series: Materials Science and Engineering **995** (2020) 012006 doi:10.1088/1757-899X/995/1/012006
enter the amount of petrol in the keypad. After swiping the card, the keypad is used to enter the fuel prices. The microcontroller triggers the relay driver for that specific time period when the desired number is entered on the keypad.
The relay output is related directly to the petrol pump. Then the fuel is pumped to sub tank from the main tank with the help of flow sensor the sub tank is consists of measuring scale so, the customer can able to view the proper quantity of the fuel it help to eliminates the possibility of all forms of cheating. At the same time the camera analysis the sticker which has pasted at the surface of fuel tank in the vehicle and send the signal to the microcontroller. After receiving the signal to robotic arm from the microcontroller. The robotic arm connects the filling hose to the vehicle's fuel tank triggered by the microcontroller. To activate the pumping motor for the specific time to perform our proposed work, a relay driver circuit is used. The quantity of fuel from the tank will be dispatched at the same time to the petrol tank of the vehicle and the corresponding cost will be shown on the LCD panel. The fire sensor detects and sends a control signal to the microcontroller in the event of a fire emergency occurring at the fuel station. The microcontroller shuts down the entire process and activates the relay to which the fire sprinklers are linked. In order to prevent fire at the station, the fire sprinklers spray water on the specific device. In order to save time and manpower, this project can also be easily implemented as an automated gas station.

2. Literature Survey

Automobile Fuel Pump Control System Using Embedded System [2]:
This paper is intended for the design of the Atmega-2560 microcontroller for Anti-Theft Control System for Automobiles, the Arduino mega-2560 Development Board and the relay switching circuit for the electronic fuel pump controller for pumping fuel to the vehicle engine. The main purpose of this project is to prevent a vehicle from being stolen. We are introducing a novel automotive anti-theft monitoring device that seeks to prevent a vehicle from being stolen. This device uses an embedded chip with an optical slot sensor or, during insertion that senses the key. This is accompanied by the device present in the vehicle asking the user via the Fingerprint Scanner to enter a unique password. A finger print consists of the password. If the user fails to enter the correct password and the location is monitored using a GPS module, a text message is sent to the owner. A warning about unauthorised use is also sent to the owner. In addition, the vehicle's fuel injector is deactivated so that the driver can not start the vehicle by any means. This method helps to take quick steps in an effort to rob. The design is powerful and simple.

SMART AUTOMATIC PETROL PUMP SYSTEM [3]:
The main objective of the project is to build a device capable of automatically deducting the quantity of petrol dispensed from the RFID technology based user card. In our everyday lives, liquid dispensing systems in various places such as offices, bus stands, train stations, gas pumps are very commonly found. Here we are going to present the latest fuel dispensing device of the era that is meant to be run using RFID technology with a prepaid card. The project primarily aims to build a prepaid card using RFID technology for the fuel bunk system and also the fuel dispensing system. At present, the petrol stations are manually run. Such petrol pumps need more human resources and are time-consuming. It is very expensive to put gas stations in distant areas to provide customers with excellent facilities. The use of an unmanned power pump, which takes less time to operate and is reliable and can be mounted anywhere, solves all these issues. The customer has to pay through the electronic clearing system to make use of the service.

3. SYSTEM OVERVIEW

The automation of the gas station is a mandatory activity to be carried out due to the dramatic rise in the number of vehicles. To provide quick access and save time consumption in the petrol station, an automated petrol station using ATMEGA2560 is proposed. Our work reduces the time as well as the use of manpower, therefore removing the likelihood of all kinds of cheating. For the convenience of the customer, automated billing allows swiping of credit / debit cards as well as cash depositor in the fuel counter itself. This provides real-time details about the density of the fuel.

IOP Conf. Series: Materials Science and Engineering **995** (2020) 012006 doi:10.1088/1757-899X/995/1/012006

3.1 Arduino Microcontroller

Arduino software is an open-source, Arduino is used for rapid prototyping, it has inbuilt library for many application. Easy-to-use both hardware and software. In this project, the Arduino board helps to Control all the peripherals which are interfaced. It controls all the parts through coding.

3.2 ATMEGA 2560

ATMega2560 is an arduino family based microcontroller board. Compared to all other arduino board ATMega2560 has most number of I/O port in it.

3.3 IR Sensor

Infrared technology is a broad range of wireless technology. Sensing is the main field of this sensor and is used remotely for TV. The infrared is categorized into three regions within the electromagnetic spectrum, which are the near infrared region, the mid infrared region and the far infrared region.
Near infrared region 700 nm to 1400 nm used for fibre optic communication
Mid infrared region 1400 nm to 3000 nm used for Heat sensing element
Far infrared region 3000 nm to 1 mm used for Thermal image processing

3.4 Display

One of the liquid crystal displays (LCD-16 * 2) is used in this prototype. You can monitor the LCD display with the liquid crystal library. It is quite because it has inbuilt library feature in it to use arduino interfacing of LCD.
To wire your LCD screen to your board, connect the following pins:
LCD RS pin connected to digital pin 12
LCD Enable pin connected to digital pin 11
LCD D4 pin connected to digital pin 5
LCD D5 pin connected to digital pin 4
LCD D6 pin connected to digital pin 3
LCD D7 pin connected to digital pin 2

3.5 Servo Motor

Servo motors are a single type of motor that can be operated by the control signal from the controller, motor angle or motor shaft, often control the direction of the servo motor. In many automation applications, this type of servo motor is used and even in the robotic arm, etc. Based on the function, various servo motors can be categorised according to size or torque. With Arduino's support, the standard size of the servo motors can be driven directly without the need for an external power supply or driver IC. Usually, servo motors come with arms that are attached to the object required for movement, such as metals or plastic.

3.6 Keypad

In this proposed method, the keypad is used to enter the cost for petrol after swiping the card and also used in case amount.

3.7 DC Motor

A motor is an electric machine that transforms mechanical energy into electrical energy. The dc motor is used in this proposal to pump the fuel from the main tank to the sub tank and fill the vehicle tank too though.

3.8 Relay

One type of mechanical switch is a relay. The electric current flows through the relay coil and the coil is converted into an electromagnet, so that the circuit can be automatically closed. In order to be used in high-end industrial application products, relays have relays for their efficient operation. With proper coding, the microcontroller relay can be controlled automatically.

3.9 Camera

To image the location of the fuel tank in the car, the camera is mounted near the robotic arm. On the surface of the fuel tank the colour labels are pasted.

IOP Conf. Series: Materials Science and Engineering **995** (2020) 012006 doi:10.1088/1757-899X/995/1/012006

3.10 Flow sensor

The flow sensor is used to calculate the amount of fuel from the fuel reserve that is being filled into the fuel tank. Between the pump and the nozzle, the flow sensor is fixed. For continuous readings, one of its wires is linked to the controller.

4. METHODOLOGY OF INTELLIGENCE SYSTEM

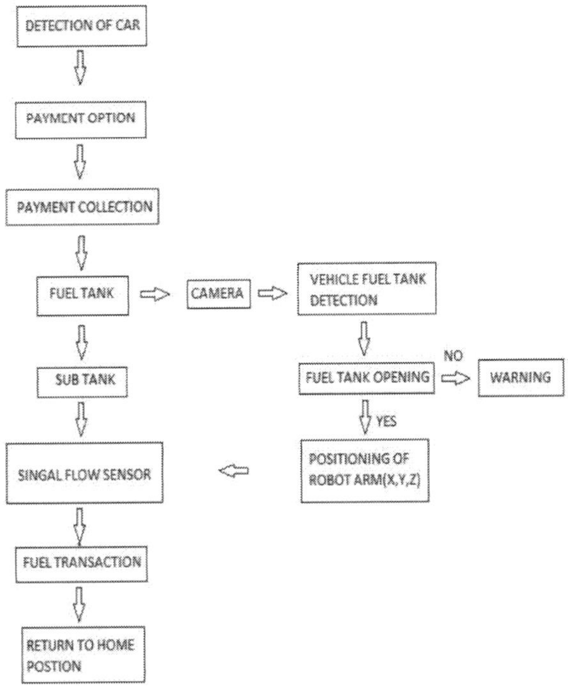

Fig 1: methodology of intelligence system in petrol station

When entering the parking lot, the car is parked and per the marker signals. The fuel is transferred by the customer from the main tank to the sub tank according to volume specifications. By capturing the sticker that has been pasted on the fuel box surface, the fuel loading side of the vehicle is filmed by the camera. The photos for more pro- cessation will go to the controller (Parikh et al. 2013). In the meantime, the image sent by camera will be processed to obtain the fuel cap centre stage. The fuel tank tracking is based using a signal mark. The marks are seen on the fuel tank in such a way that their coincidence becomes the subject of the fuel tank (Sheth et al. 2010). The centre will still remain the same, as seen from an angle, and even rural can be used. The controller quickly collects the four points and uses algorithms to locate the middle (Chauhan et al. 2011). This will assign the X, Y and Z coordinates of the fuel cap that is fed to the controller. According to this, the mechanism's drives would align the robotic arm to the fuel cap core. The arm travels in the Z direction now so that the fuel cap is pierced. If the arm is penetrated, the pump will be triggered to supply the gasoline. The pump stops when the reading of the flow sensor matches the controller's measured value. The retraction of the z axis is then carried out and the robot returns to its home location.

IOP Conf. Series: Materials Science and Engineering **995** (2020) 012006 doi:10.1088/1757-899X/995/1/012006

5. BLOCK DIAGRAM

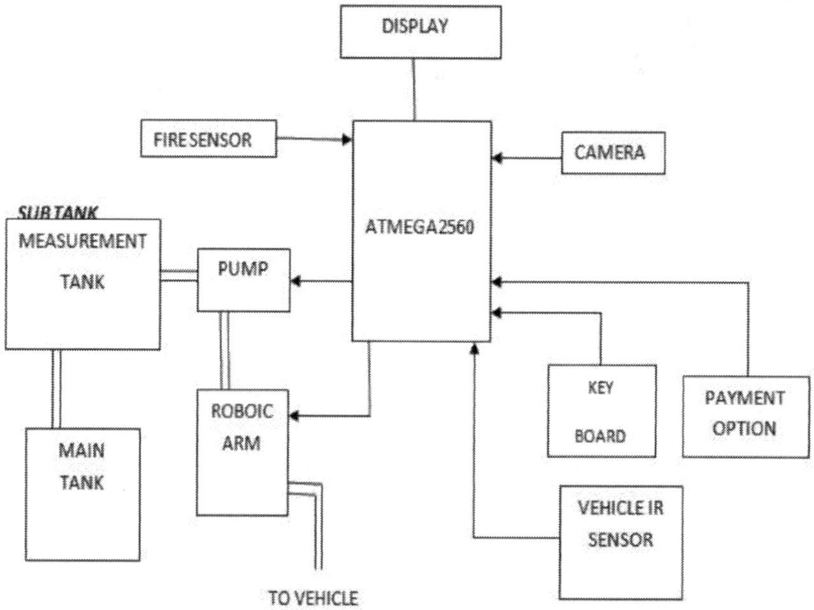

Fig 2: Block diagram of microcontroller based automated petrol station

5.1 Detection and User Interface Module

In Detection and User Interface Module we had illustrated about the detection of the vehicle, identification of the card (debit/credit), cash detection and the user interface of 4 × 4 matrix keyboard. Vehicle is detected by using a IR sensor module, when a vehicle entered in to the bay in the petrol station the IR senses the vehicle and send the control signal to the microcontroller, as soon as the signal received by the microcontroller responses to the user to enter the desired among in the keypad.

When the desired mount is entered in to the keypad for example say Rs.50. then the user is asked for the next step of payment. The payment process is carried out by providing an cash /card unit .As per the preference of the user can choose either cash or card .then the final step of the user is carried out by using a enter button finally the user enters all the necessary data required for filling the petrol.

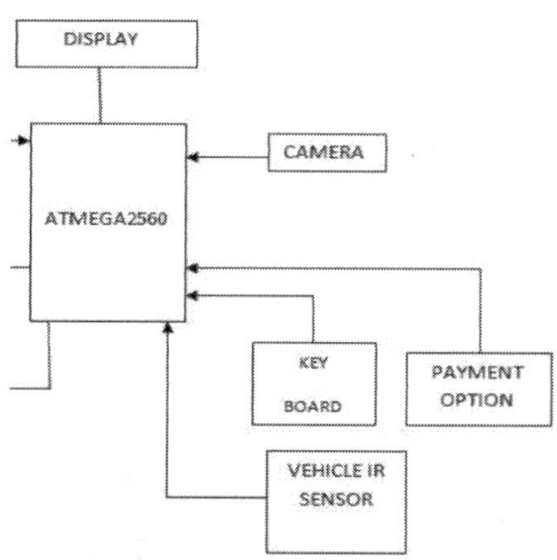

Fig 3: Detection ar ce Module

IOP Conf. Series: Materials Science and Engineering **995** (2020) 012006 doi:10.1088/1757-899X/995/1/012006

5.2 Fuel Dispensing Module

The fuel dispensing module consists of a main tank as well as sub tank (measurement tank). The main tank is used for deposition of fuel in the underground, when the fuel is less in the sub tank it pumps the petrol from the ground to the measurement tank this is carried out by using a submersible motor. The sub tank (measurement tank) is responsible for measurement of fuel .the sub tank consists of measurement scale which is placed at the edge of the tank for accuracy and for avoiding cheating in the petrol station. When the desired amount is entered in the keyboard a solenoid valve activates and the sub tank dispenses fuel to the fuel tank of the vehicle when the desired amount of fuel is filled then the pump stops the desired amount of fuel can be verified by viewing the measurement tank.

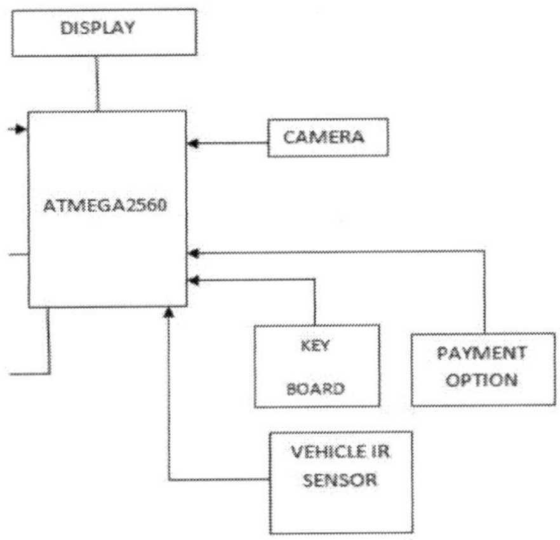

Fig 4: Fuel Dispensing Module

5.3 Robotic arm

Robotic arm is used to minimize the work of men. The automatic controlled "Robotic Arm" will fix the filling hose to the fuel tank of the vehicle which is triggered by the microcontroller. This will give the fuel cap, which is fed to the controller, the X, Y and Z coordinates. According to this, the mechanism's drives would align the robotic arm to the fuel cap core. The arm travels in the Z direction now so that the fuel cap is pierced. If the arm is penetrated, the pump will be triggered to supply the gasoline. The pump stops when the reading of the flow sensor matches the controller's measured value. The retraction of the z axis is then carried out and the robot returns to its home location.

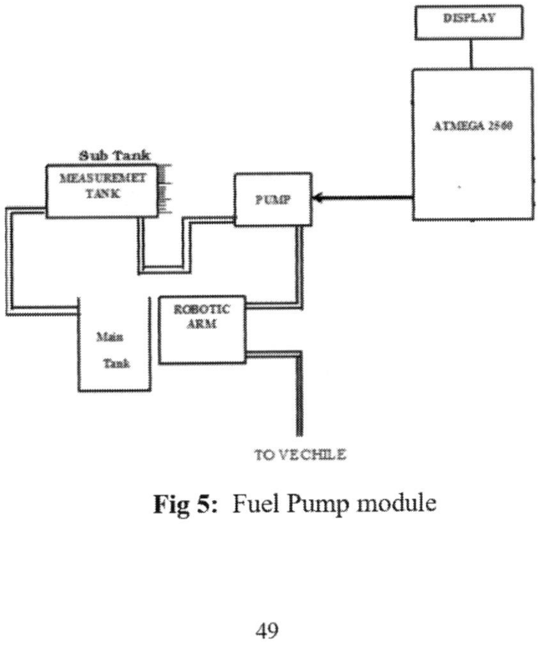

Fig 5: Fuel Pump module

IOP Conf. Series: Materials Science and Engineering **995** (2020) 012006 doi:10.1088/1757-899X/995/1/012006

6. Results and Discussion

Fig 6: Prototype Model of Microcontroller Based Automated Petrol Station

The Fig 6 shows the proposed working model of the microcontroller based automated petrol station. The model consists of 3 modules which are interfaced with arduino board along with displaying unit.

Fig 7: Ideal State of Automatic Petrol Station

Describes in Detection and User Interface Module we had illustrated about the detection of the vehicle, identification of the card (debit/credit), cash detection and the user interfacing of 4 × 4 matrix keyboard.

Fig 8: Entry of the Vehicle

Vehicle is detected by using a IR sensor module, when a vehicle entered in to the bay in the petrol station the IR senses the vehicle and sensed the control signal to the microcontroller, as soon as the signal received by the microcontroller responses to the user to enter the desired amount in the keypad.

Fig 9: Entering of Desire Amount

Fig 10: Payment mode Via Card

Fig 11: Payment mode Cash

Fig 9,10,11 Describes working of the proposed model, the desired amount is entered in to the keypad for example say Rs.50, and then the user is asked for the next step of payment.

The payment process is carried out by providing an cash /card unit .As per the preference of the user can choose either cash or card .then finial step of the user is carried out by using a enter button finally the user enters all the necessary data required for filling the petrol.

Fig 12: Dispensing of Desire Amount of Fuel

The main tank is used for deposition of fuel in the underground, when the fuel quantity is enter by the customer in the sub tank it pumps the petrol from the ground to the measurement tank this is carried out by using submersible motor. The sub-tank (measurement tank) is responsible for measurement of fuel.

The sub tank consists of measurement scale which is placed at the edge of the tank for accuracy and for avoiding cheating in the petrol station. When the desired amount is entered in the keyboard a solenoid valve activates and the sub tank dispenses fuel to the fuel tank of the vehicle when the desired amount of fuel is filled then the pump stops the desired amount of fuel can be verified by viewing the measurement tank.

Fig 13: Sticker pasted on the Fuel tank

IOP Conf. Series: Materials Science and Engineering **995** (2020) 012006 doi:10.1088/1757-899X/995/1/012006

At the same time the camera track the fuel tank by using the tracking sticker which has been placed on the surrounding the fuel tank in the vehicle. And send the accurate position location to the microcontroller the microcontroller commend the robot arm to move toward the x,y,z axis to fill the fuel in the fuel tank with the help of flow sensor. After complete the process the robot arm move towards its home position.

7. Conclusion

Our project is a novel idea which can be a remedy for conversion of today's manual system into automated system for fuel distribution. Microcontroller based automatic petrol station can be cost effectively utilized by increasing the robotic arm outlet. The microcontroller based automated petrol station utilizes cash/card swipe for fuel consumption which is user-friendly to customer. This project can be widely applied replacing existing system with minimum cost of replacement.

References

[1] Pradeep Raj, "Increasing accidents in the unmanned level crossing of the railways", 2012.

[2] Automobile Fuel Pump Control System Based on Embedded System Security International Journal of Computer Technology and Electronics Engineering (IJCTEE) Volume 3, Issue 2, April 2013By Eng.Homam Al-Bakri

[3] Aishwarya Jadhav, Lajari Patil, Leena Patil, A. D. Sonawane smart automatic petrol pump system International Journal of Science Technology and Management (IJSTM) Vol.No.6,Issue No.04, Apirl 2017

[4] Sheth, S., Kher, R., Shah, R., Dudhat, P., & Jani, P. (2010). Automatic sorting system using machine vision. Multi-Disciplinary International Symposium on Control, Automation & Robotics,. doi:10.13140/2.1.1432.1448.

[5] Chauhan, V., Sheth, S., Hindocha, B. R., Shah, R., Dudhat, P., & Jani, P. (2011). Design and development of a machine vision system for part color detection and sorting. In Proceedings of Second International Conference on Signals, Systems & Automation (ICSSA) (pp. 90–93). doi:10.13140/2.1.1628.7526.

[6] Xishi Wang, Ning Bin, and Cheng Yinhang, "A new microprocessor based approach to an automatic control system", International Symposium on Industrial Electronics, pp. 842-843, 1992.

[7] Jeong Y., Choon-Sung Nam, Hee-Jin Jeong, and Dong Shin, "Train Auto Control System based on OSGi", International Conference on Advanced Communication Technology, pp.276-279, 2008.

[8] Atul Kumar Dewangan, Meenu Gupta, and Pratibha Patel, "Automation of Railway Gate Control Using Micro-controller, International Journal of Engineering Research & Technology, pp.1-8, 2012.

[9] Parikh, P. A., Joshi, K. D., & Sheth, S. (2013). Color guided vehicle—an intelligent material handling mechatronic system. In Proceedings of the 1st International and 16th National Conference on Machines and Mechanisms (iNaCoMM2013), IIT Roorkee, India, pp. 628–635.

[10] Gunyoung Kim, Kyungwoo Kang, "Railway Gate Con- trol System at Railroad-Highway Grade Crossing in Korea".

[11] J. Banuchandar, V. Kaliraj,P. Balasubramanian, S. Deepa, N.Thamilarasi "automated unmanned railway level crossing system" International Journal of Modern Engineering Research (IJMER) Vol.2, Issue.1, Jan-Feb 2012 pp-458-463 ISSN: 2249-6645.

[12] R.AnujTyagi,Er. GoutamArora "synopsis on automatic railway crossing system" S.D. Institute of Technology & Management Israna-(Panipat). Evaluation of Structural and Biological systems V, SPIE (2006).

[13] GuoXie, XinhongHei, Hiroshi Mochizuki, Sei Takahashi and Hideo Nakamura."Formal Analysis of Automatic Train Protection and Block System for Regional Line Using VDM++" IJR International Journal of Railway Vol. 5, No. 2 / June 2012, pp. 65-70.

[14] QiaoJian-hua; Li Lin-sheng; Zhang Jing-gang; "Design of Rai Surface Crack-detecting System Based on Linear CCD Sensor" IEEE Int. con on Networking, Sensing and Control, 2008.

[15] K.Vijayakumar, S.R. Wylie, J. D. Cullen, C.C. Wright, A.I.AIShamma'a,"Non invasive rail track detection system using Microwave sensor", Journal of App. Phy., 2009.

[16] Bowler, J. R., "Eddy contemporary intera ultimate crack: I. The bold problem," J. Appl. Phys., Vol. 75, 8128–8137, 1994.

IOP Conf. Series: Materials Science and Engineering **995** (2020) 012006 doi:10.1088/1757-899X/995/1/012006
[17] Tagaki, T., M. Hashimoto, H. Fukutomi, M. Kurokawa, K. Miya, and H. Tsuboi, "Benchmark models of vortex tide taxing for steam generator tube: research and geometric analysis," Int. J. Appl. Electrom, Vol. 4, 149–162, 1994.
[18] Tagaki, T., H. Huang, H. Fukutomi, and J. Tani, "Numerical evaluation of correlation between crack amount and current contemporary taxing signals by a identical abstain simulator," IEEE Trans. Magn. Vol. 34, 2582–2584, 1998
[19] Specogna, S. and F. Trevisan, "Voltage obsessed coils contained by a discrete numerical come near to to 3d eddy-currents," Proceedings of the 11th intercontinental IGTE meeting on geometric Fields Calculation, 81–84, 2004.
[20] Tonti, E., "Algebric topology and computational electromagnetism," Proceedings of the 4th Int. Workshop on stimulating and captivating Fields , 284–294, 1998.
[21] Dodd, C. V. and W. E. Deeds, "Analytical solutions to eddy-current probe-coil problems," J. Appl. Phys, Vol. 39, 2829–2838, 1968.
[22] Lewis, A. M., "A academic fashion of the comeback of an eddy-current prod to a surface-contravention metal low energy crack in a unexciting assess piece," J. Phys. D.: Appl. Phys, Vol. 25, 319–326, 1992.
[23] Ferraris, G., "Rotazioni elettrodinamiche prodotte for each mezzo di correnti alternate," Atti dell'Accademia delle Scienze di Torino, Vol. 23, 360, 1888 (in Italian Language).
[24] Rodger, D., P. J. Leonard, and H. C. Lai, "Interfacing the all-purpose 3D A- ψ sense with a gauzy expanse conductor model," IEEE Trans. Magn, Vol. 28, No. 2, 1115–1117, 1992.
[25] Tsuboi, H. and T. Misaki, "Three dimensional laboratory analysis of whirlpool present giving out by the boundary bit reasoning by vector variables," IEEE Trans. Magn, Vol. 23, 3044–3046, 1987.
[26] Weissenburger, W. D. and U. R. Christensen, "A set of contacts network manner to evaluate whirlpool currents on conducting surfaces," IEEE Trans. Magn, Vol. 18, No. 2, 422–425, 1982.
[27] Weisteinn, W., "Euler angles," MathWorld, a wolfram web resource, available online:http://mathworld.wolfram.com/EulerAngles.html.

IOP Conf. Series: Materials Science and Engineering **995** (2020) 012007 doi:10.1088/1757-899X/995/1/012007

Vibration Based Piezoelectric Energy Harvesting - A Review

E.L.Pradeesh[1], S.Udhayakumar[2], M.G.Vasundhara[3], V.Vadivel Vivek[4]

[1, 4]Assistant Professor,Department of Mechatronics, Bannari Amman Institute of Technology, Sathyamangalam, India
[2]Assistant Professor, Department of Mechanical Engineering, PSG College of Technology, Coimbatore, India
[3]Assistant Professor, Department of Mechanical Engineering, Malnad College of Engineering, Hassan, India
pradeesh.dmt@gmail.com, udhaya_mech@yahoo.com, vasundharamg@gmail.com, vadivelvivek@bitsathy.ac.in

[*] Corresponding Author: Tel (+91)9944820144 Email ID: pradeesh.dmt@gmail.com

Abstract: In this digital race, electronic equipment has been integrated into human beings as a part of their body. Some electronic equipment is connected by wires, while some are self-powered by batteries. Today the ultra-low-power smart electronic gadgets and smart wireless sensor devices need an unlimited battery for enhancing the performance. In a remote area such as forests and hill areas, conventional charging methods of batteries by wire is not possible. Supplying power through wires is difficult. To overcome this, a sustainable solution is energy harvesting. The renewable sources for energy harvesting are light, heat, wind, tidal, motion, and vibration. Researchers have more interest in harvesting energy through mechanical vibration due to its abundant availability. This paper reviews the work about piezoelectric crystals and their role in energy harvesting, simulation software used, energy harvesting circuits and storage devices.

Keyword: Energy Harvesting; Piezoelectric materials; Simulation software; Harvesting circuit; Storage unit

1. Introduction

The process of collecting functioning or unwanted energy from the sources of natural and human-made things from the environment could be called energy harvesting. This system is used to charge batteries for self-powered portable devices or wireless sensor networks [1]. Non-renewable energy has a great impact on the environment, but piezoelectric material is renewable. The piezoelectric material could able to convert mechanical energy into electrical energy. It received the most attention of researchers due to the direct conversion of applied stress into electricity. Due to the compactness and smartness of piezoelectric material, it has a major role in portable electronic applications. The applications are power generation in a highway [2] and human bodies are to monitor breathing, blood flow and power pacemaker [3,4]. The portable devices such as wireless sensor networks in automotive, railways, aerospace, temperature and humidity sensing systems can be powered using piezoelectric materials [5]. This paper focuses on the review of literature related to piezoelectric materials and their role in energy harvesting, simulation software used, energy harvesting circuits and storage devices [6].

1.1 Sources of Energy Harvesting

Harvesting of energy can be broadly classified into two types, Micro and Macro energy harvesting. In micro harvesting the generation of power in m/μ volts, but in macro harvesting the production of power in kilo/mega volts [7], [8,9]The energy sources for micro harvesting are vibration, motion, heat, etc. The sources for Macro harvesting are windmills, solar farms, tidal, etc. Table 1 shows a detailed summary of the sources of energy harvesting [7,8].

Table 1. Sources of energy harvesting

Type of energy harvesting	Energy source	Conversion mechanism	Energy level
Micro energy harvesting	Vibration	Piezoelectric, Electromagnetic, Electrostatic	100-330μW/cm^3
	Ambient radiation	Electromagnetic	1μW/cm^2
	Ambient light	Photovoltaic	100 μ-100 mW/cm^2
	Thermal	Thermo-electric	60 μ W/cm^2
	Fluid flow	Small size Turbines	100- 500 mW/cm^3
Macro energy harvesting	Windmills	Alternator	1-10 kW
	Tidal	Alternator	1kW-1.2 MW
	Solar farms	Photovoltaic, Large scale Turbines	10 kW – 500 MW

2. Power Generation Through Piezoelectric Configuration

2.1 Piezoelectric Effect

In 1880 Pierre Curie and Jacques discovered the direct piezoelectric effect. When the mechanical strain is applied to a piezoelectric material, it tends to electrically polarize. The applied strain was proportional to polarization. Later in 1881 Jacques and Curies discovered inverse piezoelectric effect, in which applied polarization to the piezoelectric material can deform its size. [10] Figure 1 illustrates the direct and inverse piezoelectric effects. Figure 1. (a) Shows the piezoelectric material under no-load conditions. If an external force compresses the piezoelectric material then there is a change in dipole moment, which causes the same polarity poling voltage to appear between the two nodes as shown in Figure 1 (b). If the material is stretched, then piezoelectric material will produce the opposite polarity as shown in Figure 1 (c). When the opposite polarity of poling voltage is applied to the piezoelectric material then it will shorten as shown in Figure 1 (d). When the same polarity of poling voltage is applied to the piezoelectric material then it will expand as shown in Figure 1 (e). The piezoelectric material will expand and shorten by applying the alternative voltage and the frequency of material is based on the applying voltage as shown in Figure 1 (f).

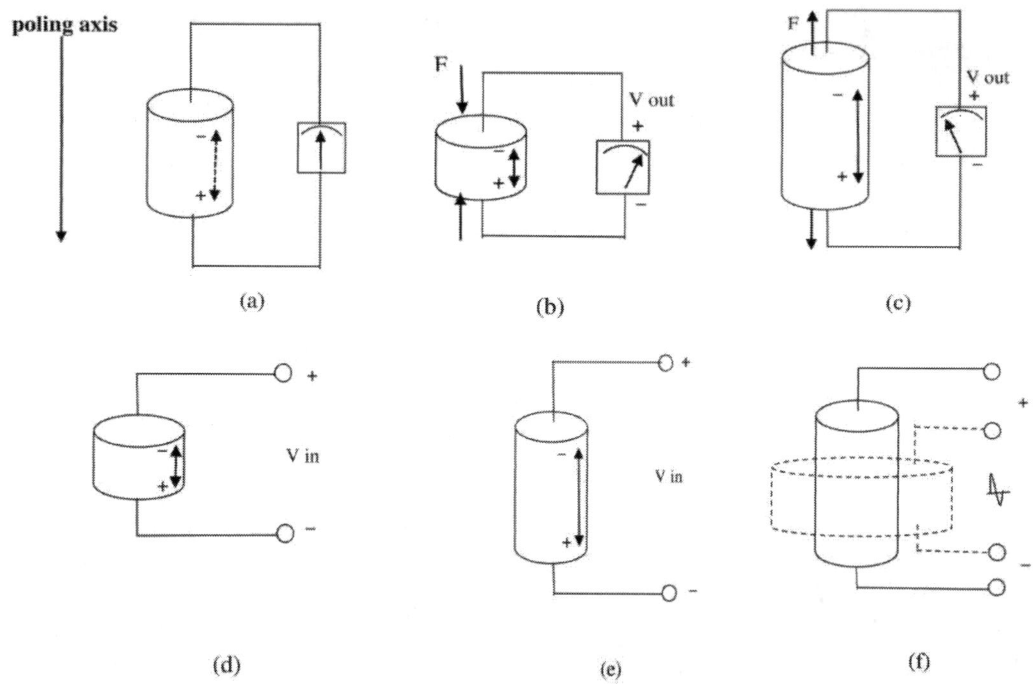

Figure 1: Direct and Inverse piezoelectric effect

2.2 Sources of Mechanical Vibration

When energy hits a medium, it can cause vibrations, the common sources of vibration are drilling machine, milling machine, bearing bed testing machine, automotive, household things, public transports, etc. [11]. Table 2 shows the sources of vibration with their acceleration and frequency levels [12,13].

Table 2. Sources of mechanical vibration

Vibration source	Equipment	Acceleration (m/s^2)	Frequency (HZ)
Industries	Base of machine tool	10	70
	Drilling machine	0.93	178
	Lathe	1.36	68
Household things	Washing machine	0.82	62
	Dryer	3.5	121
	Microwave oven	0.49	40
	Refrigerator	0.14	110
Automotive	Car engine	12	200
	Truck engine	1.98	37
	A/C Compressor	2.14	59
	Car instrument panel	3	13
Public places	Door frame	3	125
	Windows to near busy road	0.7	100
	HVAC vents in office buildings	0.2-1.6	62

2.3 Piezoelectric Materials

Piezoelectric materials can be obtained from the natural and synthetic method. The material is available in the form of crystal, ceramics and polymers. The naturally occurring crystals are Quartz, Sugarcane, Rochelle salt, Topaz, Tourmaline Group Minerals, Berlinite (AlP04), and dry bone. The man-made ceramics include Barium titanate (BaTi03), Lead titanate (PbTi03), Lead zirconate titanate (PZT), Lithium tantalate (LiTa03), Lithium niobate (LiNb03), Sodium tungstate (Na2W03), Potassium niobate (KNb03), Ba2NaNb505, Pb2KNb5015. An excellent example of Polymer is Polyvinylidene fluoride (PVDF) [14]. Anton and Sodano suggested that piezoelectric materials can be arranged in different ways through altering the electrode pattern, stress direction, changing the direction of poling and varying the piezoelectric materials [15]. They also reviewed that the selection of materials has a major influence on harvesting functions and performance. PZT is highly brittle and PVDF is flexible. Lee et al. investigated Monolithic PZT and concluded that it is vulnerable to fatigue failure during continuous load. The PVDF film coated with electrodes can withstand the fatigue crack damage to the electrode [16]. Sodano et al. investigated that Micro Fiber Composite (MFC) and Quick Pack (QP) Interdigitated Electrode (IDE) are low capacitance devices for energy harvesting. The Monolithic PZT, MFC, QP has greater flexibility for energy harvesting [17]. Sarker et al. developed a MEMS-based energy harvester using C0MS0L Multiphysics software. The comparison between PZT and PVDF was done by varying the size of the micro cantilever beam. The generated voltage for PZT is 0.4V and for PVDF is 0.2V. From the simulation results, they have concluded that an increase in the length of a cantilever beam can decrease the resonant frequency [18]. Hassan et al. converted ultra-low sound frequency into electrical energy using a piezoelectric material. In this experiment, PZT was used to harvest energy from a loudspeaker at different distances and then converted into electrical energy. They have found that 27mV was produced at 80 dB of sound at the frequency of 62 Hz at 100 mm distance using coupling mode 31. Using coupling mode 33, authors obtained a generation of 90 mV at 103 dB of sound at 374 Hz. Finally, they concluded that an increase in the distance could lead to decrease in voltage generation [19]. Cook-Chennault et al. 2008 reviewed that generative and non-generative power supply systems for Microelectromechanical system (MEMS), by using piezoelectric energy harvesting system and found that PZT performance is not significantly influenced by temperature [20]. M.Taware and Deshmukh reviewed that PZT with a thickness of 0.28mm on the length of 11 mm beam with proof mass at the end can produce 375 μW at 120Hz. The generated power is used to demonstrate a radio transceiver [21]. [22,23] The summary of various piezoelectric materials is shown in Table 3.

Table 3. Summary of various piezoelectric materials

Author	Material	Dimensions	Power/ voltage	Application	Advantages/ Disadvantages
K A Cook Chennault et al (2008)	PZT Rectangular structure	1 x 1 x 1.8 cm^3	4.8 mW	Knee implant	Stacked PZT can produce more power
[24] Mohammadi et al (2003)	Piezofiber Composite	34 x 11 x 5.85 mm	120 mW	MEMS	Increased flexibility
Sodano et al (2004)	Composite of MFC & QP IDE,	-	-	Recharging of batteries	PZT proved 6.8 % more efficient for vibration excitation
K A Cook Chennault et al (2008)	PVDF bimorph windmill	60 x 20 x 0.5 mm^3	10.2 mW	Windmill	33- mode have higher oscillating frequencies
Lee et al (2005)	Monolithic PZT	-	-	To power transducers	Most commonly used. Fatigue crack during continuous loading

K A Cook Chennault et al (2008)	PZT – 5 A membrane	Area = 1 cm^2 Thickness = 9 μm	2.3μW	Biomedical	Insufficient power production
Lee et al (2005)	PVDF film coated	-	-	To power transducers	More resistance to Fatigue Failure
[25] Churchill et al (2003)	Piezofiber Composite	130 x 13 x 0.38 mm	7.5 mW	Wireless sensor networks	Increased flexibility
K A Cook Chennault et al (2008)	PVDF bimorph	-	8.4 mW	Shoe insert	Layer thickness influences energy harvesting
Kim et al (2005)	Cymbal transducer	1.8 x 0.4mm thick x 29 mm of diameter	52 mW	MEMS	Withstand high force applications

2.4 Coupling Modes

The coupling modes of piezoelectric material with a substrate could affect the performance of energy harvester. Several authors investigated [15, 26-28] that 33 mode and 31 mode are the most influencing coupling modes if piezoelectric material in energy harvesting. When the applied force is perpendicular to the poling direction then it is called as mode 31. When the applied force is same as the poling direction then it is called as mode 33. Figure 2. Shows the coupling modes of piezoelectric materials. Mode 31 is commonly employed in the harvesting of energy because, a small force at lower vibration levels, it can produce efficient results than mode 33.

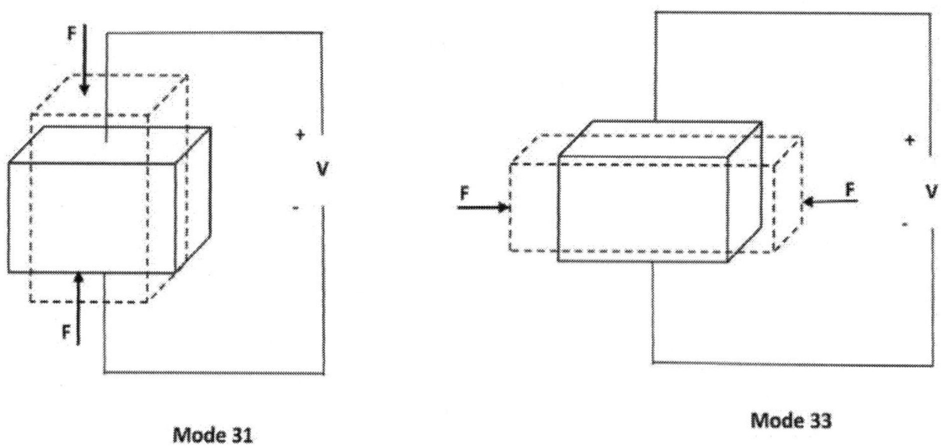

Figure 2: Coupling Modes of Piezoelectric Materials

2.5 Various Geometries of Energy Harvester

Energy harvesting through a piezoelectric material can improve by optimization of piezoelectric geometry. The rectangular cantilever beam is the most commonly used geometrical configurations. Mateu and Moll analytically compared the rectangular and triangular (the small end is free and the large end is clamped) cantilever beams. It was proven that the triangular cantilever beam can produce a maximum deflection and higher strain equated to a rectangular cantilever beam with the same base and height [29]. When a piezoelectric material is subjected to higher strain and deflection, the output generated is also high. The authors concluded that the rectangular cantilever beam produces less compared triangular cantilever beam. S. Roundy and Wright suggested that the strain in a trapezoidal based cantilever beam is uniformly distributed, whereas it's non-uniform in a rectangular cantilever beam [30-32]. They also presented that the trapezoidal beam was able to generate twice the energy as that of a rectangular cantilever beam. Baker 2005 proved that the triangular trapezoidal cantilever beam generated 30% more energy than the rectangular cantilever beam, through experimental studies [33]. Rami Reddy et al. modelled and experimentally compared the rectangular beam with and without a cavity. The cavity of the substrate tends to shift the neutral axis of energy harvester away from the surface which increases the strain and voltage [34]. They found that the beam with a cavity can produce 75% more voltage than the beam without a cavity. Reddy et al. experimentally compared the trapezoidal and rectangular cavity beam with the conventional rectangular beam. From the analytical and experimental model, they found that the trapezoidal cavity beam produces 97.5% and 108% more voltage compared with the rectangular cavity beam and the beam without cavity [35]. Usharani et al. designed a wideband energy harvester which can vary the resonant frequency range by varying the length of the beam by introducing step section in the beam. From the experiment, they concluded that step thickness of 2 mm beam reach resonance condition at 5.52 Hz while the beam without step section reach resonance at 21.24 Hz and the voltage is also 4.55 times higher compared to the beam without step section [36]. J. Park et al. designed the piezoelectric energy harvester which can be vibrated by any rotary motion of mechanical devices. From the experiment, authors conclude that their design can produce 160% more power than the tapered beam [37]. S P Matova et al. numerically and experimentally analyzed the effect of size of tapered beams. From their study, they showed that the beam with long and slender taper has a positive effect on power performance and the beam with short and wide does not affect power generation [38]. Hosseini and Hamedi developed the formula to calculate the resonance frequency of V-shaped trapezoidal beam using the Rayleigh-Ritz method and they validate their work using ABAQUS simulation [39]. Goldschmidtboeing and Woias considered the different beam shapes and sizes for piezoelectric energy harvester using the RR method and they validate the result experimentally. Finally, the authors conclude that the shapes have an only small influence on energy but have more effect on excitation amplitude and the triangular shaped with a tip mass performed well compared to rectangular beam [40]. Salmani et al. have developed and proved the mathematical modelling for exponentially tapered piezoelectric energy harvester. While tapering the beam, the natural frequency of the beam increases due to the thinner substrate layer and adding a tip mass at the free end. Table 4 shows a summary of several piezoelectric geometries investigated [41].

Table 4. Summary of several piezoelectric geometries investigated

Author	Piezoelectric configuration	Advantages / Disadvantages
Mateu and Moll (2005)	Rectangle and triangle beam	Triangle beam produces more strain and power compared to rectangle beam
Roundy et al (2005)	Trapezoidal cantilever beam	Trapezoidal cantilever beam can produce more energy than rectangular beam.
Baker et al (2005)	Rectangular and trapezoidal cantilever beam	Rectangle beam produce 30 % less power
A R Reddy and et al	Rectangular beam with and without cavity	Beam with cavity has produce 75% more power.
R Usharani and et al	Introduction of step section in the beam	The beam with step section can produce 4.55% more voltage than the beam without step section.
J park and et al	Variable geometry	Their design can produce 160% more power that the taper beam
S P Matova and et al	The ratio of length and width on tapered beams	The beam with long and slender taper have positive effect on power performance and the beam.
A R Reddy and et al	Rectangular cantilever beam with trapezoidal cavity and rectangular cavity	Trapezoidal cavity beam can produce 97.5% and 108% more voltage compared with the rectangular cavity beam and the beam without cavity

[42] E L Pradeesh & S Udhayakumar	Rectangular beam compared with triangle and various tapered beams with single piezoelectric material	Inverted taper beam produced 47.39 % & 58.3 % more power than other beam
[43] E L Pradeesh & et al.	Rectangular beam compared with triangle and various tapered beams with multiple piezoelectric material	Inverted taper beam produced 46.15 % & 76.30 % more power than rectangle and triangle cantilever beam

3. Various Simulation Software used in Energy Harvesting

Saadon and Sidek designed and reviewed the different types of cantilever beam excited at various resonance frequencies. The geometry of the cantilever beam, vibration frequency, and centre level of resonance frequency was simulated using ANSYS [28]. Sarker et al. developed a micro cantilever beam using C0SM0L Multiphysics. From the various analysis, they found the optimum placement of PZT and silicon mass on the beam [18]. Diyana et al. modelled and simulated comb-shaped piezoelectric beam structure using C0MS0L Multiphysics. The piezoelectric patches and structure were modelled in C0MS0L Multiphysics, MATLAB was used to make mathematical works. The natural frequency of beams was compared using both software. The power output of piezo with and without tip mass was simulated using C0MS0L Multiphysics [44]. Boban et al. optimized energy harvesting by piezo transduction mechanism. NI-PXI workstation was interlinked with LabVIEW software to measure and monitor the output voltage. They have harvested noise and converted it into electrical energy [45]. Prabhakar and Krishna analyzed rectangular cantilever beams at different conditions using Modal analysis in ANSYS. The comparison of resonant frequency for different beam mounding methods was made, and the results had only slight variations compared to experimental work [46]. Motter et al. developed non-controlled rectifier circuits for energy harvesting using a piezoelectric transducer. All tests are performed using MATLAB Simulink setup. The use of the software is to ensure the accuracy of numerical simulations and experimental verifications [47]. Nandish and Hosamani used Multisim software to simulate and analyze the results of the hardware circuit. The simulated result has a close agreement with experimental work [48]. Kundu and Nemade developed a mathematical model using MATLAB and FEM model using C0MS0L Multiphysics. Finally, the short circuit resonance frequency for C0MS0L Multiphysics and MATLAB are 100.0 Hz and 99.80Hz [48,49]. E.L.Pradeesh & S.Udhayakumar used C0SM0L Multiphysics to analysis the optimal location of piezoelectric material over the length of the beam. The effect of proof mass material and shapes were also analyzed. From the numerical analysis, they observed that piezoelectric material located near to the fixed produced more power compared to other locations [50].

4. Energy Harvesting Circuit

The piezoelectric materials can produce AC voltage as an output [51]. The generation of AC voltage is about in μ / m volts, so the AC has to be changed into DC for storage or direct use of electronic equipment [23, 52-55]. Figure 3. Shows a Full wave bridge rectifier circuit, which is used to convert AC to DC.

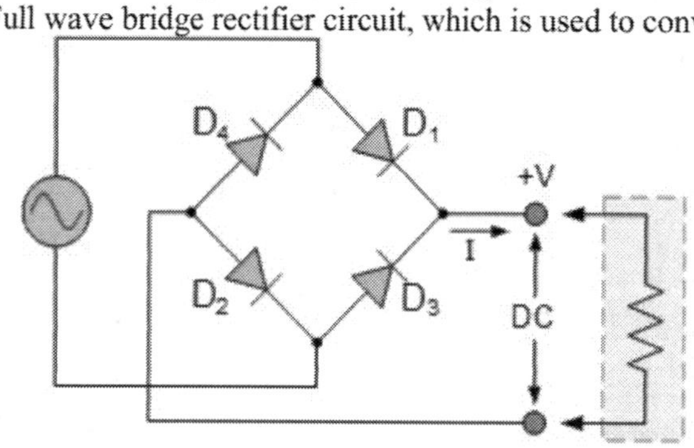

Figure 3: Full Wave Bridge Rectifier Circuit

The basic circuit of energy harvesting is shown in Figure 4, ambient vibration is caused by surrounding environments and a piezoelectric element which is used to harvest the energy from vibration [56]. Step up circuit is used to boost the power of piezoelectric [57]. The signal conditioning circuit consists of several Dedicated Integrated Chips to provide a regulated voltage for the storage of power in the storage unit [58].

IOP Conf. Series: Materials Science and Engineering **995** (2020) 012007 doi:10.1088/1757-899X/995/1/012007

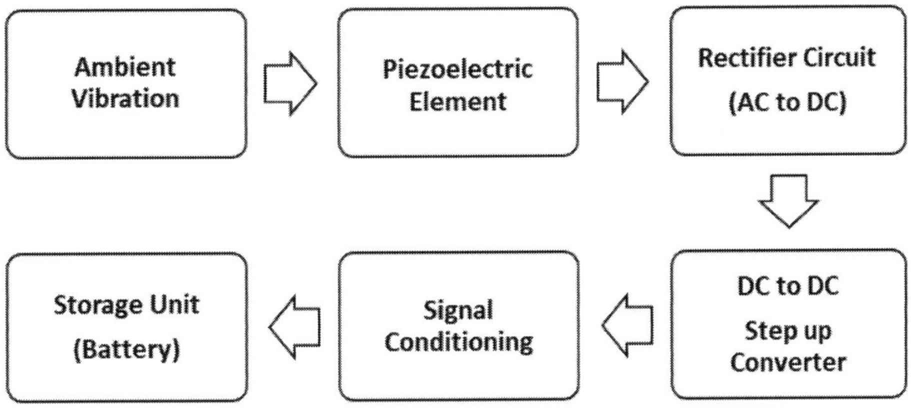

Figure 4: Basic Energy Harvesting Circuit

Yamuna. M. B et al. investigated that the voltage generated from piezoelectric has some ripples, this could be avoided using rectification and filtering. Rectifier circuit converts AC to DC and the regulator circuit removes ripples and maintains the same value of DC. They use MCP73862 charge management controllers used to charge Lithium-ion batteries [59]. Linear Corporation Technology of USA developed a device LTC 3588 series for standalone energy harvesting which is shown in Figure 5. It is completely optimized energy harvesting solution for high impedance sources of piezoelectric elements. It can be interfaced with a piezoelectric element, storage device, load and microcontroller. By using this unit the rectifier circuit, step-up circuit and signal conditioning unit can be eliminated. This unit can also be used to operate ultra-low-power electronic equipment without batteries [60].

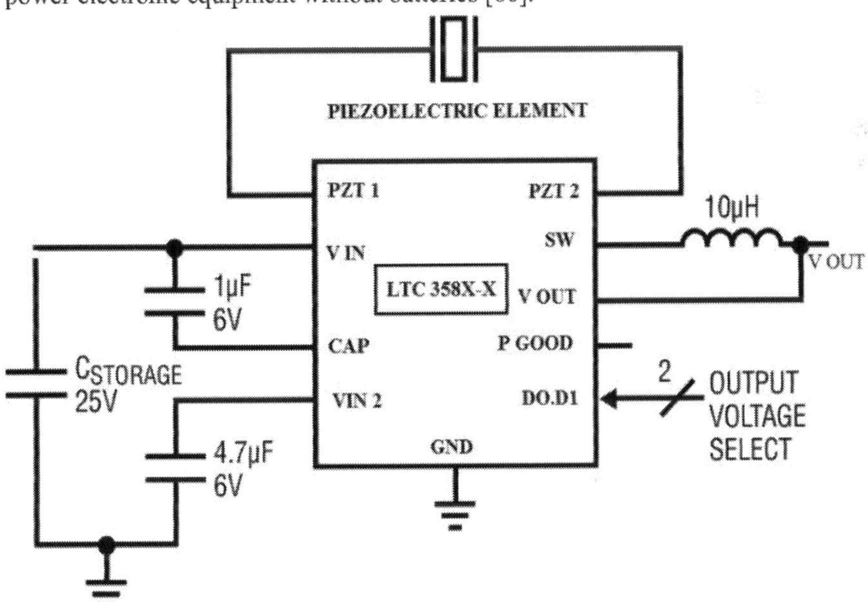

Figure 5: Standalone Energy Harvester Circuit

5. Storage Devices

Storage devices are used to store and power the electronic equipments, the storage devices are a capacitor [61,62] and the battery. Typically, batteries are in different types, Nickel Metal Hydride battery (NiMH) and Lithium-Ion battery (Li-ion) are used in piezoelectric harvesting. Sodano et al. compared Nickel Metal Hydride battery and a capacitor for storage of power from PZT. They found that the charging and discharging rates of the capacitor are quick and the capacitor can't store much power like batteries.

Sodano et al. use a Nickel Metal Hydride battery instead of Lithium-Ion batteries because NiMH has a high current charge density and Li-ion batteries need a voltage regulator or charge controller [17]. Guan and Liao investigated that the super capacitor has high efficiency of 95 %, while Li-ion batteries have 92 % of efficiency and NiMH batteries have only 65% of efficiency. The lifecycle of the super capacitor is unlimited, but batteries can withstand only 300-1000 life cycles. The self-discharge rate of Lithium-ion batteries about 95 % in 30 days. The Nickel Metal Hydride battery has a self-discharge rate of 70% in 30 days and super capacitor can withstand only 65% in 30 days [63]. The conclusion was that a super capacitor can perform better than the batteries in the terms of longer lifetime, charging and discharging efficiencies, but it has poor withstand capacity compared to batteries.

6. Conclusion

Vibration based piezoelectric energy harvesting was reviewed in this paper. It is one of the lasting power source solutions for transportable microelectronic devices by harvesting vibration. The piezoelectric energy harvester is clean and greener technologies can protect the environment. This system is less maintenance and cost-efficient to generate power. However, the real application of energy harvester is still limited and it does not use commercially by consumer electronic equipment. The limitation can overcome by improving geometric configuration, selection of multiple piezoelectric patches, development of efficient smart circuits and selection of better storage devices. Vibration is everywhere around us and we look forward that self-powered smart devices will exist in our life shortly.

References

[1] Liu Z and Li L 2012 Modeling of Energy harvesting device with segmented piezoelectric layer *Procedia Eng.* **47** 470–3
[2] Kokkinopoulos A, Vokas G and Papageorgas P 2014 Energy harvesting implementing embedded piezoelectric generators-The potential for the Attiki Odos traffic grid *Energy Procedia* **50** 1070–85
[3] Trolier-Mckinstry S, Griggio F, Yaeger C, Jousse P, Zhao D, Bharadwaja S S N, Jackson T N, Jesse S, Kalinin S V. and Wasa K 2011 Designing piezoelectric films for micro electromechanical systems *IEEE Trans. Ultrason. Ferroelectr. Freq. Control* **58** 1782–92
[4] Deterre M, Lefeuvre E, Zhu Y, Woytasik M, Boutaud B and Molin R D 2014 Micro blood pressure energy harvester for intracardiac pacemaker *J. Microelectromechanical Syst.* **23** 651–60
[5] Kumar S and Jain Y K 2015 Simulation of Circular-Shaped PZT-5H Sensor for Train Measurement Using COMSOL Multiphysics *IEEE Sens. J.* **15** 4380–7
[6] Alper Erturk D J I 2011 *Piezoelectric Energy* (A John Wiley and Sons, Ltd., Publication)
[7] Sravanthi Chalasani S and Conrad J M 2008 A survey of energy harvesting sources for embedded systems *IEEE SoutheastCon 2008* 442–7
[8] Dikshit T and Shrivastava D 2010 Energy Harvesting via Piezoelectricity. *BIJIT - BVICAM's Int. J. Inf. Technol.* **2** 265–70
[9] Wang L and Yuan F G 2008 Vibration energy harvesting by magnetostrictive material *Smart Mater. Struct.* **17**
[10] Dhingra P, Biswas J, Manipal C E T, Prasad A and Meher S S 2012 Energy Harvesting using Piezoelectric Materials *Spec. Issue Int. J. Comput. Appl. Int. Conf. Electron. Des. Signal Process. 2012* 38–42
[11] Khalid A, Redhewal A K, Kumar M and Srivastav A 2015 Piezoelectric Vibration Harvesters Based on Vibrations of Cantilevered Bimorphs : A Review *Mater. Sci. Appl.* **6** 818–27
[12] Roundy S, Wright P K and Rabaey J 2003 A study of low level vibrations as a power source for wireless sensor nodes *Comput. Commun.* **26** 1131–44
[13] GAO; R X and CUI Y 2005 Vibration-Based Sensor Powering for Manufacturing Process Monitoring *Namrc Thirty-Three* pp 1–8
[14] Song J, Zhao G, Li B and Wang J 2017 Design optimization of PVDF-based piezoelectric energy harvesters *Heliyon* **Volume 3,** 1–18
[15] Anton S R and Sodano H A 2007 A review of power harvesting using piezoelectric materials (2003-2006) *Smart Mater. Struct.* **16**
[16] Lee B-S, He J-J, Wu W-J and Shih W-P 2006 MEMS generator of power harvesting by vibrations

using piezoelectric cantilever beam with digitate electrode **6169** 61690B

[17] Sodano H A, Inman D J and Park G 2005 Comparison of Piezoelectric Energy Harvesting Devices for Recharging Batteries *J. Intell. Mater. Syst. Struct.* **16** 799–807

[18] Sarker M R, Mohamed A and Mohamed R 2016 Develop A Vibration Based MEMS Piezoelectric Energy Harvester Using Micro Cantilever Beam *Int. J. Appl. Eng. Res. ISSN 0973-4562* **11** 3421–6

[19] Hassan H F, Idris S, Hassan S and Rahim R A 2014 Acoustic Energy Harvesting Using Piezoelectric Generator for Low Frequency Sound Waves Energy Conversion *Int. J. Eng. Technol.* **5** 4702–7

[20] Cook-Chennault K A, Thambi N and Sastry A M 2008 Powering MEMS portable devices - A review of non-regenerative and regenerative power supply systems with special emphasis on piezoelectric energy harvesting systems *Smart Mater. Struct.* **17**

[21] M.Taware S and Deshmukh S P 2013 A Review of Energy Harvesting From Piezoelectric Materials *J. Mech. Civ. Eng.* **4** 43–50

[22] Kim S, Clark W W and Wang Q M 2005 Piezoelectric energy harvesting with a clamped circular plate: Analysis *J. Intell. Mater. Syst. Struct.* **16** 847–54

[23] Yu H, Zhou J, Deng L and Wen Z 2014 A vibration-based MEMS piezoelectric energy harvester and power conditioning circuit *Sensors (Switzerland)* **14** 3323–41

[24] Mohammadi F and Khan A 2003 Power generation from piezoelectric lead zirconate titanate fiber composites *Mater. Res.* **736** 1–6

[25] Churchill D L, Hamel M J, Townsend C P and Arms S W 2003 Strain energy harvesting for wireless sensor networks *Smart Struct. Mater. 2003* **5055** 319–327

[26] Park J C, Park J Y and Lee Y P 2010 Modeling and characterization of piezoelectric d33-Mode MEMS energy harvester *J. Microelectromechanical Syst.* **19** 1215–22

[27] Sirohi J and Chopra I 2000 Fundamental understanding of piezoelectric strain sensors *J. Intell. Mater. Syst. Struct.* **11** 246–57

[28] Saadon S and Sidek O 2011 A review of vibration-based MEMS piezoelectric energy harvesters *Energy Convers. Manag.* **52** 500–4

[29] Mateu L and Moll F 2005 Optimum Piezoelectric Bending Beam Structures for Energy Harvesting using Shoe Inserts *J. Intell. Mater. Syst. Struct.* **16** 835–45

[30] Roundy S 2005 On the effectiveness of vibration-based energy harvesting *J. Intell. Mater. Syst. Struct.* **16** 809–23

[31] Dergisi F B, Rahmatian M A, Rahmatian M R and Rahimzadeh H 2015 A review of studies on low-level vibrations as a source of electric power generation *Özel Sayı Sci. J.* **36** 1–7

[32] Roundy S and Wright P K 2004 A piezoelectric vibration based generator for wireless electronics *Smart Mater. Struct.* **13** 1131–42

[33] Baker J 2005 Alternative Geometries for Increasing Power Density in Vibration Energy Scavenging *3rd Int. Energy Convers. Eng. Conf. 15 - 18 August 2005, San Fr. Calif.* 1–12

[34] Rami Reddy A, Umapathy M, Ezhilarasi D and Gandhi U 2016 Improved energy harvesting from vibration by introducing cavity in a cantilever beam *J. Vib. Control* **22** 3057–66

[35] Reddy A R, Umapathy M, Ezhilarasi D and Uma G 2015 Cantilever Beam with Trapezoidal Cavity for Improved Energy Harvesting *NTERNATIONAL J. Precis. Eng. Manuf.* **16** 1875–81

[36] Usharani R, Uma G, Umapathy M and Choi S 2017 Design of high output broadband piezoelectric energy harvester *J. Mech. Sci. Technol.* **31** 3131–42

[37] Park J, Lee S and Kwak B M 2012 Design optimization of piezoelectric energy harvester subject to tip excitation *J. Mech. Sci. Technol.* **26** 137–43

[38] S P Matova, MRenaud, MJambunathan Mg and R V S and Holst 2013 Effect of length / width ratio of tapered beams on the performance of piezoelectric energy harvesters *SMARTMATERIALS Struct.* **22** 1–8

[39] Hosseini R and Hamedi M 2016 An investigation into resonant frequency of trapezoidal V-shaped cantilever piezoelectric energy harvester *Microsyst. Technol.* **22** 1127–34

[40] Goldschmidtboeing F and Woias P 2008 Characterization of different beam shapes for piezoelectric energy harvesting *J. MICROMECHANICS MICROENGINEERING* **18**

[41] Salmani H, Rahimi G H and Kordkheili S A H 2015 An Exact Analytical Solution to Exponentially Tapered Piezoelectric Energy Harvester *Shock Vib.* **2015** 1–13

[42] Pradeesh E L and Udhayakumar S 2019 Investigation on the geometry of beams for piezoelectric energy harvester *Microsyst. Technol.* **25** 3463–75

[43] Pradeesh E L, Udhayakumar S and Sathishkumar C 2019 Investigation on various beam geometries for piezoelectric energy harvester with two serially mounted piezoelectric materials *SN Appl. Sci.* **1** 1–11
[44] Diyana N H, Muthalif A G A, Fakhzan M N and Nordin A N 2012 Vibration energy harvesting using single and comb-shaped piezoelectric beam structures: Modeling and simulation *Procedia Eng.* **41** 1228–34
[45] Boban B, Babu S K S, Satheesh U and Devaprakasam D 2014 *Energy Harvesting By Optimized Piezo Transduction Mechanism* (Coimbatore)
[46] Prabhakar T N and Krishna D V 2016 Numerical Vibration Analysis of Rectangular Beams for Different End Conditions *Int. J. Recent Innov. Trends Comput. Commun.* **Volume: 4** 98–101
[47] Motter D, Lavarda J V, Dias F A, Da S and Member S 2012 Vibration Energy Harvesting Using Piezoelectric Transducer and Non- Controlled Rectifiers Circuits *Abcm* **xxxiv** 378–85
[48] Nandish B M and Hosamani B 2014 A Review of Energy Harvesting From Vibration using Piezoelectric Material **3** 1607–11
[49] Kundu S and Nemade H B 2016 Modeling and Simulation of a Piezoelectric Vibration Energy Harvester *Procedia Eng.* **144** 568–75
[50] Pradeesh E L and Udhayakumar S 2019 Effect of placement of piezoelectric material and proof mass on the performance of piezoelectric energy harvester *Mech. Syst. Signal Process.* **130** 664–76
[51] Ali W G and Ibrahim S W 2012 Power Analysis for Piezoelectric Energy Harvester *Energy Power Eng.* **04** 496–505
[52] Wang W, Yang T, Chen X and Yao X 2012 Vibration energy harvesting using a piezoelectric circular diaphragm array *IEEE Trans. Ultrason. Ferroelectr. Freq. Control* **59** 2022–6
[53] Eltamaly A and Addoweesh K 2016 A Novel Self Power SSHI Circuit for Piezoelectric Energy Harvester *IEEE Trans. Power Electron.* **32** 1–1
[54] Singh P E, Bhumi S and Monika R 2015 Real Time Battery Charging System by Human Walking 225–32
[55] https://www.electronics-tutorials.ws/diode/diode_6.html Accessed 12 October 2020
[56] M.Hanif M.Ramli, M.HazwanM. Yunus, Cheng Yee Low A J 2014 Scavenging energy from human activities using piezoelectric material *Procedia Technol.* **15** 827–31
[57] Shevtsov S and Flek M 2016 Random Vibration Energy Harvesting by Piezoelectric Stack Charging the Battery *Procedia Eng.* **144** 645–52
[58] R. Amirtharajah A P C 1997 Self-powered low power signal processing *IEEE Symp. VLSI Circuits, Dig. Tech. Pap.* **30** 25–6
[59] Yamuna. M. B K S S 2014 Design of Piezoelectric Energy Harvesting and Storage Devices *Int. J. Adv. Res. Electr. Electron. Instrum. Eng.* **3** 10945–53
[60] www.linear.com. Accessed 12 October 2020
[61] Porhonen J, Rajala S, Lehtimaki S and Tuukkanen S 2014 Flexible piezoelectric energy harvesting circuit with printable supercapacitor and diodes *IEEE Trans. Electron Devices* **61** 3303–8
[62] M.Hanif M.Ramli,M.Hazwan M. Yunus, Cheng Yee Lowa A J 2014 Scavenging energy from human activities using piezoelectric material *Procedia Technol.* **15** 827–31
[63] Guan M and Liao W-H 2006 On the energy storage devices in piezoelectric energy harvesting *Smart Struct. Mater. 2006 Damping Isol. Febrary 27, 2006 - March 1, 2006* **6169** SPIE; ASME

Modelling of Robotic Single Peg-In-Hole Assembly Using ADAMS/MATLAB Co-simulation

P Nagarajan[1], S Vigneshwaran[1], K Yuvaraj[1], A Mohammed Ismail[1] and S Raghavendra Prabhu[1]

1 Assistant Professor, Bannari Amman Institute of Technology, Department of Mechatronics, Erode, Tamilnadu, India.

nagarajanp@bitsathy.ac.in, vigneshwaran@bitsathy.ac.in, yuvarajk@bitsathy.ac.in, mohamedismail@bitsathy.ac.in, raghavendraprabhu@bitsathy.ac.in

Abstract. A virtual assembly model of robotic single peg-in-hole assembly environment is developed to determine contact force using ADAMS/MATLAB Co-simulation in this work. The existing dynamic models of the insertion process are incapable of replicating the real assembly scenarios due to its complicated mathematical expressions and handling variety of dynamic parameters. In order to address this problem, a co-simulation model is proposed to study the dynamic behaviour of a planar robot in executing an assembly process. Initially, the physical robotic assembly environment with 3-R planar manipulator is built in SolidWorks and it is imported to ADAMS environment for the analysis. A Cartesian trajectory for the end-effector is given as input for executing the assembly process. Since simulation in ADAMS is executed with joint values, inverse kinematic calculations of the manipulator are employed in MATLAB environment to convert the given Cartesian descriptions into their corresponding joint descriptions. Further, the joint values are properly exchanged between these different software environments using co-simulation model to simulate the intended assembly process. Besides, the contact forces at peg end are analysed under defined lateral error condition.

Keywords: Peg-in-Hole, Robotic Assembly, 3-DOF robot manipulator, ADAMS/MATLAB Co-Simulation, Contact force analysis.

1. Introduction

A peg-in-hole insertion task is considered as a typical and primitive assembly task by many researchers [1, 7]. The peg-in-hole assembly has its application in various industrial sectors like camshaft assembly, flange assemblies in automobile industry, IC assemblies in electronics industries etc. [11]. Such assembly processes are complicated as the aligning and the insertion phases are path and force constrained. Henceforth, the process is automated using robotic manipulators due to its higher accuracy, flexibility and reflex to the dynamic environment.

The robotic manipulators require sensors such as vision, force sensors to sense and localize the hole component with respect to the peg component. In this regard, the assembly operation is influenced by pose estimation of the mating components and positional error in the robotic manipulator. The positional error creates misalignments between the mating parts along their lateral and angular axes. Further, the misalignments in the peg component leads to large reactive forces by establishing contacts with the hole component. Successful assembly depends on how precisely the reactive forces are managed throughout the insertion depth [8].

These reactive forces are predominately controlled by introducing passive or active compliance devices such as RCC and force control techniques. In order to control compliance devices, the insertion process needs to be modelled as either quasi-static assembly or dynamic assembly. Since the peg is inserted at a very low speed, the insertion process is considered as static in quasi-static insertion modelling [7]. Due to this approach, inertial effects of the parts are negligible in static condition. In reality, the static condition is not replicating actual assembly scenario. The mass of the mating parts and the speed of insertion influence the insertion mechanics. In this regard, the dynamic modelling of the insertion including the acceleration and the inertial effects are developed by few researchers [1, 5, 7, 11].

Shahinpoor and Zoohor [11] have developed a dynamic model of the insertion process with an assumption of having insertion depth as a known function of time. The insertion speed is calculated for the given insertion time and the dynamic model is developed using the insertion speed. Trong et al. [7] developed a dynamic model for compliantly supported chamfered peg-in-hole assembly including the factors like inertia, gravity, passive compliance, location of compliance center and insertion speed. Trong et al. presented the influence of insertion force and tilt angle with respect to time for the insertion process. Du et al. [1] derived an extensive dynamic model using the framework of Trong et al. to investigate the insertion process of a compliantly supported chamfered peg-in-hole assembly. They have also presented the insertion forces, tilt angle of the peg, lateral deviation of the peg and insertion depth with respect to time. Xia et al. [9] have proposed a dynamical insertion model of a peg-in-hole including the elastic contact mechanics in order to allow small deformations at all the instances of contact during insertion. Since the traditional rigid body dynamic models are not exhibiting enough deformation of the mating parts. Xia et al. have developed a dynamic model by considering the limitation of elastic deformation that cannot destroy the assembly parts.

The above proposed dynamic models were quite complicated and are computationally difficult to solve since they involve huge computation with different parameters like dimensions and material of the mating parts, mass, insertion speed, insertion depth etc., The dynamic models are also having inconsistent solutions and vary for the different values of the parameters. In order to overcome the above concern, the capabilities of a multibody dynamic simulation tool known as ADAMS, Automated Dynamic Analysis of Mechanical Systems have been utilized in this work. The software is capable of performing a virtual prototype analysis for a mechanical system with an interactive graphical interface. ADAMS solver module is based on the Euler-Lagrange technique similar to the previous dynamic models developed by the researchers. ADAMS has the capability to represent the motion of the joints as a function of time and it is incapable of addressing the Cartesian space inputs. However, the assembly task is executed with Cartesian space information. In order to simulate the assembly process in ADAMS, the joint angles are to be determined using inverse kinematic model of the manipulator in other numerical computing software such as MATLAB. Consequently, the resultant joint angles are imported to the ADAMS environment for the intended simulation. This process of controlling the multibody system developed in ADAMS through other platform is termed as ADAMS co-simulation. MATLAB being a tool, with strong computation capability and capability to interface with ADAMS is used in this work. Several researchers have performed co-simulation using the above mentioned tools for various applications such as robot based welding, virtual bionic hand model and space exploration rovers [3, 9, 6, 4]. ADAMS/MATLAB Co-simulation offers a virtual kinematic and dynamic analysis of robotic applications before the implementation in an actual robot. Guojun et al. [3] have experimented the Co-simulation for robot welding application. The welding coordinates are converted to joint space and the trajectory is planned using robotic toolbox of MatLab. The computed angles are implemented in ADAMS using the in-built step function. Haitao et al. [4] has established a virtual prototyping model for dynamic analysis of 2 DOF robot arm and the PD control system is developed using ADAMS/Control and MATLAB/Simulink. Zhiming et al. [10] has developed an intelligent bionic hand for virtual prosthetic prototyping using ADAMS and PD control using MATLAB.

On this consideration, this work is intended to develop a virtual model of 3-R planar robotic single peg-in-hole assembly using ADAMS/MATLAB co-simulation and to carry out contact force analysis of the intended insertion process. This paper is organized as follows: Section 2 depicts the single peg-in-hole robotic assembly environment adopted in this work, Section 3 presents the ADAMS/MATLAB Co-simulation of robotic assembly model, Section 4 discusses the dynamic insertion analysis results of the assembly from co-simulation and Section 5 provides the conclusion of this paper.

2. PROPOSED ROBOTIC PEG-IN-HOLE ASSEMBLY ENVIRONMENT

A robotic peg-in-hole assembly with position uncertainty is an intricate operation. In this regard, this work adopts a single peg-in-hole robotic assembly environment as shown in Figure 1. The assembly task is assumed to be performed in planar work space. A 3-R planar robotic manipulator is used in this work dexterously to assemble a rigid peg in a rigid hole. In this considered environment, the peg component is mounted on the end-effector and inserted into a static chamfered hole component. A positive clearance (clearance fit) between peg and hole is considered in this work. The assembly process is carried out in two phases: Alignment phase (rotation and translation take place) and insertion phase (only translation). In the alignment phase, the co-ordinate frames of the hole 'H' and peg 'P' components are aligned. The insertion task has been executed in insertion phase. The poses of the peg and hole component are assumed to be known using vision sensor. A offset distance above the hole axes is introduced to avoid the abrupt collision between the mating parts and to ensure minimal angular misalignment during the insertion. ${}^{B}T_{P}$ Pose of the peg with respect to robot base 'B' is first aligned with ${}^{O}T_{B}$ the pose at the considered offset location 'O'. The angular misalignment between the axes is minimum at the end of aligning phase.

In the alignment phase, the peg frame $P_s\left(X_{ps}, Y_{ps}, Z_{ps}\right)$ at the initial stage is first translated and then rotated with respect to the base frame $B\left(X_B, Y_B, Z_B\right)$. Further, frame P is also rotated for the required intermediate point orientation φ_i followed by translation to $I_i\ \left(X_i, Y_i, Z_i\right)$. The pose of the intermediate points I_i change according to defined trajectory followed between the initial peg pose and the offset pose. The homogeneous transformation for this alignment phase is given as,

$$T_A = T_i \cdot R_i \cdot R_B \cdot T_B \tag{1}$$

Where, T_B is the translation matrix between the peg frame and the base frame.

R_B is the rotation matrix to align the peg frame and the base frame.

IOP Conf. Series: Materials Science and Engineering **995** (2020) 012008 doi:10.1088/1757-899X/995/1/012008

R_i is the rotation matrix to rotate the peg frame to the intermediate point orientation.

T_i is the translation matrix to translate the peg frame to the intermediate point position.

As the orientation of the offset and the hole component are same, only translation T_I is carried out in the insertion phase.

In the insertion phase, the peg is translated from the defined virtual offset ${}^{O}T_P$ to ${}^{H}T_O$ i.e. pose of the hole base with respect to the virtual offset. The total assembly process is therefore defined as,

$$ {}^{H}T_P = {}^{H}T_O \, {}^{O}T_B \, {}^{B}T_P \quad (2) $$

Where, ${}^{H}T_P$ is the pose of peg with respect to the hole. At the end of the assembly process, the pose of peg with respect to the robot base ${}^{B}T_P$ and the pose of hole base with respect to the robot base ${}^{B}T_H$ will be collinear.

$$ {}^{B}T_H = {}^{B}T_O \, {}^{O}T_H \quad (3) $$

Where, ${}^{B}T_H$ and ${}^{O}T_H$ are known.

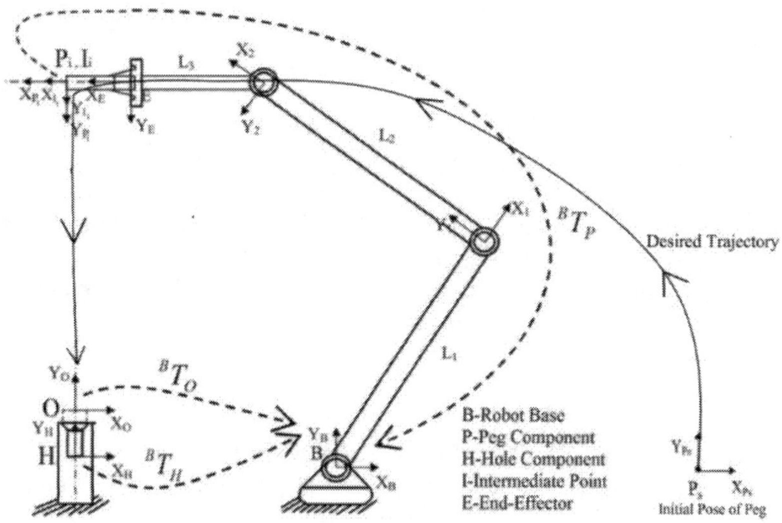

Figure 1. Robotic Peg-in-Hole Assembly environment

3. ADAMS/MATLAB CO-SIMULATION OF ROBOTIC ASSEMBLY MODELLING

A Simulation of robotic assembly aids in understanding, prototyping and analyzing the environment. MSC ADAMS being dynamic analysis software is used to test and perform the simulation and contact force analysis during the constrained motion of the assembly task. Figure 2 represents the steps involved in ADAMS/MATLAB Co-Simulation modelling of a single peg-in-hole robotic assembly process.

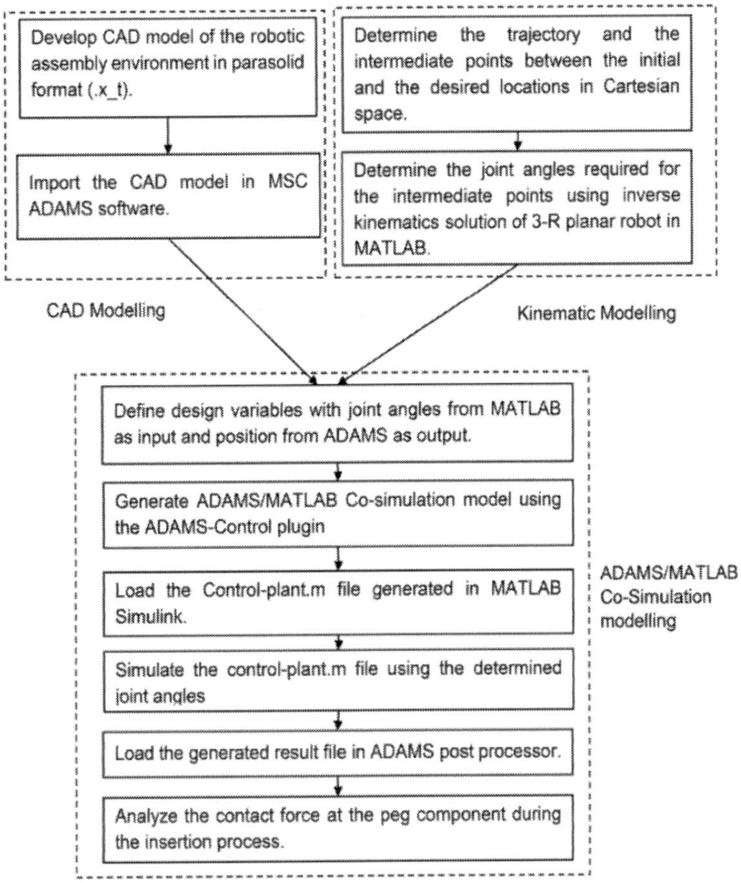

Figure 2. ADAMS/MATLAB Co-Simulation modelling of assembly environment

3.1. CAD Modelling of 3-R Planar Robotic Assembly

The 3-DOF robot manipulator & mating parts are modelled and assembled in SolidWorks. The CAD model is converted into Parasolid.x_t file format in order to transfer the exact 3-D solid or surface data. Then the file is imported into the ADAMS / View environment. The CAD assembly model is illustrated in Figure 3.

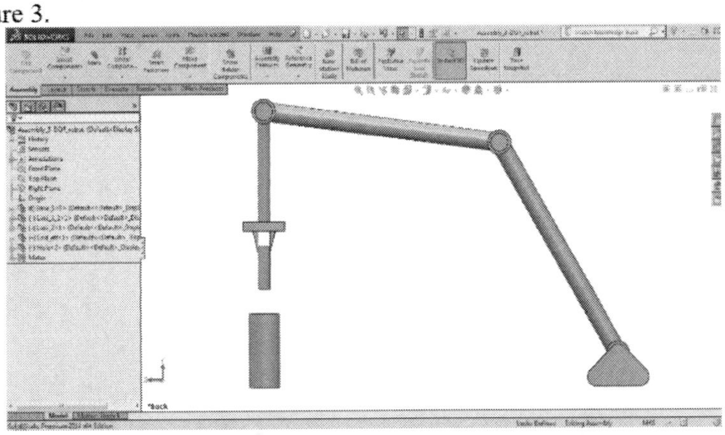

Figure 3. CAD Model of 3-R robotic assembly

3.2. *Kinematic Modelling of the Robotic Assembly*

The robotic assembly process is carried out in two phases: alignment phase and insertion phase. In the alignment phase the peg component is translated and rotated to an offset distance above the hole surface. The offset distance is provided to prevent the collision of the peg surface with the hole chamfers or the collision of end-effector with the hole surface. At the completion of this phase the hole and peg axes are aligned with zero/minimum misalignment. In the insertion phase, the peg component is only translated along the insertion axes since the axes are aligned in the previous phase. In this work, a simple straight line trajectory is assumed to be followed by the robotic arm during the alignment and the insertion phase as shown in Figure 4.

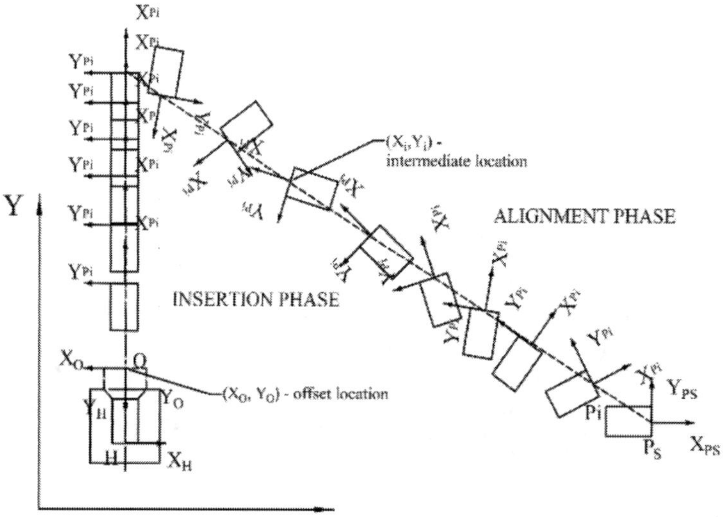

Figure 4. Sequence of Peg positions during the alignment phase and insertion phase

The straight line trajectory is generated as,

$$\begin{aligned} x(t) &= x_p + (x_O - x_P)\cdot t \\ y(t) &= y_p + (y_O - y_P)\cdot t \quad \forall 0 \le t \le 1 \\ \varphi(t) &= \varphi_p + (\varphi_O - \varphi_P)\cdot t \end{aligned} \tag{4}$$

Corresponding translation and rotation matrices as in equation 1 are given as,

$$T_i = \begin{bmatrix} 1 & 0 & 0 & x(t) \\ 0 & 1 & 0 & y(t) \\ 0 & 0 & 1 & 0 \\ 0 & 0 & 0 & 1 \end{bmatrix}$$
$$R_i = \begin{bmatrix} \cos(\varphi(t)) & -\sin(\varphi(t)) & 0 & 0 \\ \sin(\varphi(t)) & \cos(\varphi(t)) & 0 & 0 \\ 0 & 0 & 1 & 0 \\ 0 & 0 & 0 & 1 \end{bmatrix} \tag{5}$$

A 3-R planar robot arm moving in X-Y plane and rotation about Z-axis as shown in Figure 2 is adopted in this work. The Denavit-Hartenberg (D-H) parameters of this planar arm are given in table 1.

Table 1. D-H parameters of 3-R planar arm

Link	Joint Offset (b_i)	Joint Angle (θ_i)	Link Length (l_i)	Twist Angle (α_i)
1	0	θ_1	l_1	0
2	0	θ_2	l_2	0
3	0	θ_3	l_3	0

Overall transformation of the end-effector to base frame is given by

$$ {}^0T_3 = \begin{bmatrix} c_{123} & -s_{123} & 0 & l_1c_1 + l_2c_{12} + l_3c_{123} \\ s_{123} & c_{123} & 0 & l_1s_1 + l_2s_{12} + l_3s_{123} \\ 0 & 0 & 1 & 0 \\ 0 & 0 & 0 & 1 \end{bmatrix} \tag{6} $$

where, $c_i = \cos(\theta_i)$, $s_i = \sin(\theta_i)$. $\theta_{12} = \theta_1 + \theta_2$ *and* $\theta_{123} = \theta_1 + \theta_2 + \theta_3$. For the given joint angle values the end effector pose is calculated as

$$ px = l_1c_1 + l_2c_{12} + l_3c_{123}, $$
$$ py = l_1s_1 + l_2s_{12} + l_3s_{123}, $$
$$ \varphi = \theta_1 + \theta_2 + \theta_3. $$

Whereas in assembly, the robotic manipulator moves along the desired trajectory in the Cartesian coordinates. The Cartesian values are converted to joint angles using the inverse kinematics solution of the manipulator.

$$ \cos(\theta_2) = \frac{(px - l_3c\varphi)^2 + (py - l_3s\varphi)^2 - l_1^2 - l_2^2}{2l_1l_2} $$
$$ \sin(\theta_2) = \pm\sqrt{\left(1 - \cos(\theta_2)^2\right)} $$
$$ \theta_2 = a\tan 2(\sin(\theta_2), \cos(\theta_2)) \tag{7} $$

Two solutions for θ_2 is possible as there are two values for $\sin(\theta_2)$corresponding to the elbow up and elbow down configuration.

$$ \sin(\theta_1) = \frac{(l_1 + l_2c_2)(py - l_3s\varphi) - l_2s_2(px - l_3c\varphi)}{(px - l_3c\varphi)^2 + (py - l_3s\varphi)^2} $$
$$ \cos(\theta_1) = \frac{(l_1 + l_2c_2)(px - l_3c\varphi) + l_2s_2(py - l_3s\varphi)}{(px - l_3c\varphi)^2 + (py - l_3s\varphi)^2} $$
$$ \theta_1 = a\tan 2(\sin(\theta_1), \cos(\theta_1)) $$
$$ \theta_3 = \varphi - \theta_1 - \theta_2 \tag{8} $$

The inverse kinematic solution is programmed in MATLAB to yield the joint angles for all three joints at various time instants corresponding to the intermediate points generated in the desired trajectory.

IOP Conf. Series: Materials Science and Engineering **995** (2020) 012008 doi:10.1088/1757-899X/995/1/012008

3.3. Co-Simulation Model

Co-simulation using ADAMS and MATLAB/SIMULINK software helps to build mechanical system in ADAMS and control it with the input information from MATLAB/SIMULINK. The control of the ADAMS model can be ensured by the proper data exchange between the two programs. The output function variable defined in the ADAMS model forms the input to the MATLAB control system and the input variable described by the MATLAB/Simulink application is the input to the ADAMS mechanical system. This forms a closed loop between ADAMS and MATLAB. The steps followed for the co-simulation are as follows:
The required motion constraints are added for the considered 3-DOF planar robotic assembly system. The base of 3-DOF robot is fixed on the ground. The revolute joints are created in order to establish rotational relationship between the respective links. Then, the element variables for the mechanical system are assigned to provide motion in terms of joint angles about rotational axis (Z-axis). This is a significant step on the Co-simulation.
The function variables that read the joint angles are given as the input and the position variable is given as the output in the ADAMS/Control Plugin. The ADAMS/MATLAB interface helping file is created as (.m) file format in the working directory. This interface file is opened in MATLAB/Simulink, to represents the physical element as a Simulink element. Figure 5 depicts the Co-simulation model with mechanical system, the input and output variables as blocks.

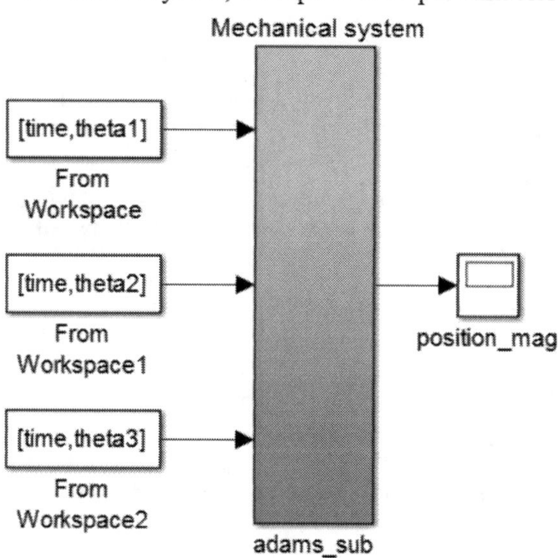

Figure 5. Co-simulation model with input and output variables

The adams_sub block in Figure 5 represents the robotic assembly system. The Joint angles determined from the inverse kinematic solution are read from the MATLAB workspace and forms the input for the ADAMS mechanical block. Subsequently the insertion simulation is performed for the determined joint angles at the specific time instants. The reactive forces produced between the peg and hole during the insertion is analyzed by loading the simulation results in ADAMS post processor.

4. RESULTS AND DISCUSSION OF DYNAMIC ANALYSIS

The developed co-simulation model of the robotic peg-in-hole assembly is simulated and analyzed for estimating the contact force. Dynamic analysis of the insertion process is performed in MATLAB/Simulink using the generated control plant file. The contact force during the insertion process is influenced by the material properties of the mating components. The material for the mating

IOP Conf. Series: Materials Science and Engineering **995** (2020) 012008 doi:10.1088/1757-899X/995/1/012008

parts and the manipulator are taken as steel and aluminium. The Dynamic insertion of 3-DOF robotic assembly simulation is illustrated using the mating component parameters used by Du et al. [1].

- Peg diameter, $d = 19.8\,mm$, Hole diameter, $D = 20\,mm$, Lateral error ξ_0 = -1mm,
- Length of manipulator links: $L_1 = 400mm$, $L_2 = 400mm$, $L_3 = 200mm$,
- Initial peg location : [x=1000 mm, y=0 mm, z=0 mm], Offset location : [x=600 mm, y=300 mm, z=0 mm], Final location : [x=600 mm, y=120 mm , z=0 mm],
- Insertion velocity: 18 mm/sec, Alignment time = 10 sec, Insertion time =10 sec.

The simulation results file in Simulink is imported in ADAMS post processor to interpret the contact force results. Experiments are carried out with and without lateral error in the peg component. The contact forces generated during the insertion process for both the cases are shown in Figure 6.

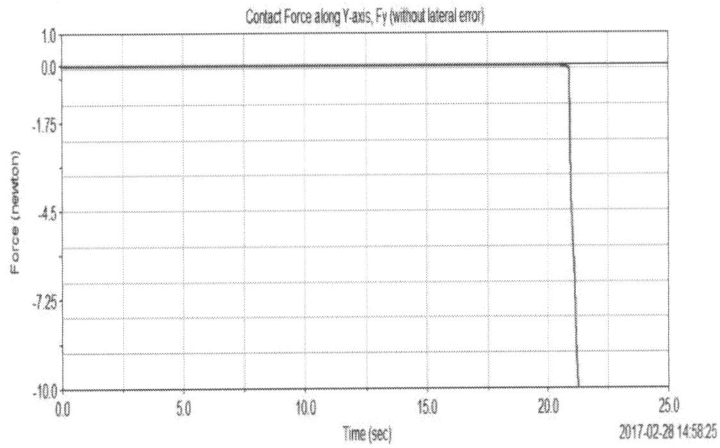

(a) Without lateral error

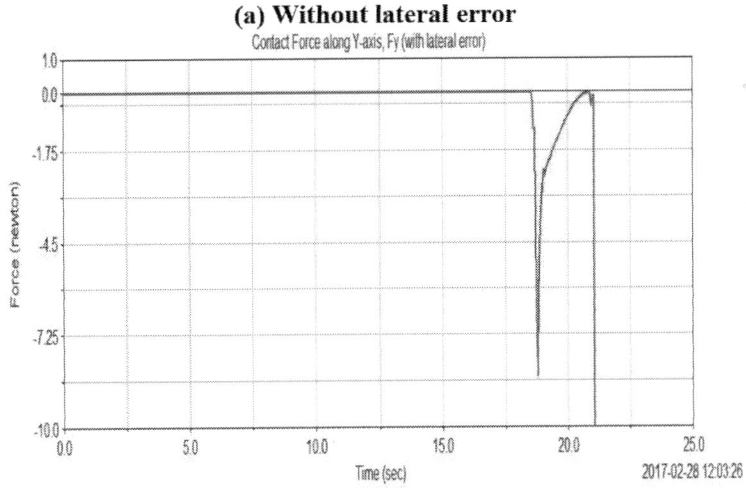

(b) With Lateral error -1mm.

Figure 6. Contact force generated at the bottom of insertion depth.

When the contact occurs between the peg component and the hole, reactive forces are developed. It is evident from Figure 6 (a), that in case of no lateral error there are no contact force until the completion of the insertion depth. If the insertion force is exerted further, there is a steep increase in the contact force which ensures the completion of the assembly process. Figure 6 (b) shows the contact forces generated during insertion in case of 1mm lateral error along X-direction. It is observed that there is an initial increase in the reactive force to a maximum of 8.23N when the first contact is established on the chamfer.

The peg slides inside and establishes one point contact along the chamfer region and leads to a minimal the lateral error on applying forces continuously along the insertion axis. It is evident that there is a decrease in the reactive forces and it becomes zero when there is no lateral error. After the completion of the insertion depth there is a sudden increase in the reactive forces due to restricted motion. Any kind of reactive forces at the peg end are transferred to the robot links and further increases the force at the joints. Figure 7 gives the element force profile at joint 1 of the robotic manipulator with and without lateral error conditions. A constant force of 22.32 N is developed at the joint until the initial contact is established. It is evident from Figure 7 (b) on introduction of lateral error the joint force increase at the chamfer contact and reduces while sliding into the hole. At the completion of the insertion process, the contact between the surface of peg and hole leads to a significant increase in joint forces. The results of the contact forces at various stages can be used to set the threshold values in force control techniques or select RCC devices.

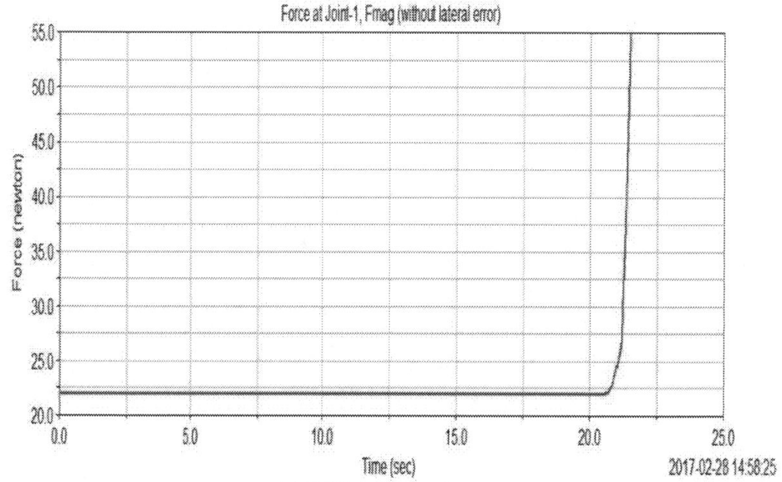

(a) Without lateral error

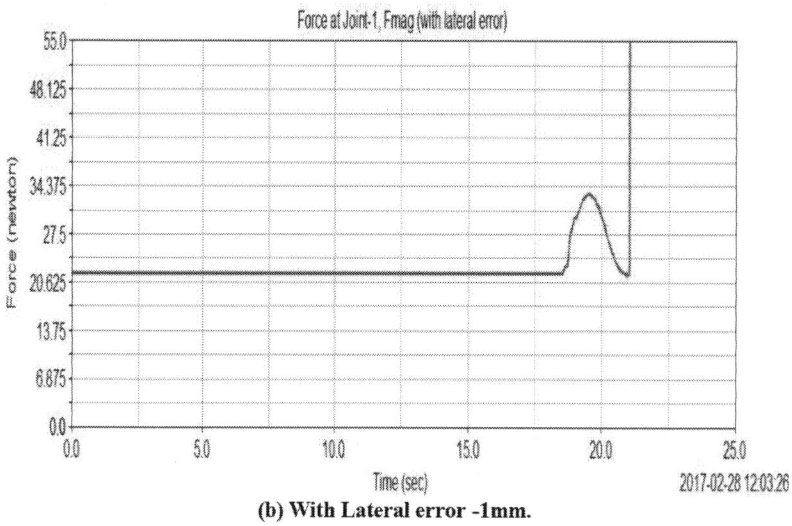

(b) With Lateral error -1mm.

Figure 7. Reactive forces generated on the Joint 1 of 3-R planar robot

5. CONCLUSION

The experimentation of the robotic peg-in-hole environment requires a costly experimental setup and requires precise evaluation. To address this issue, a simulation environment using ADAMS-Multibody dynamics solver and MATLAB for developing an experimentation scenario. A virtual model of 3-R planar robotic single peg-in-hole assembly has been proposed using ADAMS/MATLAB co-simulation to carry out contact force analysis of the intended assembly process. The CAD model of the robotic manipulator is modelled in SolidWorks and it is imported to ADAMS platform. Since ADAMS works only with joint description in simulating the motions, the inverse kinematic calculations are performed in MATLAB environment to handle Cartesian coordinates of assembly descriptions. Consequently, these software are integrated using co-simulation model. The simulation results of the assembly with lateral error have also been analyzed for estimating contact forces of peg component during its insertion. The proposed co-simulation of robotic peg-in-hole assembly can be used to design the compliant devices in executing the real-time assembly.

REFERENCES

[1] Du, K.L. Zhang, B.B. Huang, X. and Hu, J. 2003. "Dynamic analysis of assembly process with passive compliance for robot manipulators", In *Proceedings of the IEEE International Symposium on Computational Intelligence in Robotics and Automation*, (Kobe, Japan, July 16-20, 2003), pp. 1168-1173, DOI=https://doi.org/10.1109/CIRA.2003.1222162.

[2] Gouasmi, M. Ouali, M. Fernini, B. and Meghatria, M. 2012, "Kinematic Modelling and Simulation of a 2-R Robot Using SolidWorks and Verification by MATLAB/Simulink", *International Journal of Advanced Robotic Systems*, Vol 9, Issue 6.

[3] Guojun, W. Linhong, X. Fulun, H. and Xia, Z. 2009. "Kinematics simulation to Manipulator of Welding Robot Based on ADAMS", In *Proceedings of the International Workshop on Intelligent Systems and Applications*, (Wuhan, China, May 23-24, 2009), DOI= https://doi.org/10.1109/IWISA.2009.5072932.

[4] Haitao, L. Yuwang, L. Zhengcang, C. and Yuquan, L. 2013. "Co-simulation Control of Robot Arm Dynamics in ADAMS and MATLAB", *Research Journal of Applied Sciences, Engineering and Technology*, 6(20), pp. 3778-3783.

[5] Haskiya, W. Maycock, K. and Knight, J.1999. "Robotic assembly: Chamferless peg-hole assembly", *Robotica,* Vol. 17, pp. 621-634.

[6] Srividhya, G. Sharma, G. and Suresha Kumar, H.N. 2013. "Software for Modelling and Analysis of Rover on Terrain", In *Proceedings of Conference on Advances in Robotics*, (Pune, India, July 04-06, 2013), DOI=https://doi.org/10.1145/2506095.2506112.

[7] Trong, D.N. Betemps, M. and Jutard, A. 1995. "Analysis of dynamic assembly using passive compliance", In*Proceedings of the IEEE International Conference on Robotics and Automation*, (Nayoga, Japan, May 21-27, 1995), pp. 1997-2002, DOI=https://doi.org/10.1109/ROBOT.1995.525556.

[8] Whitney, D.E. 1982. "Quasi-static assembly of the compliantly supported rigid parts", *Transaction of the ASME Journal of Dynamic Systems, Measurement and Control*, Vol. 104 No. 1, pp. 65-77.

[9] Xia, Y. Yin, Y. and Chen, Z. 2006. "Dynamic analysis for peg-in-hole assembly with contact deformation", *The International Journal of Advanced Manufacturing Technology*, Vol. 30, pp. 118-128.

[10] Zhiming, Y. Yantao, T. Zhuojun, X. and Yang, L. 2014. "Co-simulation and Control Algorithm of Intelligent Bionic Hands with Multi-degree of Freedom", In *Proceedings of the 9th IEEE Internationalon Industrial Electronics and Applications*, (Hangzhou, China, 9-11 June 2014,June 9-11, 2014), pp. 639 – 644, https://doi.org/10.1109/ICIEA.2014.6931242.

[11] Zohoor,H. and Shahinpoor,M. 1991. "Dynamic analysis of peg-in-hole insertion for manufacturing automation", *Journal of Manufacturing Systems*, Vol. 10, pp 99-108.

Development of Low Cost Pellet Loading and Unloading AGV

T Vignesh[1], M Bhuvaneswari [2], A Mohamed Ismail[3], R Aravindh [4], R R AshwinRaaj [5], J BeniRijo Singh [6] and G Saktheeswaran[7]

Department of Mechatronics Engineering
[1,2,4,5,6]Sri Krishna College of Engineering and technology, Coimbatore – 641008
[3] ,Bannari Amman Institute of Technology, Sathyamangalam, Erode - 638401
vigneshtt@skcet.ac.in , bhuvaneswarim@skcet.ac.in,mohamedismail@bitsathy.ac.in ,16eumt010@skcet.ac.in, 16eumt013@skcet.ac.in , 16eumt014@skcet.ac.in

Abstract. This paper explains the steps and processes and components used in the fabrication of an Automated Guided Vehicle (AGV) for loading and unloading pellets in a tank producing industry. In this work, an AGV based on indoor mapping mechanism is proposed for the purpose of material handling efficiently in confined space. The AGV has four geared DC motors at the base for the wheels, the top of the AGV has a set of four rollers which are coupled by gears with one main motor, a pallet will be placed on the rollers in which the pellets will be collected. The AGV will collect the pellets on the pallet from the cutter initially. Once the pellet is filled it is detected by level sensor, and the AGV moves to the next location by avoiding obstacles with the help of the camera. The roller on top will start to rotate on reaching the required spot and this will transfer the pallet on the roller.

Keywords- AGV; indoor mapping; material handling;rollers; camera;

INTRODUCTION

An automated guided vehicle is the remote operated or automatic programmed machine which is used to transfer the material in the industry. It is the automated transporting material from one place to another with low cost manner. Using AGVs in the industry it is increase the flexibility and production time also due path is easily selected and reconfigured to accommodate new locations [1]. In olden days wires is buried under the earth and magnetic strips/ electric cable also used for the navigation, and global position system, laser techniques [2]. In the market many Unit loads AGVs is available [3]. Compare to other system this AVGS system more flexible and faster one, other system more complex and slower. [4]. The material handling is the wide process starting from the storing of raw material to the product manufacturing, it is taken so much time and cost too. To minimize the cost the process should be faster and labour cost is to be reduced. AVGS have some advantages:Reduction in Labour Charge, Increase in productivity, accuracy, High reliability, Material handling ability because to programmability, integrated operation of all AGVs [5]
There are many method is there to control the AVGS, one of the best method is the path following, based on the programming the robot is follows the path. The robot is take the decision based on the instruction is given in the programming. The predefined data store in the system once the robot move from the one place to another place it follows the path, the path data acquired from the sensors it sent to the processor and it compare to the stored data based on that the decision is taken. [6]
Most common guidance technologies in AGV are

- Landmarked based navigation system
- Behaviours based navigation system

- Vision based navigation system

1. MODELLING AND CALCULATION

1.1 Design of AGV

The AGV has a lower frame, upper frame, columns, rollers, and a tray[7]. The lower frame has four wheels coupled to geared DC motor. The columns are supported between the lower and upper frame. The upper frame has four rollers which are coupled by gears and sprockets. A tray will be placed on the top of the roller. The dimensions of AGV are length is 50cm , the breadth is 40cm, the height is 40cm.

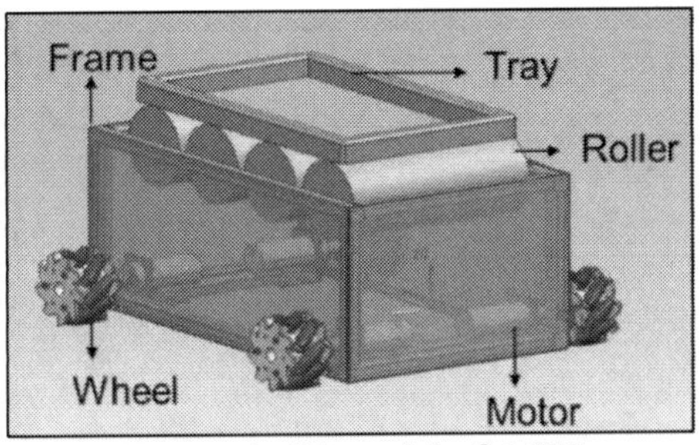

Figure 1. 3D model of AGV

1.2 Method

The wheels in the lower frame are coupled to a motor. Once the tray is filled with pellets the motors will start to rotate resulting in the movement of AGV. The AGV moves in the path which is already given to the AGV by means of indoor mapping. By this method the AGV will avoid obstacles[8]. After reaching the destination the IR sensor will sense it and starts the rotation of the roller, so the tray on the roller will be placed on the required spot. Then the AGV will follow its own path to the starting location.

1.3 Material Strength Selection:

Material selection equation shown in equation 1

$$\frac{\mu}{I} = \frac{\sigma}{y} \tag{1}$$

μ- Load
σ-Stress
I-Moment of Inertia
y - Length

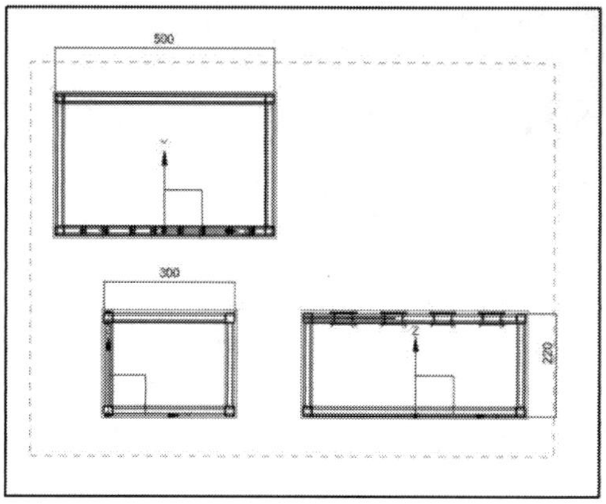

Figure2. Frame Design

The material strength calculation is done to find the correct material that is required to fabricate the lower frame, upper frame and the column. The material is found by comparing the yield stress of the materials.

Motor Selection for wheels

$$F = \mu * N \quad (2)$$

Were

F - Friction

μ- Coefficient of friction

N - Speed

=0.8*24.5

F (friction) = 19.6N

Total load for one wheel (F) = F= (N/4)+ F(friction)

Total load (F) =98/4+19.6 = 44.1N

$$T = F * r \quad (3)$$

Were

T – Torque

F – Force

R – Radius

=44.1*0.05

T =2.21N

N (speed) = (10*60)/2*3.14

N=95.5 rpm

$$A = \frac{v}{r} \quad (4)$$

Were

A - Angular velocity

V – Velocity

R - radius

v=d/t

$$P = T * A \tag{5}$$

Were

P – Power

$$P = \frac{2\pi N}{60} \tag{6}$$

Were

N – speed

v =15/30 = 0.5m/s

Angular velocity=0.5/0.05==10rad/sec

Power =2.21*10=22.1W

Motor selection for Rollers

Friction (F) = 7.81N

Total Load = 27.44N

Torque = 0.82N

Speed = 31.52rpm

Power = 5.4W

Column Calculation

Crippling load by Euler's equation:

$$P_{cr} = \frac{n\pi^2 EA}{\left(\frac{l}{k}\right)^2} \tag{7}$$

P_{cr} = 11.03N

Slenderness ratio = l/k

l/k = 257.5

The boundary condition is satisfied and this column is safe.

2. ELECTRONICS COMPONENTS

2.1 Level Sensor:

This sensor is used to find the level of the pellets collected in the tray. Once the sensor senses the signal will pass and the motor in the wheels will start and the AGV will move[9,10].

IOP Conf. Series: Materials Science and Engineering **995** (2020) 012009 doi:10.1088/1757-899X/995/1/012009

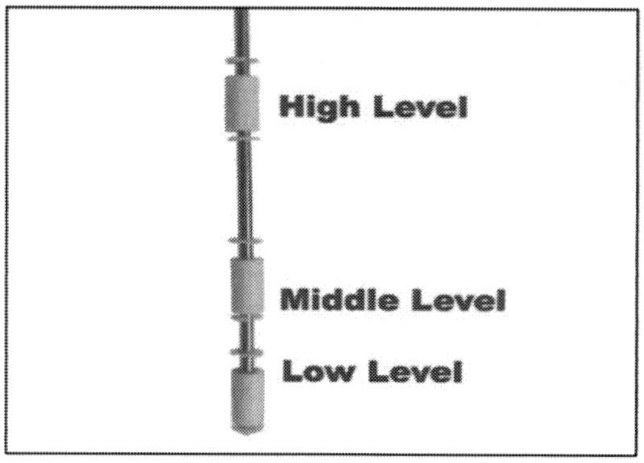

Figure 3. Level Sensor

2.2 IR (InfraRed) Sensor:

The IR sensor is used to detect the AGV near the end location [11,12]. After the AGV reaches the end point, the IR sensor senses and the roller will start to rotate and the tray will move and land in therequired spot.

Figure 4. IR Sensor

2.3 Circuit Diagram:

The Arduino controller board use for the prototype purpose. The level sensor and IR sensor also connected to the arduino. The IR sensor used to find the obstacle and level sensor used to find the level of pallets. The circuit diagram of the AGV shown in figure 5.

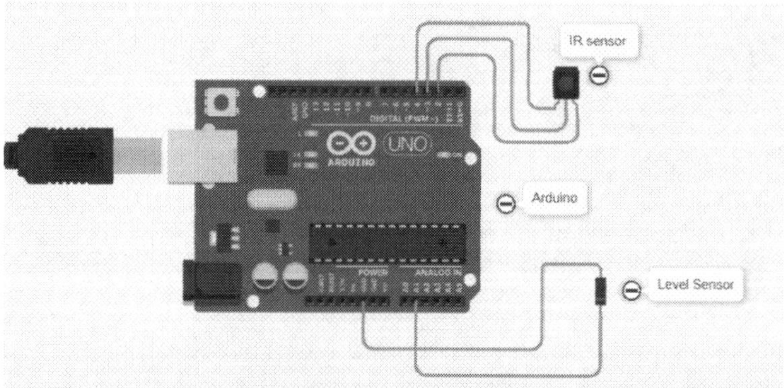

Figure 5. Circuit Diagram

IOP Conf. Series: Materials Science and Engineering **995** (2020) 012009 doi:10.1088/1757-899X/995/1/012009

3. RESULT AND DISCUSSION

The proposed model is in absolute working condition. The AGV is cost efficient and weighs less. The AGV starts from the cutter after loading the pellet and moves through the industry by avoiding the obstacles and unloads the pellets in the hopper. After unloading it returns back in the same path and stops in the starting point. The cycle time is perfectly achieved in its to and fro path. Using the AVG in the industry it is increase the productivity. Through designing this AGV we have provided the basic functions like path following and collision avoiding. And the main function, transportation of goods from one station to another station. The listed factors are considered while design the prototype model.

- Delivery Speed
- Vehicle speed adjustment
- Path flexibility
- Layout changes
- Collision avoidance
- Reducing labour cost
- Running cost reduced compare to conveyor system
- Add more sensors
- Continues cycle of working

By comparing the results of this project and the KhosroBijanrostami, "Design and Development of an Automated Guided Vehicle for Educational Purposes", shows that the efficiency of this project increased.

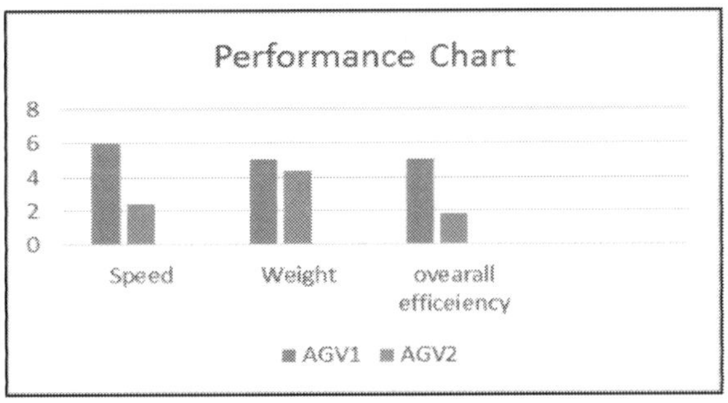

Figure 6. Comparison

Figure 7. Fabricated AGV

IOP Conf. Series: Materials Science and Engineering **995** (2020) 012009 doi:10.1088/1757-899X/995/1/012009

4. CONCLUSION

In this project, design and fabrication of the pellet loading and unloading AGV is done. The AGV has four geared DC motors at the base for the wheels, the top of the AGV has a set of four rollers which are coupled by gears with one main motor. The AGV completes its given function of loading and unloading. The industry layout is fed into the AGV and thus the AGV moves in the exact same path and returns in the same path. Thus the functioning of the AGV is as expected.

References

[1] Tuan Le-Anh, M.B.M. and DeKoster, "A Review Of Design and Control of Automated

[2] Guided Vehicle Systems", European Journal of Operational Research, Volume 171, pp. 1-23, 2006.

[3] KhosroBijanrostami, "Design and Development of an Automated Guided Vehicle for Educational Purposes", Master of Science Thesis, Institute of Graduate Studies and Research, Eastern Mediterranean University, Gazimağusa, North Cyprus, 2011.

[4] G. A Koff, "Basics of AGVS," in Proc. SME ULTRATECH, vol. 1, September 1986

[5]. Apple J. M., 1972, Material Handling Systems Design (New York: Wiley).

[6] E. Shigley and C. R. Mischke, Mechanical Engineering Design, McGraw Hill, 198

[7] Alexander Lange, Martin Albert, Karl-Heinz Siedersberger and Klaus Bengler, "Ergonomic Design of the Vehicle Motion in an Automated Driving Car", Procedia Manufacturing, Vol. 3, pp. 2761-2768, 2015.

[8] Robert Arifin and Pius J. Egbelu, "Determination of vehicle requirements in automated guided vehicle systems: A statistical approach", Production Planning & Control, Vol. 11, pp. 258-270, 2010.

[9] Fang, Q.; Xie, C. A Study on Intelligent Path Following and Control for Vision-Based Automated Guided Vehicle. In Proceedings of the Fifth World Congress on Intelligent Control and Automation, Hangzhou, China, 14–18 June 2004; pp. 4811–4815.

[9] Sabikan, S.; Sulaiman, M.; Salim, S.N.S.; Miskon, M.F. Vision Based Automated Guided Vehicle for Navigation and Obstacle Avoidance, In Proceeding of the 2nd International Conference on Engineering and ICT, Malacca, Malaysia, 18–20 February 2010

[10] . Sun, Y.; Cao, Q.; Chen, W. An Object Tracking and Global Localization Method using Omnidirectional Vision System. In Proceedings of the Fifth World Congress on Intelligent Control and Automation (WCICA 2004), Hangzhou, China, 15–19 June 2004; pp. 4730–4735

IOP Conf. Series: Materials Science and Engineering **995** (2020) 012009 doi:10.1088/1757-899X/995/1/012009

[11]. Lotharschulze ,SebasitianBehiling " Automated Guided Vehicle System: a Driver for Increased Business Performance" IMECS 2008 19-21 March 2008 hongkong .

[12] K KishorDesgine Of Automated Guided Vehicel , IJARC Volume 3, Issue 1, January-April (2012)

Predicting the Emissive Characteristics of an IC Engine Using DNN

Pravin M C[1], Mukilan M[2], Vishnu Prakash G,[3] Nithish P[4], Monish Kanna B[5], Logesh E[6].

1 Assistant Professor, Bannari amman Institute of Technology,Department of Mechatronics, sathyamangalam, Erode, Tamil Nadu , India.

2, 3, 4, 5, 6 Students, Bannari amman Institute of Technology,Department of Mechatronics, sathyamangalam, Erode, Tamil Nadu , India.

Abstract. Biodiesel is the new form of automotive fuel that the world is now concerning and several researches are going on for the production of an efficient form of Bio-Diesel because of the fact that Diesel and Petrol are going to be exhaustible in nearly 60 years. In order to produce an efficient fuel, it is inevitable to calculate the emission characteristics concerning the fuel. This project deals with the efficient and intelligent way of analyzing and calculating the engine emission characteristics of Bio-diesel operated IC engines. A Machine Learning based model using TensorFlow library has been developed using python programming for the calculation of emission characteristics such as Carbon-monoxide (CO) and Carbon-dioxide (CO_2) of an IC engine upon injection of Bio-diesel as fuel in different proportions. These investigations and data-sets are considered for a four stroke internal combustion engine. In this Machine Learning model TensorFlow library has been used for the better visualization of the results and error rectification. The results of the developed TensorFlow model are then compared with an existing Fuzzy model for the same application. The results predicted by this model clearly are in good correlation with the actual values which depicts that this method is effective and the total error of the developed model was found to be ±0.02 which is comparatively lower than that of the existing Fuzzy model. Concludingly, the Machine learning model using TensorFlow was found to be the best model for the calculation of engine emission characteristics of Bio-diesel operated IC engines as it offers more visualization tools and better predictive analysis.

Keywords: Biodiesel, Carbon-monoxide, Carbon-dioxide, TensorFlow, TensorFlow.

1. Introduction

Diesel is our primary mode of fuel for almost all automotive vehicles along with Petrol. But the fact is that Diesel and Petrol are going to be extinct in future. Many alternative fuels have been developed to replace fossil fuels such as Diesel and Petrol in order to make the automotive sector keep moving. Out of those alternative fuels developed, in this project Biodiesel is mainly addressed as an efficient and eco-friendly alternative for Diesel. Bio- diesel (Pongamia pinnata) is a form of Diesel obtained from plants and animals and it is typically made by chemically reacting lipids such as soybean oil with alcohol. For concluding Bio-diesel as a better alternative to Diesel, we should calculate the IC Engine performance characteristics such as Brake Thermal Efficiency (BTE) produced for different ratios of hydrogen injection and Brake specific energy consumption by considering the emission characteristics such as carbon monoxide and carbon - dioxide. It is prepared from edible and non-edible vegetable oils which considerably reduce the CO2 emission when compared to diesel and also it acts as promising fuels in the upcoming decades.

R.S Karthick et al [1] developed a dataset by analyzing IC engines using biodiesel as a fuel. These engine parameters of the Bio-diesel engine should be effective so that it may serve as a better alternative for the Diesel. From the dataset obtained loads and bends are considered as input meanwhile emission of CO and CO2 emission are to be considered as output for this prediction model.

In this project, an intelligent model has been developed from machine learning using TensorFlow for the calculation of the above-mentioned engine parameters and engine efficiency for the usage of Bio-diesel as a fuel. A Machine Learning based model using TensorFlow library has been developed using python programming for the calculation of emission characteristics such as Carbon-monoxide (CO) and Carbon-dioxide (CO_2) of an IC engine upon injection of Bio-diesel as fuel in different proportions. This investigation and data-sets are considered for a four-stroke internal combustion engine. In this Machine Learning model TensorFlow library has been used for the better visualization of the results and error rectification.

Visualization of a dataset is one of the important aspects for this proposed intelligent model. Visualization of a dataset using tensor flow is referred from [2] as a base for this prediction model. Fuzzy logic model for the same dataset has been developed in [1]. Result of this intelligent model is compared with a fuzzy intelligent model and the best intelligent model is discovered.

2. Proposed Machine Learning Model

2.1. *Tensor Flow*

TensorFlow is an open source library for machine learning and deep learning developed using python programming, it uses differential calculus and dataflow programming as a base. TensorFlow is developed by Google. The proposed model uses TensorFlow version 2.x to build up the intelligent system for the calculation of efficiencies.

2.2. *Implementation*

All the coding and the implementation of the proposed machine learning model is done by using Google Collaboratory. Google colaboratory uses an online cloud based Graphical Processing Unit(GPU) and Tensor Processing Unit(TPU) for developing intelligent machine learning and deep learning models. The proposed machine learning model uses a Tensor Processing Unit(TPU) to predict the emission characteristics of various proportions of Biodiesel operated IC engines upon hydrogen injection. Experiments were carried out using four-stroke, single cylinder and stationary compression ignition engines at a constant speed of 1500 rpm with the rated power of 5.2 KW. The emission characteristics such as Carbon monoxide(CO), Carbon dioxide(CO_2) are measured with varying blends and loads using AVL gas analyser. Therefore for developing the model, the parameters such as Loads and Blends of Diesel and Biodiesel are considered as inputs and the emission parameters such as CO and CO_2 are considered as outputs. The source of data sets were taken experimentally using a four-stroke, single cylinder ignition engine at a constant speed of 1500 rpm with the rated power of 5.2 KW. The data sets were considered as different input and output parameters for developing the machine learning model. The implementation of the algorithm of the model is shown as follows.

IOP Conf. Series: Materials Science and Engineering **995** (2020) 012010 doi:10.1088/1757-899X/995/1/012010

2.2.1. Python Libraries Used

There are many libraries available in python for machine learning. In order to train the model for prediction of unseen values, structuring of the data, mathematical modeling and visualization of the output are inevitable. These are achieved by the following libraries.

Table 1: Python libraries used

S.No.	Library used	Function
1.	tensorflow	faster numerical computing
2.	Pathlib	creating new path
3.	matplotlib	plotting 2D graphs
4.	Numpy	computation
5.	Pandas	data manipulation and analysis
6.	Seaborn	statistical data visualization

The data-sets required for the model are imported from the Google drive location from which it is already stored. The parameters such as Loads in the range of 0 - 100 and blends of Diesel and Biodiesel in the range of 0 - 100 are considered as inputs to the model and emission characteristics of CO and CO_2 are taken as output from the model.

2.2.2. Visualization Of Inputs

The input data-sets are visualized using a seaborn python library in order to get an overall idea on how to design the model as shown in the figure 2.1.

In the figure "e_co" and "e_co2" represents the emission characteristics of CO and CO_2 respectively.

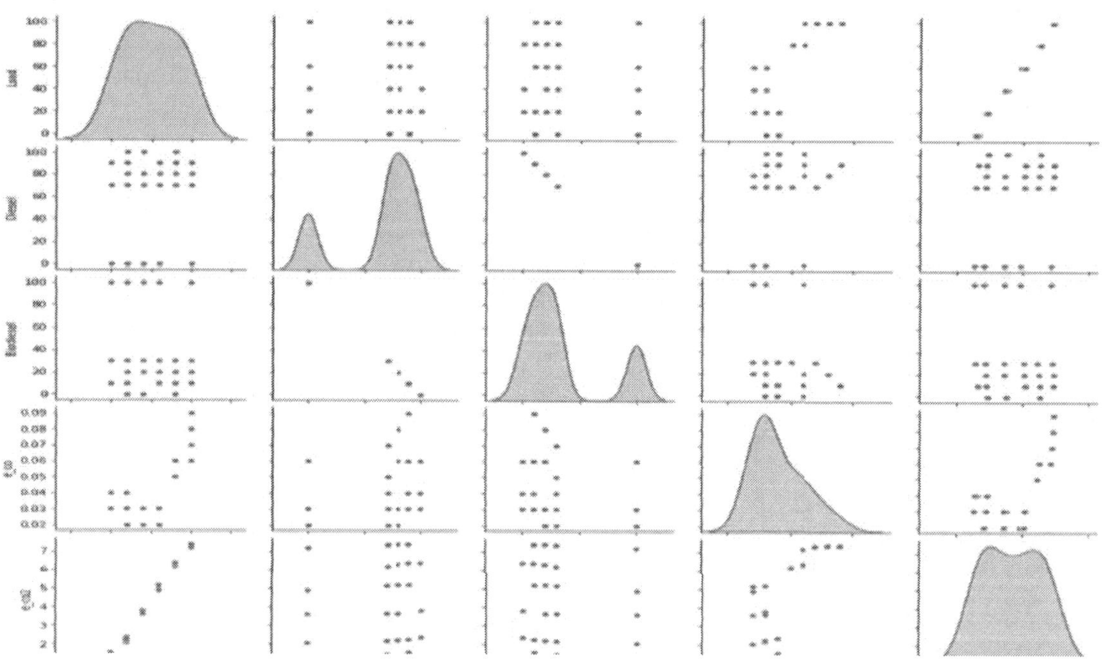

Figure 1: Visualization of input data-set

2.2.3. *Splitting The Data*

Sometimes due to the over training of the model leads to over-fit of the training data(*over-fit - performing very good on training data set and performing very poor on new unseen data set). To avoid this we are splitting the available data into two sections, one is "training_data_set" which is then divided into "training_data_set" and "validation_data_set" and another one is "test_data_set". The "validation_data_set" is a minimal subset of "training_data_set" in which values are taken randomly from the "training_data_set" automatically for the verification of the model and error rectification. First we are training the model with the training_data_set and the trained model is verified and error is rectified by using the "validation_data_set". Then the performance of the model is tested with unseen "test_data_set". Finally, from the predicted output we will find the error and correlation of the model with the actual values(true value - predicted value). Normally the train : test data set ratio is 8:2.

2.2.4. *Calculating Statistical Data*

Statistical data such as mean and standard deviation of the datasets provided is found by using the "describe()" function. These statistical data is helpful in normalizing the data in order to achieve the results effectively.

Table 2: Calculating the statistical data

Inputs	**Mean**	**Standard deviation**
Load	50.833333	33.868886
Diesel	65.416667	35.628416
Bio-diesel	34.583333	35.628416

2.2.5. *Extracting The Labels (Output Values)*

Emission of CO and CO_2 is the output or predictable value in this model. So it is necessary to extract it from the training set and label it separately. Both the CO data and CO_2 data are labeled separately and this label is later used to train the model as well as predicting and differentiating the outputs generated from the model.

2.2.6. *Normalizing The Data*

Data handled by the model can vary in range; this can cause huge error while predicting the output. So it is necessary to scale down those data by using normalization techniques. Even Though the input data does not vary, normalization will give better output. Here, we are normalizing the input parameters using their statistical parameters (Mean, standard deviation) in the range of -1 to 1. Normalization is done by using,

$$\text{Normalized data} = (\text{Data} - \text{Mean}) / \text{Standard deviation}$$

2.2.7. *Building The Model*

Here a function is defined for model building. The deep neural network model built here consists of three layers.

- First layer - Input layer
- Second layer - Hidden layer
- Third layer - Output layer

In the first layer there are 64 units which means 64 neurons, the hidden layer also has 64 neurons, on the third layer there is only one neuron because of the model is a single output model. In every layer we are using "relu" activation which means if the input value is less than zero it replaces that value with 0 and the rest will remain the same. For building the model we need to define the input shape by the "input_shape" parameter. In tensorflow there are a lot of optimizers which contain the inbuilt algorithms to reduce the error automatically. Here, we use the RMS prop optimizer (Root Mean Square) which will reduce the Mean Square Error (MSE) and Mean Attribute Error (MAE). For this optimizer the initial learning rate is given as 0.001. Then after setting all the instructions, the model is compiled using "model.compile". The built model will work commonly for calculating both CO and CO_2 emission by calling the model by using different data-sets and parameters for CO and CO_2 respectively.

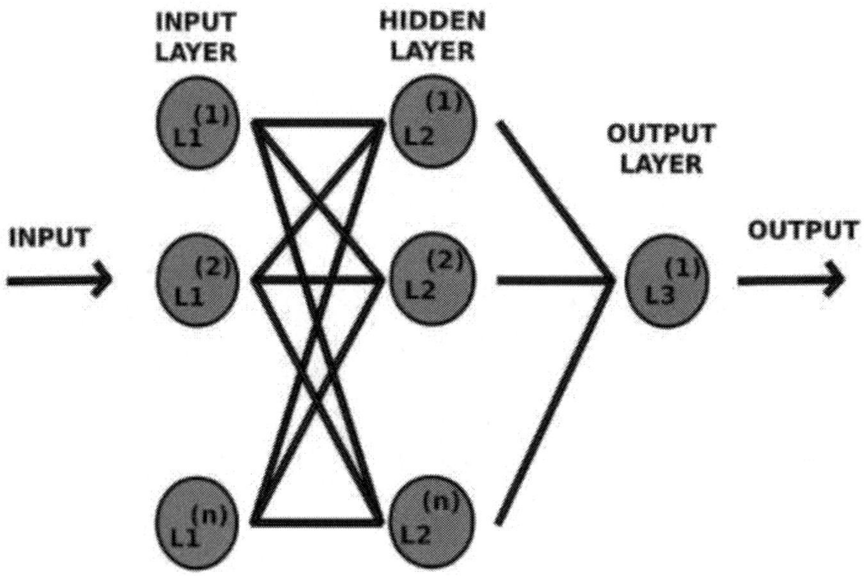

Figure 2: Neural network representation

2.2.8. *Calling The Model*

Two models are created for CO and CO_2 and they are called for execution. Finally the models are tracked for their route in different layers. By using the "summary()" function we can print the performance of every layer for the CO and CO_2 model as shown in the table 2.3. It explains the input and output shapes of the layers.

Table 3: Tracking the CO and CO_2 model's route

Layers (type)	Output shape	Parameters
Layer 1 (Dense)	(None, 64)	256
Layer 2 (Dense)	(None, 64)	4160
Layer 2 (Dense)	(None, 1)	65
Total parameters : 4, 481		
Trainable parameters : 4, 481		
Non-Trainable parameters : 0		

2.2.9. *Training The Model (For Co)*

Training of the model is done between normalized training datasets and training labels of CO and CO_2. Here what happens is the training datasets is split into training and validation datasets and the training is done in an iterative manner by using epochs (iterations), then the validation loss of the model is calculated and monitored by using an "early stopping callback()" function which is an inbuilt function in the keras module of tensorflow. The early stopping callback function monitors the validation loss produced in the model and stops the training of the model where the validation loss becomes a constant value upon iterations. Hence overfitting of the model is prevented by using this function.

2.3. *Error Analysis*

2.3.1. *Error Analysis For Co And Co_2*

Here we are plotting a graph between epochs (iterations) and MAE (Mean Attribute Error) and also between epochs and MSE (Mean Square Error) by using the values of validation loss and training loss obtained from the early stopping function in the of the training model.

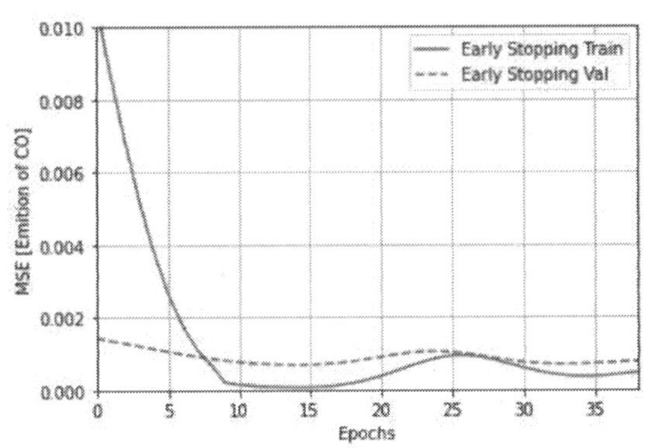

Figure 3: Epochs vs MAE for CO

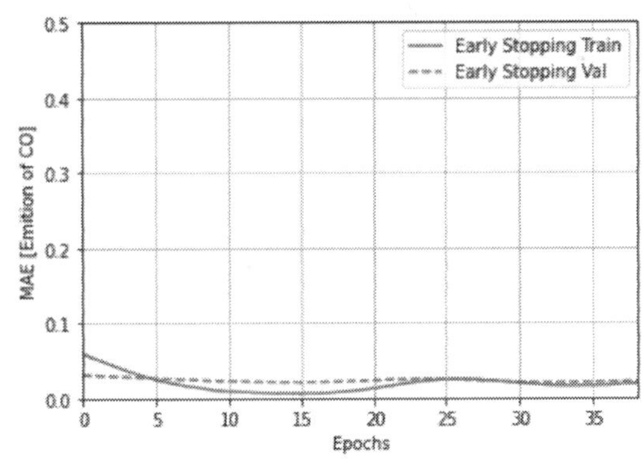

Figure 4: Epochs vs MSE for CO

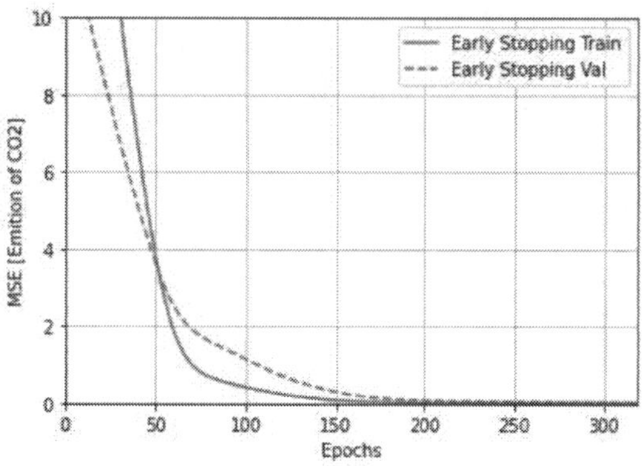

Figure 5: Epochs vs MSE for CO_2

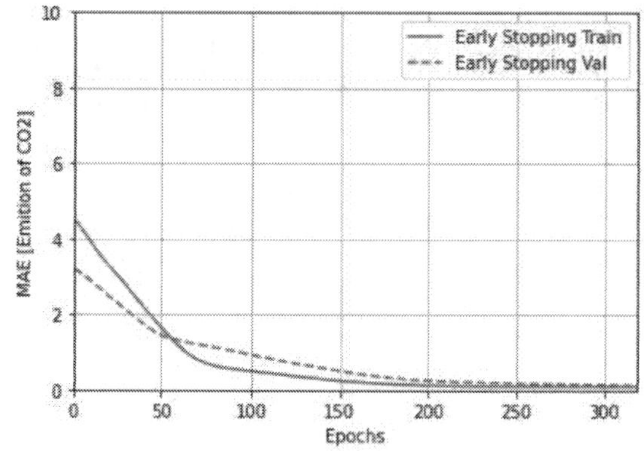

Figure 6: Epochs vs MAE for CO_2

From the plots shown, the straight line represents the training losses and the dotted line represents the validation losses of both the CO and the CO_2 models. It is clearly seen that the validation losses and the training losses of both the CO and CO_2 models nearly coincide and the error gets reduced up to iterations. The total error thus produced from the CO model is 0.034 and for the CO_2 model is 0.074. So, from these graphs it is concluded that the models developed show good coherence with the original values and the error is also minimal and acceptable.

3. Results

3.1. Tolerance Level

The tolerance level for CO2 and CO emission characteristics are plotted below. The maximum tolerance for CO2 and CO emission is more or less than ± 0.002 that means that this model is pretty good at predicting the unseen data values.

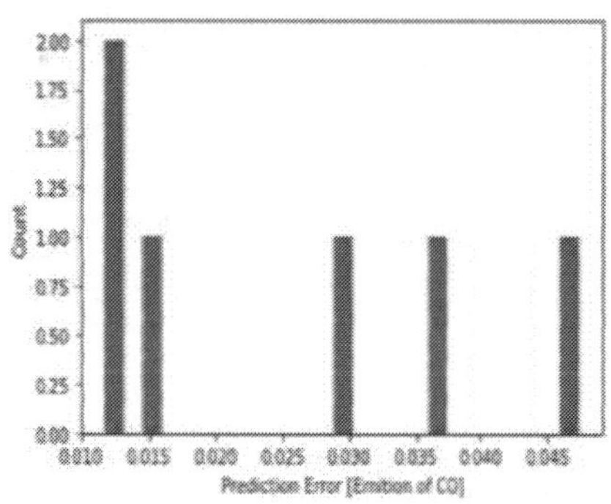

Figure 7: Tolerance graph for CO

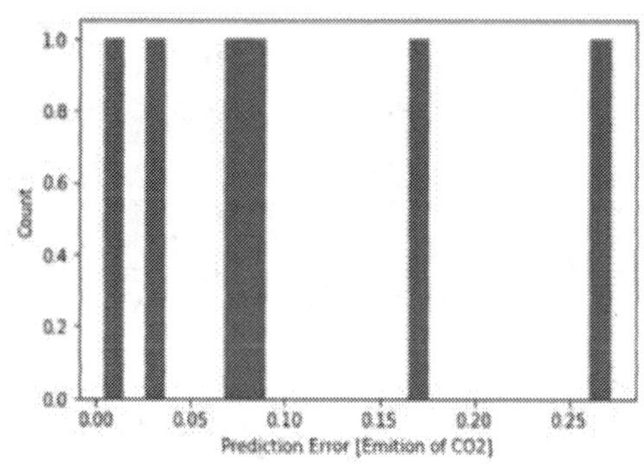

Figure 8: Tolerance graph for CO2

Figure 7 and 8 depicts a tolerance graph of CO and CO2. The above developed graph is plotted based on prediction error (x axis) and count (y axis). The prediction error for CO ranges from 0.010 to 0.045 randomly meanwhile the prediction error for CO2 ranges from 0 to 0.25. From the graph it is clear that the tolerance of CO and CO2 emission is 0.002.

3.2. *Prediction Model*

Above mentioned dataset is divided as training dataset and test dataset. Hitherto we have used a training data set and executed our model. In order to depict our intelligent model as the best we have to prove that the original value should be more or less equal to the prediction value. For that a prediction graph is developed as it is shown in the figure 3.3 and 3.4.

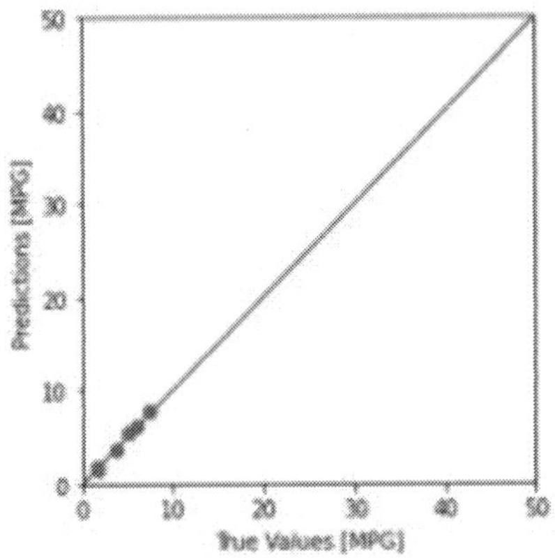

Figure 9 : Prediction graph for CO2

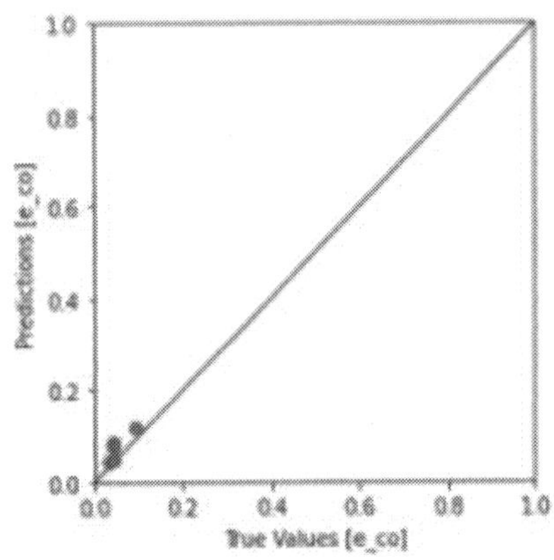

Figure 10 : Prediction graph for CO

4. Conclusion

For an efficient model the prediction model should lie in the center line, if the deflection is more its efficiency is less and its error is more. In our model its deflection is very low so its efficiency is high. Its efficiency is approximately 99.5 and its error is 0.05.

Error of this intelligent model using TensorFlow is calculated by using a formula described below.

Error 1= CO prediction - CO label

Error 2= CO2 prediction - CO2 label

After calculating Error 1 and Error 2, overall error for this intelligent model is calculated. That overall error is approximately 0.05.

When compared to a fuzzy intelligent model its error is 0.91 [1]. By comparing these two values it is clear that the efficiency of intelligent models using TensorFlow is better than the efficiency of intelligent models using fuzzy logic.

From the above described results it is clear that, Comparing the results obtained from the two models, it is found that both the models seem to be equally effective and intelligent. Bio-diesel as an IC engine fuel offers better efficiency and produces good engine knocks upon hydrogen injection. Hence, concludingly Biodiesel can be used as a better alternative for Diesel.

REFERENCE

[1] Karrthik, R.S., Baskaran, S. and Raghunath M., (2019), "*Development of intelligent model to predict the characteristics of biodiesel operated CI engine with hydrogen injection*", Advances in Computational Design, *Vol. 4, No. 4 (2019) 367-379* DOI:https://doi.org/10.12989/acd.2019.4.4.367.

[2] Wongsuphasawat, K., Smilkov, D., Wexler, J., Wilson, J., Mane, D., Fritz, D., ... Wattenberg, M. (2018). Visualizing Dataflow Graphs of Deep Learning Models in TensorFlow. IEEE Transactions on Visualization and Computer Graphics, 24(1), 1–12. doi:10.1109/tvcg.2017.2744878

[3] Berber, A., Tinkir, M., Gultekin, S. and Celikten, I. (2011), "Prediction of a diesel engine characteristics by using different modeling techniques", *J. Phys. Sci.*, **16**, 3979-3992.

[4] Oliveira, A., Santos, E., Botelho, G., Valente, O. and Sodre, J. (2013), "Hydrogen electronic injection system for a diesel power generator", *J. Green Energy,* **38**(19), 7986-799.

[5] https://doi.org/10.1016/j.ijhydene.2013.04.118.

[6] Parlak, A., Islamoglu, Y., Yasar, H. and Egrisogut, A. (2006), "Application of artificial neural networks to predict specific fuel consumption and exhaust temperature for a diesel engine", *J. Appl. Thermal Eng.*, **26**(8-9), 824-828. https://doi.org/10.1016/j.applthermaleng.2005.10.006.

[7] Rai, A., Satheesh Kumar, N., Pai, S. and Rao, S. (2012), "Fuzzy logic based prediction of performance and emission parameters of a LPG-Diesel dual fuel engine", *J. Appl. Thermal Eng.*, **38**, 280-292.

[8] https://doi.org/10.1016/j.proeng.2012.06.036.

[9] Sakthivel, G., Ilangkumaran, M., Syamkumar, U., Vasudevan, M., Venkatesh, S. and Viswanathan, B. (2013), "Development of fuzzy logic model to predict the engine performance of fish oil biodiesel with diethyl ether", *J. Ambient Energy,* **47**, 1-33. https://doi.org/10.1080/01430750.2013.829786.

[10] Sakthivel, G. and Ilangkumaran, M. (2014), "Neural network (ANN) approach to predict the engine performance of fish oil biodiesel with diethyl ether using back propagation algorithm", *J. Ambient Energy,* **78**, 9-40.

[11] Saravanan, N., Nagarajan, G. and Narayanasamy, S. (2008), "An experimental investigation of hydrogen enriched air induction in a diesel engine system", *J. Renewable Energy,* **33**, 1761-1765.

[12] https://doi.org/10.1016/j.ijhydene.2007.12.065.

[13] Saravanan, N. and Nagarajan, G. (2009), "Performance and emission study in manifold hydrogen injection with diesel as an ignition source for different start of injection", *J. Renewable Energy,* **34**(1), 328-334. https://doi.org/10.1016/j.renene.2008.04.023.

[14] Tasdemir, S., Saritas, I., Ciniviz, M. and Allahverdi, N. (2011), "Artificial neural network and fuzzy expert system comparison for prediction of performance and emission parameters on a gasoline engine", *J. Expert*

[15] Amarpuri, Lakshay, et al.(2019) "Prediction of CO 2 emissions using deep learning hybrid approach: A Case Study in Indian Context." *Twelfth International Conference on Contemporary Computing (IC3)*.

[16] Schürholz, D., Kubler, S., & Zaslavsky, A. (2020). "Artificial intelligence-enabled context-aware air quality prediction for smart cities." *Journal of Cleaner Production*, 121941.

Detection of Soft Sensor Fault Using EKF Algorithm for Two Tank Interacting System

K. Sneha[1], P. Vaishnavi[2], M. Ganesh[3], C. Jagadeesh[4], A. Hariharan[5]

[1,2] Assistant Professors, Department of Electronics and Instrumentation Engineering, Bannari Amman Institute of Technology, Sathyamangalam, Tamilnadu, India.

[3,4,5]Students, Department of Electronics and Instrumentation Engineering, Bannari Amman Institute of Technology, Sathyamangalam, Tamilnadu, India.

sneha@bitsathy.ac.in, vaishnavi@bitsathy.ac.in, ganesh.ei17@bitsathy.ac.in, jagadeesh.ei17@bitsathy.ac.in, hariharan.ei17@bitsathy.ac.in

Abstract. This paper deals with the identification of fault acting in the highly non-linear two-tank interacting system. This two-tank interacting system is reported with a sensor fault that is acting on the system which makes the system to be highly unstable.For detecting the fault, the estimation algorithm is used. Here the Extended Kalman Filter algorithm is used for identifying the sensor fault acting in the system. In the EKF algorithm residual is generated that indicates the presence of the fault. The residual that is generated gives the difference between the actual and the predicted measurement. This residual generation shows the magnitude of fault that is present in the system and also the nature of the fault that is acting in the system. This EKF based detection of fault indicates the occurrence of fault at the instant of time.

Keywords:State Estimation, Residual Generation, Two Tank Interacting System, Extended Kalman Filter, Fault Identification, Additive Fault.

1.Introduction

Nonlinear, dynamic structures are nowadays all over in our lives.They're in the Computers, ships, planes, and vehicles.Composite structures, such as chemical reactors, reactors in nuclear power plants and fractionating columns, are used in every industry.They work consistently, creating our lives more secure and more fun before these systems malfunction.Faults in complex structures are occurrences which rarely occur in unpredicted situations.The fault is a change in the characteristics property of the system. In dynamic systems, it is difficult to anticipate and avoid faults. Faults can cause accidents that lead to damages to human and huge economic loss.

A number of examples are: the explosion ofnuclear power plant Chernobyl in Ukraine on 26 April 1986. It is not possible to eliminate the occurrence of faults in complex systems, to avoid the effects of faults, or at least to mitigate their severity.In order to minimize the occurrence of accidents, the most important step is to introduce fault detection and diagnosis method.

Model- oriented identification of faults in dynamic systems has gained much recognition over the past decades in the form of research and real-world technology studies. The problem in model-oriented fault detection and diagnosis concerns the consistency of the model that defines the monitoring system's behavior. The model-based FDD involves the detection and diagnosis of fault on a system by using methods that derive characteristics from available signals (i.e. known inputs and measurements) and mathematical model processes. Fault are identified by generating the residual that shows the difference between the actual and the predicted state measurements.A number of residuals can be generated, based on the individual faults that occur at different system locations. The examination of each residual leads to fault diagnosis until the threshold is surpassed. For non-linear systems, a large range of model- oriented FDD approaches, e.g., based on observer approach, parity space approach, and system recognition method. Here for this system, the Extended Kalman Filter algorithm is considered for identification of a fault.

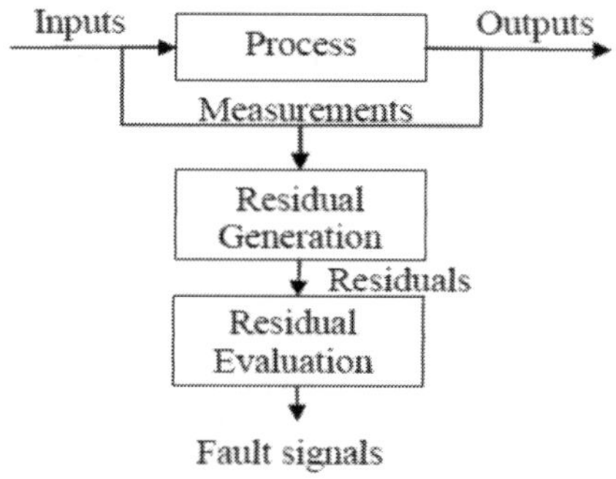

Figure 1. Model Based Residual Generation.

Here the model that is considered is two-tank system which is highly non-linear. The two-tank interacting system has the valve that is connecting between the two tanks and hence if any of the operating parameters get changed, this automatically changes the dynamics of the second tank or vice versa. Here in this system, the height of the tanks has to be held at a constant level say 4m and 3m respectively. Here the fault is introduced to the system as the additive representation of fault. The additive fault is widely used to represent sensor, actuator and component fault in the system. Just in state-space representation of the process, the additive sensor fault is added. The residual generation block produces the residual signal by comparing the system's input and output. The output produced will be , under no fault condition usually zero or close to zero. The method used to measure residuals is called generation of residuals. In order to remove fault symptoms from the system, such a technique is used, with the fault symptom reflecting the residual signal. The process of managing multiple faults lies in the simultaneous occurrence of similar fault effects generated by single faults. For correct residual evaluation, it is therefore necessary to gain the right residual structure. The residual assessment stage tests residuals for the faults and then uses a decision rule-base to decide where there were some faults.Whcich is shown in Figure 1and this produces the residual vectors which is given in the equation below,

$$r = y - \hat{y} \tag{1}$$

Residuals may also be produced by methods of identification of changes, as noise, fluctuations and other unknown signals, calculated and measurable signals and other unknown signals are present.

Many scholars have studied sensor fault tolerant problem. In the following, we will analyse the contribution of some of these papers.

In [1] the author explains about the detection and isolation of fault by using fuzzy logic system and state estimation algorithm. The 20% sensor fault is added to the two tank systems. The detection of fault is implemented by using both the technqiues and the performance is validated based on the effectiveness of fault detection.

This paper [2] demonstrates estimation of state by using EKF algorithm and also controlling the two tank system by using model predictive controller. The MPC controller strategy focused on basic use of some kind of system model to predict the variable governed over the horizon of finite time. From this paper, the internal state estimation of the system is learned and also the detection of systems sensor fault.

This paper [3] shows the detection , isolation and compensation of fault in three tank interacting system. Here the results shows the PI controller output with faults and the fault-tolerant output of the system. The adaptive observer whose function is to approximate both the undetermined performance and the inaccessible state variable for the calculation can be used to extend the work.

In [4] explains the fault tolerant control system. Here the first method is to estimate the fault by using LPV observer. Second method is to detect the fault by using fault decision scheme and to compensate the effect of fault is the third method.

In [5], the detection and fault tolerant control system of two tank is implemented by using observer based method and its performance is evaluated to check for its efffectivness in detecting and isolating sensor fault in two tank system.

The work by Marta Capiluppi, Andrea Paoli [6] deals with the implementation of fault-tolerant control strategy for two-tank system. Here the FTC allows the permissible amount of fault in the system and thus making the system stable.Here the detection of sensor fault is learned.

2. Process description

The two interacting tank device is shown in Figure 2 consists of two tanks which are interacting in nature that is connected with the help of cross section valve. In the process the nonlinearity is introduced by the valves. The flow rate of tanks areQin_1 and Qin_2 are defined as inputs for the dynamic model, whereas the two quantities$L_1(t)$ and $L_2(t)$ i.e. the fluid height in the tank is considered as outputs. The following differential equations are defined in 1 and 2.
For Tank 1,

$$S_1 \frac{dL_1}{dt} = Qin_1 - c1\sqrt{l_1 - l_2} \qquad (2)$$

For Tank 2,

$$S_2 \frac{dL_2}{dt} = Qin_2 + c1\sqrt{l_1 - l_2} - \quad b2\sqrt{l_2} \qquad (3)$$

Here the valve coefficient of the tank1 and tank are $c_1 = m_1 n_1 \sqrt{2h}$ and$c_2 = m_2 n_2 \sqrt{2h}$. The system parameter for two tank is given in Table 1.

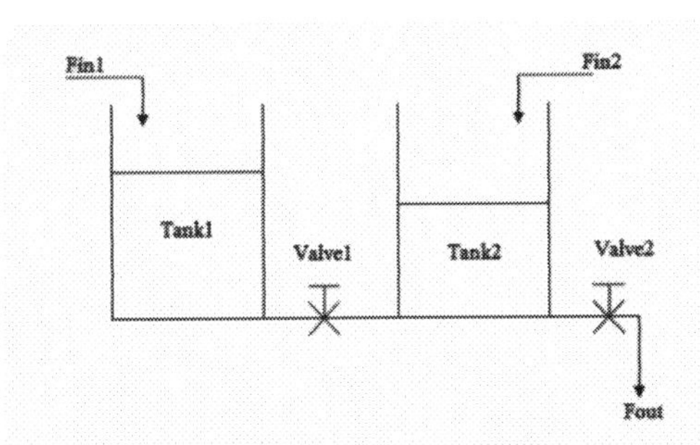

Figure 2. Two Tank Interacting System

Table 1. System Parameter for Two Tank System

SYSTEM PARAMETER	NOTATIONS
Tank1 and Tank2 area	S_1, S_2
Gravity-induced acceleration	h
Maximum tank height	L_{max}
Cross section area of linking pipes	m
Co-efficient of connecting pipes	n
Height of the tanks	L_1, L_2

3. State estimation algorithm

State Estimation is a method to determine the state of a system by extracting measurement data from inaccuracies and errors.

3.1. Extended kalman filter

The most commonly used nonlinear state estimation technique employed in recent decades Such methods require The term "higher order" is more than a direct linearization of the nonlinear system. These techniques include second order kalmanfiltering ,iterated kalman filtering , sum-based kalman filtering , and grid-based kalman filtering. These filters have ways of reducing the inherent linearisationfaults. Typically, they have better predicted performance They do so at the risk of higher complexity and computational costs than the EKF.

The simple agenda of the EKF comprises state estimation of a discrete-time non-linear dynamics systems, defined as

$$C_k = U(c_{k-1}, f_k, e_k) \tag{4}$$

Where,

Cindicates the estimation of state at thestep $k-1$.

f_k indicates the input vector, and

e_k indicates thenoise in the processThe normal multivariate distribution with covariance is assumed to be drawn from the zero mean. Q_k .

$$e_k = M(0, J_k) \tag{5}$$

At time, an observation (or measurement) b_k of the true state c_k is made accordingly as,

$$b_k = s(c_k, m_k) \tag{6}$$

Where,

c_{k-1}indicates the estimation of state at timestep k, and

IOP Conf. Series: Materials Science and Engineering **995** (2020) 012011 doi:10.1088/1757-899X/995/1/012011

m_k indicatesThe measuring noise known as Gaussian white sound with zero mean and covariance R_k.

$$m_k = M(0, I_k) \tag{7}$$

The function u and $s$describes an n-dimension vector function and they are presumed to be identified. The function uis used to measure the approximate state from the previous data like the function s is used to calculate the estimated measurement from the previous predicted state. Every time the jacobian matrix is estimated and in extended kalman filter these matrices are used. This procedure is most important for linearising the non-linear state.

EKF consist of two phases: predicting and updating. The predicted phase uses the state estimate of the preceding the step time to make the state estimate at the current time step. Measurement at the same time is used for updating phase to refine this prediction in order to attain a better accuracy in the estimation of the state, for the same current phase time.

3.2. Extended kalman filter algorithm

3.2.1. Prediction state and covariance equation

Estimation of State

$$\hat{c}_{k|k-1} = u(\hat{c}_{k-1|k-1}, e_k, 0) \tag{8}$$

Estimation of Covariance

$$K_{k|k-1} = U_k K_{k|k-1} U_k^T + J_k \tag{9}$$

Here the following Jacobians describe the State transition matrices:

$$U_k = \frac{\partial b}{\partial c} | \hat{c}_{k|k-1}, f_k \tag{10}$$

3.2.2. Correction state and covariance equation

Residual measurement

$$\tilde{b}_k = A_k - s(\hat{c}_{k|k-1}, 0) \tag{11}$$

Covariance measurement

$$H_k = S_k K_{k|k-1} S_k^T + I_k \tag{12}$$

Kalman gain

$$P = K_{k|k-1} S_k^T + H_k^{-1} \tag{13}$$

Estimation of State

$$\hat{c}_{k|k} = c_{k|k-1} + P_k \tilde{b}_k \tag{14}$$

Estimationof Covariance

$$K_{k|k} = (R - P_k S) K_{k|k-1} \tag{15}$$

Here, the following Jacobians describe the measurement matrices:

$$S_k = \frac{\partial s}{\partial c} | \hat{c}_{k|k-1} \tag{16}$$

4. Simulation results

4.1. Identification of fault using EKF

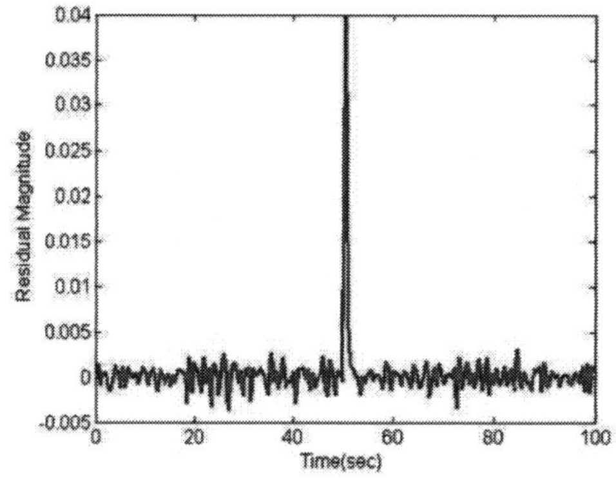

Figure 3. Tank 1 Residual Plot

From the Fig 3, it is observed that the residual is generated which indicates the fault in the system. Here at time t=50 sec, 10% sensor fault that is added to the system is shown through residual generation. This magnitude change indicates the percentage of fault that is added to the tank1.

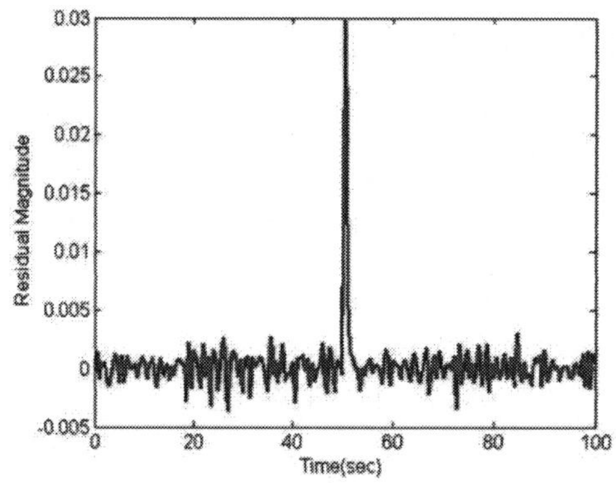

Figure4. Tank 2 Residual Plot

From the Fig 4, it is observed that the residual is generated which indicates the fault in the system. Here at time t=50 sec, 10% sensor fault that is added to the system is shown through residual generation. This magnitude change indicates the percentage of fault that is added to the tank2.

5. Conclusion

In this paper, the identification of sensor fault in two tank interacting system has been implemented by using Extended Kalman Filter algorithm. The study on EKF algorithm has been made and the simulation results show that the algorithm is successful in identifying the sensor fault that occurs in both tank1 and tank2 for the detection of faults.

References

[1] PandiyanManikandan and Mani Geetha 2014 Takagi Sugeno fuzzy expert model based soft fault diagnosis for two tank interacting system Archives of Control Sciences Vol. 24 pp.271-287.

[2] Geetha M Jovitha Jerome Devatha V 2013Design of State Estimation Based Model Predictive
Controller for a Two Tank Interacting System Elseiver Procedia Engineering Vol. 64 pp. 244-253.

[3] Mondher Amor TaoufikLadhari Salim Hadj Said and FaouziM'Sahli2020Actuator Fault-Tolerant Control Applied to Three-Tank System Mathematical Problems in Engineering.

[4] Ali Abdullah and Mohamed Zribi 2013 Sensor-Fault-Tolerant Control for a Class of Linear
Parameter Varying Systems With Practical Examples IEEE Transactions On Industrial Electronics Vol. 60 pp.5239-5251.

[5] Marcel Luzar and Marcin Witczak2012Model-based sensors fault detection for two-tank system, IFAC Proceedings Vol. 48 pp.1102-1107.

[6] Marta Capiluppi Andrea Paoli 2005Distributed fault tolerant control of the two-tanks system benchmark IEEE Conference on Decision and Control Vol. 48 pp. 205-211.

[7] MiralChangela and Ankit Kumar 2015 Designing a Controller for Two Tank Interacting System
International Journal of Science and Research Vol. 4 pp. 589-593.

[8] Reshmi Banerjee 2015 Water Level Controller by Fuzzy Logic International Journal of Innovative Research in Advanced Engineering Vol. 2 pp. 250-263.

[9] NirjharBhattacharjee and Binoy Krishna Roy 2010 Fault detection and isolation of a two non-interacting tanks system using partial PCA International Conference on Industrial Electronics, Control and Robotics pp. 137-14.

[10] Didier Theilliol Hassan Noura and Jean-Christophe Ponsart 2002 Fault diagnosis and accommodation of a three-tank system based on analytical redundancy ISA Transaction Vol. 41 pp. 365-382.

IOP Conf. Series: Materials Science and Engineering **995** (2020) 012012 doi:10.1088/1757-899X/995/1/012012

Design And Development Of An Application For Database Maintenance In Inventory Management System Using Tkinter And Sqlite Platform

K Yuvaraj[1], G M Oorappan[2], K K Megavarthini[3], M C Pravin[4], R Adharsh[5] and M Ashwath Kumaran[6]

1, 2, 3, 4 Assistant Professor, Bannari Amman Institute of Technology, Department of Mechatronics Engineering, Sathyamangalam, Tamilnadu, India.
5, 6 UG Scholar, Bannari Amman Institute of Technology, Department of Mechatronics Engineering, Sathyamangalam, Tamilnadu, India.

yuvarajk@bitsathy.ac.in, oorappan@bitsathy.ac.in, megavarthini@bitsathy.ac.in, pravin@bitsathy.ac.in, adharsh.mc18@bitsathy.ac.in, ashwathkumaran.mc18@bitsathy.ac.in

Abstract: The challenges in classical way of inventory log are human mathematical ability, paper maintenance, untidy process of editing and chances of missing the data of invoice and outward. The proposed idea is to develop an inventory management application using Tkinter and SQLite platform that creates a way towards progressively from traditional applications to a fully connected and customized inventory management system. This application uses password encryption to ensure the data privacy between the administrators and other workers of the industry. The system utilizes the integration of Tkinter and SQLite for the effective Graphical User Interface interfaced with Relational Database Management System. This customized application enables the administrator to add a new employee or remove the existing employee, he can update the stock quantity, and can back up the invoice history. Due to the login feature, the details of the product outward such as who withdrew and when it was withdrawn are automatically stored in the database. This digitization of inventory management system increases flexibility, reliability, smart storage, resource utilization, easy access to product location and warehouse management. It would be one time investment no further investment needed if at all any problem or errors occur in the app by human mishandling. The application saves us a lot of time because it uses SQL in it which is structured language, the data can be retrieved soon without any time delay and the human errors are avoided due to its structured nature.

Keyword: Internet of things, Cloud computing, Cyber-physical systems, Structured Query Language (SQL)

Introduction

Warehouse management is the act of organising and controlling everything within your warehouse and making sure it all runs in the most optimal way possible. As one of the essential things in Industry 4.0, the smart factory utilizes the recent advanced technology such as Internet of things (IoT), Cloud computing, Cyber-physical systems and automated machines to real-time monitor of physical process, information storage, visualization through virtual system and optimal decision making to improve the production and quality. When it comes to a product manufacturing unit, the process should also deals with effective product part management in the warehouse with progressive view towards present and future demands.

In the smart factory, the resources should be effectively managed with the help of software based tools. Warehouse management systems support warehouse staff in performing all the processes required to handle all of the major and minor warehouse tasks such as receiving, inspection and acceptance, put-away, internal replenishment to picking positions, packing, and value added services, order assembly on the shipping dock, documentation and shipping.

A warehouse management system also helps in directing and validating each step such as capturing and recording all inventory movement and status changes to the data file. With the help of our python based Inventory system, we can effectively manage all the warehouse related tasks such as resource tracking, stock control, purchase order-log maintenance, admin-user specified task based commands most of all the manual inventory maintenance problems such as inaccuracy of data, protection of data books(inventory details) can be tackled out.

1.0 Literature Review

Aaditya Tirodkar, Amogh Singh, Nida Parkar and Vimal Negi stated that the inventory management application is a real-time relational database capable of dealing with huge warehouse of an industry. The application could be applied for the tracking of the inventory of a warehouse, or to control the distribution of inventory among several warehouse of a huge firm. Anyhow, the application restocks datum and enables warning message of below par inventory at any area at an accurate c programming language. The objective is to minimize the stress of monitoring preferred to address all shop upkeep. The inventory management application is completely computer based system developed on Python period on Tkinter GUI using Python shell. WAMPserver is a software of windows operating system mainly based on installation and configuratuion of structured query language database server, scripting language, php MyAdmin and SQLite.

The ultimate objective of the work is to broaden inventory control system software code in which every information realted to the the inventory of the employee can be recieved. This computer application is mainly based at the management of inventory of a user. The software consists of modern organisation profile, inward data, purchase data and the existing stock that are supplied within the industry.

Nasuha Lee Abdullah, Ooi Chun Wei and Rosnah Idrus proposed that across several industries including manufacturing industries, poor inventory management causes big issues. The problems may include product delivery delay and/or shortages, inventory overstock and/or stock-out, tied up cash flow, and other issues that risk the profitability of the firm. A sample named company which deals with wide range of product from handheld products such as multi-meter and thermal imager, RF measurement products such as power meter and power sensor, power supply, oscilloscope, spectrum analyzer, network analyzer, etc., which provide a total solution to wireless communication for aerospace, defense and semiconductor market. In that company in order to handle each department a separate ERP has set up which takes care of The ERP is used mainly for supply chain management, finance, accounting, and asset management.

Lingyun Wei and Xueqing Yu proposed that the stock administration device is a real time stock database management system equipped for managing huge inventories of a company. The ultimate goal is to broaden the existing stock management system with a software in which all the facts concerning the inventory management is carried out. This software is clearly made to progress towards material management in the inventory. The aim is to reduce the strain of monitoring rather than to address all shop upkeep. The product Inventory control system is a standalone application designed using Tkinter which is a python's open source GUI (graphical User Interface) platform. All the data are processed and maintained using a private SQLite server. The application contains all the modern day inventory control features for product In-Out, Income information and also alerts when there is stock less than safer levels. Each new stock is uploaded to the software with the product's details such as supplier, price, stock entry date and time. Being a web-based app the login page is created in a way to manage and prevent the inventory from misuse and threads.

So there is a need for an inventory which handles large quantities of items which is complex to manage. In order to manage inventory, the company must predict the quantity of input foods at a previous time. Studies shows that data mining using sequential patter analysis is an effective approach to identify repeated pattern of input foods purchased. However the sequential pattern analysis should be involved in food based processing industries.

1.1 Problem Statement

The advanced development of automation provides automatic control of every process in a smart factory. The challenges in smart manufacturing system are lack of development in inventory since it is maintained manually through a logbook. In order to abide with a smart factory setup the inventory system should also be improved accordingly. The top level management cannot visually monitor the material available at the instant of time.

In the existing system the usage of mySQL heavily affects the processing speed of the process as it requires additional server for connection establishment with the database. The amount of memory required for processing is very large.

1.2 Proposed System

The proposed system is to create an app layout using python and User Interface development using Tkinter and to acquire real time physical information on the inventory materials, data processing and database management for prediction of future supply needs. It should be followed by the Interfacing Raspberry pi, setting up the working environment on Raspberry Pi. In order to improve the inventory, our proposed system suits well in all aspects for every level of the management to tackle out the problems till date.

2.0 Methodology

The python shell and SQLite are needed to be installed to the microcontroller. Once the installation process is completed, SQLite is needed to be interfaced with the python using the structured query language. The interfacing should be followed by the tkinter libraries should be accessed using python functions.

The beginning step of the project is to create a graphical user interface using the appropriate tkinter functions. In this project tkinter serves as the front end application. After the front end is completely developed, the back end process should be intiated. The queries in SQLlite should be used as per the requirement with the help of structured query language and python

2.1 Hardware Requirements

- Raspberry pi 3
- Power cable with adaptor
- Monitor

2.1.1 Raspberry pi

Figure 1. Raspberry pi

The raspberry Pi is a low cost computer that can easily suits with all generalized plug and play devices such as mouse, keyboard, HMI port etc. It is a capable device that can be programmed with the trending high level languages like python and R programming. It is capable of doing almost all the works that can be done with a computer.

2.1.2 Power Cable with Adaptor
Cables with adaptor are required for the powering up of the microcontroller, which operates at the voltage of 5v.

2.1.3 Monitor

IOP Conf. Series: Materials Science and Engineering **995** (2020) 012012 doi:10.1088/1757-899X/995/1/012012

In order to view the output of the controller a displaying device is needed. So a display monitor is connected to the HDMI port of the raspberry pi. Also to make the user friendly interface it is needed.

2.2 Software Requirements:

- SQLite
- Python Shell

2.2.1 SQLite

SQLite is a Simple, intuitive and fast software for simple implementing relational database management system. As always SQLite is an open source application and you could download it on SQLite web site for free. Unlike MySQL it doesn't require any addition sever connection to run. The latest version of SQLite is SQLite3.It almost requires no additional configurations as in other structured query languages as every configuration are already done in the system by default.

2.2.2 Python Shell

PowerShell is a task-based command-line shell and scripting language built on .NET. PowerShell helps system administrators and power-users rapidly automate tasks that manage operating systems (Linux, macOS, and Windows) and processes. PowerShell commands let you manage computers from the command line. PowerShell providers let you access data stores, such as the registry and certificate store, as easily as you access the file system. PowerShell includes a rich expression parser and a fully developed scripting language. Here it is used to run the python code.

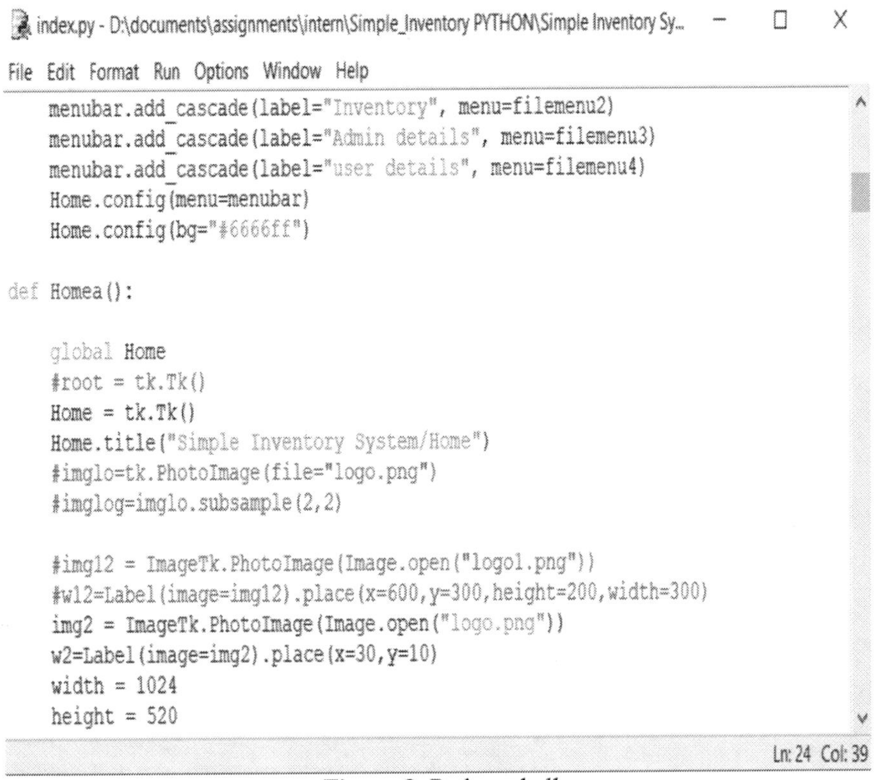

```
index.py - D:\documents\assignments\intern\Simple_Inventory PYTHON\Simple Inventory Sy...
File Edit Format Run Options Window Help
    menubar.add_cascade(label="Inventory", menu=filemenu2)
    menubar.add_cascade(label="Admin details", menu=filemenu3)
    menubar.add_cascade(label="user details", menu=filemenu4)
    Home.config(menu=menubar)
    Home.config(bg="#6666ff")

def Homea():

    global Home
    #root = tk.Tk()
    Home = tk.Tk()
    Home.title("Simple Inventory System/Home")
    #imglo=tk.PhotoImage(file="logo.png")
    #imglog=imglo.subsample(2,2)

    #img12 = ImageTk.PhotoImage(Image.open("logo1.png"))
    #w12=Label(image=img12).place(x=600,y=300,height=200,width=300)
    img2 = ImageTk.PhotoImage(Image.open("logo.png"))
    w2=Label(image=img2).place(x=30,y=10)
    width = 1024
    height = 520
Ln: 24 Col: 39
```

Figure 2. Python shell

3.0 Design and Implementation

Even in the small scale industries thousands of components are used every week, such that it becomes a difficult task for the store keeper to manage the inventory stock manually. Though there are some existing inventory management supplications, they are built with mySQL which requires complicated software setup as it involves in creation of additional servers for database connection establishment

For the purpose of overcoming the software complexities we have interfaced tkinter with SQLite, which results in the simple and elegant Graphical user interface enabling the customer to use it an easier way.

3.1 Block Diagram

The blocks depict the general warehouse management system. This diagram views on the entire warehouse rather than specific task based maintenance system such as information system, material In-Out section, loading and unloading.

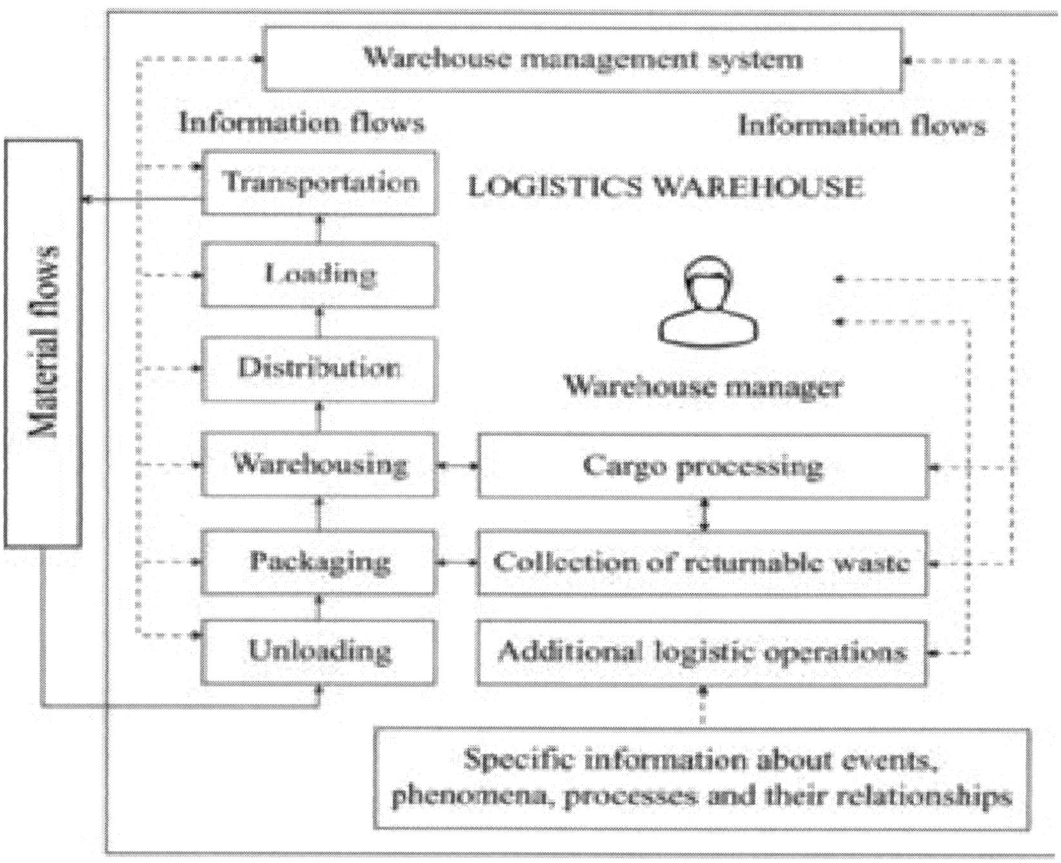

Figure 3. Block diagram

Our software works as a generalised platform which is highly customisable and suits with every maintenance mechanisms like 5S as the storage locations of the product is also mentioned in the product log. So the user can easily locate the location of components without any hazzle.

3.2 Working

The software mainly deals with all the possible ways to improve an inventory system in an industry where traditional system is followed which progress lesser productivity. Our product has individual features on the same software for different users. If the logged account is a user then the software shows only limited but highly required features. The features include making entry while taking out a product from inventory, checking for any product in the inventory, getting rank number and bin number of the product searched, making

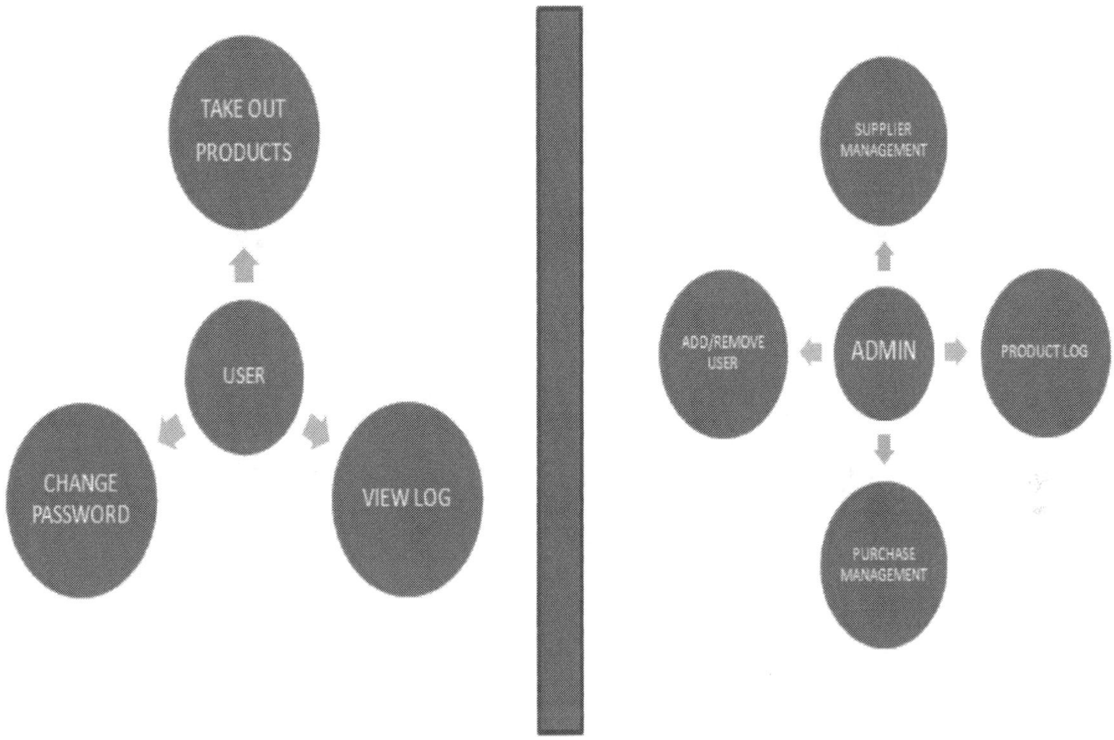

Figure 4. Admin/user accessibility

The software has twin dedicated login options, one is for the user and other is for the administrator. In our tkinter based framework user and admin can login with their own credentials which ensure hassle free login sessions.

On user's dashboard all the options are provided expect decision making which is made through top level management described as admins.

The Users can login with their provided user-id and password. They are provided with options for making entry which is must while taking a product in and out; also they can view the available list of items in the inventory. The admin has the ultimate power to add an admin and can able to add/remove items, made purchase decisions. While coming to products the system follows a dedicated private database Management system on SQLite. It ensures that when a product is taken from any rack by any user in the inventory, it deducts from the number of items in the stack. When the number of items reaches a low count the system automatically sends alert to managers of the inventory and the unit to ensure product availability.

IOP Conf. Series: Materials Science and Engineering **995** (2020) 012012 doi:10.1088/1757-899X/995/1/012012

Figure 5. Application outlook

The admin can add a new admin/user or remove the existing one and he has the access to the invoice history dating back to last 10 years. He can add a stock when inward occurs and he also has the provision to make changes in the product location. The admin is allowed to access the whole database.

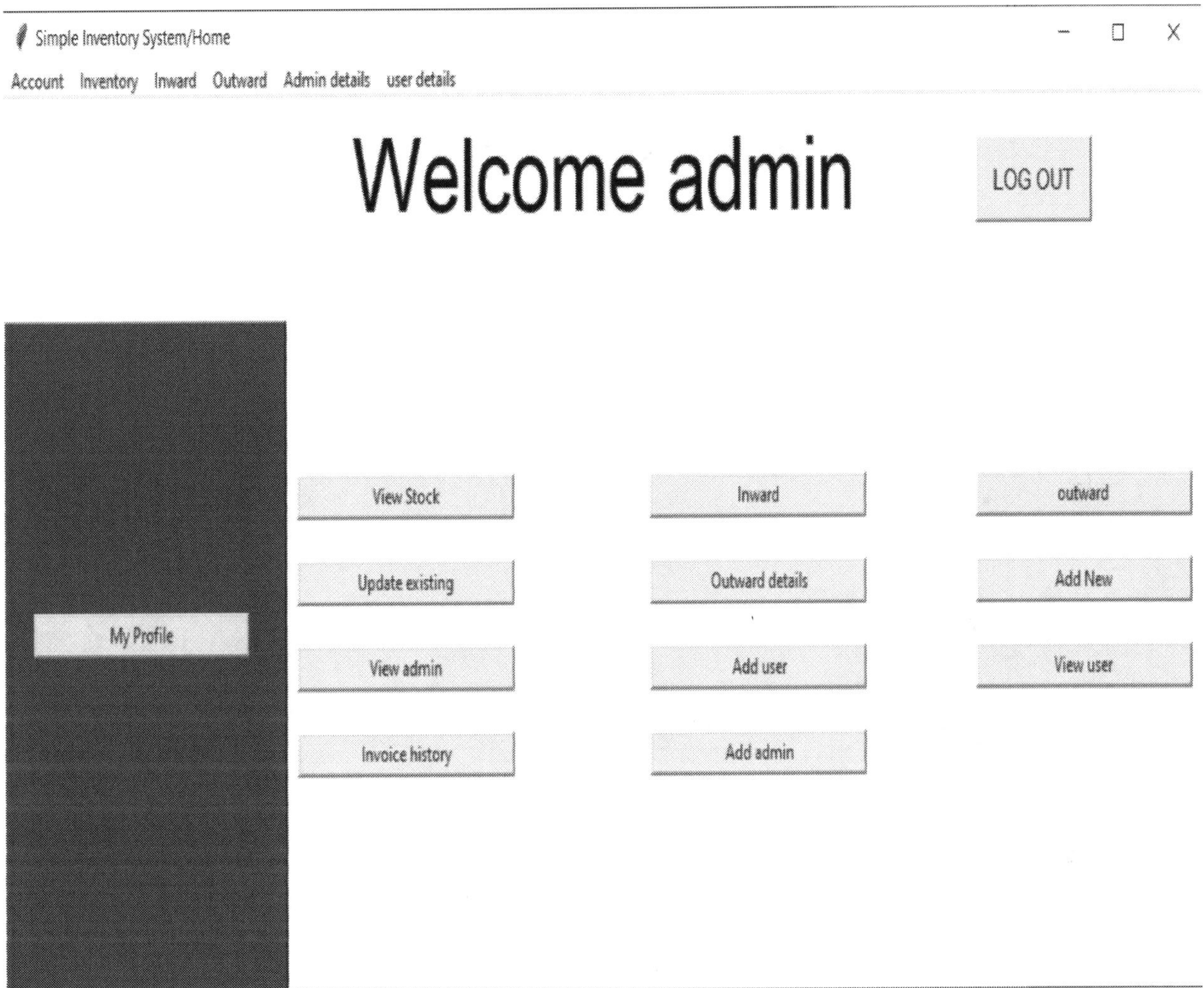

Figure 6. Admin dashboard

On the other side user can only fill outward details or check the existing stock in the warehouse. He is given with the provision to change his password.

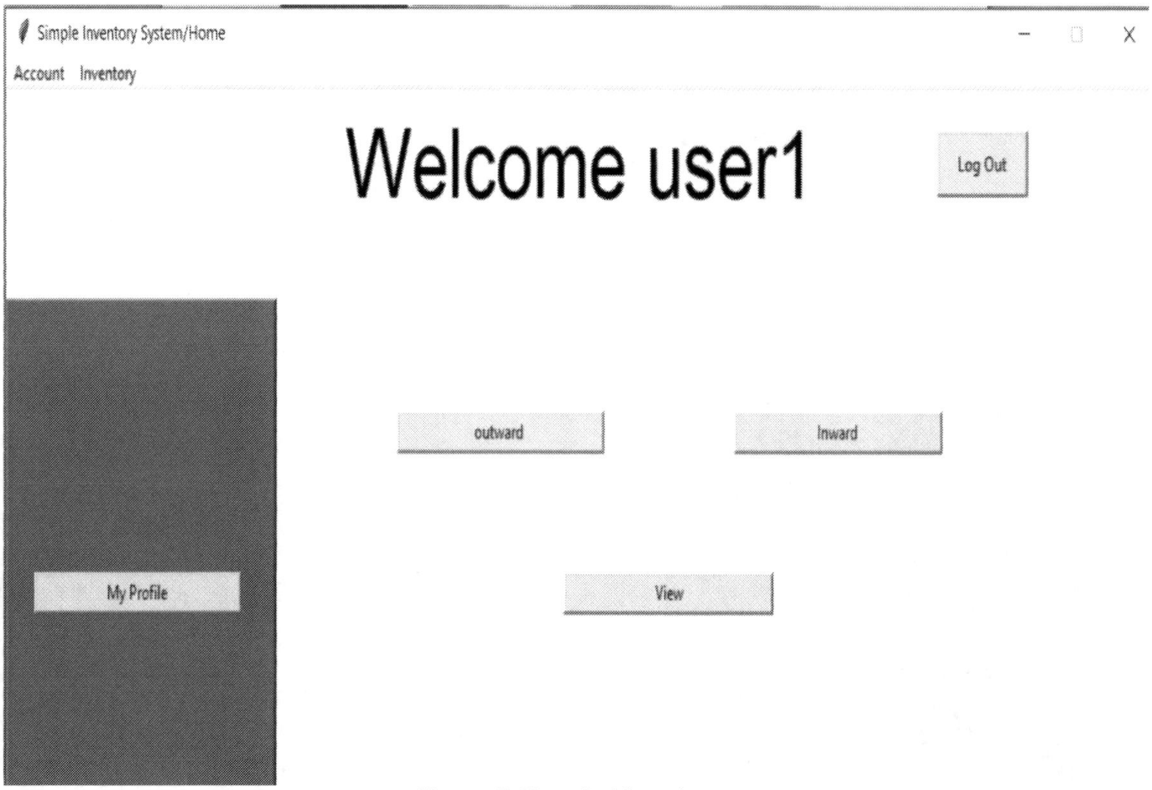

Figure 7. User dashboard

The outlook of the app is enhanced with the aesthetic feature available in tkinter such as inserting images, colours and fonts.

The "view **stock**" option can be used to view the components available in the warehouse and the exact storage location in warehouse as storage rack and bin numbers are also displayed. The user can also search for a particular product using the Search menu.

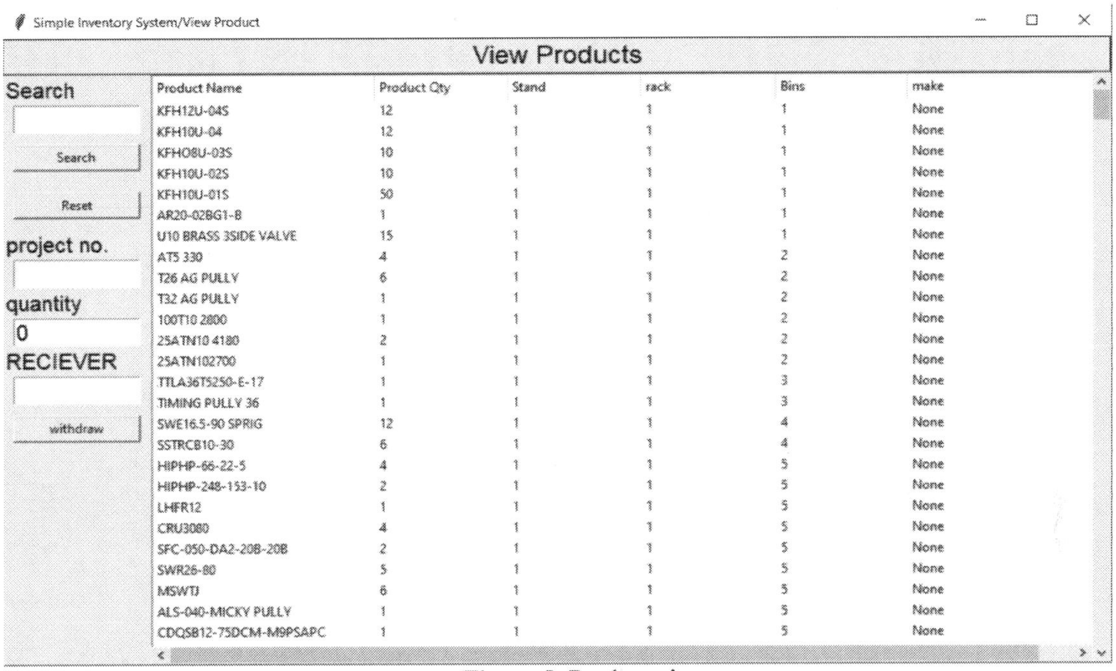

Figure 8. Product view

The store keeper can add the additional products upon purchasing with the help of the "add new" option.

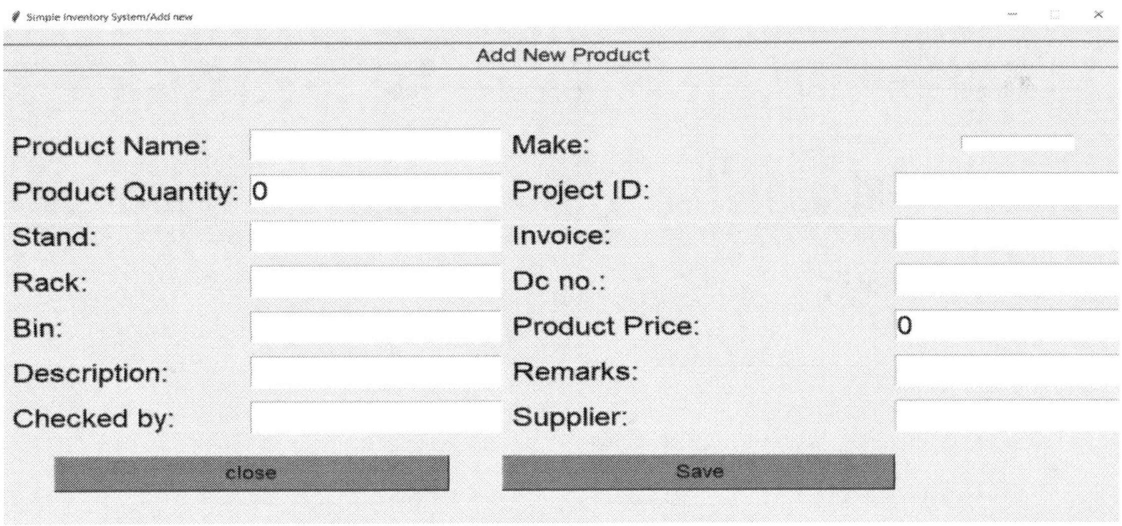

Figure 9. Updating products

If the admin is in the need of adding new user or new admin, he can use the "add user" or "add admin" option respectively.

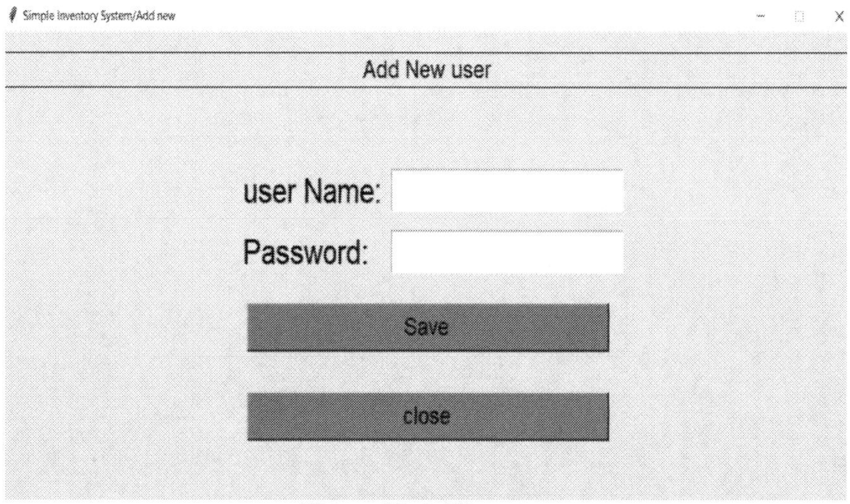

Figure 10. Adding user

To search for a particular user or admin, the option "view user" or "view admin" can be clicked.

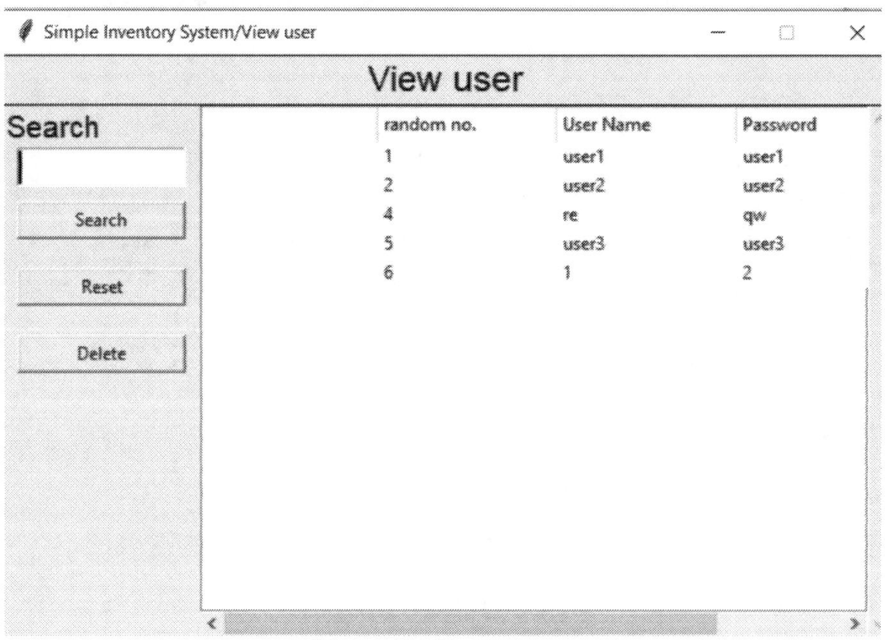

Figure 11. Viewing user

In case the store keeper wants to know the particulars of person who has withdrawn the components, the "outward details" option would help him. It delivers the details such as person name, time, date and quantity withdrawn.

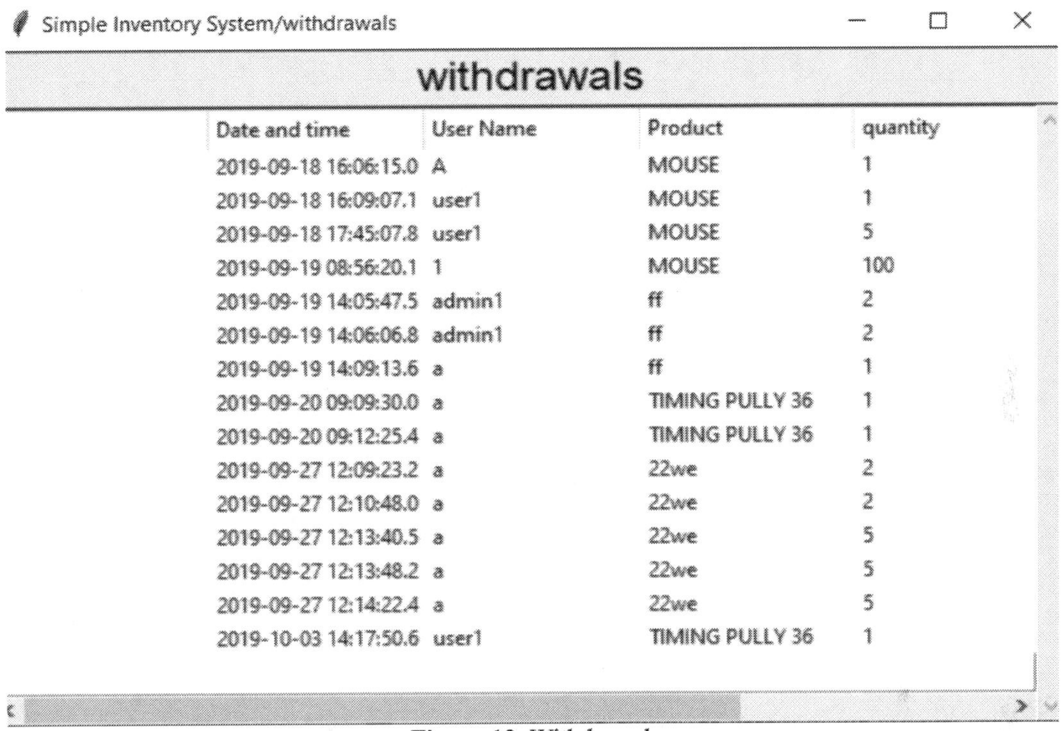

Date and time	User Name	Product	quantity
2019-09-18 16:06:15.0	A	MOUSE	1
2019-09-18 16:09:07.1	user1	MOUSE	1
2019-09-18 17:45:07.8	user1	MOUSE	5
2019-09-19 08:56:20.1	1	MOUSE	100
2019-09-19 14:05:47.5	admin1	ff	2
2019-09-19 14:06:06.8	admin1	ff	2
2019-09-19 14:09:13.6	a	ff	1
2019-09-20 09:09:30.0	a	TIMING PULLY 36	1
2019-09-20 09:12:25.4	a	TIMING PULLY 36	1
2019-09-27 12:09:23.2	a	22we	2
2019-09-27 12:10:48.0	a	22we	2
2019-09-27 12:13:40.5	a	22we	5
2019-09-27 12:13:48.2	a	22we	5
2019-09-27 12:14:22.4	a	22we	5
2019-10-03 14:17:50.6	user1	TIMING PULLY 36	1

Figure 12. Withdrawals

Encryption is the most wanted thing in the digital world, so we have an option "change password" under the "my profile" menu to change one's password when they are logged in.

Figure 13. Changing password

On completing your task, the "log out" option is provided for switching the user or to end ones ongoing process and exit.

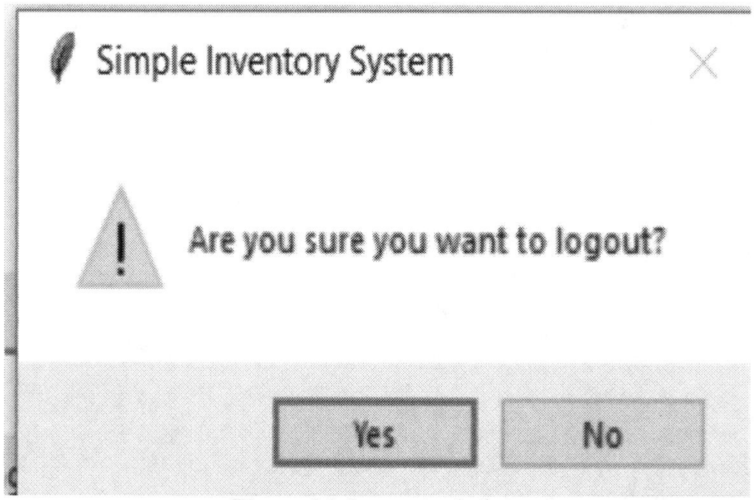

Figure 14. Log out window

4.0 Conclusion and Future Scope

4.1 Conclusion

The development of the application is effectively designed and implemented. This system will be helpful for the large scale warehouse management. It can save the time and reduce errors by the relational database management system. As an addition benefit the implementation of internet of things (IoT) is also implemented in our project such that utmost authority of the company can access the inventory status right from his mobile from anywhere around the world.

4.2 Future Scope

The proposed application can be further added with enhanced security features with the interfacing of RFID scanner and fingerprint sensor; so that even the username and password is known other account can't be accessed. Further an additional feature that enables the store keeper to receive an email if a product quantity goes below the pre-set value. Such that deficiency in store supply never happens which saves the halt time. Upon adding these two features, this application would attract the app market for sure.

IOP Conf. Series: Materials Science and Engineering **995** (2020) 012012 doi:10.1088/1757-899X/995/1/012012

References

[1] Aaditya Tirodkar, Amogh Singh, Nida Parkar and Vimal Negi, 'Inventory Management Software for Windows in Python', IOSR Journal of Engineering (IOSR JEN) ,ISSN (e): 2250-3021, ISSN (p): 2278-8719, PP 42-48.

[2] Alan D. Moore and B. M. Harwani, 'Python GUI Programming - A Complete Reference Guide: Develop responsive and powerful GUI applications with PyQt and Tkinter'.

[3] Alexey Marques Espíndola, Douglas Bezerra Beniz, 'Using tkinter of python to create GUI for scripts in LNLS'.

[4] Bryan Johnson, 'SQL AND Python Programming'.

[5] Chih-Chin Liang, (2013) 'Inventory Prediction in a Food processing and Distribution Company', 2013 5th International Conference on Service Science and Innovation, 978-0-7695-4985-9.

[6] Lingyun Wei and Xueqing Yu,(2018) 'Inventory management in e-commerce supply chain with lateral trans-shipment and quick response', fifth International Conference on Industrial Engineering and Applications, 978-1-5386-5748-5.

[7] Nasuha Lee Abdullah and Ooi Chun Wei, Rosnah Idrus,(2017) 'Extended ERP for Inventory Management: The case of a Multinational Manufacturing Company', fifth International Conference on Research and Innovation in Information Systems (ICRIIS), DOI: 10.1109/ICRIIS.2017.8002489.

[8] Stuart Hall, 'how to create a database in python using sql lite 3'.

[9] Thomas Nield, 'Getting Started with SQL: A Hands-On Approach for Beginners'.

Formulation of lagrangian equation and dynamic analysis of 2 - DoF Industrial robot using Roboanalyzer

Megavarthini K K[1], Arun Kumar K [2], Nagarajan P [3] Mohammed Ismail A [4], Karrthik R S[5] and Raghul Kumar S[6]

1, 5 Assistant Professor, Bannari Amman Institute of technology, Department of Mechatronics Engineering, Sathyamangalam, Tamilnadu, India.
2 1 UG student, Bannari Amman Institute of technology, Department of Mechatronics Engineering, Sathyamangalam, Tamilnadu, India.

megavarthini@bitsathy.ac.in,
arunkumark@bitsathy.ac.in,karrthik@bitsathy.ac.innagarajanp@bitsathy.ac.in,
mohamedismail@bitsathy.ac.in, raghulkumar.mc17@bitsathy.ac.in

Abstract. Industrial robot is extensively used in all major manufacturing and assembly line throughout the world. In order to properly actuate the robot links, it is necessary to calculate the torque required at link joints. In this paper we consider a two degrees of freedom robot manipulator having two rotary joints J_1 and J_2. This is a classic example of inverse dynamics problem, inverse dynamics in robotics used to calculate torque or force required at joints from known values of link mass, velocity, and acceleration and we choose lagrangian formulation to solve equation of motion. In addition, Robo analyzer software we plot the torque values for predefined value of centre of gravity and link mass

Keywords: acceleration, calculus of variation, degrees of freedom, dynamic analysis Industrial robot, lagrangian, torque

1. Introduction

The usage of the robotic arms are everywhere from medium scale sector to large scale sector in order to have 24/7 production the arms are widely used. Industrial robots have been extensively used in manufacturing industries after the uprising in computer science technology, faster calculation, precise movement and the support from academia helped the rise of industrial robot. A range of industrial robot was designed to suit the application of the customer such as serial manipulator and parallel manipulator to name the few. Design had been always a major constraint in developing any product, industrial robot designers faced the same concern as different customer requires dissimilar application robots to gratify their needs [2][3][5][6].

2. Literature survey

Robot manufacturer started building a variety of degrees of freedom robots to suit various application , For example a pick and place operation need only 3 dof robot but a painting job robot requires 6 dof robot[1][4]. Degrees of freedom (dof) is the primary design parameter for industrial robot and is defined as number of independent motion a particular system can make or number of variables required to define the position and orientation of robot end effector .The industrial robot plays an essential role in reducing expenditure of production and haste delivery of products to its customers. By accepting the current situation even medium scale industries rely on industrial robots. The robot manufacturer had faced a lot of issue regarding selecting proper motor for the applications.

To decide on proper motors we should be acquainted with the payload capacity and other parameters such as acceleration and mass of the robot manipulator. In order to discover the solution, we need to solve the inverse dynamics problem to find the equation of motion of the system. Inverse dynamics is defined as finding the required torque (rotary joint) and force (prismatic joint) of an industrial robot [9]. There are a variety of methods in classical mechanics to solve the system such as Newton - Euler, Lagrangian, Hamiltonian and Featherstone formulation. Each method have its own advantages and disadvantages. When we gaze into Newton Euler method [10], it is incredibly simple to solve the equation of motion for simple system such as single degrees of freedom but as the system becomes complex solving the equation of motion is very time consuming by using Newton - Euler. At this point we call for lagrangian and Hamiltonian formulation. Both these method uses the kinetic and potential energy of the individual links of the robot to solve the inverse dynamics, former uses the difference of kinetic and potential energy and later uses addition of kinetic and potential energy[7][8].

3. Calculus of Variation

It is concerned with maximizing and minimizing quantities called functional, which depend not on one or more independent variables, but on one or more functions. These quantities are typically definite integrals involving some function in the integrand which must be chosen to extremize the integral.

For every continuously differentiable function η such that η(x1) = η(x2) = 0 it is possible to represent any member of {Y} by an expression of the form

$$Y(x)= y(x)+\epsilon\, \eta(x) \qquad x_1 \leq x \leq x_2 -$$

Whose derivative are given by

$$Y'(x)= y'(x)+\epsilon\, \eta'(x) \qquad x_1 \leq x \leq x_2$$

replacing y and y' by Y and Y' respectively we obtain the integral

$$I(\epsilon)=\int_{x1}^{x2} f(x,Y,Y')dx$$

$$I'(0)=0$$

$$\frac{dI}{d\epsilon}= I'(\epsilon)=\int_{x1}^{x2}\left[\frac{\partial f}{\partial Y}\frac{\partial Y}{\partial \epsilon}+\frac{\partial f}{\partial Y'}\frac{\partial Y'}{\partial \epsilon}\right] dx = \int_{x1}^{x2}\left[\frac{\partial f}{\partial Y}\eta +\frac{\partial f}{\partial Y'}\eta'\right] dx$$

Which upon setting ϵ=0 and making use of I'(0)=0

$$I'(0)=\int_{x1}^{x2}\left[\frac{\partial f}{\partial Y}\eta +\frac{\partial f}{\partial Y'}\eta'\right] dx$$

Applying integration by the parts to the last term in this integral we obtain

$$I'(0\partial f)=\frac{}{\partial y'}\eta|^{x2}_{x1}+ \int_{x1}^{x2}\left[\frac{\partial f}{\partial y}-\frac{d}{dx}\left(\frac{\partial f}{\partial y'}\right)\right]\eta\, dx = \int_{x1}^{x2}\left[\frac{\partial f}{\partial y}-\frac{d}{dx}\left(\frac{\partial f}{\partial y'}\right)\right]\eta\, dx=0$$

$$\frac{\partial f}{\partial y}-\frac{d}{dx}\left(\frac{\partial f}{\partial y'}\right)=0 \quad \text{.............(a)}$$

Equation 'a' is known as Euler- Lagrangian equation

A. Equation of motions for the 2 DoF robot arm.The two links have distributed masses that requires moments of inertia on the calculation of kinetic energy. We calculate the velocity of the centre of masses of link 2 by differentiating its position. Fig 1 shows 2-Dof industrial robot.

Fig 1 shows two degrees of freedom industrial robot
(Courtesy: IIT Delhi, Department of Mechatronics)

$$x_D=l_1C_1+0.5l_2C_{12}\rightarrow \dot{x}_D=-\ l_1\ S_1\theta\text{-}0.5l_2\ S_{12}\,(\dot{}_1+\dot{}_2)\ \ldots\ldots(1)$$

$$\gamma D=l_1\ S_1+0.5\quad l_2\ S_{12}\rightarrow\ \dot{}_D=-\ l_1\ C_1\dot{}+0.5l_2\ C_{12}\,(\dot{}_1+\dot{}_2)\ldots\ldots(2)$$

Therefore, the total velocity of the centre of mass of link 2 is:

$$v_D{}^2=\dot{x}^2{}_D+\dot{}^2{}_D=\dot{}_1{}^2(l_1{}^2+0.25\ l_2{}^2+l_1\ l_2\ C_2)+\dot{}_2{}^2\ (0.25\ l_2{}^2)+\dot{}_1\dot{}_2(0.5l_2{}^2+l_1\ l_2\ C_2)\ldots\ldots(3)$$

The total kinetic energy of the system is the sum of the kinetic energies of links 1 and 2

$$K=K_1+K_2=[\tfrac{1}{2}I_A\dot{\theta}_1{}^2]+[\tfrac{1}{2}I_D(\dot{\theta}_1+\dot{\theta}_2)^2+\tfrac{1}{2}m_2v_D{}^2]=[\tfrac{1}{2}(\tfrac{1}{3}m_1l_1{}^2)\dot{\theta}_1{}^2]+[\tfrac{1}{2}(\tfrac{1}{12}m_2l_2{}^2)(\dot{\theta}_1+\dot{\theta}_2)^2+\tfrac{1}{2}\ m_2v_D{}^2]\ldots\ldots(4)$$

Substituting equation 3 in 4

$$K=\dot{\theta}_1{}^2(\tfrac{1}{6}\ m_1l_1{}^2+\tfrac{1}{6}\ m_2l_2{}^2+\tfrac{1}{2}C_2l_2\ m_2l_1+\tfrac{1}{2}\ m_2l_1{}^2)+\dot{\theta}_2{}^2(\tfrac{1}{6}\ m_2l_2{}^2)+\dot{\theta}_1\dot{\theta}_2(\tfrac{1}{3}\ m_2l_2{}^2+\tfrac{1}{2}\ C_2l_2\ m_2l_1)\ldots\ldots(5)$$

The potential energy of the system is the sum of potential energies of the two links:

$$P=m_1g\tfrac{l1}{6}S_1+m_2g(l_1S_1+\tfrac{l2}{6}S_{12})\ldots\ldots(6)$$

Lagrangian for two link robot will be

$$L=K\text{-}P=\dot{\theta}_1{}^2(\tfrac{1}{6}\ m_1l_1{}^2+\tfrac{1}{6}\ m_2l_2{}^2+\tfrac{1}{2}\ C_2l_2\ m_2l_1+\tfrac{1}{2}\ m_2l_1{}^2)+$$

$$\dot{\theta}_2{}^2(\tfrac{1}{6}\ m_2l_2{}^2)+\dot{\theta}_1\dot{\theta}_2(\tfrac{1}{3}\ m_2l_2{}^2+\tfrac{1}{2}\ C_2l_2\ m_2l_1)\text{-}\ m_1g\tfrac{l1}{6}S_1\ \text{-}m_2g(l_1S_1+\tfrac{l2}{6}S_{12})\ldots\ldots(7)$$

Taking the derivatives lagragian we get two equation of motions:

$$T_1=(\tfrac{1}{3}m_1l_1{}^2+m_2l_2{}^2+\tfrac{1}{3}m_2l_2{}^2+C_2l_2\ m_2l_1)\ddot{\theta}_1+(\tfrac{1}{3}\ m_2l_2{}^2+\tfrac{1}{2}\ C_2l_2\ m_2l_1)\ddot{\theta}_2-(m_2l_1l_2S_2)\dot{\theta}_1\dot{\theta}_2\ -$$

$$(\tfrac{1}{2}\ m_2l_1l_2S_2)\dot{\theta}_2{}^2+(\tfrac{1}{2}m_1+m_2)gl_1C_1+\tfrac{1}{2}m_2gl_2C_{12}\ \ldots\ldots(8)$$

$$T_2=(\tfrac{1}{3}m_2l_2{}^2+\tfrac{1}{2}C_2l_2\ m_2l_1)\ddot{\theta}_1+(\tfrac{1}{3}m_2l_2{}^2)\ddot{\theta}_2+(\tfrac{1}{2}m_2l_1\ l_2S_2)\dot{\theta}_1{}^2+\tfrac{1}{2}m_2gl_2C_{12}\ldots\ldots(9)$$

Analysis of Inverse dynamics is shown in table 1

Link no:	Joint offset(m)	Joint angle (deg)	Link length(m)	Twist angle(deg)	Centre of gravity (m)	Mass (kg)
1	0	0-180	0.2	0	0.1	1
2	0	0-90	0.2	0	0.1	1

Table 1: Analysis of Inverse dynamics

The below figure 2- 3 shows the inverse dynamic analysis for link 1 and 2 (Torque) and the inverse dynamic analysis for link 1 and 2 (Acceleration).

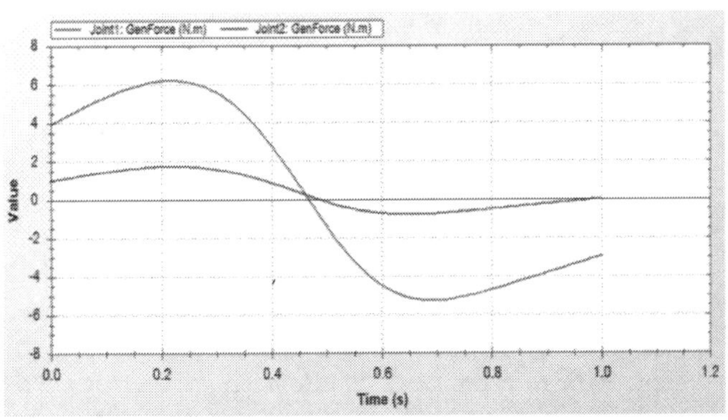

Fig 2 shows inverse dynamic analysis for link 1 and 2 (Torque)

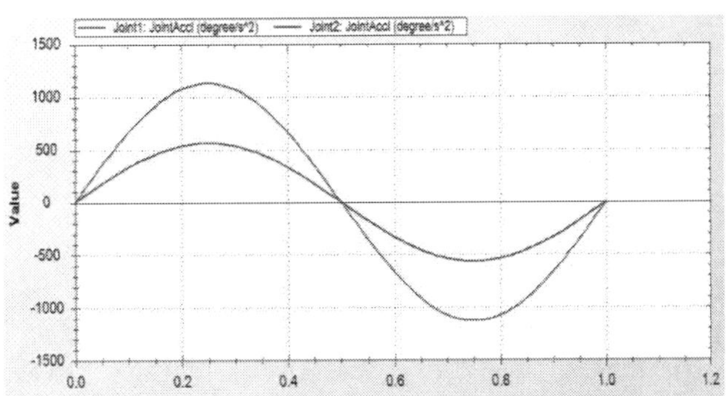

Fig 3inverse dynamic analysis for link 1 and 2 (Acceleration)

4. Results andDiscussion

Fig 2 shows the value of torque required for link 1 and link 2 . It is clear that link 1 need more positive torque ~6 Nm compare with link 2 which in turn require less positive torque of 2 Nm

$$T_1= (\frac{1}{3}m_1l_1^2+ m_2l_2^2+\frac{1}{3}m_2l_2^2+ C_2l_2\, m_2l_1)\ddot{\theta}_1+(\frac{1}{3}\, m_2l_2^2+\frac{1}{2}\, C_2l_2\, m_2l_1)\ddot{\theta}_2-(m_2l_1l_2S_2)\dot{\theta}_1\dot{\theta}_2\ \ -(\frac{1}{2}$$

$$m_2l_1l_2S_2)\dot{\theta}_2^{\,2}+(\frac{1}{2}m_1+m_2)gl_1C_1+\frac{1}{2}m_2gl_2C_{12} \text{ ....} \quad (10)$$

$$T_2=(\frac{1}{3}m_2l_2^{\,2}+\frac{1}{2}C_2l_2\, m_2l_1)\ddot{\theta}_1+(\frac{1}{3}m_2l_2^{\,2})\ddot{\theta}_2+(\frac{1}{2}m_2l_{12}S_2)\dot{\theta}_1^{\,2}+\frac{1}{2}m_2gl_2C_{12} \text{ ....} \quad (11)$$

Table 2: shows comparison of torque values for link1 and link2 using lagrangian formulation and simulation software Robo analyzer

Link	Torque Values (simulated), Nm	Torque values (Theoretical), Nm
1	6	4.112
2	2	1.046

5. Conclusion

It is not essential to apply lagrangian formulation for solving inverse dynamics problem, there are other methods such as Newton- Euler, Hamiltonian and Featherstone method to obtain the equation of motion. The similar method shall be used for higher degrees of freedom system however the complexity of solving it increases theoretically. It is far better to use simulated software to gain the result provided if the simulation does consider centrifugal and Coriolis Effect. In future studies, researchers can check the theoretical results of a range of configured industrial robot using lagrangian and hamiltonian formulation

Acknowledgment
The authors would like to express gratitude Bannari Amman Institute of Technology for providing computing infrastructure to complete the research work and IIT Delhi, Department Of Mechatronics to make use of images from software Robo analyzer.

References

[1] Armstrong, B., Khatib, O. and Burdick, J., 1986, April. The explicit dynamic model and inertial parameters of the PUMA 560 arm. In Proceedings. 1986 IEEE international conference on robotics and automation (Vol. 3, pp. 510-518). IEEE.

[2] Flores, P., Ambrosio, J., Claro, H.C.P., Lankarani, H.M., Koshy, C.S.: "A study on dynamics of mechanical systems including joints with clearance and lubrication". Mechanism and Machine Theory 41, (2006) Mar, 247–261.

[3] Chang, C.O., Nikravesh, P.E.: "An adaptive constraint violation stabilization method for dynamic analysis of mechanical systems." Journal of Mechanisms, Transmissions, and Automation in Design 107, 488–492 (1985)

[4] Hwang, Yunn Lin, Jung Kuang Cheng, and Van Thuan Truong. "Dynamic Analysis and Control of Industrial Robotic Manipulators." In Applied Mechanics and Materials, vol. 883, pp. 30-36. Trans Tech Publications, 2018.

[5] Roger Boudreau, Jérémie Léger, Hakim Tinaou, and André Gallant ; " Dynamic analysis and optimization of a kinematically redundant planar parallel manipulator "Transactions of the Canadian Society for Mechanical Engineering, (2018) Apr, 42(1): 20-29,

[6] Chembrammel, Pramod, and ThenkurussiKesavadas. "A new implementation for online calculation of manipulator Jacobian." PloS one 14, no. 2 (2019): e0212018.

[7] Ivan Virgala, FrantišekŠimčák, ZdenkoBobovský;" Dynamic Analysis of Manipulator Arm for 6-legged Robot"American Journal of Mechanical Engineering, November 12. 2013, 1 (7), pp 365-369.

[8] Rajeevlochana CG, Jain A, Shah SV, Saha SK. Recursive robot dynamics in RoboAnalyzer. In

Machines and Mechanisms (Proc. of the 15th Nat. Conf. on Machines and Mechanisms), edited by: Bandopadhyay, S., Gurunathan, SK, and Ramu, P., Narosa Publishing House, New Delhi 2011 (pp. 482490).

[9] Slotine JJ, Weiping L. Adaptive manipulator control: A case study. IEEE transactions on automatic control. 1988 Nov; 33(11):995-1003.

[10] Urrea, C. & Pascal, J. (2017). Parameter identification methods for real redundant manipulators. Journal of Applied Research and Technology, 15, 320–331. [12].Virendra, Patidar, R. T. (2016). Survey of robotic arm and parameters. In 2016 International Conference on Computer Communication and Informatics, ICCCI, 1–6.

Intelligent and Stable Next Generation HetNets with Self-aggregated Framework

Anita Sethi[1], Manoj Bisht

[1]ICFAI University, Department of Computer Science, India.

seep4g@gmail.com

Abstract. The malfunctioning of critical infrastructure causes high-quality deterioration, financial effect and loss of life in a town in the present scenario. It is essential to check the cost-effectiveness and reliability of the quality of critical infrastructures in the pandemic situation. Sensitive monitoring of an infrastructure is costly but necessary, in which fault identification and corrective action are needed at service levels to avoid cascading failures with minimal communication latency. With enhanced expense, resource, and process performance, the growing population in smart cities demanded a high living standard, rigorously straining energy, city services, and infrastructure. Intelligent houses, intelligent transport and smart buildings linked with intelligent networks are three pillars of the smart city initiative. In this paper, an improved framework for enhanced Quality of Service with higher performance in a 5G network is presented. A detailed description of reconfigurable intelligent Smart networks and reconfiguration analysis at various layers are also discussed. SDN and NFV, with the help of an intelligent SDN controller, provide the context for managing a multi-objective heterogeneous network. For self-optimization and self-healing, multiple clustering heuristics are useful for managing these networks at the various locations of smart projects.

Keywords: 6th Generation (6G), Software-defined networking (SDN), Self-Optimized Network (SON), quality of experience (QoE), HetNets

Introduction

Present challenge in ad hoc automation for business units, automation, security, activities, developers, and infrastructure provides accessible, efficient, and agentless applications. The efficiency of network and resource utilization for Smart projects in smart cities can be estimated with its demands generated by the customers. The essential attributes of future 6G communication are deep connectivity, intelligent connectivity, Holographic and Pervasive Connectivity. In order to support various business models, a combination of artificial intelligence and wireless mobile communication networks provides an integrated, quick, and efficient open-platform. In order to serve diverse and demanding business models with the help of AI techniques, the future 6G network architecture is increasingly heterogeneous, massive and multi-objective with self-optimization feature. In indoor applications with deep coverage to maximise indoor access requirements, deep networking is used. Deployment and design of future wireless technologies to take into account ITU scenarios of both deep coverage requirements as well as objects and people [1]. The integration of various technologies in the 6G system with the capability of intelligence, dynamic, self-aggregation includes the interconnection of all aspects and IIoTs. A challenge for the next-generation network is the aggregation of various technologies and dynamic versatility and help to achieve complex and varied scenarios and market demands.

1. Background

The goal of 4G when studying human behaviours was people-centered contact, a case of an indoor scenario that uses a wireless communication device of several generations. The need for current industrial applications as well as Smart projects of Smart Societies with mobile and intelligent networks is higher speed communication. An in-depth view of the intellectual architecture for the self-organized network that is intended as the primary technology to fulfil both conventional domain applications of networking as well as novel paradigms of nanoscale communication. The creation and management of highly challenging business model services in the current industry needs an intelligent Smart Network with zero latency. It is difficult to implement and efficiently manage heterogeneous infrastructures with multiple network networks managed by multiple administrative domains [2].

The digitalization process has allowed smarter technology to proactively manage applications, users and systems, to make maximum use of resources and to make physical space more effective for business operations. The Internet of Things and Robotics play an important role in the smart grid or e-healthcare solutions of Smart Cities that increase the gross domestic product with minimal running costs. In Figure 1, the implementation of Smart City was shown to rely on various domains and infrastructures, such as IoT, mobile phones or Big Data Cloud, with analytical steps in real-time scenarios [3]. Traffic congestion is a major problem that causes decreased cost of productivity, loss of life quality, and contamination of the atmosphere. In urban buildings, energy and resource use are strongly occurring, and smart home protection and automation systems are more in demand. Numerous technologies used in the construction of a smart project using various technologies are seen in the Figure 1.

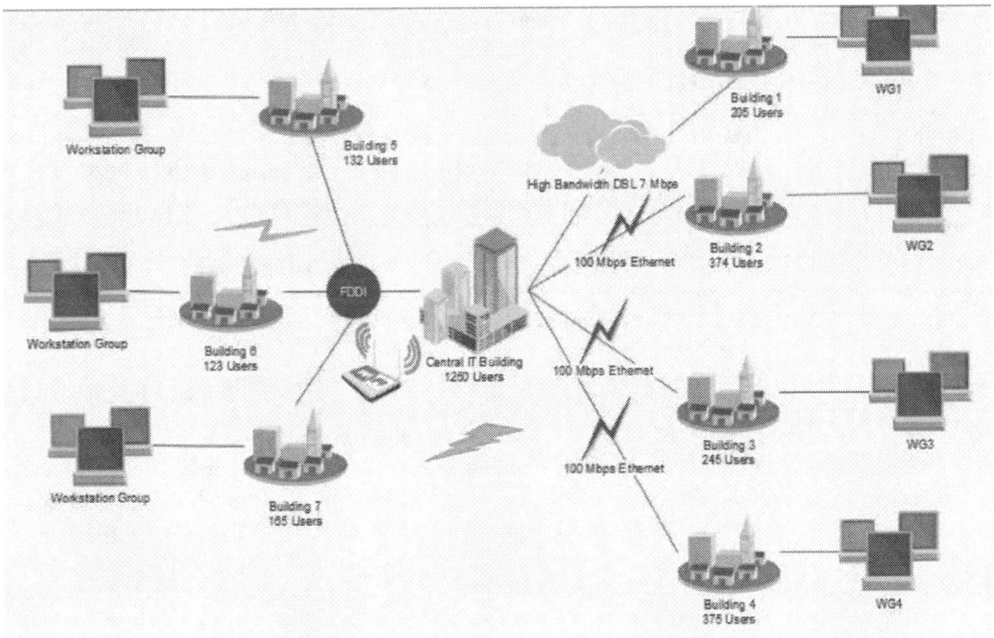

Figure 1.

2. Research and Technical challenges of 6G communications

Until 6G communication systems can be deployed successfully, many technical issues must be solved. Some of the problems are discussed briefly below.

2.1. Connectivity by backhaul of high bandwidth, Interference and Spectrum Regulation

There will be very tightly wired connection between heterogeneous massive IoT devices in 6G networks. Furthermore, these access networks are diverse and widely distributed and two access networks have high data accessibility for various types of users. In order to connect access nets to a core network to support client high data rating services, 6G backhaul networks will have to manage the huge volume of data; if not, a bottleneck will be created. Optical fibre and FSO networks are possible substitutes for high bandwidth backhaul links, so the overall ability for these networks can hardly be increased by data depth. It is necessary to use the 6G spectrum effectively, including spectrum sharing schemes and creative specimen management methods, due to interference problems and the lack of bandwidth. Effective spectrum management is important for maximum resource usage with QoS maximisation. In 6 G, researchers must deal with issues such as how to swap spectral shares and how to handle the spectrum phase in heterogeneous networks at the same frequency. The possibility of cancelling interference with traditional methods, such as simultaneous interference cancellation and consecutive interference cancellation, must also be investigated by researchers.

2.2. The complexity of 3D network resource management and utilization

The 3D networking spread vertically which introduced a new dimension. Multiple opponents can also intercept information which can deteriorate the overall machine efficiency significantly. New techniques are also important for resource management, routing protocol, mobility management, and optimization and multi-access technologies. Timing involves a new configuration of the network.

2.3. The performance of high energy efficient processing elements

There will be several unique features involved in the 6G system. The introduction of new features should be possible for devices such as smartphones, mobiles and processing elements. It is extremely difficult to support 1 Tbps, XR, artificial intelligence, and advanced sensing [8] using individual devices with communication features. The 5G devices will not support only a few of the 6G features and performance enhancement in 6G devices will also increase costs. Millions of devices are going to be connected to 5G, and we have to make sure that they are all 6G technology compatible. Mobile ultra-wide-scale networking has become an integral part of the use of world electricity. Not only does it produce huge carbon emissions, it also takes up a large part of operating expenses. 6G networks will have extraordinarily high performance, very high bandwidth, and an exceedingly high number of omnipresent wireless nodes in the future, which are likely to face unimaginable challenges in terms of energy demand. Spectrum efficiency and spectrum capacity are growing, but the efficiency can be significantly improved. The pervasive and dense sensors of wireless sensing networks packed with output and living space intensify two problems with energy consumption. The enormous number of sensors will lead to high overall energy usage, and it is also a challenge to supply energy for ubiquitous deployment conveniently and efficiently.

2.4. Heterogeneous hardware requirements for Autonomous Wireless Networks

The 6G regime knowingly permits AI-based automation technology, including UAV, Autonomous Vehicles & Industry 4.0. Convergence of various heterogeneous subsystems, including independent systems, interoperable processes, system systems , computer training, independent clouds, system machines and heterogeneous wireless systems, is required to make wireless systems self-sufficient [7]. Thus, the overall development of the system becomes complicated and challenging. To create a truly autonomous driverless system, 6G researchers need extra effort as they need to develop fully automatic driving vehicles that perform better than human-controlled cars. A broad range of heterogeneous forms of communication attributes, such as frequency bands, communication topologies, service delivery, and so on, are included in 6G. In addition, in hardware environments, portable terminals and access points would be substantially different. Massive MIMO technology is being further upgraded from 5G to 6G, and more complex architectures may be required. The communication protocol and algorithm design will also be complicated with Machine Learning and AI technologies. In addition, there are different hardware architectures for different communication systems.

2.5. Frequency bands of THz, Beam forming and Sub-mmWave (THz) frequency modelling

High THz frequencies provide high data rates, while atmospheric absorption and high propagation loss properties challenge THz bands for data transmission over relatively long distances [5], we need to provide a new design for the transceiver architecture of the THz communication network. The transceiver must run at elevated frequencies and we must ensure that very wide bandwidths are used to the maximum extent possible. Another barrier to THz contact is just a very slight gain and an open area with distinct THz band antennas. It is also important to address the health and security communications infrastructure problems with THz bands. mmWave and submmWave (THz) propagation properties are subject to environment conditions, so the effects are absorbent and dispersive [6]. The state of the environment is changing continuously and is therefore very unpredictable. Therefore, this band's channel architecture is very complex and there is no perfect channel model for this band. Hence the estimation of the beam power of the sub-mmWave, i.e. the THz band, is difficult.

3. Vision of 6G and future core technologies

The primary requirement is the flexibility of 6G wireless networks to handle huge amounts of data and high-data accessibility per machine. The objective of 6G (2030-) would be to serve the information needs of society, so 6G ought to be the demand that 5G is unable to serve and that need to be further developed [9]. Based on the demand that 5G will satisfy and the growth trend in the 6G vision of other related areas. Smart Homes, Towns, and Villages can be enabled by ultra-quick Internet connectivity with 10-11 Gbps data speeds. S2S networking makes it possible to access 6 G networks to space for global connectivity.

3.1. Intelligent Connectivity

Artificial Intelligence (AI) is one of the best ideas today, and almost all areas are exploring the use of such AI technologies. The convergence of cellular mobile communication networks and AI to make AI a better-enabling network is also becoming an inevitable development. At present, attempts are being made to use AI technology in 5G, but the current 5G and AI combination can only be seen as optimising traditional network design via AI instead of using a new smart communication network. Second, in the 5G network, the use of AI technology is comparatively late [10,11]. We have just recently begun our research and attempts to apply AI technology to a 5G network, and the 5G network architecture has already been finalised. Secondly, while AI technologies are rapidly emerging and have shown strong capacity in some areas, in other sectors they are still exploratory. Research studies into the combination of AI and wireless communication technologies have just begun and a long-term research phase is required before actual technologies come to fruition. The trend for AI indicates, however, that the next decade will have a technical maturity.

3.2. Dynamic Network Slicing in 3D networking

The 6G solution combines ground and aerial networks to enable vertical expansion of connectivity to users through 3D base stations, UAVs and low orbit satellites. The incorporation of new altitude dimensions and related degrees of freedom varies significantly from 3D networking to conventional 2D networks. At the same time, dynamic network slicing allows a service provider to enhance the efficient delivery of any service to a wide range of clients, cars, equipment and industries across dedicated virtual networks. It is one of the most important management elements when a wide variety of users are connected to a number of heterogeneous networks in 5G communication systems[6,7].

3.3. Universal Connectivity

Concerning the connectivity plight of customers, the 5G system would dramatically broaden the space and form of data on the Internet of Things anytime and anywhere. The range of activities would massively extend the geographical space for connectivity, including detectors in the deep, deep sea and high altitude human/unmanned aircraft, autonomous robots in harsh environments, smart remote control, etc.. Besides, human aerospace activity is also increasingly expanding with the rapid advancement of technologies in the fields of aerospace engineering, deep-sea exploration, and other sectors, and the enhancement of survival capability in certain extreme natural environments [11,12]. For instance, in 2030–2040, people will reach the deep space more likely than specific communication needs to be limited to a few skilled scientific exploratory fields, and signs of human activity on earth would likely be more popular and more frequent between satellites and spacecraft.

3.4. Blockchain

Blockchain is a distributed architecture to handle vast volumes of data in future communication networks. A distributed directory is a database that is spread over several nodes or processing elements. For every node, an identical copy of the ledger is replicated and saved. P2P manages the blockchain without a server or centralized authority. Data is stored and organized into blocks on a blockchain that is interlinked and cryptographically encrypted. The blockchain technology therefore can provide a range of facilities, such as software interoperability, mass data traceability, autonomous IoT-systems interactions, and reliability for 6G communications system mass connectivity.

3.5. Optical wireless technology with FSO backhaul network

In comparison to RF-based communications for every future mobile-to-access network, optical wireless systems are intended for 6G communications, as well as network-to-backhaul/fronthaul network services for such networks. Established technologies for OWC technologies include visual light connectivity, automatic camera information exchange, and light fidelity and FSO optical band communication. Companies are working to enhance the reliability of these innovations and to overcome their problems. Using LiDAR, which is a revolutionary technology for high-resolution 6 G 3D mapping, optical wireless technology-based communication can provide low latency, very high data rates, and secure communications [13, 6]. FSO is an excellent technology for providing 6G backhaul connectivity over a very long distance of more than 10,000 km. FSO promotes high-power backhaul connectivity for remote and non-remote areas such as sea, ground, underwater or isolated islands; FSO also promotes BS cellular connectivity. Huge multiple inputs and multiple output techniques: the application of the MIMO technique is one of the main techniques for increasing spectral efficiency.

4. Innovative taxonomy of STATE OF ART

Content-centered networking (CCN) was initially considered for solving the problems of existing Internet Protocol (IP) communication. The current host-centered internet architecture has been changed by CCN into a data-centered architecture. In the form of interest packets, the conventional CCN uses the broadcasting system to submit requests for content. However, today's broadcast / flooding of the packet in either direction is not ideal for large-scale networks because this creates the network bottle. To remedy this shortcoming, a new CCN approach was proposed in order to find out from local controllers if the content is available on the local network [16], software defined network controller. We also introduced an improved SDN-based CCN forwarding scheme to decrease both existing vulnerabilities and network traffic. The security problems facing the network and its devices are very significant with the ongoing growth in various technological scenarios. Due to the wireless existence of WSN, its protection is a major concern. Some WSN applications are highly critical in nature and result in a catastrophe as a result of the intrusion.

As computers are changed by a few generations, they are disposed of in failure, but they can become very powerful when all of them are put together and share each other's computing power. A cluster is one type of parallel computer system made up of a group of separate, interconnected computers that function together as a single integrated system. Clusters use a communication interface, and fragments of code must be explicitly configured to use the distributed hardware underlying them. More computing nodes based on needs will be included in the cluster. In the design of parallel clusters, we mainly use open source and Linux-based system components [17]. A parallel cluster is a way that similar computers can be combined and can do a job by sharing the workload. The function is shared in partnership by master and slave nodes in a group collaboration. Parallel computing clustering enables a community of hardware-setting computers to work together to solve various computational problems. In a cluster computing environment, resources such as memory, storage, and CPU core are combined into a single device.

5. Heterogeneous Network Scenario

Mobile ad hoc networks (MANETs) and hybrid optical-wireless networks have historically been taken as distinct and separate, without any synergic management solutions, often referred to as FiWi. The first main focus of FiWi is on the distribution of mobile nodes in extremely complex and infrastructure-less settings, the second offers high-bandwidth and low-latency connectivity to cellular nodes. Recent developments and penetration of SDN techniques have also encouraged the adoption of versatile SDN monitoring and control for the MANET and FiWi sectors, including the use of software-defined networking. It is possible to identify VANET applications as Emergency & Safety apps and other on-demand and entertainment apps. Protection technologies are linked to the emergency broadcast of any accident and traffic conditions that have occurred. There are two kinds of security messages; periodic messages contain substantial data used to minimize hazardous situations [18a]. The current location, path, vehicle speed, and other information transmitted in V2V contact by other vehicles. These messages can often lead to a problem with broadcast storms. An incident led notification takes precedence over the current location, the incident, and time of the vehicle that could be used in an emergency.

It is difficult to supervise the building or area when there are a wide building or area and fewer guards. In this case, the protection of and life of guards is more fragile, so we present with this article an idea of a modifiable vehicle for protection and surveillance. This vehicle is modifiable and is useful in the fields of protection, monitoring, and security. As a car, it is mobile in nature and can be remotely controlled from a base station. Since the vehicle can be changed, it can be fitted with any desired module, in which case the vehicle is installed with a module consisting of a camera, a sensor, and a siren. When the sensors sense a moving target, the camera shifts in that direction. If an intruder or suspicious activity is detected, prompt action can be taken to prevent a problem [19a]. The main goal of this mobile device is to minimize the risk to life and to protect a significant portion of the city or building with limited manpower. The mobile system's slogan is one-time work and all-time use.

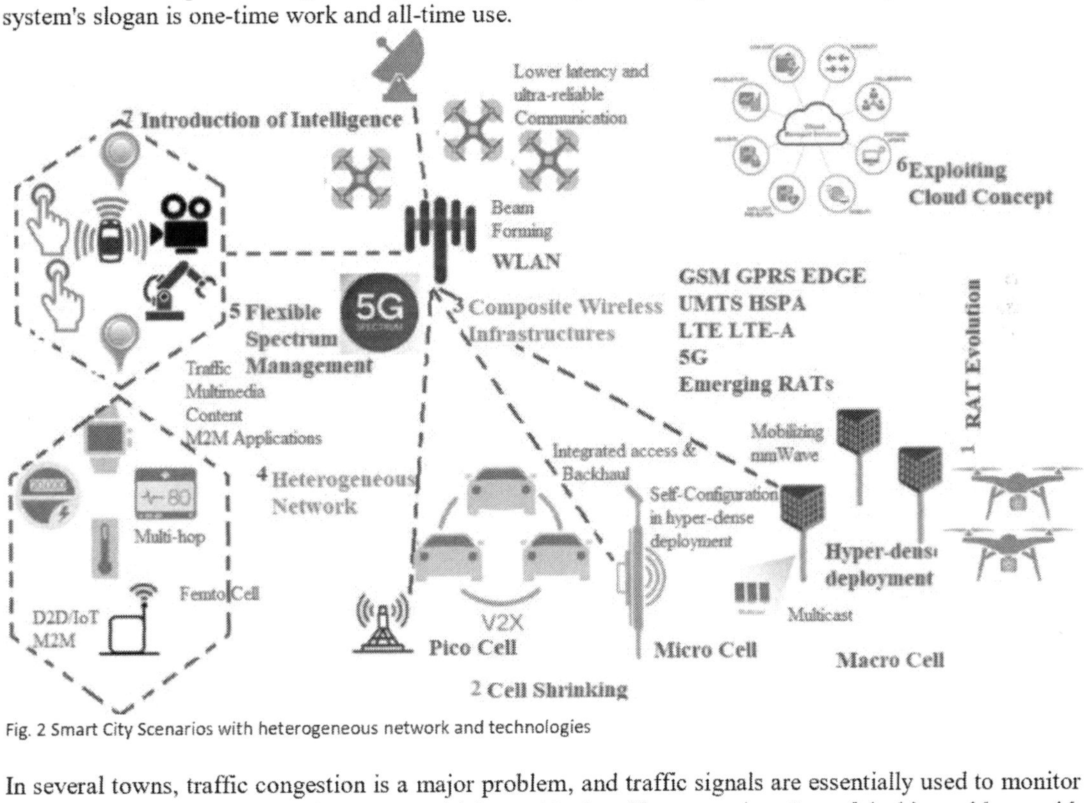

Fig. 2 Smart City Scenarios with heterogeneous network and technologies

In several towns, traffic congestion is a major problem, and traffic signals are essentially used to monitor the movement of weak law enforcement vehicles and bad traffic congestion. One of the big problems with many cities is that we cannot extend the current infrastructure further, so better traffic management is just one solution open to us. The efficacy of the traffic control system depends on its ability to adapt to the conditions of real-time traffic. Conventional traffic control systems, however, are not capable of doing this. Whatever the high or low traffic density, the signals are all timed and run only at that time [20a]. This leads to increased road congestion, contributing again to substantial air pollution, increased safety risk, and negative impacts on the economy and overall living standard. Advance the Internet of Things (IoT)-based traffic management framework in this need for expected innovations. The control of traffic lights is suggested and planned to help the decision-making of traffic offices. Based on the density of the vehicles, the device can detect the congestion level of each road at the crossing.

A challenging research field in Intelligent Transportation Systems (ITS) is vision-based automatic detection and monitoring of on-road vehicles. Several research studies are available in the literature in this area of research using various machine learning (ML) techniques. This article proposes a method for on-road vehicle detection and tracking under varying weather conditions by changing the configuration of a single-look vehicle (YOLO), a recently established deep-learning computer vision object detection model. A single convolutionary neural network, CNN, predicts the identification of binding boxes and the probabilities for classes in YOLO models together in a complete evaluation from the full images. A single, convertible neural network is used to detect vehicles [21a]. The detection mechanism can be interpreted as end-to-end systems since the full detection is achieved through a single network. 16 convolution layers of YOLO and 2 fully connected layers were used at the end of this work. In various weather conditions, the vehicles have been found. To assess device performance, two separate public datasets, namely CDNet 2014 and LISA 2010 were used.

Owing to the increased number of cars, road accidents also cause thousands of lives to be put at risk. This resulted in the need for controlled vehicle contact between the various vehicles in the network. The key issue related to VANET is its complex existence. The network is constantly evolving, creating high mobility and difficulty in controlling all the nodes within the network [22a]. However, researchers have identified various routing protocols to resolve this question, but the problem arises when it comes to network stability. In this paper, a stable routing algorithm called 'CP-PSO' is proposed and optimized using the PSO optimization technique. In terms of better throughput, high packet delivery ratio, and low energy consumption, the Matlab simulator is used for simulation, and results are shown. Pedestrian crossing may be a part of the road where the pedestrians have the right to cross the road. The intersections with cross-country tracks are also the maximum distance in many regions as a kilometer away. A foot traffic study that can significantly help reverse and prevent pedestrian accidents is a very important part of pedestal crossing. The main goal is to detect the footpath in the signal path and build an intelligent wagon to prevent fatal deaths and traffic. The use of ultra-sensors for vehicle detection and crash avoidance.

6. PROPOSED SYSTEM MODEL

The increased use of mobile devices, machine learning, and cloud computing provides an inconvenient basis for multi-domain IoT growth. Numerous processing elements, specifically IoT sensors that are linked to a server cluster and machine learning technology, comprise the proposed model system. The cloud provides services such as software, data collection and network server computing, open databases for easy and flexible use of resources across the internet. The cloud data centre is also a major component of the device being proposed, and is used to store the vast amount of sensory data generated by the sensors. The following are briefly mentioned.

6.1. Cloudlet

It offers cloud services on mobile devices such as intelligent phones, tablets and IT sensing, as well as servers connected for job storage and processing [12-13a]. Cloudlet is able to reduce the response time of applications running on IoT sensors as well as the hosting of cloud computer services through the use of low latency and high bandwidth wireless networking [14a]. This reduces long-distance contact latency delays between the IoT sensing system and the cloud data centre.

6.2. Compatibility and global vision with OpenFlow

OpenFlow was developed to show the feasibility of forming a centralised SDN control and administration system controller. For example, load monitoring, the quickest way to find the routing network, and the provision of a network flow control are several OpenFlow features currently available. New OpenFlow extensions, however, are needed to allow advanced technologies, such as advanced bandwidth allotment strategies based on traffic preview schemes as discussed. SDN controllers should be given a global perspective on the MANET topology (lightweight coordination between the MANET islands). In real-time, traffic flows inside the MANET domain could also be tracked.

6.3. MANET-domain SDN-based traffic control system

Smartphones and tablets consume 40% of data traffic and 99% of signal traffic is generated [15a]. Therefore, to ensure a global and precise backhaul (optical network) traffic status, SDN capabilities need to be strengthened for real-time traffic management in the MANET networks. In terms of SDN controllers, the global network view should be capable of handling topology, up-/down-link traffic flows and specifications, permitted and acceptable traffic per domain, and traffic and latency monitoring. In fact, it will allow them to gain full knowledge of the current network status and traffic of upcoming applications, allowing them to make proper traffic management decisions.

7. Conclusion

For hundreds of IoT devices, a smart urban network needs a secure system as well as a stable data storage system. A detailed analysis is provided of self-optimization, intelligent heuristic, virtualization of network functions, and heterogeneous networks and their possible advantages, use cases, and B5G research challenges. Heterogeneous network scenarios used in B5G and their specifications are discussed with the key drivers of B5G applications and technologies. The expanded vision of Smart intelligent networks lists a range of service groups used in a heterogeneous network with their network efficiency goals and specifications are specified which helps in research. Modern technical challenges with intelligent resource allocation heuristics for heterogeneous network and sensing applications are raised in virtualization function. A brief overview is given of heterogeneous ad hoc networks and recent developments in uncoordinated vMAC and modern IoT random access. AI-enabled B5 G network architecture and resource allocation approaches are addressed with optimization and the role of edge and cloud computing in the next-generation network. Some related performance metrics for self-optimization and self-configuration in smart network topologies with extended SDN controllers are addressed.

References

[1] Long Zhang, Hui Zhaoz, Shuai Hou, and Haitao Xuy, A Survey on 5G Millimeter Wave Communications for UAV-Assisted Wireless Networks, IEEE Access, vol. 7, no. 2, pp. 2241-2263, Sep. 2019.
[2] K. David and H. Berndt, "6G Vision and Requirements: Is There Any Need for Beyond 5G?," in IEEE Vehicular Technology Magazine, vol. 13, no. 3, pp. 72-80, Sept. 2018, doi: 10.1109/MVT.2018.2848498.
[3] H. Zhang, W. Quan, H. Chao and C. Qiao, "Smart identifier network: A collaborative architecture for the future internet," in IEEE Network, vol. 30, no. 3, pp. 46-51, May-June 2016, doi: 10.1109/MNET.2016.7474343.
[4] Tekbiyik, Kursat, A. R. Ekti, G. Kurt, A. Gorcin and H. Yanikomeroglu. "A Holistic Investigation on Terahertz Propagation and Channel Modeling toward Vertical HetNets." *arXiv: Information Theory* (2020): n. pag.
[5] T. Nakano, Y. Okaie, S. Kobayashi, T. Hara, Y. Hiraoka and T. Haraguchi, "Methods and Applications of Mobile Molecular Communication," in Proceedings of the IEEE, vol. 107, no. 7, pp. 1442-1456, July 2019, doi: 10.1109/JPROC.2019.2917625.
[6] S. Samarakoon, M. Bennis, W. Saad and M. Debbah, "Federated Learning for Ultra-Reliable Low-Latency V2V Communications," 2018 IEEE Global Communications Conference (GLOBECOM), Abu Dhabi, United Arab Emirates, 2018, pp. 1-7, doi: 10.1109/GLOCOM.2018.8647927.
[7] Halim Yanikomeroglu. 2018. Integrated Terrestrial/Non-Terrestrial 6G Networks for Ubiquitous 3D Super-Connectivity. Proceedings of the 21st ACM International Conference on Modeling, Analysis and Simulation of Wireless and Mobile Systems MSWIM'18, 3–4. DOI:https://doi.org/10.1145/3242102.3242148

[8] Z. Zhou, H. Liao, B. Gu, K. M. S. Huq, S. Mumtaz and J. Rodriguez, "Robust Mobile Crowd Sensing: When Deep Learning Meets Edge Computing," in IEEE Network, vol. 32, no. 4, pp. 54-60, July/August 2018, doi: 10.1109/MNET.2018.1700442.

[9] E. C. Strinati, S. Barbarossa, J. L. Gonzalez-Jimenez, D. Ktenas, N. Cassiau, L. Maret, and C. Dehos, "6G: The next frontier: From holographic messaging to artificial intelligence using sub terahertz and visible light communication", IEEE Vehicular Technology Mag., vol. 14, no. 3, pp. 42–50, Sep. 2019.

[10] C. Jiang, H. Zhang, Y. Ren, Z. Han, K.-C. Chen, and L. Hanzo, Machine learning paradigms for next-generation wireless networks, IEEE Wireless Commun., vol. 24, no. 2, pp. 98–105, Apr. 2017.

[11] R. Li, Z. Zhao, Z. Xuan, G. Ding, C. Yan, Z. Wang, and H. Zhang, Intelligent 5G: When cellular networks meet artificial intelligence, IEEE Wireless Commun., vol. 24, no. 5, pp. 175–183, Oct. 2017.

[12] M. G. Kibria, K. Nguyen, G. P. Villardi, O. Zhao, K. Ishizu, and F.Kojima, Big data analytics, machine learning, and artificial intelligence in next-generation wireless networks,‖ IEEE Access, vol. 6, pp.32328–32338, 2018.

[13] K. David and H. Berndt, 6G vision and requirements: is there any need for beyond 5G, IEEE Vehicular Technology Magazine, vol. 13, no. 3, pp. 72-80, Sept. 2018.

[14] Sethi A., Jain S. K., Vijay S. (2020) Secure Self Optimizing Software-Defined Framework for NB-IoT Towards 5G, Procedia Computer Science Volume 171, 2020, Pages 2740-2749 https://doi.org/10.1016/j.procs.2020.04.298

[15] N. Kato, Z. M. Fadlullah, B. Mao, F. Tang, O. Akashi, T. Inoue, and K. Mizutani, "The deep learning vision for heterogeneous network traffic control: Proposal, challenges, and future perspective", IEEE Wireless Communications, vol. 24, no. 3, pp. 146–153, Dec. 2017.

[16] H. Sun, X. Chen, Q. Shi, M. Hong, X. Fu, and N. D. Sidiropoulos, Learning to optimize: Training deep neural networks for wireless resource management, available online arXiv:1705.09412, May 2017.

[17] P. Du, S. Nazari, J. Mena, R. Fan, M. Gerla, and R. Gupta, Multi-path TCP in SDN-enabled LEO satellite networks, in Proc. IEEE Mil. Commun. Conf. (MILCOM), Nov. 2016, pp. 354–359.

[18] Sethi A., Vijay S., Gupta V. (2020) Mobility and QoS based cross-network for ICN Vehicular Communication, Procedia Computer Science Volume 171, 2020, Pages 897-906 https://doi.org/10.1016/j.procs.2020.04.097

A prototype of Fixed Wing UAV for delivery of Medical Supplies

R Pavithran R[1], V Lalith V[1], Naveen C[1], Sabari S P[1], Ajay Kumar MST[1], V Hariprasad[1]

[1] Department of Aeronautical Engineering, Bannari Amman Institute of Technology, Sathyamangalam, Tamilnadu, India.

hari.aero007@gmail.com

Abstract. A prototype Unmanned Aerial Vehicle (UAV) is designed for the purpose of last mile delivery to supply medicines and vaccines. This work aims to ascertain that UAVs can be employed to decrease transportation times, increased power efficiency, and improved safety for transportation of necessary products. The prototype model is design to deliver 700 g maximum payload up to 5 km with an endurance time of 30 minutes. The model is fabricated with a wing span of 1.5 m, Aspect ratio 7, and contains NACA 6 series airfoil shape as its cross section. During the transportation of payload, it has to fly beyond the line of sight which requires an autonomous Pixhawk flight controller installed in the model. Autonomous module aids in the safe flight avoiding the obstacles during the operation. To comply with government regulations for flying the model cruises at an altitude of 350 feet. After reaching the desired GPS destination, the payload is delivered to the target customer via parachute. However, the model requires a runway length of 200 m for its take-off and landing operation.

Keywords: Fixed Wing; UAV; Medical Supply; Pixhawk; Parachute

1. Introduction

The use of drones is proposed for number of domains as it can provides a faster and reliable solution for many transportation issues[1]. Due to an exponential increase in number of road vehicles and the dense traffic pose a challenge for the timely supply of life saving medicines and supply. Hence the use of UAV is proposed mainly for the major task of medical supply transmissions from one point to another. This medical supply generally comprises of blood packets antidotes and even in the transportation of organs which is very difficult in case of road transportation vehicles[2]. To control the spread of diseases and reduce the growing rates of mortality due to lack of adequate health facilities, special attention needs to be given to health care in rural areas. Most people lose their life due to a lack of blood and vaccines. It is difficult to travel inside a rural environment from urban areas because of gridlock as well as some of the rural environments are far away from the city which requires medical supply in a timely manner. Air transportation has many advantages over other means of transportation in the medical supply[3]. An autonomous fixed-wing unmanned aerial system that has higher range capability can travel faster even in rural areas[4] and can carry payloads of crucial components such as blood samples, vaccines, tablets, etc. The fixed wing UAV properly equipped with the autonomous module can fly beyond the line of sight[5]. Once they move beyond the line of sight, with the help of GPS signals it reach the destination and return back using predetermined waypoints. The payload is kept in a bay which is completely enclosed inside the fuselage of the model[6–8].

2. Design of Fixed Wing UAV

The design and calculation of the medical supply UAV happens in three phase namely conceptual design, preliminary design and detailed design. The conceptual design involves listing a variety of configurations that meet the required mission profile and mission requirements in terms of aerodynamics, propulsion, flight performance, structural and control systems. Further fundamental aspects such as fuselage shape, wing configuration and location, engine size and type are determined during the conceptual design process. In preliminary design process, UAV is further designed to fit the required design parameters of the mission requirements. In this design process component level testing, CFD to estimate wing performance, lift drag estimation, weight estimation, structural design, control surface design, and structural materials are finalized for fabrication. Finally, in the detailed design process, weight of the selected components and electronics is optimized to maximize endurance and ease of operation.

2.1. Mission Profile

The drawing the mission profile (Figure 1) for the proposed application for Fixed Wing UAV provides insight on the actual implementation. From the mission profile the important milestones are highlighted with actual flight envelop (Table 1). From the mission profile the aircraft is outlined to have seven phase in its overall flight. Take-off of the UAV happens on the runway, followed by the climb, the model cruises at an altitude of 350 feet allowed by DGCA regulations, it descend to the altitude of 100 feet to drop the package during using parachute, it again ascents to its cruise altitude during its return flight, it finally descend back to ground followed by ground run before it comes to halt. In the flight envelop, during take-off and landing phase is operated manually the remaining is flight phase is managed by the autonomous flight controller module.

Table 1. Mission Requirements

Criteria	Requirement
Max wingspan	1.5 m
Gross Take-off weight	2.3 kg
Max Endurance	30 minutes
Max Flight speed	12-15 m/s
Payload weight	700 g
Service ceiling (altitude)	150-350 m
Operational range	5 km

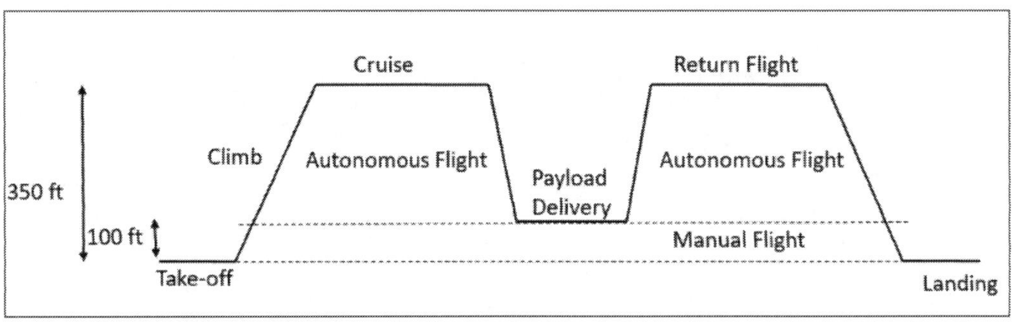

Figure 1: Mission Profile

2.2. Airfoil and Wing Planform

The design of the wing begins with the selection of an airfoil as it is the primary means of generating sufficient lift and minimizing drag effects. The airfoil is selected based on the Reynolds number, coefficient of lift (C_l), coefficient of drag (C_d) and coefficient of moment (C_m) as per the mission requirement. The lift curve slope should be primarily focused while judging the airfoil. Based on above parameters NACA 63-412 was selected as the airfoil for the aircraft. NACA 63-412 has a maximum C_l of 1.2 at an angle of attack 12 degree with a maximum thickness of 12% at 34.9% of chord (Figure 2). This is followed by the selection of an appropriate wing planform. The planform of a wing contributes significantly to its aerodynamic performance, particularly in terms of 3D drag due to wingtip vortex shedding. A detailed analysis led to the involvement of washout and winglets on a combined rectangular tapered planform.

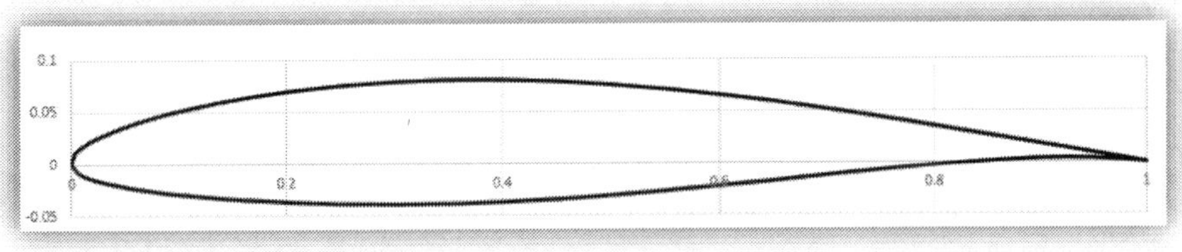

Figure 2: NACA 63-412 Airfoil

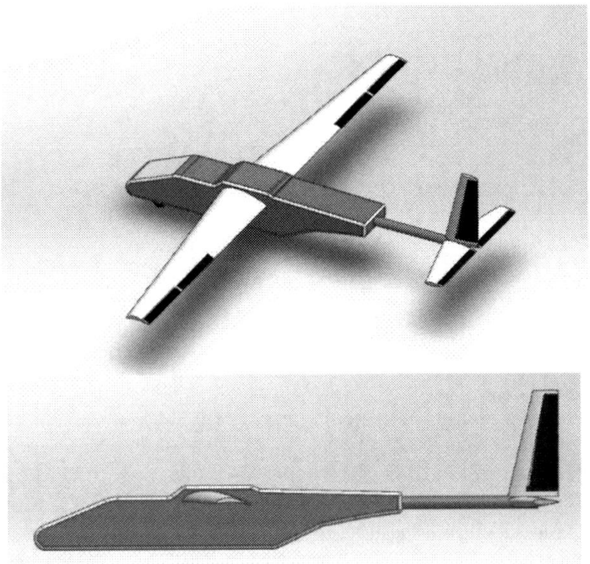

Figure 3: Conceptual Design of Model using CATIA

2.3. Fuselage

The fuselage is the body of the fixed-wing radio controlled plane. It provides space for cargo, controls, accessories, passengers, and other equipment. Fuselage must produce minimum drag and it must have space for all the electronics and propulsion system. From an aerodynamic point of consideration, it is important that the fuselage is streamlined so as to facilitate the airflow with minimum drag. The tail assembly and the propulsion system are attached to fuselage and it bears the all load during the land. Hence the fuselage should have a good structural integrity and it is measured as 1 X 0.12 X 0.10 m^3 (Figure 3). The centre of gravity (CG) of the UAV plays an important role in the flight stability, hence the mass distribution and CG balancing was carried out using the weights of individual components. The electronics, payload and structural components are positioned in the fuselage at different locations and the CG of the model is balanced by altering their locations, so as to locate the CG near the aerodynamic centre.

2.4. Empennage

The empennage for the prototype is located towards the rear end of the model. The empennage contains both horizontal and vertical stabilizers which provides the longitudinal and lateral stability for the plane. It contributes to the overall aircraft stability by controlling the pitch and yaw of the model. Wing position decides the effective location of the empennage. The tail efficiency is reduced when it encounters the wake generated from the wing's trailing edge. Wake creates a reduced pressure region around its centreline for a particular wake width, which depends on the tail moment arm and the minimum drag coefficient of the wing. So the tail has to be positioned outside the wake region for the maximum estimated angle of attack of the flight. The tail configuration was chosen to be of the inverted 'T' type due to its structural strength and the horizontal stabilizer is fixed on the fuselage. Both horizontal and vertical stabilizers are designed to have a cross section of NACA 0012.

2.5. *Electronics*
The propulsion system calculations are made with the assumption as a glider model, with thrust to weight ratio as 0.4. Using Static thrust equation propeller dimension is estimated as 11 X 4.5 which provides the necessary thrust for the model. The propeller is driven by an 850 kv brushless DC motor (BLDC) supported by 30 Amp electronic speed controller. The electronic speed controller is positioned such that it is sufficiently air cooled to withstand the heat generated throughout the flight envelop. A 2.4 GHz transmitter is used to assist the model during take-off and landing process while the autonomous flight is supported by a Pixhawk flight controller and 433 MHz telemetry. The power is supplied to all the electronic components using a 6500 mAh Lithium polymer battery (Table 2).

Table 2. Propulsion Specification

Criteria	Requirement
BLDC motor	850 kv
Electronic Speed Controller	30 Amp
Propeller	11 X 4.5
Battery	6500 mAh (4s)
Transmitter/Receiver	2.4 GHz
Autonomous	Pixhawk
Telemetry	433 MHz

2.6. *Autonomous Module Installation and Calibration*
The autonomous module comprises of a pixhawk flight controller board, Pitot-static tube, GPS module, telemetry, power module and PPM encoder. The entire autonomous module is controlled by the ground control unit which is Mission Planner. The ground control unit transmits the signals to the model using ground and air telemetry. Take-off and landing operations of the model are controlled manually by the pilot. The pixhawk board is placed in the centre of gravity location and it should be free from interference of electromagnetic waves and permanent magnet. The receiver is connected with PPM encoder which converts the PWM (Pulse Width Modulation) signals to digital signals.

Initially the pixhawk is installed in the model for calibration. Through telemetry the flight controller connects with mission planner software. The fixed-wing configuration in chosen manually in the mission planner software and then it uploads the necessary packages to the controller. Once the data is uploaded, the control surfaces in the model are calibrated to ensure proper working for the inbound transmitter signals. The model can be flown in any one of the following three autonomous modes.

- Normal
- Stable
- Fully autonomous

In normal mode, the model is controlled manually. During stability mode, the stability of the model is alone autonomous. In fully autonomous mode the flight controller takes the entire control based on the ground control unit. After completing this process, the model is ready to fly in autonomous mode.

2.7. *Parachute design*
The parachute is designed to ensure the safe delivery of payload (Figure 4). The parachute is made of nylon fabric to withstand the drag force experienced during the descend phase. The nylon fabric is designed to have a hemispherical surface in its fully extended condition. The strings attaching the payload bay and parachute fabric is arranged and kept together inside fuselage in the manner that they do not tangle during the drop. The surface area of the parachute is designed so that the payload descends at a rate of 3 m/s.

3. Fabrication
To fabricate the fuselage, the bulkheads are made from plywood of thickness 0.3 cm. The remaining structural members are made of duraform sheets of thickness 3 mm and 5 mm. The bulkhead provides the necessary shape to the fuselage. Fuselage is divided into three sections: the first section will contain a propulsion system, the second section is allotted to payload, and the third section contains the autopilot system. The landing gear is attached to the bottom of the fuselage to support ground run during take-off and landing. The prototype has a conventional tail contains horizontal and vertical stabilizers. They are made as separate sections using the duraform sheet and attached to the fuselage.

The wing contains the essential parts such as spar, ribs and stringers. Initially the ribs are cut to the required airfoil sections. A circular aluminium rod of 10 mm diameter is used as spar. The ribs are placed at a distance of 6 inch apart and covered using a 3 mm duraform to complete the wing. The wing screwed

IOP Conf. Series: Materials Science and Engineering **995** (2020) 012015 doi:10.1088/1757-899X/995/1/012015

to the fuselage and the servos rods are connected to the control surfaces (Figure 5). The entire fuselage and wing structure is covered by vinyl sheets to reduce the skin friction drag and provide a smooth finish.

Figure 4: Parachute attached to payload

Figure 5: Prototype model with electronics installed

4. Flight Test

The prototype model that need to deliver the medical supply is initially tested for the proper working of control surfaces and propulsion system. The pixhawk module attached with the model is calibrated to the ground control unit to ensure the safe operation during autonomous flight. Payload to be delivered is attached to the parachute system and installed in the payload bay. The payload bay is operated using a servo. The servo opens and closes the payload bay based on the input from the pixhawk. Flight path is instructed to the model through waypoints from the ground control unit to the installed pixhawk using telemetry. Payload drop location is provided to pixhawk as GPS waypoint location.

A fully charged battery is replaced in the model before the flight. The take-off of the model is performed manually until it reaches the altitude of 100 feet and ensured of obstacle free flight. The autonomous module is engaged for the rest of the flight. The model then climbs to 350 and follows the waypoints towards the destination. As soon as the flight close to the drop location it descends to 100 feet altitude and the servo is enabled by the pixhawk to release the payload. Payload using the parachute is descends at 3 m/s and safely landed on the ground (Figure 6). The parachute delivery has to be planned on an open area to avoid the landing on nearby building and trees due to cross wind. The model then ascents to 350 feet and continues its flight using the waypoint for the return flight. Once it reaches the landing spot and acquires the telemetry connection, it will be intimated by the ground control station. The model is then safely landed and prepared for the next flight.

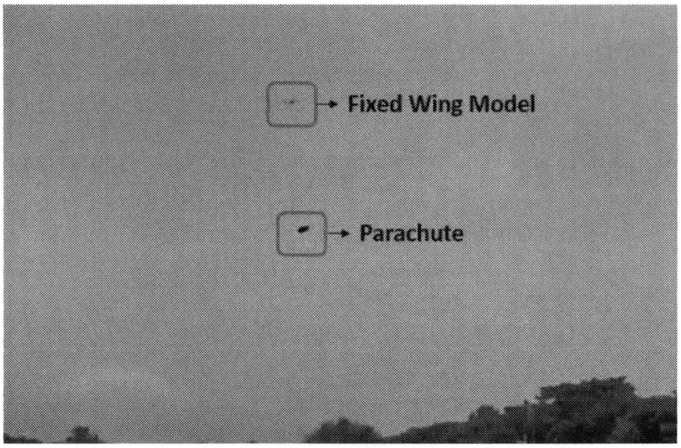

Figure 6: Prototype Test Flight delivering Payload using Parachute

5. Conclusion

The following are concluded from the prototype development

1. Fixed wing UAV is designed and tested for carrying the payload of 700 kg
2. Working of the autonomous module is successful for cruise flight condition
3. Parachute can decent the payload at a rate of 3 m/s to provide safe landing
4. Take-off and landing is to be performed manually by pilot to ensure obstacle free flight
5. Atmospheric wind direction will affect the flight endurance and payload delivery

References

[1] Dündar Ö, Bilici M and Ünler T 2020 Design and performance analyses of a fixed wing battery VTOL UAV *Engineering Science and Technology, an International Journal* **23** 1182–93

[2] Yan C, Xiang X and Wang C 2020 Fixed-Wing UAVs flocking in continuous spaces: A deep reinforcement learning approach *Robotics and Autonomous Systems* **131** 103594

[3] Jayathunga S, Owari T and Tsuyuki S 2018 The use of fixed–wing UAV photogrammetry with LiDAR DTM to estimate merchantable volume and carbon stock in living biomass over a mixed conifer–broadleaf forest *International Journal of Applied Earth Observation and Geoinformation* **73** 767–77

[4] Kim H J, Kim M, Lim H, Park C, Yoon S, Lee D, Choi H, Oh G, Park J and Kim Y 2013 Fully Autonomous Vision-Based Net-Recovery Landing System for a Fixed-Wing UAV *IEEE/ASME Transactions on Mechatronics* **18** 1320–33

[5] Kang Y and Hedrick J K 2009 Linear Tracking for a Fixed-Wing UAV Using Nonlinear Model Predictive Control *IEEE Transactions on Control Systems Technology* **17** 1202–10

[6] Euston M, Coote P, Mahony R, Kim J and Hamel T 2008 A complementary filter for attitude estimation of a fixed-wing UAV *2008 IEEE/RSJ International Conference on Intelligent Robots and Systems* 2008 IEEE/RSJ International Conference on Intelligent Robots and Systems pp 340–5

[7] Beard R, Kingston D, Quigley M, Snyder D, Christiansen R, Johnson W, McLain T and Goodrich M 2005 Autonomous Vehicle Technologies for Small Fixed-Wing UAVs *Journal of Aerospace Computing, Information, and Communication* **2** 92–108

[8] Beard R W, McLain T W, Nelson D B, Kingston D and Johanson D 2006 Decentralized Cooperative Aerial Surveillance Using Fixed-Wing Miniature UAVs *Proceedings of the IEEE* **94** 1306–24

IOP Conf. Series: Materials Science and Engineering **995** (2020) 012016 doi:10.1088/1757-899X/995/1/012016

Dielectric Fluid Parameter Optimization in Machining of Composite Material using WEDM Process

K L Senthil Kumar[1] and S K Dhinesh [1]

[1] Department of Mechatronics, Bannari Amman Institute of Technology, Sathyamangalam, Erode, Tamil Nadu, India.

senthilkumarkl@bitsathy.ac.in, dhineshmemts@gmail.com

Abstract: Wire Electric Discharge Machining (WEDM) has become most popular among the non-conventional machining processes because of precise machining and accuracy of parts. The process is preferable for accurate machining of complex geometries in hard materials such as Aluminium Silicon Carbide metal matrix composite. In this work, an attempt is made to analyze and optimize the dielectric fluid parameters in machining of Al/SiCp 10% work material. The input variables such as volume flow rate, flow velocity and chemical composition of dielectric medium used in the WEDM are chosen for this. Taguchi's L9 orthogonal array is used to conduct experiments in WEDM machine and ANOVA results have confirmed the influence of chosen parameters on the output responses. Computer simulation model of the WEDM tank with dielectric fluid, work piece and nozzle is presented using Fluent software. The Fluent model is analyzed based on the maximum removal of debris in the work material. Optimization is carried out to maximize the Material Removal Rate (MRR) and minimize surface roughness. The results obtained from Taguchi's optimization and Fluent analysis have been validated by carrying out tests on WEDM. The study shows that volume flow rate and flow pressure can adequately influence MRR and surface roughness.

Keywords: Composite material, Material Removal Rate, Surface Roughness, Dielectric Fluid, CFD

1. Introduction

WEDM is a non-conventional, thermo-electric process in which the material from work piece is eroded by a series of discharge sparks between the work and tool electrode immersed in a liquid dielectric medium and widely used to produce dies, punches and moulds, finishing parts for aerospace and automotive industry and surgical components [1-3]. The minute arcs created in numerous numbers within very short duration, locally evaporates the metal which are cooled down and flushed away by the flowing di-electric fluid. The method is practised by the industries to produce complex shapes that are either difficult or uneconomical through conventional machining processes. WEDM process is proven its ability in machining hard but conductive materials and reproducing complex shapes [4].

Industries like aerospace, automotive and marine are increasing the usage of aluminium based composite materials which are usually very difficult to machine using turning and milling operations. [5,6]. Such composite materials, even though can be conveniently machined using WEDM process, still needs to be tested for improving the machining efficiency and apply flushing effectively [7].

2. Literature Review

Material removal rate (MRR) increases with increasing flushing pressure and becomes constant. Tool wear reduces to certain extent, up to optimal value is reached, and surface finish also showed the same trend [8]. Various dielectric fluids were used and their effect on EDM was revealed. Study concludes that tap water, as dielectric medium, can enhance MRR. As far the surface quality is considered, hydrocarbon oils performance was inferior to that of deionized water [9].

As water, based dielectrics are economic and environment friendly, they must be investigated further as an option of hydrocarbon oil for performance improvement of EDM with different additives and for machining of materials like composites and carbides [14].

Dhakar et al. [10] have used two-phase dielectrics like water-air, EDM oil-air and Glycerine-air mixtures as dielectric medium in near dry EDM process and discovered that high viscosity fluids perform better than that of low viscosity. They also reported that MRR for glycerine-air mixture was enhanced by 300% compared to other mixtures used in the study.

Boron carbide powder varied from 1-15 gpl and mixed along with EDM oil was tried as dielectric medium in machining Ti-6Al-4V alloy [11]. Results revealed that with an increase in B4C concentration had improved the MRR, tool wear rate and surface finish and however concentration beyond 15 gpl of B_4C powder resulted in reduced MRR. The authors have also tried with dielectric fluid of surfactant and graphite powder addition to the EDM oil [12]. They concluded that adding surfactant and graphite powder responded positively as far as MRR and TWR were concerned, whereas the quality of machined surface and Recast Layer Thickness (RLT) had come down. Taguchi method was used to determine the optimal concentration of dielectric mixture.

Singh et al. [13] have reviewed the effect of different dielectric fluids on MRR and TWR. Effects of gaseous, liquid, liquid plus powder particles, vegetable oil based dielectrics were reviewed and reported that vegetable-based oil dielectric had comparatively good sustainability index. Further, it was suggested that Jatropha oil could be a best alternative for hydrocarbon based dielectric fluids. The review which was considering researches of past 36 years in this field, have not reported a good number of works carried out to know the effect of dielectric fluid parameters on MRR and TWR. From the above literature survey, it is understood that a gap exists in predicting the effect of controlling parameters of dielectric medium on the common outputs of machining. Even there is a very short report on wire EDM, a variant of EDM process. Hence, an effort was made to study how the dielectric fluid parameters such as fluid pressure, volume flow rate, ion exchanger resin composition effect on MRR and surface quality in wire EDM of aluminium matrix composites.

3. Experimental Design

This research is carried out on machining of composite material in WEDM with brass electrodes. Dielectric parameters such as volume flow rate, flow pressure and resin composition are varied to investigate their effects on the Material Removal Rate (MRR) and finish of the cut surface. De-ionized water, used as dielectric fluid and injected through nozzle, enhances debris ejection in the gap up to a certain extent. Work materials are prepared by reinforcing 10% SiC particles in Al6061 alloy matrix, through stir cast route and design of experiments is applied to conduct the experiments. This section elaborates these in detail.

3.1 Preparation of work materials

Induction furnace is used to melt the Al6061 alloy at 700°C. Preheated (800°C) SiC particles of 40 μm are presented into the vortex of molten aluminium, created by rotating stirrer at 110 rpm. Stirring is continued for 10 minutes before pouring into the preheated (350°C) metal mould and allowing it to solidify. The microstructure of the work material cast shows that SiC reinforcement is mixed with aluminium matrix (Fig. 1)

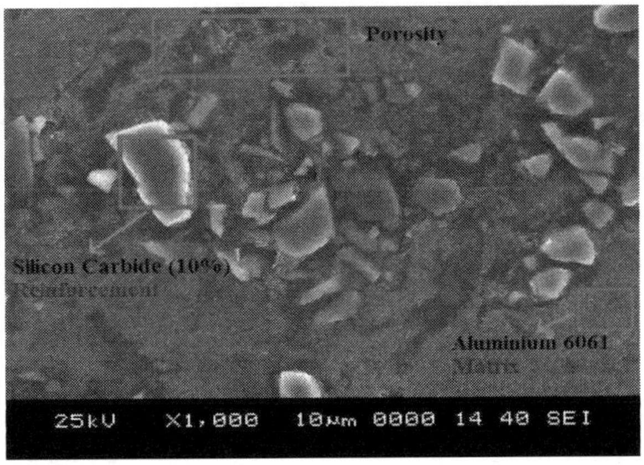

Figure 1. Microstructure of MMC.

3.2 Design of Experiments

From the literature and opinion of the experts in this filed, dielectric fluid parameters namely fluid pressure, flow rate and ion exchanger resin composition are selected for the analysis. Levels of each variable are fixed with the fluid circulation system capabilities. Since three levels are decided (Table 1) for each parameter a total of 3 x 2 = 6 degrees of freedom (dof) is needed to conduct the experiments. Taguchi's L9 orthogonal array, which can accommodate up to four parameters at three levels has 8 DOF is the suitable design to conduct the experiments. Three trials on each trial is conducted to reduce the experimental error.

After deciding the factors and their levels (table1), L9 orthogonal array (OA) was chosen to plan the experimental layout as shown in table 2. This OA can accommodate upto four three level factors. The levels for each factor are selected based on the past researches and the equipment capability. Material removal rate (MRR) and the surface roughness are the response variables used. All the nine experiments are replicated two times (totally three times) and complete randomization is followed in the order of conducting the trials. t, the optimization of the observed values was determined by comparing the standard analysis and analysis of variance (ANOVA) which was based on the taguchi method.

Table 1. Dielectric medium parameters and levels.

Control Parameter	Level 1 Minimum	Level 2 Intermediate	Level 3 Maximum	Response measured
Volume Flow Rate (lpm)	10	12	14	Material Removal Rate (g/min)
Pressure (bar)	2	2.5	3	
Ion exchanger resin composition (kg)	10	15	20	Surface Roughness (μm)

Design of experiments for volume flow rate, Pressure and resin composition were done by using MINITAB V16 software. Table 3 indicates the work conditions of EDM used while conducting experimental tests.

Table 2. Layout of experiments with L_9 OA.

Trial No	Volume Flow Rate (lpm)	Pressure (bar)	Ion exchanger resin composition (kg)
1	1	1	1
2	1	2	2
3	1	3	3
4	2	1	2
5	2	2	3
6	2	3	1
7	3	1	3
8	3	2	1
9	3	3	2

Table 3. Working Conditions

Work Condition	Description
Work piece	Aluminium 6061 MMC with SiC (10%)
SiC Particle size	40 (μm)
Tool Material	Brass
Wire Diameter	0.25 mm
Dielectric fluid	Deionized water

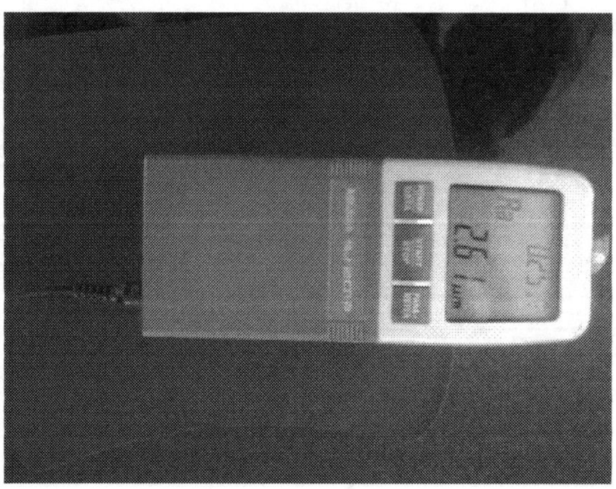

Figure 2. Surface roughness measurement using surftest meter

The weight of the work piece before (wj_b) and after (wj_a) conducting each trial are measured and time taken for each trial (t) is also noted down and finally the MRR is computed as below [4]:

$$MRR = (wj_b - wj_a)/t \text{ -------------------- } (1)$$

Surface roughness is measured with Mitutoyo tester (figure 2) with 0.8 mm cut off length. Experiments are conducted and data are obtained from experiment. Table 4 shows the L9 Orthogonal array with replaced parameters and levels and output result after conducting experiments. MRR are taken as outputs and signal to noise ratios were calculated individually.

Table 4. L_9 Observed data and S/N ratio for material removal rate.

S.NO	Parameters			Material removal rate (g/min)	Signal to noise ratio (db)
	Volume Flow Rate (lpm)	Pressure (bar)	Ion exchanger resin composition (kg)		
1	10	2.0	10	1.300	2.279
2	10	2.5	15	1.567	3.900
3	10	3.0	20	1.733	4.778
4	12	2.0	15	2.500	7.959
5	12	2.5	20	2.300	7.235
6	12	3.0	10	1.177	1.413
7	14	2.0	20	3.633	11.206
8	14	2.5	10	2.167	6.716
9	14	3.0	15	2.400	7.604

Table 5 shows the L9 Orthogonal array with replaced parameters and levels and output result after conducting experiments. Surface roughness is taken as output and signal to noise ratio were calculated individually.

Table 5. Observed data and S/N ratio for surface finish

S.No.	Parameters			Surface Roughness (μm)	Signal to noise ratio (db)
	Volume Flow Rate (lpm)	Pressure (bar)	Ion exchanger resin composition (kg)		
1	10	2.0	10	2.610	-8.3328
2	10	2.5	15	2.540	-8.0967
3	10	3.0	20	2.730	-8.7233
4	12	2.0	15	2.513	-8.0050
5	12	2.5	20	2.710	-8.6594
6	12	3.0	10	2.450	-7.7833
7	14	2.0	20	2.503	-7.9704
8	14	2.5	10	2.297	-7.2220
9	14	3.0	15	2.397	-7.5922

IOP Conf. Series: Materials Science and Engineering **995** (2020) 012016 doi:10.1088/1757-899X/995/1/012016

4. Analysis of Data

In order to understand the influence of selected factors, that is to know the main effects of parameters, an analysis of variance (ANOVA) is done. Further the observed data are subjected to optimization using Taguchi's technique.

4.1 ANOVA for material removal rate

The ANOVA for MRR is formulated as shown in Table 6 and the percentage contribution of each of the process parameter is used to understand the degree of influence of various factors.

Table 6. ANOVA for MRR.

Factor	DOF	SS	Adj SS	MSS	F Test	P (%)
Volume flow rate (lpm)	2	2.2156	2.2156	1.1078	39.8800	46.4627
Pressure (bar)	2	0.7622	0.7622	0.3811	13.7200	15.2008
Ion exchanger resin composition (kg)	2	1.6156	1.6156	0.8078	29.0800	33.5564
Error	2	0.0556				4.780
Total	8	4.6489				100

From the results it is evident that volume flow rate of the dielectric fluid influences highly in the removal of material followed by ion exchange resin composition and fluid pressure. This supports the principle of wire EDM in which the material removal is due to movement of ionized particles.

4.2 ANOVA for Surface Roughness

ANOVA for surface roughness is also formulated as shown in then Table 7 and found that the flow rate of the dielectric fluid highly influences the surface roughness.

Table 7. ANOVA for surface roughness.

Factor	DOF	SS	Adj SS	MSS	F Test	P (%)
Volume flow rate (lpm)	2	0.2571	0.2571	0.1285	84.4543	84.354
Pressure (bar)	2	0.0271	0.0271	0.0135	8.8947	10.521
Ion exchanger resin composition (kg)	2	0.0018	0.0018	0.0009	0.5861	0.912
Error	2	0.0030				4.213
Total	8	0.2890				100

From the data obtained from the experiments, result was analyzed using ANOVA method. The ANOVA results for various parameters for SR were shown in table 7. The analysis of the above ANOVA results show that the calculated F values and percentage contribution of each factor were obtained are as volume flow rate influences majorly for SR, secondly flow pressure and finally Ion exchanger resin ratio which influences the surface roughness in the work piece. For SR in the above rank table for SR is validated by percentage contribution in ANOVA table and optimized parameters from Taguchi are validated by ANOVA.

Delta and rank table formulated for MRR and surface finish supplements the ANOVA results (refer table 8 & 9).

Table 8. Response table for MRR

Level	Volume flow rate (lpm)	Pressure (bar)	Resin Composition (kg)
1	3.652	7.148	3.469
2	5.535	5.950	6.488
3	8.509	4.598	7.739
Delta	4.857	2.550	4.270
Rank	**1**	**3**	**2**

Table 9. Response table for SR.

Level	Volume flow rate (lpm)	Pressure (bar)	Resin Composition (kg)
1	-8.384	-8.103	-7.779
2	-8.149	-7.993	-7.898
3	-7.595	-8.033	-8.451
Delta	0.789	0.110	0.672
Rank	**1**	**3**	**2**

4.3 CFD Analysis

CFD analysis had done for parameters such as volume flow rate and pressure of dielectric medium in WEDM. Fluent software is used for analysis.

4.3.1 Geometric Modeling.

Geometric models of WEDM tank with dielectric fluid shown in figure3, work piece and nozzle shown in figure4 were modeled by using Pro-E software. Geometric models were modeled based on the actual size of tank, work piece and nozzle for debris analysis.

4.3.2 Meshing of Model.

Geometric model is meshed by using ANSA software. Triangular mesh is used to create surface mesh of the model and total count of triangular mesh is 116100 shown in figure 3. After surface meshing, T- Grid software is used to create triangular volume mesh in the model. Outer boundary and inner boundary of model are connected by using prism mesh and prism mesh used is to capture the perfect physics in model. Finally, it is *simulated* by CFD software for maximum removal of debris.

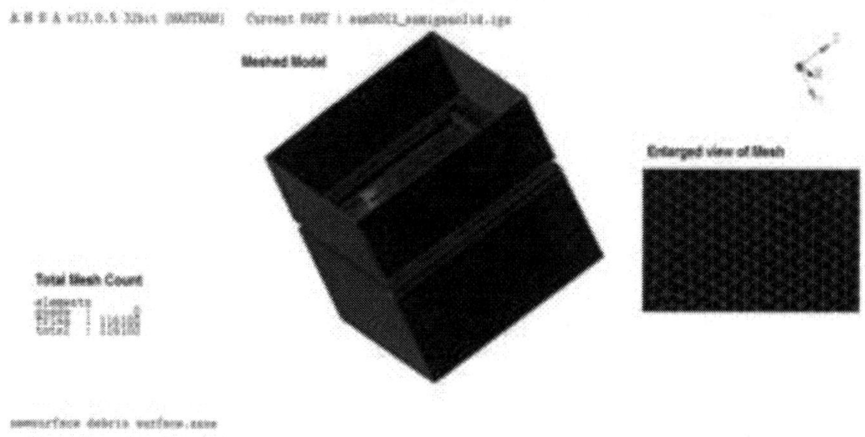

Figure 3. Fluent model.

4.3.3 Flushing Analysis in Fluent.

A simulation is made to simulate the behaviour of dielectric fluid in the working gap during the machining process with the aid of commercially available FLUENT V13. The volume removal rate of debris, which is an indirect measure of material removal rate, is notified for each trial with different parameter settings. Then, their results were compared with all the parameters and the parameters those influence maximum removal of debris, are presented.

5. Results and Discussion

In this section, results obtained from the ANOVA for MRR, ANOVA for surface roughness and CFD analysis by Fluent are discussed with analysis results. Taguchi analysis had done for WEDM dielectric parameters, graphs and rank of parameters are discussed below

5.1 Signal to Noise (S/N) ratio graph

The main effects plot on S/N ratios of MRR and surface roughness are shown by figures 4 and 5 respectively. Regardless of the output response variable the maximum S/N is preferred always. In that aspect, by comparing figures 4 and 5 it could be understood that the optimal process parameter values those produce a better MRR with minimum roughness are 14 lpm for volume flow rate, 2.0 bar pressure and 20 kg for ion exchanger resin composition.

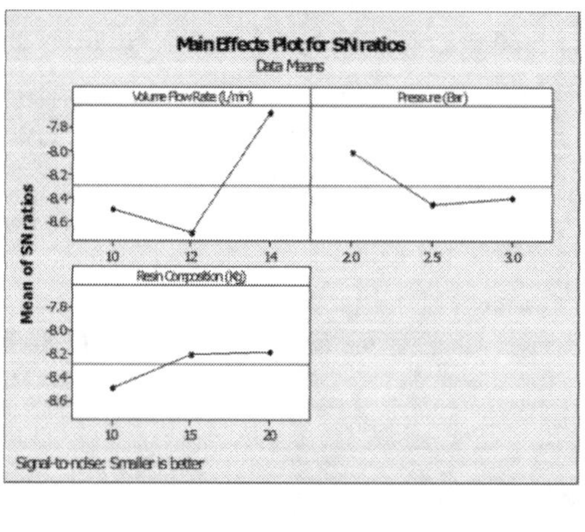

Figure 4. S/N graph for MRR.

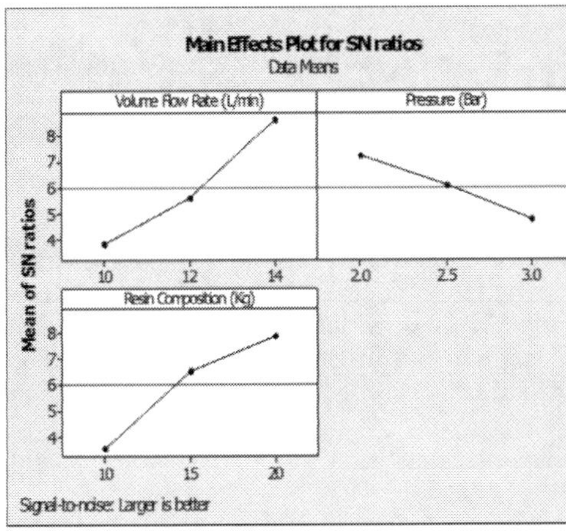

Figure 5. S/N graph for surface roughness

5.2 CFD Analysis by Fluent

In Fluent software, analyses were done for various dielectric medium parameters. Parameters such as volume flow rate and flow pressure are taken as input boundary conditions. According to the input boundary conditions the analyses were done by Fluent for all parameters and the best parameter, which has maximum removal debris, are 14 lpm volume flow rate and 2 bar pressure. For this parameter the amount of debris, which was removed from work piece, are 599.0 kg/m^3 as depicted in figures 6 and 7. Debris removal time from work piece for the best parameter is also shown in figure 8. This result coincides with that of Taguchi's S/N ratio analysis results as far as the volume flow rate and pressure of the dielectric fluid are concerned. It means that the parameter values those are responsible for maximizing MRR has resulted in maximum debris removal.

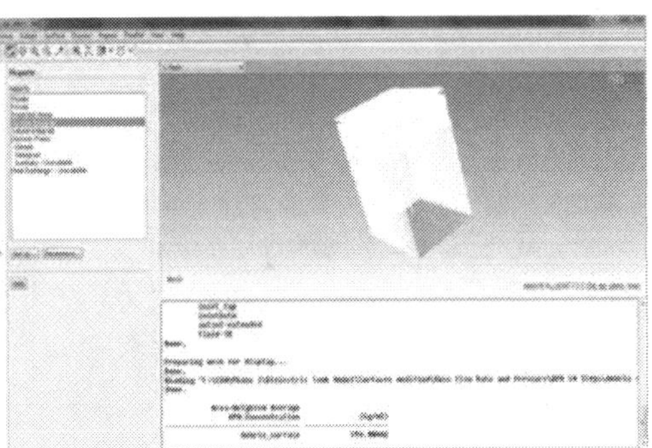

Figure 6. CFD analysis of debris removal rate.

Figure 7. CFD analysis of debris removal time.

5.3 Confirmation experiment:

The prediction of the response values can be done using the following relation:

$P = \mu + (A3-\mu) + (B1-\mu) + (C3-\mu)$ -- (2)

$P = 2.631$ g/min for MRR

$P = 2.533$ μm for surface roughness

Where 'P' is the predicted value of the concerned response and 'μ'is the mean value of total responses. A confirmation test is conducted with the optimum values to ensure the reliability of results obtained through the above methods. Three trials are conducted with the parameters values of 14 lpm volume flow rate, 2.0 bar pressure and with 20 kg ion-exchange resin. An average material removal rate of 2.756 g/min and surface roughness value of 2.62 μm are obtained. These values are closely matching with the predicted values.

6. Conclusion

In this work wire electro discharge machining of Al based composite with 10% SiC is carried out. WEDM dielectric medium parameters such as volume flow rate, flow pressure and ion exchanger resin composition are taken as input variables and analysis of data is done for maximizing MRR and minimizing surface roughness. Taguchi optimization method and fluent analyzed model of WEDM dielectric tank are applied to the data collected.

Optimized results of dielectric medium parameters for maximizing MRR and minimizing surface roughness are as follows:

- Maximum MRR is obtained for a volume flow rate of 14 lpm (which contributes 46.46%), ion exchanger resin composition of 20 kg (33.55%) and fluid flow pressure of 2 bar (15.20%)
- Minimum surface roughness is obtained for a volume flow rate of 14 lpm (which contributes 84.35%) followed by fluid flow pressure (10.52%) and ion exchanger resin composition
- CFD analysis by fluent showed the optimized debris removal concentration of 599.0 kg/m^3 happens at 14 lpm volume flow rate and 2 bar pressure which also results with a minimum time (0.034 seconds) of debris removal in work piece.
- Confirmation test is conducted to verify the same and the results were found to be satisfactory.
- Results also reveal that volume flow rate of dielectric fluid plays a vital role in maximizing the removal of material and minimizing surface roughness.

7. References

[1] Kathiresan, M. and Sornakumar, T., (2010), "EDM studies on aluminum alloy-silicon carbide composites developed by vortex technique and pressure die casting" Journal of Minerals & Materials Characterization & Engineering, Vol. 9, No.1, pp.79-88

[2] Tomadi, S.H., Hassan, M. A. and Hamedon, Z (2009), "Analysis of the Influence of EDM Parameters on Surface quality, material removal rate and electrode wear of tungsten carbide", International MultiConference of Engineers and Computer Scientists, Vol II, 978-988-17012-7-5

[3] Mohammed Amri Lajis, H.C.D. Mohammed Radzi and A.K.M. Nurul Amin (2009), "The implementation of Taguchi method on EDM process of tungsten carbide" European Journal of Scientific Research ISSN 1450-216X Vol.26 No.4

[4] Sushant Dhar, Rajesh Purohit, Nishant Saini, Akhil Sharma and Hemath Kumar, G., (2007), "Mathematical modeling of electric discharge machining of cast Al–4Cu–6Si alloy–10 wt.% SiCP composites", Journal of Materials Processing Technology 194,24–29

[5] Lauwers, B., Kruth, J. P. Liu, W., Eeraerts, W., Schacht, B. and Bleys, P., (2004), "Investigation of material removal mechanisms in EDM of composite ceramic materials", Journal of Materials Processing Technology 149,347–352

[6] Velusamy Senthilkumar and Bidwai Uday Omprakash (2011), "Effect of titanium carbide particle addition in the aluminium composite on EDM process parameters", Journal of Manufacturing Processes 13, 60–66

[7] Lei Li, Lin Gu, Xuecheng Xi and Wansheng Zhao (2010), "Influence of flushing on performance of EDM with bunched electrode", International Journal of Advanced Manufacturing Technology, DOI 10.1007/s00170-011-3357-8

[8] Makanzi, M. M. and Ikua, B. W., (2012), "A review of flushing techniques used in electrical discharge machining", Proceedings of the 2012 Mechanical Engineering Conference on Sustainable research and innovation, Vol. 4, 3-4 May 2012, pp: 162-165.
[9] Chakraborty, S., Dey, V. and Ghosh, S. K., "A review on the use of dielectric fluids and their effects in electrical discharge machining characteristics", Precision Engineering, Vol. 40, 2015, pp: 1-6.
[10] Krishnakant Dhakar, Akshay Dvivedi and Amit Dhiman, "Experimental investigation on effects of dielectric mediums in near-dry electric discharge machining", Journal of Materials Science and Technology, Vol. 30, No. 5, 2016, pp: 2179-2185.
[11] Murahari Kolli and Adepu Kumar, Effect of Boron carbide powder mixed into dielectric fluid on electro-discharge machining of titanium alloy", Procedia: Material Science, Vol. 5, 2014, pp: 1957-1965.
[12] Murahari Kolli and Adepu Kumar, "Effect of dielectric fluid with surfactant and graphite powder on EDM of titanium alloy using Taguchi method", Engineering Science and Technology, an International Journal, Vol. 18, 2015, pp: 524-535
[13] Kuldeeep Singh, Gianender, Ajay Kumar Agarwal, Ajit and Rinki Yadav, "Effects of dielectric fluids used on EDM performance: A review", International; Journal of Engineering Technologies in Engineering research, Vol. 5, No. 10, 2017, pp: 10-16
[14] Sanghai, C. R. and Acharya, G. D., "Effect of various dielectric fluids on performance of EDM: a review", Trends in Mechanical Engineering & Technology 2016; 6(1): 55–71

Experimental Investigation of Automobile radiator using Tungsten trioxide Nano-fluid

K S Hanumanth Ramji[1], J Vinoth kumar[2], and A Amar Karthik[3]

[1,2] Assistant Professor, Department of Mechanical Engineering, Bannari Amman Institute of Technology-Erode, India.

[3] Professor, Department of Mechanical Engineering, Bannari Amman Institute of Technology-Erode, India.

hanumanthramjiks@bitsathy.ac.in, vinothkumar@bitsathy.ac.in , amarkarthik@bitsathy.ac.in

Abstract. Conventional fluids (ethylene glycol and H_2O) heat transfer performance results to be poor when incorporated with an automobile radiator. An experimental study was performed to explore the enactment of automobile radiator using H_2O and WO_3 nanofluid as operational fluids. Analysis and comparison were carried out for the heat transfer rate and outlet temperature of fluid for various flow rate ranging from $0.2m^3/hr$ to $0.5\ m^3/hr$.Preperation of nanofluid was done for water with 0.0167%, 0.034% and 0.0667% volume concentration of WO_3 nanoparticle. The inlet condition of fluid was set at 50℃ as the impact for the variation of temperature conditions on system performance were minimal. On experimentation, the maximum increase in the heat transfer rate was by 4.18% on comparison with base fluid.

Keywords: Nanofluid, Radiator, Heat transfer rate, WO_3, Performance

1. Introduction

Heat dissipation using conventional coolant fluid is employed in most of the engineering systems. At present nanofluids (base fluid dispersed with nanoparticle) outperforms significantly in most of the heat transfer systems. Enough precaution need to be carried out to prevent fouling, sedimentation and increase of pressure drop.To increase the performance of the heat transfer system needs either increase in size of the system or increase in performance of cooling fluid. Due to structural constraint it's good for increasing the performance of the coolant. Peyghambarzadeh et al. [1] investigated the radiator performance using Al_2O_3/Water nanofluid for rate of flow range 2-5 lit/min and coolant inlet temperature range 37-49℃. They concluded an increase of 45% in rate of heat transfer for 1% volume concentration nanofluid on comparison with water and minimal effect on performance for variation in inlet temperature of coolant. Sharma et al. [2] conducted experimental study on a radiator using Al/water nanofluid of 0.2% volume concentration and 0.3 volume concentration with inlet coolant fluid temperature range from 45℃-55℃. They resulted with a 23.563% increase in coefficient of heat transfer for volume concentration of 0.3%. Selvam et al.[3] summarized the enhancement of the Nusselt number for volume concentration of 0.5%. GnP/(water/EG) nanofluid by about 88% at 35℃ and 90% at 40℃. Oliveira et al. [4] studied the application of MWCNT (Multi-Walled Carbon Nanotube)/water of 0.05-0.16 weight.% in an automotive radiator for coolant inlet temperature range 50℃-80℃ and resulted in 17% decrement in heat transfer for 0.16 weight % nanofluid. Ali et al. [5] investigated the performance of a radiator using ZnO/Water nanofluid of 0.01%-0.3% volume concentration for inlet temperature 45℃-55℃ and concluded with an enhancement of 46.5% by using 0.2% volume concentration. nanofluid. Ali et al. [6] conducted experimental study on a radiator using 0.06, 0.09, &0.12 volume concentration MgO / water nanofluid and resulted with a 31% increase in the heat transfer rate for 0.12 volume concentration %.Naraki et al.

[7] studied the application of CuO/water nanofluid of 0.04-0.4 volume concentration % and resulted in 8% enhancement in overall heat transfer coefficient for using 0.4 volume concentration % nanofluid. Hussain et al.[8] investigated the performance of a radiator using SiO_2/Water nanofluid of 1%-2.5% volume concentration for inlet temperature 60℃-80℃ and the coolant flow rate of 2-8 lit/min. They concluded with an enhancement of 46% by using 2.5% volume concentration nanofluid. Ahmed et al.[9] conducted experimental study on a radiator using TiO_2/water nanofluid of 0.1% volume concentration, 0.2% volume concentration and 0.3 volume concentration with inlet coolant fluid temperature range from 20℃-80℃ and found the maximum overall heat transfer coefficient of 2050 W/m^{2}K 23.563% for 0.3% volume concentration Palaniapan et al. [10] studied the application of Flyash nanofluid of about 2% volume concentration in a radiator for Reynold number range 4000-8000 and resulted in 43.7% increase in heat transfer at Reynold number 8000 and 2% volume concentration nanofluid. From the literature there is no study for automobile radiator using tungsten trioxide nanofluid. Tungsten Trioxide nanoparticles are a category of semiconductors that are broadly used in photo thermal systems, based on the awareness of the authors, this type of particles have not been castoff yet in automobile radiator as the additive to working fluid. Hence, the current paper targets to analyse the result of using WO_3/water nanofluids at various concentrations on the performance of a radiator.

2. Heat transfer rate analysis:

Calculation of heat transfer rate in radiator is obtained by using Eq. (1) & Eq. (2)

$$Q = \dot{m}c_p\left(T_i - T_o\right) \quad (1)$$

$$Q = hA(T_b - T_w) \quad (2)$$

The nanofluid's specific heat capacity can be calculated by using Eq. (3) [11]

$$c_{p,nf} = \frac{(1-\varphi)\rho_{bf}c_{p,bf} + \varphi\rho_n c_{p,n}}{(1-\varphi)\rho_{bf} + \varphi\rho_n} \quad (3)$$

The nanofluid's density is calculated from the following Eq. (4) [11]

$$\rho_{nf} = (1-\varphi)\rho_{bf} + \varphi\rho_n \quad (4)$$

The nanofluid's Thermal conductivity is calculated from the following Eq. (5) [11]

$$k_{nf} = \frac{k_p + (n-1)\,k_{bf} - \varphi(n-1)(k_{bf} - k_p)}{k_p + (n-1)\,k_{bf} + \varphi(n-1)(k_{bf} - k_p)} \quad (5)$$

The nanofluid's viscosity is calculated using Eq. (6) [11]

$$\mu_{nf} = \mu_{bf}(1 + 2.5\,\varphi) \quad (6)$$

The Nusselt number for the entire radiator can be calculated using the Eq. (7) [11]

$$Nu_{exp} = \frac{h_{exp}\,x\,D_h}{k} \quad (7)$$

Volume fraction of the nano particle is calculated by using Eq.(8)

Density = mass / volume (8)

Initially, the rate of heat transfer in the radiator setup is analysed using H_20 as working fluid. As the properties of H_20 are standard, the specific capacity of heat, density, thermal conductivity and viscosity are taken from data books for corresponding working temperatures. Input parameters such as mass flow water(m),inlet fluid temperature(T_i) ,outlet fluid temperature(T_o) and bulk mean temperature(T_b) are measured using 4 RTD Pt 100 type temperature sensors with ±0.1 °C accuracy. On substituting these parameters in Eq.(1) we obtain the rate of heat transfer. Further the calculated rate of heat transfer along with the surface area(A),bulk mean temperature(T_b) and the surrounding temperature(T_w) are substituted in Eq.(2) to obtain heat transfer coefficient(h).For nanofluid in the radiator setup, the properties-specific heat capacity, density, thermal conductivity and viscosity are found out using Eq.(3),Eq.(4),Eq.(5) and Eq.(6).

3. Preparation of nanofluid and experimentation

3.1. WO_3 nanofluid synthesis

Without extra processing and of diagnostic grade, all chemicals substances were used as received. Supplier of the WO_3 nanoparticle: Nano Research labs, Jharkhand, India. The size of nanoparticle is of the range of 40-50 nm and 99.9% purity. The spherical shaped nanoparticles possess a lower value of thermal conductivity than the cylindrical shaped nanoparticles same way the thermal conductivity of smaller size elements is greater than the thermal conductivity of larger size particles[14]. Nanofluids can be prepared using 1 step and 2 step method. Due to ease in preparation, nanofluid is prepared using 2-step method .Initially, grinding of WO_3 nanoparticles are done by means of a mortar to prevent or minimize agglomeration. Later, the addition of WO_3 nanoparticles with distilled water by an appropriate mass fraction and the mixture is manually agitated for a minimal period of 10-min. The prepared mixture is placed in the ultrasonic bath (ANALAB, India) under sonication with a frequency of 30 kHz, 1000 watts power during a period of 4 hours[12]. Three different nanofluids of 0.0166%, 0.0334% and 0.06667% volume concentration is prepared. Finally, stability testing of nanofluid is carried out using Zeta potential analyzer.

3.2. Nanofluid stability

To avoid nanofluid sedimentation during the trials, by using ultrasonication the nano-WO_3 and water is mixed. Two methods have been proposed for uniformly distribute nanoparticles in nanofluids and to prevent sedimentation; using surfactants and changing nanofluids pH [15].For all temperatures in the experimental procedure ,the stability of nanofluid is found to be maintained.(Fig. 1).

Figure 1. Ultrasonicator

3.3. Experimentation

The experimental setup comprises an automobile radiator with a fan, a reservoir, connectors, rotameter, pump, thermometers, a heater, small fuses, tubes and expansion resource. 4 RTD Pt 100 type sensors(temperature) with ±0.1 °C accuracy offer tremendous exactness over an extensive range of temperature (from −200 to +850 °C) are used to quantify the inlet and outlet temperature of the fluid used as coolant and temperature of the radiator. The experiment was conducted in Government college of Engineering-Salem, India. The circulation of operational fluid into the radiator is done through pump (500W).On circulation, the working fluid enters the radiator initially and then it arrives a tank with 20 L capacity and continues the process. The constant temperature bath is utilized to cool the gain of heat obtained from the radiator. Flow rate adjustment (three way valve),a rotameter for the desired flow rate upto 4 l/min and a pressure indicator for water and nano fluid. The flow rate of the working fluid is limited to 0.5 m^3/hr above which it ceases to be in laminar flow regime[9]. Radiator specification , test condition for experimentation and properties of the nano particle WO_3 is shown in Table 1. and Table 2.The Schematic layout of experiment apparatus is showed in Fig. 2.

Table 1. Radiator specifications.

S.No	Parameter	Value (cm)
1	Radiator length	38.4
2	Radiator width	33
3	Radiator height	2.2
4	Width of the radiator tube	0.2
5	Fin length	1
6	Distance between adjacent fins	0.1
7	Fin thickness	0.008

Table 2. Experiment test condition

Parameter	Water based nanofluid
Nanoparticle concentration (volume %)	0-0.0667%
Nanoparticle	WO_3
Inlet temperature (K)	323
Volumetric flow rate (m^3/hr)	0.2 to 0.5
Reynold number	200-600

Table 3. Nanoparticle property.

S.No	Parameter	Value
1	Density	7160 kg/m^3
2	Thermal conductivity	1.63 W/m-K

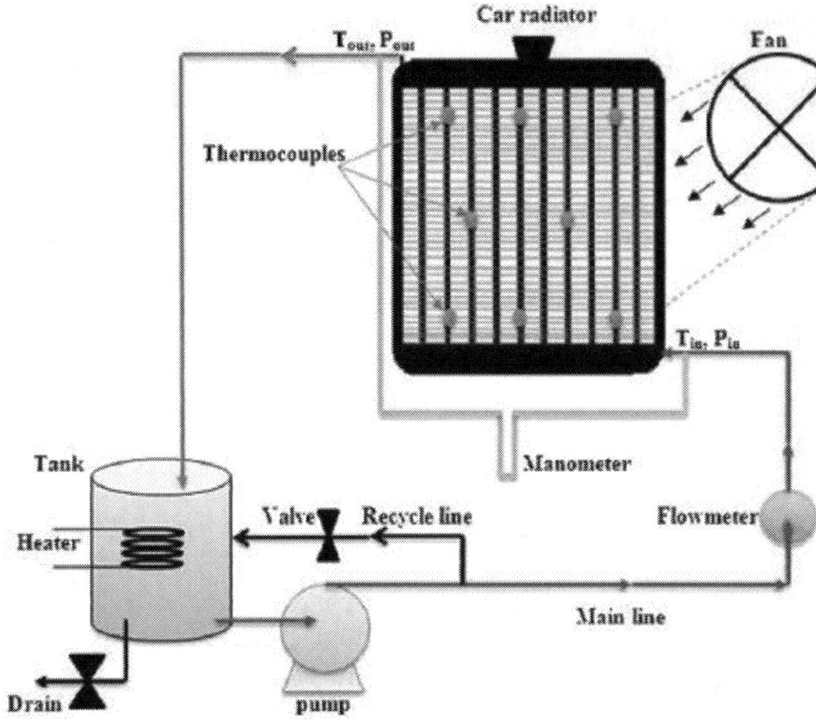

Figure 2. The Schematic layout of experiment apparatus

4. Results and Discussion

In this study, experimentation is carried out using WO_3 nanofluid with concentrations 0.0167, 0.0334 and 0.0667 volume concentration % for various rates of flow from 0.2 to 0.5 m^3/hr as working fluids. Volume concentration are limited to 0.0667% since further increase in volume concentration tends to increase pumping power for nanofluid drastically. However, may professionals settle that an prime enactment can be attained at a little nanoparticle volume element of less than 1% [13].Figure 3 it depicts an increase of the rate of heat transfer with the rise in rate of flow and nanoparticle concentrations. An increase in the coefficient of heat transfer is resulted at high flow rates due to mixing fluctuation produced by the effect of diffusion and the disordered movement of the nanoparticles. A surge in the heat transfer was found by a surge in volume concentration of nanoparticle addition with H_2O. The figure 2 shows outlet temperatures of radiator, T_{out}, as a utility of volume flow rate circulating in the radiator. It can be observed that accumulation nanoparticles to base fluid tends to reduce radiator outlet temperature. It can be concluded as for the equal mass flow rate in a heat transfer system tends to reduce fluid outlet temperature thus enabling improvement in thermal enactment of the system. Figure 2 also displays depreciation in the rate of heat transfer for rise in the volume flow rate passed through the system. The reason would be due to rise in the fluid velocity impacted by the rise in the volume flow rate. Thus, the fluid would have minimal duration for relating itself with the air circulated through fan and results in the rise of the exit temperature of coolant. It will be illustrious that all the statistics in Figure 2 were found for 54 C of radiator fluid inlet temperature.

IOP Conf. Series: Materials Science and Engineering **995** (2020) 012017 doi:10.1088/1757-899X/995/1/012017

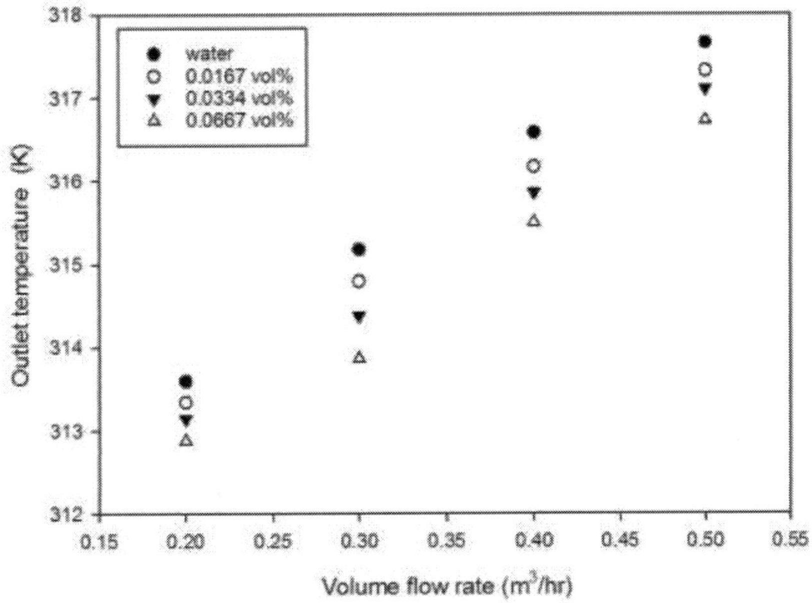

Figure 3. Outlet temperature of coolant fluid based on variation of volume flow rate

Figure 3 illustrates the heat transfer enhancement in a radiator is obtained due to the use nanofluids instead of base fluid. It can be perceived that the ratio of thermal energy transfer by nanofluid to base fluid increases with volume flow rate irrespective of nanofluid concentrations. In addition, the impact of volume flow rate is insignificant at higher concentration nanofluid. Also, Figure 4 displays that for an addition of 0.0667 volume concentration % WO_3 nanoparticle to water mixture, the heat transfer rate was found to increase by 4.18% on comparison with the heat transfer rate on using water. More over suggestions by the researcher on enhancement of heat transfer due to the significant factor Brownian motion is detailed.

Finally, the replacement of conventional cooling fluid by nanofluid in car radiators, would surge the rate of heat transfer. An increased rate of heat transfer is depicted the addition of nanoparticle with the bae fluid on using with the radiator Thus results the possibility of a reduction in radiator size as well as reduced weight. So an improved fuel economy and reduced emission to an environment would be achieved.

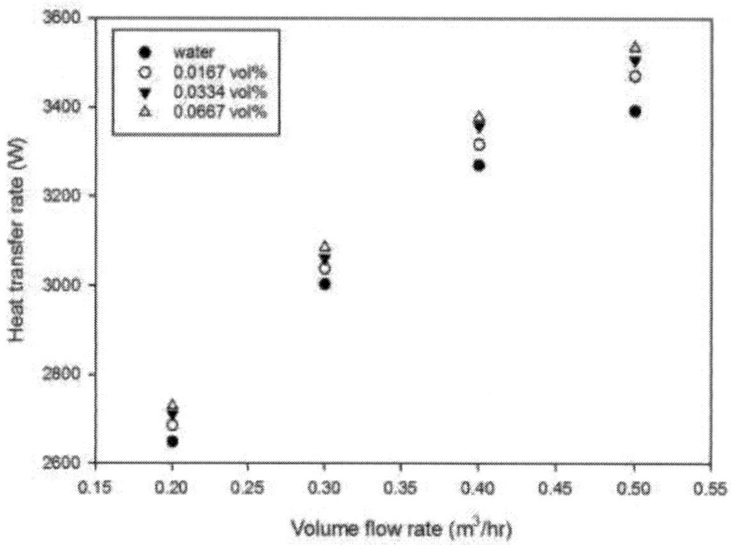

Figure 4. Heat transfer rate of coolant fluid based on variation of volume flow rate

5. Conclusion:

In this study, the impact of WO_3 /Water nanofluids in a radiator was examined. On evaluation with water, the rates of heat transfer enhanced significantly for the usage of (WO_3 /Water) nanofluid on evaluation to the water. In the preeminent condition, the maximum improvement in the rate of heat transfer was obtained to be 4.18% for 0.067 volume concentration % of (WO_3 /Water) fluid. The investigational outcomes exhibited that heat transfer rate enhanced with the rise in flow rate of (WO_3 /Water) fluid. Also, the surge in volume concentration amplified the heat transfer rate. The outcomes specifies that utilizing (WO_3 /Water) fluid can surge the rate of heat transfer. Hence, providing hopeful means for scientists to improve extremely compact and effective automotive radiators. The weight reduction of radiator in turn reduces the consumption of fuel . Moreover, it can offer effective heat transfer and maximized the lifespan of sections of a vehicle.

Acknowledgment

This research was supported by the mechanical engineering department that was provided Government college of engineering, Salem.

References

[1] Peyghambarzadeh S M, Hashemabadi S H, Jamnani M S and Hoseini S M, Improving the cooling performance of automobile radiator with Al_2O_3/water nanofluid, Appl. Therm. Eng. (2011) 1833–1838.

[2] Sharma S, Fabricating an experimental setup to investigate the performance of an au- tomobile car radiator by using aluminum / water nanofluid, J. Therm. Anal. Calorim.9 (2018).

[3] Selvam C, Lal D M and Harish S, Enhanced heat transfer performance of an automobile radiator with graphene based suspensions, Appl. Therm. Eng. 123 (2017) 50–60.

[4] Oliveira G A, Cardenas Contreras E M and Bandarra Filho E P, Experimental study on the heat transfer of MWCNT/water nanofluid flowing in a car radiator, Appl. Therm. Eng. 111 (2017) 1450–1456.

[5] Ali H M, Ali H, Liaquat H, Bin Maqsood H T and Nadir M A, Experimental investiga- tion of convective heat transfer augmentation for car radiator using ZnO-water nanoflu- ids, Energy. 84 (2015) 317–324.

[6] Ali H, Azhar M, Saleem M, Saeed Q and Saieed A, Heat transfer enhancement of car radiator using aqua based magnesium oxide nanofluids, Therm. Sci. 19 (2015) 2039– 2048.

[7] Naraki M., Peyghambarzadeh S M, Hashemabadi S H and Vermahmoudi Y, Parametric study of overall heat transfer coefficient of CuO/water nanofluids in a car radiator, Int. J. Therm. Sci. 66 (2013) 82–90.

[8] Hussein A M, Bakar R A and Kadirgama K, Case Studies in Thermal Engineering Study of forced convection nanofluid heat transfer in the automotive cooling system, Case Stud. Therm. Eng. 2 (2014) 50–61.

[9] Ahmed S A, Ozkaymak M, Sözen A, Menlik T and Fahed A, Improving car radiator performance by using TiO2-water nanofluid, Eng. Sci. Technol. an Int. J. (2018) 996– 1005.

[10] Palaniappan B and Ramasamy V, Thermodynamic analysis of fly ash nanofluid for auto- mobile (heavy vehicle) radiators, J. Therm. Anal. Calorim. (2018).

[11] Devireddy Sandhya, Mekala Chandra Sekhara Reddy and Veeredhi Vasudeva Rao, Improving the cooling performance of automobile radiator with ethylene glycol water based TiO_2 nanofluids

[12] Minea A A, Advances in New Heat Transfer Fluids: From Numerical to Experimental Techniques, CRC Press, 2017.

[13] Sidik N A C,.Yazid M N A W M, Mamat R., Recent advancement of nanofluids in engine cooling system, Renewable Sustainable Energy Rev. 75 (2017) 137–144.

[14] Duangthongsuk W and Wongwises S, Measurement of temperature-dependent thermal conductivity and viscosity of TiO_2-water nanofluids, Exp. Therm Fluid Sci. 33 (4) (2009) 706–714.

[15] Sözen et al. A, Utilization of fly ash nanofluids in two-phase closed thermosyphon for enhancing heat transfer, Exp. Heat Transfer 29 (3) (2016) 337–354.

IOP Conf. Series: Materials Science and Engineering **995** (2020) 012018 doi:10.1088/1757-899X/995/1/012018

Development of a system to assist in library to automate the book handling process

V Vadivel Vivek[1], E L Pradeesh[2], N Natarajan[3], M C Pravin[4], S Raghavendra Prabhu[5]

[1,2,4,5] Assistant Professor, Bannari Amman Institute of Technology, Department of Mechatronics, Sathyamangalam, Erode, Tamilnadu, India.

[3] Professor, Excel Engineering College, Namakkal, Salem, Tamilnadu, India.

vadivel.vivek@gmail.com

Abstract: The library is the place where one can gain not only knowledge, but also many important aspects of life. The more and more we learn, the more and more we become wise. But nowadays, the books we used to read in library are not returned to their place by us. Instead we drop or return the book in one place. The library staffs are the person, who handles the dropped books and stack it in their allotted place. It is a long process when it comes to multiple floors and multiple books and it is a more time consuming also. This work is aimed at automating the book return process by creating a new mechanism. The main objective of this paper is to return book to its place without any human involvement and to reduce the book handling time and reduce repeated to and fro motion. In order to automate this process a control system is required. The control system is implemented through some physical device; it is usually based on mathematical logic to get desired counts. Arduino mega is used to implement the required control for this project with the help of RFID technology. CAD design is made and the required robot for automation is fabricated. Hence by developing an autonomous robot the time required for handling the book is reduced and the book is handled safely. The material handling cost is also reduced with less man power.

Keywords: RFID, library, automation, material handling, less man power

1. Introduction

Sree Lakshmi et al used radio frequency identification (RFID) to automate the book handling process in the library. They have used passive RFID tags that are capable of electronically storing the information and that can be read with the help of RFID reader. By using this system, books can be issued and returned via RFID tags and the fine associated with the return of books after the due date can also be calculated using the database available with the library [1]. Smarat gosh et al developed a robotic arm working with the help of microcontroller for library management system. They have used Arduino micro controller and python programming micro controller for robotic arm to automatically place the books from the return section to their corresponding place in the rack [2]. Dhanalakshmi et al used RFID in library management system for fast transaction and for easy traceability of books. They have employed RFID readers supported with antennas at the entry of the library and at transaction systems, tha are capable of electronically storing the information and that can be read without any physical contact [3]. Yogesh Angal et al developed a robot which is capable of retrieving the book from the stored database. They have employed global positioning system for finding the location of the book. They have witnessed less man power under the supervision of library staff [4]. Sathishkumar et al used RFID and LED to identify the book. The books are identified using RFID tag and RFID, and the LED in the robot will blink if the searched book is identified [5]. Grover Priyanka et al used RFID for books transaction, their system is designed in such a way that the books are issued and returned using RFID tags, which is also capable of calculating the fine for the books returned after the due date[6]. Bansode et al implemented RFID system in Pune university library and proved that the system is a boon for the library and the time consumed for the book handing and damage of books is also reduced [7]. Faizul Nisha experimented RFID technology at Defense Science Library and found that a bouquet of benefits for libraraians, the staff and the users of the library because of automation and less time consumption [8]. Hasan nabi evaluated the performance of RFID technology in libraries by comparison with barcodes and electro magnetic strips and addressed the various challenges and issues associated with the implementation of RFID system in libraries and gave some guidelines to implement RFID system for the libraries [9]. Mayank Trivedi et al conducted a study on success of RFID technology in Indian academic libraries and they described how the RFID technology is implemented and also compared RFID with barcode[10]. Radha made a review on employment of RFID at Indian academic libraries and described some of the best practices in terms of privacy policy, collection limitation, consumer transparency and accountability [11].

It is difficult for humans to handle more number of books in a short time. Hence a robot is developed a robot to do this job. This type of robot will provide the benefits such as repeatability, easy to operate and maintain, faster material handling and low operating cost. In order to automate a process, a control system is required. The control system is implemented through some physical device; it is usually based on mathematical logic to get desired counts. RFID with ARDUINO MEGA is used to implement the required control.

The concern with book handling in library is that, they are stacked in one bin and transferred to its respective floor and sorted and arranged in the book shelf by a person in library. During this process the book may get damaged, due to poor handling or over stacking. It leads to torn pages, separated pages, sections and many more. And also, though it seems quite simple to find the book location, we couldn't find the book in its place, because it is not re stacked in its allotted space. So, we have to find the person-in-charge of that section and find its temporary location and borrow the book. In order to reduce these many small and complicated problems, we have developed a book handling robot for library. The main objective of this work is to return the book to its place without any human involvement, reduce the book handling time and to reduce the repeated to and fro motion.

RFID technology is employed in many places because of its advantages. Nowadays libraries across the world are also started using RFID for automatic check in and checkout of books and this technology is replacing the barcode technology in a steady phase. One of the main reasons for using RFID is that, it need not be visible, it can be embedded in any place and it has more details such as, position of the book, author, accession number, etc., but barcode is only an identification number. This work describes the study of employing RFID technology with a specially designed autonomous vehicle which is used for automating the book handling process.

2. Components

To automate the process, various electronic components are used. They are used to monitor the position of the robot regarding the block position and to generate and control the movement of the robot according to the requirement. The DC motor acts as the prime mover in this work which moves the robot along the respective axis. The actuation of motors is triggered and limited by the IR Sensors, that gets triggered when get stuck by the blocks on the either sides of the robot. These IR Sensor triggers are fed to the controller that decides which axis motors to be actuated or stopped at correct position using motor driver.

3. Working Methodology

The proposed system is made by assembling the required components and the CCAD model of the required autonomous vehicle is built and then it is fabricated considering the load requirements. The step by step procedure followed to make the required system to automate the book handling process is explained in detail in the following sections.

3.1. Connections

IOP Conf. Series: Materials Science and Engineering **995** (2020) 012018 doi:10.1088/1757-899X/995/1/012018

Connections are made as shown in the figure 1 below. The Arduino Mega is connected with IR Array and RFID Reader using bread board connection. The Motor is controlled by motor driver. As the RFID Reader reads any input data from RFID tag it soon detects its position and starts navigating via a path set for it. The IR Array finds any obstacle on either side, if any found it soon reroutes itself into the obstacle less path. For this the power supply is given via a battery, attached to it.

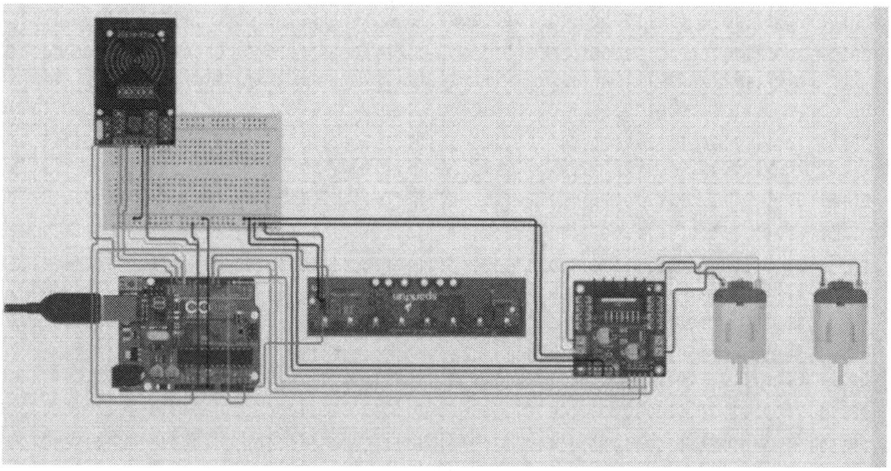

Figure 1. Connections-Line Diagram

3.2. Block Diagram

The block diagram to automate the process is shown in the figure 2 below.

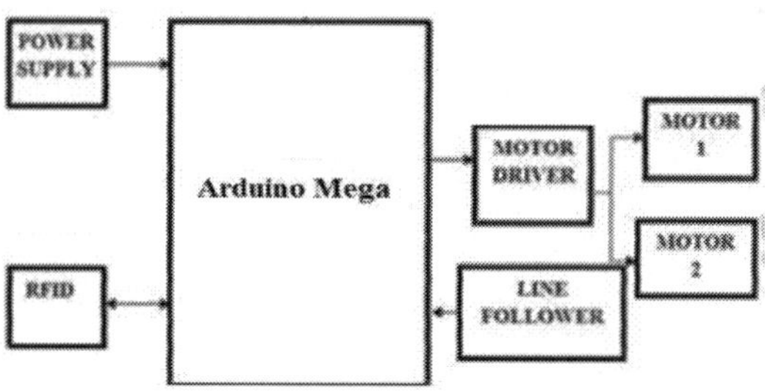

Figure 2. Block diagram of the proposed system

3.3. Working

The Line follower robot, first detects the book placed in the box via RFID. Then finds the location of the book and follows its lane via IR Array. The path is preloaded into the line follower hence the robot doesn't go to any other lanes. After reaching its location, it navigates to its destination and halts there. For now, the loading and unloading of book is done manually.

Initially the Arduino is programmed with the respected software. Then the jumper wires are connected as per the above diagram. The motor drive is used to connect the Arduino to motor. Because the motor wants high power but Arduino works with 5V power supply so the motor is connected to driver circuit.IR array is used for follow the path. Mainly the array is used for more accuracy. All are combined together to make a bot.

4. Developed Model of Library Assist Robot

Design calculations were made and a suitable model of desired robot is designed in such a way that it is capable of handling the book at its best, the figure 3 shows the CAD model of the developed system for automating the process and the figure 4 shows the actual developed system for library automation.

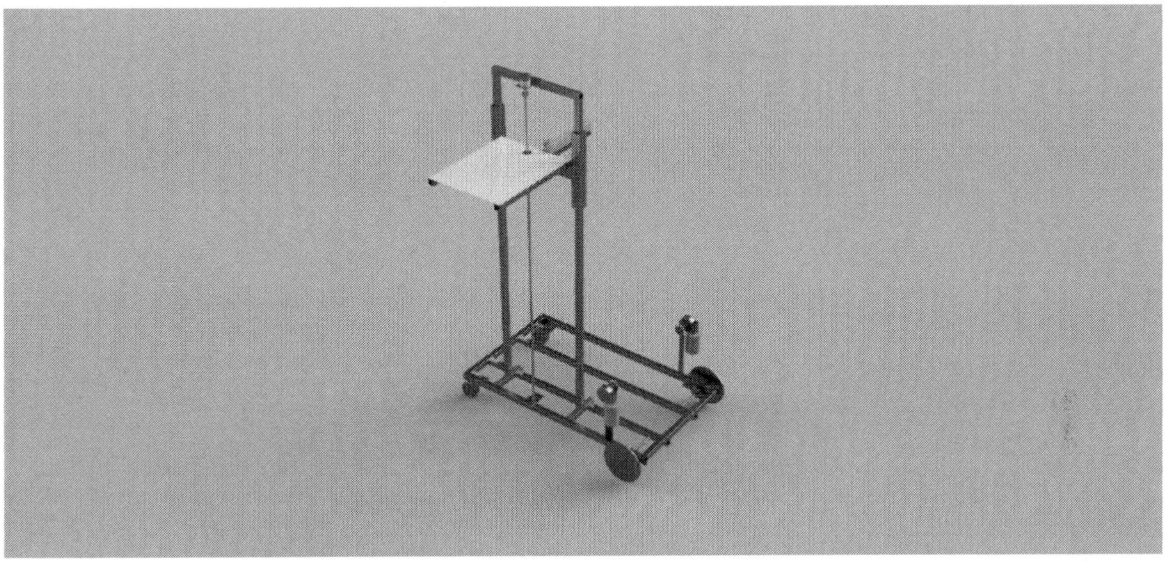

Figure 3. CAD model of library assist robot

Figure 4. Developed library assist robot

5. Experimental findings

After implementing the developed system in the library it is found that the book handling time is reduced to an extent and the books are exactly placed in its position. It is also evidenced that the damaging of books is also reduced. The figure 5 compares the time taken for handling the books before and after implementing the system, from the graph it is evident that the time taken for stacking the books after implementing the system is reduced to an extent. The figure 6 compares the number of damaged books before and after implementing the system, from the graph it is clear that the number of books damaged is reduced after implementing the system in library.

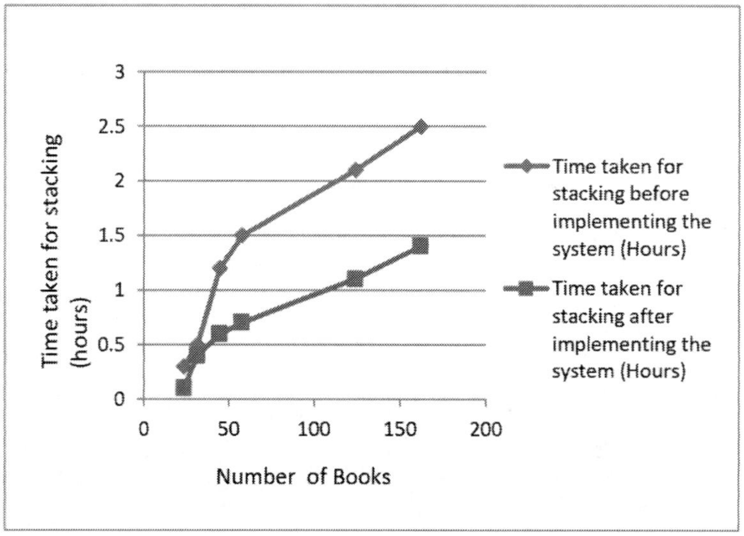

Figure 5. Time taken for stacking the books

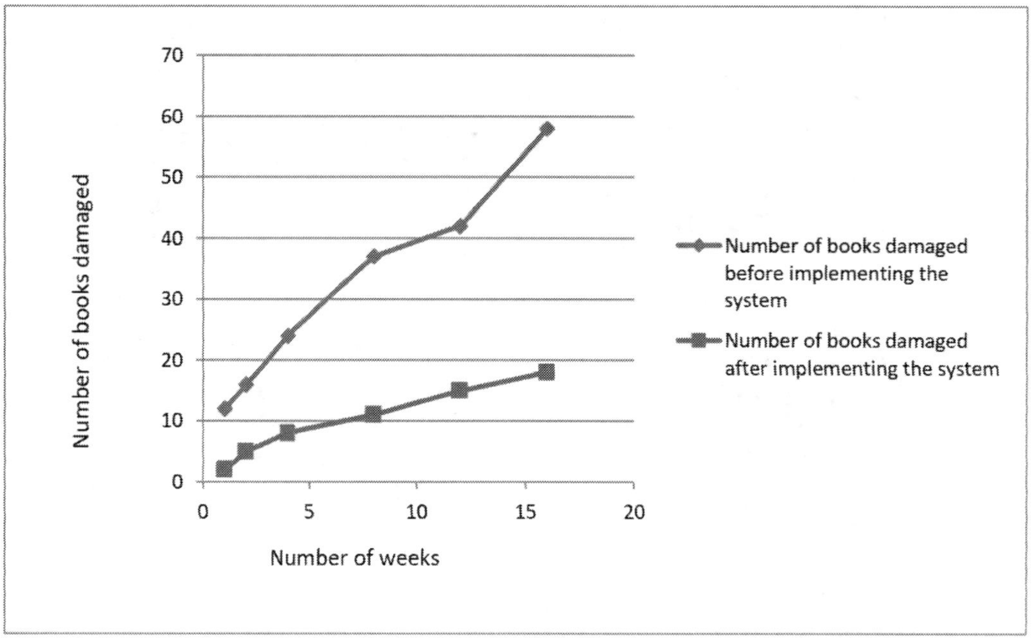

Figure 6. Number of books damaged.

6. Advantages

The main advantages because of implementing the system in the library are requirement of number of person to transfer the books can be minimized to an extent and they may be utilized for some other work. The human involvement in handling the books is reduced and the man power is utilized for some other task where it is unavoidable. There will be more accuracy and effectiveness in stacking the book exactly to its desired position and the damage of books is also reduced enormously. The time taken for stacking the books is also reduced and the valuable time of the readers is also saved as they can locate the desired book easier with minimum time as it is staked exactly in its predetermined position. The cost associated with the maintenance of book is also reduced because of less damage.

7. Conclusion

The RFID technology is one of the important user friendly technologies implemented in most of the libraries because of its effectiveness and efficiency. It is proved to better than barcode system. Even though the initial investment in implementing RFID is looking higher, after break even period it is the best method for cost efficiency and it is even suitable for medium and small libraries. By implementing this technique one can experience more savings and customer satisfaction. This technology benefitted a lot the library staff and users in provision and service. The increasing volume of books is difficult to handle with man power and it has become a tedious process for stock taking and tracking of books. Lack of skilled person for book handling is also one of the major concerns. This leads to the development of this novel technique using RFID. It automates the book handling process and saves the time of both the staff and reader. Stacking of books become easier and more number of books can be stacked in less time with minimized man power. It also prevents the misplacing and theft of books. At initial stage the cost of implementation may seems higher but it is proved to be one of the economic technology for over years. Because of more experimenting in deploying RFID in libraries for automation, this technology in near future will become affordable for all type of libraries.

8. Future Scope

In future real time robots which consist of obstacle avoidance, automatic floor crossing, searching for a particular book, stock taking with advanced image processing and shortest path finding may be developed. By fully automating the book handling inside the library, the time taken for issue and return of books may also be reduced and the missing books can be found easier and it will also prevent the theft of books.

Acknowledgment
This research was supported by the Mechatronics department, Bannari Amman Institute of Technology, Sathyamangalam.

References

[1] Sree Lakshmi Addepalli, Sree Gowri Addepalli 2014 Library Management System Using RFID Technology International Journal of Computer Science and Information Technologies, Vol. 5 (6), 6932-6935

[2] Samrat gosh, Biswarup Ganguly 2018 Robotic Arm Development for Library Management System 2nd International Conference on Computational Advancement in Communication circuit and System (ICCACCS-2018)

[3] Dhanalakshmi M and Uppala Mamatha 2009 RFID based library management system Proceedings of ASCNT 227-234

[4] Yogesh Angal, Anita Gade 2017 Development of library management robotic system, International Conference on Database Management System

[5] B.R.Sathishkumar, M.Krishnaprabha, S.Priya, M.Ragavi July 2010 Automated Library System using Android Based Robot International Journal of Recent Technology and Engineering (IJRTE) Volume-8 Issue-

[6] Priyanka Grover, AnshulAhuja July 2010 RFID based Library Management System, International Journal of Computer science and application, Vol.1, No.1

[7] Bansode, S.Y. and Desale, S.K. 2009 Implementation of RFID technology in University of Pune Library. Program: electronic library and information systems, 43(2), 202-214.

[8] Faizul Nisha 2018 Implementation of RFID Technology at Defence Science Library, DESIDOC : A Case Study 27 DESIDOC Journal of Library & Information Technology, Vol. 38, No. 1, pp. 27-33

[9] Hasan, Nabi 2014 Roadmap for RFID Implementation in Libraries: Issues and challenges. Int. J. Info. Libr. Soc., 3(1), 65-71

[10] Mayank Trivedi, Radha L June 2014 Success of RFID Technology in Indian academic libraries - A study International Journal of Information Management 3(8):51

[11] Radha Alias Nagalakshmi February 2011 Deployment of RFID (Radio Frequency Identification) at Indian academic libraries: Issues and best practice L. International Journal of Library and Information Science Vol. 3(2), pp. 34-37

Kinematics and Plane Decomposition Algorithm for Non Linear Path Planning Navigation and Tracking of Unmanned Aerial Vehicles

L Arulmurugan[1], S Raghavendra Prabhu[2], M Ilangkumaran[3], V Suresh[4], R Saravanakumar[5] and M Raghunath[6]

[1,5] Assistant Professor, Bannari Amman Institute of Technology, Department of Electronics and Communication Engineering, Sathyamangalam, Tamilnadu, India.

[2,6] Assistant Professor, Bannari Amman Institute of Technology, Department of Mechatronics Engineering, Sathyamangalam, Tamilnadu, India.

[3] Professor, K S Rangasamy College of Technology, Department of Mechatronics Engineering, Tiruchengode, Tamilnadu, India.

[4] Assistant Professor, Bannari Amman Institute of Technology, Department of Electrical and Electronics Engineering, Sathyamangalam, Tamilnadu, India.

arulmurugan@bitsathy.ac.in, maniilankumaran@gmail.com, sureshv@bitsathy.ac.in, raghavendraprabhu@bitsathy.ac.in, saravanakumar@bitsathy.ac.in, raghunath@bitsathy.ac.in

Abstract. Online planned course of the flight and obstacle avoidance are the major issues of exploitation faced in the real time implementation of Unmanned Aerial Vehicles (UAVs). The main application that are used in the UAV are path planning navigation and tracking. Our project uses two algorithms to derive the navigation control laws. The algorithms such as kinematics based algorithm and plane decomposition algorithm. The control laws rely on the two factors namely, proportionality and derivation factors. These factors guide the UAV in reaching the goal in a nonlinear path successfully. The surveillance mode and obstacle avoidance mode are taken as the important consideration in our project. In the surveillance mode UAV reaches the goal in the three dimensional space with the aid of navigation laws. In obstacle avoidance function, the obstacles are avoided in the course of its path of the flight and attains the destination with the navigation laws' assistance. The kinematics rules leads to the successful application of tracking and navigation of UAV in nonlinear path planning.

Keywords: Unmanned Aerial Vehicles (UAV), Navigation laws, Kinematics laws, obstacle avoidance, path planning.

1 Introduction

Unmanned aerial vehicles (UAV) is a particular group of aircrafts which can fly in the absence of the pilots on the aircraft. Unmanned Aerial Vehicles more normally acknowledged by their acronym UAV are essentially mobile robots used in aerial environment. The name "Unmanned Aerial Vehicles" suggests that these robots are operated by preprogrammed algorithms and therefore require very little human help at some point of operation. Nowadays the use of small UAV vehicles have increased, so is the impulse of applying them as a medium for photography and filming is gradually increasing. Unmanned plane vehicles include sensor payloads, the aircraft aspect and ground control station. They are constrained through the means of onboard digital device or through manage gadget from the floor. When it is controlled distantly from floor, its miles are referred to as RPV (Remotely Piloted Vehicle), it mandatorily needs dependable wireless media to manage. Allocated manipulate systems may be used for the discussion of massive UAVs, and may be set up aboard vehicle or also in trailers to enable the nearby closeness to UAVs which are restricted via variety or conversation abilities. One place of hobby is the potential to apply those cars as a medium for filming unique UAVs which can be in aerial navigation. Woefully, one of the maximum troubling component of this particular system is the venture of visually pursuing an aim in the live circulate from the cameras installed. Nowadays, various commercialized entities use a cell phone as their medium to show the stay circulation which sums up to the venture of locating the favored goal inside the video encase. The algorithms are translated to a smart code that's able to direct the UAV dynamically to follow a given desired course to acquire a unique purposewithvarioustiersofcourseselectionasnicelyattainingtheobstacle.

The most common movements using UAVs involve in achieving a table bound purpose, achieving a moving the purpose or following a shifting purpose at steady distance. The most common strategies used for the motive encompass imaginative and prescient primarily based sensors, audible sensors, GPS navigation machine, GPS/INS integration and included sensor gadget amongst others. Each of those techniques constitute an answer particular to the undertaking assigned to the UAV in a selected state of affairs. The two cases mentioned on this undertaking are 1) UAV following a shifting purpose in a 3 dimensional space with the objective of attaining the purpose the use of the navigation law offered within the task. 2) UAV following a moving aim in a three dimensional space inside the presence of obstacles inside the direction. UAVs require estimation when navigating in both indoor and outdoor environments. The opportunity of crashing in that aerial environment is excessive, as spaces are constrained, with many moving boundaries. Another essential aspect discussed in this undertaking deals with the idea of plane decomposition wherein a three dimensional area is split into two dimensional planes for the motive of simplifying the procedure of flight direction navigation. The derivation of the navigation law as well as a complete terminology used there in is discussed in detail so as to explain the whole procedure of UAV navigation and flight course manage. The "Obstacle Avoidance Mode" for the UAV is mentioned in detail that allows you to explain the capacity of the UAV to manoeuvre across the limitations present inside the path and effectively attain the intention the usage of an alternate path with the aid of deviating toward waypoints. This is achieved successfully by modifying the navigation parameters (proportionality and deviation factors) in the dimensional planes. MATLAB is used as a tool of choice for programming the processes and simulating it.

1.1 LITERATURE SURVEY

A cost-efficient remedy for unmanned aerial vehicle voyage in a worldwide positioning system-refused environment proposed by Shahrukh Ashraf.et.al [1] used optical flow–based set of rules with Kalman filters to offer the voyage during worldwide positioning system interruptions and worldwide positioning system environments in 10th July, 2017. This thesis proposes intelligent processes to offer the navigation potential during long GPS blackouts and through GPS-refused environments employing optical flow and inertial sensors. A Modern Approach to an Unmanned Vehicle Navigation by Artur Janowski.et.al [2] proposed, developed and tested methods for image data acquired by a single non-metric digital camera placed on mini UAV in 2016. The paper offers the approach of solving a UAV position, primarily based only on photograph statistics, as well as technique of UAV navigation supported via evolved approach and the main benefits and limitations. The method is based on photogrammetric and a modern computer vision process. An Algorithm for Autonomous Aerial Navigation Using MATLAB Mapping Tool Box by Mansoor Ahsan et.al [3] proposed aircraft elevation and leading controllers and an coherent set of rules for self-governing navigation by employing mat lab over specified areas in June 2012.

The linear administrators are assessed on non-linear aircraft model and navigation function is evolved for autonomous flight of UAV. Tracing of Compact Unmanned Aerial Vehicles by Steven Krukowski Aeronautics and Astronautics Stanford University, Stanford in 2012. In this dissertation, a photo processing subroutine is used in visually stalking compact UAVs .Matthew B. Rhudy [4] proposed Unmanned Aerial Vehicle Navigation Using Wide-Field Optical Flow and Inertial Sensors in 16th Jan,2015. In this paper, a full state construction including velocity and attitude is capable enough to approximate the aircraft ground velocity to within 1.3 m/s of a GPS recipient remedy and a disentangled construction which presumes that the slanting and vertical velocity of the aircraft are trivial and capable in managing the attitude angles within 1.4 degrees standard divergence of error for twain arrays of flight report. Alexander Cerón[5] preferred the project Vision-Based Path Finding Strategy of Unmanned Aerial Vehicles for Electrical Infrastructure Purpose in 27th Sep,2018. Here, the scrutiny is carried out by employing images encapsulated with an unmanned aerial vehicle.

The observing of power lines can be undertaken by making use of submissive sensors like cameras or active sensors namely, light detection and ranging (LIDAR) cameras, image processing techniques, and computer vision and in addition, control systems can also be availed. Hamid Teimoori Sangani[6] made a thesis on Topics In Navigation And Guidance Of Wheeled Robots in 13th July, 2009. The main theme of this thesis is to resolve the difficulty of real-time navigation and keeping track of Wheeled Mobile Robots (WMR) towards a target using range-only measurements. Christoforos Kanellakis·George Nikolakopoulos [7] Survey on Computer Vision for UAVs: Current Developments and Trends 28th April, 2016. Unmanned Aerial Vehicle was proposed by Luis Ramos [8] in 21st Nov, 2013. The main objective of the project is to create and operate UAV for target reconnaissance, navigation, taking off and landing operations. Modelling and Control of MM-UAV: Mobile Manipulating Unmanned Aerial Vehicles by MatkoOrsag [9] in 2013. This paper, however, explores the fundamental challenges in administering a mobile manipulating UAV via a commercially accessible aircraft and a light-weight prototype 3-arm manipulator. Evaluating the accuracy of vehicle tracing details acquired from Unmanned Aerial Vehicles by Giuseppe Guido [10] in Oct, 2016. A video rectification system for vehicles trajectory accession is instigated. The procedure is based on OpenCV libraries.

Yuanwei Wu [11] proposed Vision-based Real-Time Aerial Object Localization and Tracking for UAV Sensing System in 19th Mar, 2017. The propounded nudge does not need any physical or labouring commencement for tracing and also succeeds chasing execution on a great number of image concatenations. Kalyani Chopade[12] proposed A Survey: Unmanned Aerial Vehicle for Road Detection and Tracking in Nov,2015. This is for improving the performance of transportation systems .P.OmPrakasp[13] submitted a work on Navigation of Unmanned Aerial Vehicles in GPS Denied Region Using Vision Based Obstacle Detection in 2017.Here Sobels edge detection algorithm is used to discover

IOP Conf. Series: Materials Science and Engineering **995** (2020) 012019 doi:10.1088/1757-899X/995/1/012019

the dimension of the barrier of the specific encase of the video. Jay P. Wilhelm [14] published a paper on Vector Field UAV Guidance for Path Following an Obstacle Avoidance with Minimal Deviation in Aug, 2019.RuiyongZhai [15] submitted a paper Control and navigation system for a fixed-wing unmanned aerial vehicle in Nov, 2013. The entity is devised based on the inner loop and outer looptactic. Wang et al. [16] proposed the article Vision-Aided Inertial Navigation for compact Unmanned Aerial Vehicles in GPS-Denied Environments in 10th Aug, 2012. This practice hypothesize that the approximate error state is consistent and unchangeable during and the second half of one visual calculation phase.

2 EXISTING METHOD

2.1 Types of sensors in UAV navigation

The navigation law helps in planning the course of the UAV successfully so that it attains the preferred mobilising target. The numerous modes availed for aerial trailing and navigation can be sorted extensively into classes of: 1) Sensor based totally navigation method and 2) Model based totally navigation approaches. Numerous sensors are typically utilized for the cause of UAV path drafting and navigation. The typically utilized kinds of sensors consist of vision based, inertial and sonar sensors. This chapter sheds light on a number of the most commonly used techniques for UAV course planning and navigation which include imaginative and prescient based sensors, inertial sensors, included sensor systems, hierarchical control and feedback linearization.

2.1.1 Vision sensors for UAV navigation

The UAV causes an easy trouble of horizon detection that suggests a simple instance that shows the capability of imaginative and prescient primarily based sensors. The UAV is used to differentiate among the ground and the horizon the use of visible sensors. Even though this project sounds easy, it employs surprisingly superior strategies along with rudimentary picture segmentation, real-time optimization, standard multilevel altering and comments stabilization. The information of manage idea based totally concepts used in figuring out the characteristic of vision sensors are beyond the scope of this assignment, the instance serves the purpose of illustrating the advantages of imaginative and prescient based sensors utilized in UAV path making plans and navigation.

2.1.2 Inertial sensors for UAV navigation

Inertial Navigation Systems (INS) is the maximum not unusual styles of inertial sensors utilized in aerial navigation. Comparing the welfare of inertial sensors with different sensors, these sensors have a tendency to perform uniquely on the presence of outside factors which include wind course, wind pressure, interference and many others. A common place shape of INS utilized in UAVs is described within the following determine:

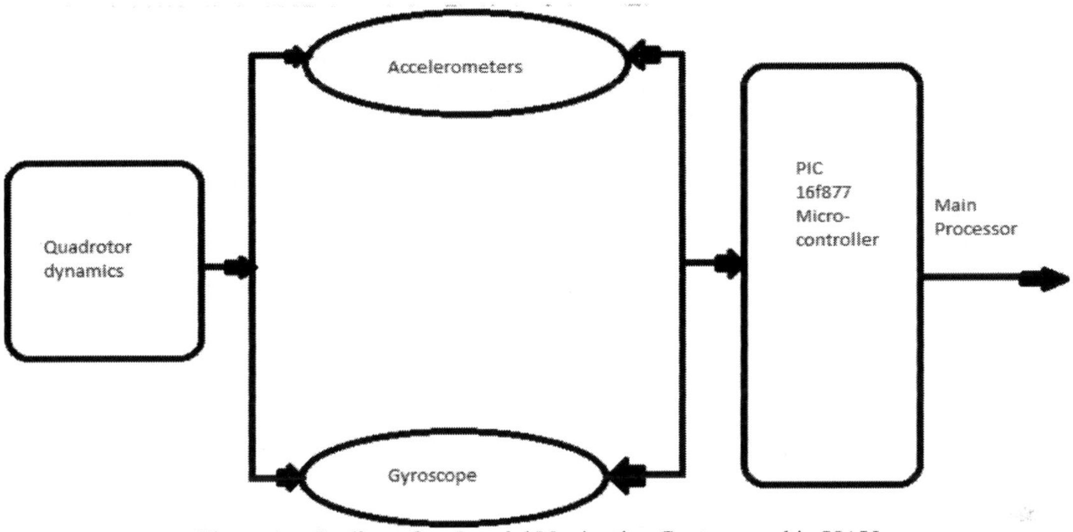

Figure 1 – Outline of an Inertial Navigation System used in UAV

As shown in the Figure 1, the INS unit is made up of normally to be had bodily additives together with quad rotors, accelerometers, gyroscopes and microcontrollers. The INS is attached to the primary processor, which gives the main processor with sensing records. However, the main disadvantage of Inertial Sensors or INS/IMU is that the placement mistakes obtained for the duration of UAV tracking compound through the years which yields faulty outcomes.

2.1.3 Integrated sensor systems for UAV navigation

In order to recoup for some of the varied misconceptions experienced by a single sort of sensors engaged in UAV navigation, dissimilar types of sensors are combined or else coalesced into a single entity to produce optimal outcomes. Several types of coalescence prevail between dissimilar types of sensors like vision/acoustic and vision/inertial.

2.1.4 Hierarchical methods for UAV navigation

One of the most common model-based approaches used for UAV navigation is the hierarchical control method in which a control system using a hierarchical organizational structure is used for path planning. A simple hierarchical control system structure is displayed in the Figure 2:

IOP Conf. Series: Materials Science and Engineering **995** (2020) 012019 doi:10.1088/1757-899X/995/1/012019

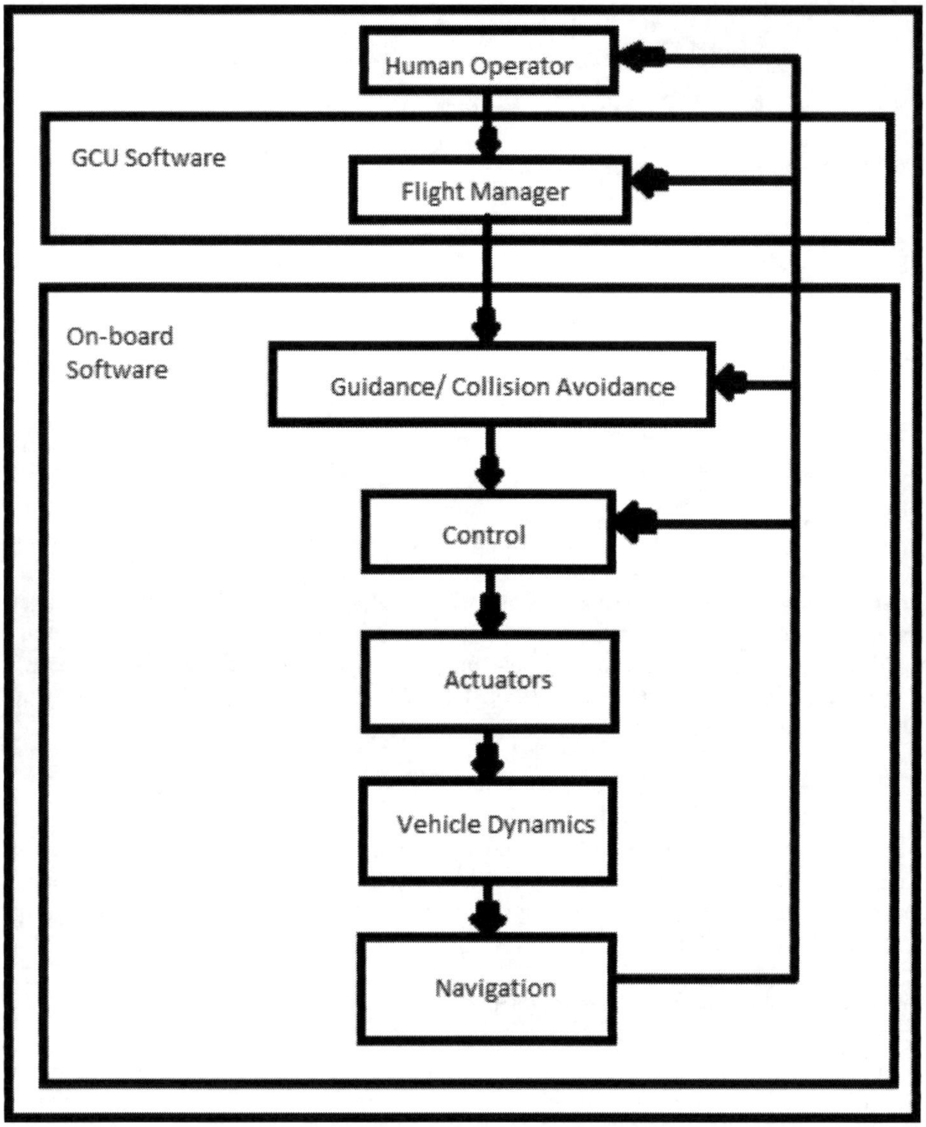

Figure 2 Hierarchical control architecture

The hierarchical architecture that integrates the Ground Control Unit (GCU) along with the on board software which involve in UAV path planning and navigation. In the above figure both blocks are linked via a human operator. In onboard software many operations such as actual path control and navigation, path estimation, collision detection and avoidance are integrated which are helpful in navigation process. Using this architecture both optimal path planning and navigation solution for UAV is employed.

IOP Conf. Series: Materials Science and Engineering **995** (2020) 012019 doi:10.1088/1757-899X/995/1/012019

2.1.5 Feedback linearization for UAV navigation

Feedback linearization is a common approach used in controlling nonlinear system. In addressing the UAV environment a nonlinear dynamic motion employed the concept of dual controller. For example, feedback linearization is applied within an inner controller and also with linear outer controller to achieve robust navigation capabilities. Linearity is not invariant under nonlinear state feedback. Therefore, it may be possible to convert nonlinear system into linear system under this transformation. This is called feedback linearization. Adaptive feedback linearization is used for altering the incorrect estimation of the non-linearity. This utilizes lyapunov stability criteria and produces a parameter update rule that treats with time for the stability of the system.

This chapter gives a brief overview of some of the most commonly used methods for navigation and tracking for Unmanned Aerial Vehicles. Modes of UAV navigation using vision sensors, inertial sensors, integrated sensor systems, hierarchical control and feedback linearization are discussed with proper examples.

3 PROPOSED METHODS

Project uses navigation law based on the kinematics system. This law allows to successfully reach the desired goal. The kinematics based rule are very advantageous compared with other type of sensors and it also very reliable for the online path planning of the UAV. To achieve the trajectory control of the UAV we used navigation criteria such as proportionality and deviation factors. This is one of the major advantages of these laws.

3.1 Definitions and kinematics modeling of UAV motion

The movement of a UAV mobilising in a three dimensional space ((x,y,z) Cartesian co-ordinate system) can be delineated with the aid of the upcoming diagram:

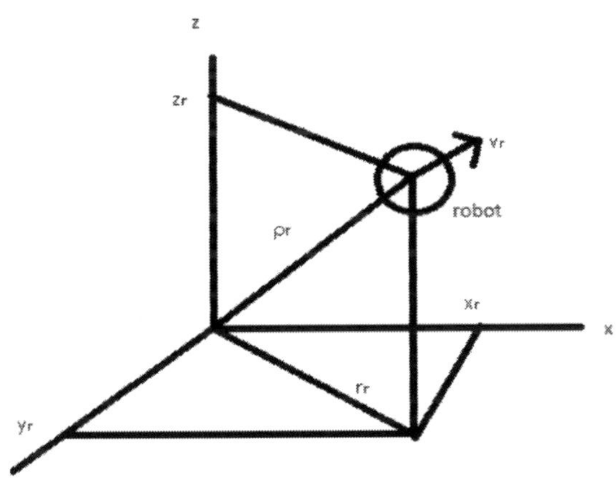

Figure 3 – Delineation of UAV motion in 3 dimensional space

The terms appearing in the figure can be defined as:

The velocity v_r of the robot in the 3 dimensional spaces.

The Euclidean distance r_r from the origin to the robot in the 3 dimensional space.
The projectionρ_r in the (x,y)plane.
The commensurating x_r x co-ordinate indicating the x position of the robot at any given time in the 3 dimensional spaces.
The commensurating y_r y co-ordinate indicating the y position of the robot at any given time in the 3 dimensional spaces
The commensurating z_rz co-ordinate indicating the z position of the robot at any given time in the 3 dimensional spaces.

3.2 Definitions of the flight path, heading and line of sight angles

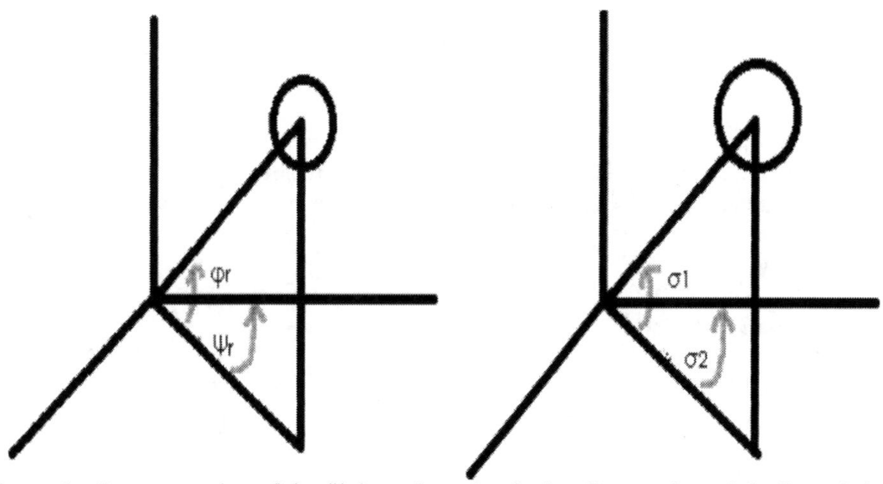

Figure 4 – Representation of the flight path angle, the heading angle and the line of sight Angles

ϕr: flight path angle (angle steering the motion of the robot in vertical direction)
ψr: heading angle (angle steering the motion of the robot in horizontal direction)
σ1, σ2: line of sight angles of the robot with respect to origin.

3.3 Plane decomposition method

In order to simplify the mathematical derivations, the 3 dimensional space is converted into 2 dimensional space: the (x,y)plane and (x,z) plane. The plane decomposition process can be explained using the following figures:

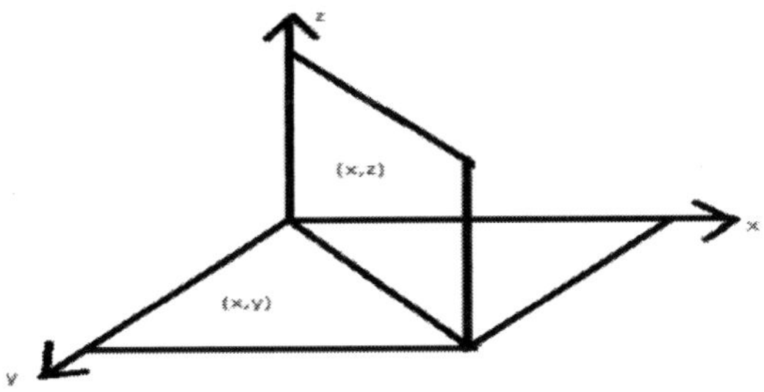

Figure 5 – Decomposition of 3 dimensional space into 2 dimensional space: (x,y) and (x,z)

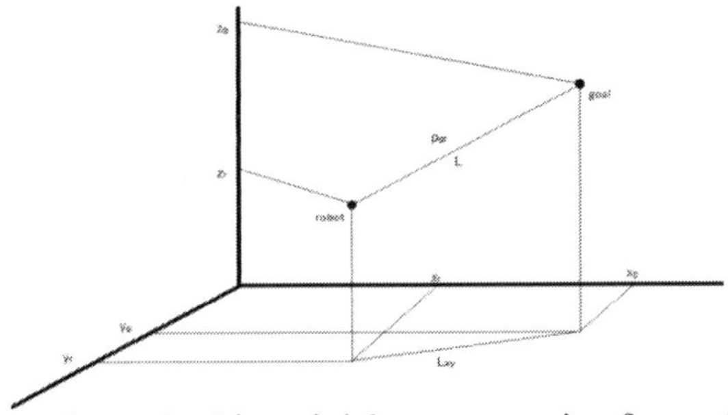

Figure 6 – Line of sight representation from robot to goal

The co-ordinates of the robot at any point in space can be defined by the following equations:
Co-ordinates in (x,y)plane: $xr= rr \cos(\sigma 2)$, $yr= rr\sin(\sigma 2)$
Co-ordinates in (r,z)plane: $rr= \rho r \cos(\sigma 1)$, $zr= \rho r \sin(\sigma 1)$

3.4 Line of sight angles in the (x,y) and (r,z)planes

The planes have been defined by decomposing the 3 dimensional space, we expound the line of sight angles for the respective planes. Let L be the line of sight and the direction of vector L is towards the goal from the robot. The projection of vector L in the (x,y)plane is denoted by Lxy. The coordinates of the robot and the goal are defined as (xr,yr,zr) and (xg,yg,zg) respectively. Taking into considerationthe coordinates of the robot and the goal, the line of sight angles in the respective planes can be defined as:
$\tan \sigma xy=ygr/xgr$ where $ygr=yg-yr$ and $xgr=xg-xr$
$\tan\sigma rz=zgr/rgr$ where $zgr=zg-zr$ and $rgr= \mathrm{sqrt}(xgr2 +ygr2)$

The above equations are valid for the case of moving agoal. In our assumed case, the goal isn't moving therefore the coordinates of the goal will remain the same; however, the line of sight angles vary withtime.

3.5 Kinematics equations and dynamic constraints

The kinematics equations defining the motion of the moving robot in a 3 dimensional space can be deduced based on the values of the various parameters discussed above. The kinematics equations of motion can be modelled as follows:

$$\dot{x}r=\upsilon r \cos(\varphi r) \cos(\psi r)$$
$$\dot{y}r=\upsilon r \cos(\varphi r) \sin(\psi r)$$
$$\dot{z}r=\upsilon r \sin(\varphi r)$$

Where, $\dot{x}r$,$\dot{y}r$ and $\dot{z}r$ are the velocity components defining the motion of robot in the 3 dimensional space.
υr is the velocity of the robot.
φr is the flight path angle (angle denoting motion in vertical plane)
ψr is the heading angle (angle denoting motion in horizontal plane).
For the sake of the simplicity, the quantity $\upsilon r \cos(\varphi r)$ can be denoted as $\upsilon 1$, resulting in the following kinematics equations in the (x,y)plane.

$$\dot{x}r=\upsilon 1 \cos(\psi r)$$
$$\dot{y}r=\upsilon 1 \sin(\psi r)$$

IOP Conf. Series: Materials Science and Engineering **995** (2020) 012019 doi:10.1088/1757-899X/995/1/012019

3.6 Obstacle Avoidance

This thesis ought to be victorious when the robot mobilizes towards a motionless or a target in motion and gradually attains it without experiencing any barriers in the path and traverse in a nonlinear path. Any obstacle in the path of the robot is treated as a sphere. As discussed earlier, the method of plane decomposition is used we use projections of the spherical obstacle in the 2 dimensional planes.
The graphical representation of obstacles in both horizontal and vertical planes is as shown below:

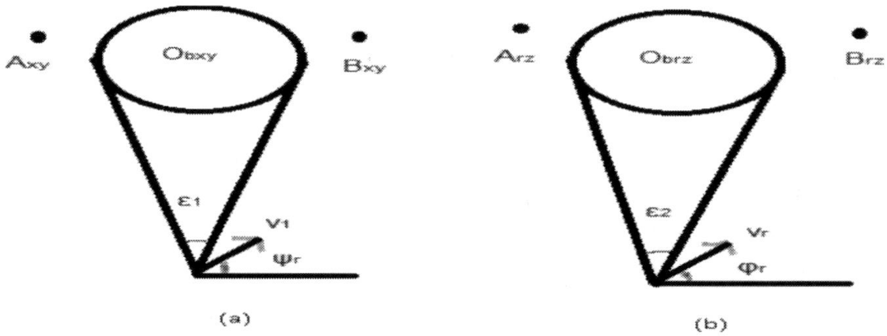

Figure 7 – Obstacles in the (x,y) and (r,z) planes

Figure 7 (a) and (b) depicts the obstacles in the (x,y) and (r,z) planes, respectively. In the (x,y) horizontal plane the velocity of the robot is defined by υ1= υr cos(φr). The angle formed by the robot with the (x,y) plane is as we know. The projection of the spherical obstacle on the horizontal plane is denoted by a circle named Obxyas shown in figure 8 (a). Now, the robot is moving towards the obstacle. At any given point in time (let us assume at time t1), the robot encounters an obstacle within its line of sight. Let us say that at time t1, the robot (assumed as a point) forms a cone with the circular obstacle in 2 dimensional (x,y) plane as shows in figure 8 (a). The angle formed by this cone is denoted by the term ε1. Axy and Bxy are the waypoints in (x,y) plane where the robot deviates to avoid collision against the obstacles. The divergence in the horizontal (x,y) plane is referred to as "Zero slope deviation".
Similarly, in the vertical (r,z) plane the velocity of the robot is denoted by υr. The angle formed by the robot with the (r,z) plane is φr. The angle formed by the collision cone in the (r,z) plane is denoted by ε2. The projection of the obstacle in the vertical plane is denoted by Obrz. The waypoints in the vertical plane are denoted by Arzand Brz. The divergence of the robot towards the waypoints in the vertical plane(r,z) is called "infinite slope deviation".

3.7 Zero deviation and infinite slope deviation

The desired heading angle's new value can be given by the equation:

Ψr_des=N1new σxy_p+c1new

Where,

Ψr_des =Desired value of heading angle to avoid collision N1new=New value of proportionality factor N1 for (x,y) plane c1new= New value of deviation factor c1 for (x,y) plane
σxy_p = Line of sight angle for waypoint in (x,y) plane

Also,the new value of the desired flight path angle can be given by the equation:

ϕr_des=N2new σrz_p+c2new

Where,

ϕr_des = Desired value of flight path angle to avoid collision N2new= New value of proportionality factor N2 for (r,z) plane c2new= New value of deviation factor c2 for (r,z) plane
σrz_p= Line of sight angle for waypoint in (r,z) plane.

3.8 Derivation parameters for Obstacle avoidance

For successfully achieving obstacle avoidance and maintaining a smooth trajectory, the values of the flight path and the heading angles at time t1 should advance smoothly to the desired values ψr_des and ϕr_des..This can be expressed by the following equations:
Ψr_t1= N1new_p (t1) + c1new + b1new ϕr_t1= σrz_p (t1) + c2new + b2new
The algorithm for obstacle avoidance can be developed by taking the values of N1new and N2new as 1.

3.9 Examples of obstacle avoidance Zero slope deviation:

This deviation is achieved in the horizontal (x,y) plane by deviating to either left or right of the obstacle. The proportionality factor N1 and deviation factor c1are changed for navigation process. The zero slope deviation can be attained by deviating towards the waypoints Axy or Bxy in the horizontal plane. The condition for collision avoidance in the horizontal plane is that the heading angle ψr should not lie in the interval [σxy-(ε1)/2, σxy+ (ε1)/2], where σxy is the line of sight angle in the horizontal plane and ε1is the collision cone width angle in the horizontal plane. Zero slope deviation can be displayed graphically in the Figure 8:

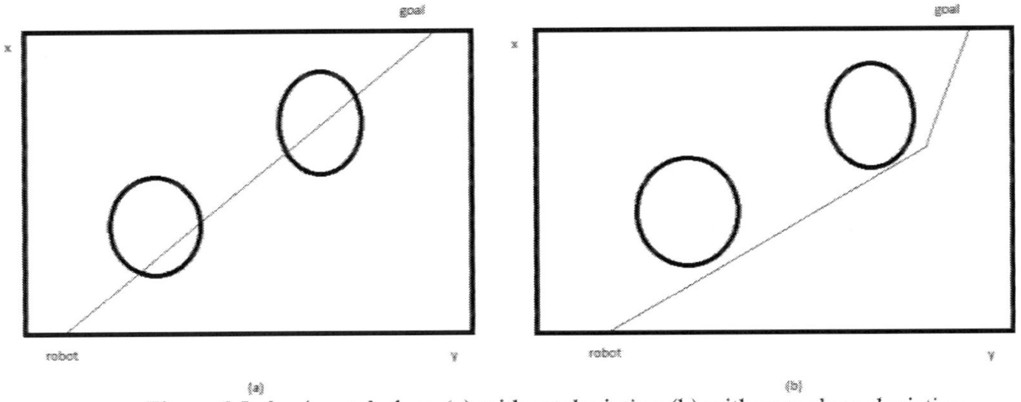

Figure 8 In horizontal plane (a) without deviation (b) with zero slope deviation

Infinite slope deviation:

This divergence is achieved in the vertical (r,z) plane by deviating to either upward or downward of the obstacle. The proportionality factor N2 and deviation factor c2 are changed for navigation process.

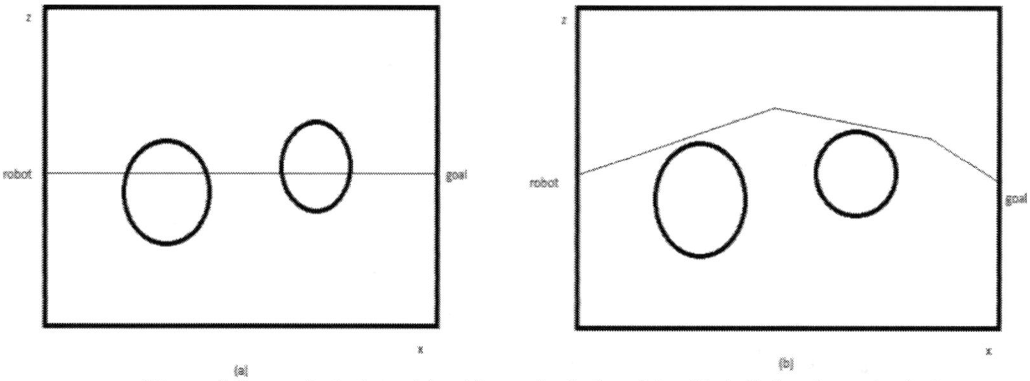

Figure 9 In vertical plane (a) without deviation (b) with infinite slope deviation

The zero slope deviation can be attained by deviating towards the waypoints Arz or Brz in the vertical plane. The condition for collision avoidance in the vertical plane is that the flight path angle φr should not lie in the interval [σrz-(ε2)/2, σrz+(ε2)/2], where σrz is the line of sight angle in the vertical plane and ε2 is the collision cone width angle in the vertical plane. Infinite slope deviation can be displayed graphically in the Figure 9:

4 RESULTS FOR SUVEILLANCE AND OBSTACLE AVOIDANCE MODE

4.1 Simulation results for surveillance mode

The starting point of the aim is given (10, 5, 3) and for the robot is given as (12, 6, 3). The robot pace is always significant than the target pace.

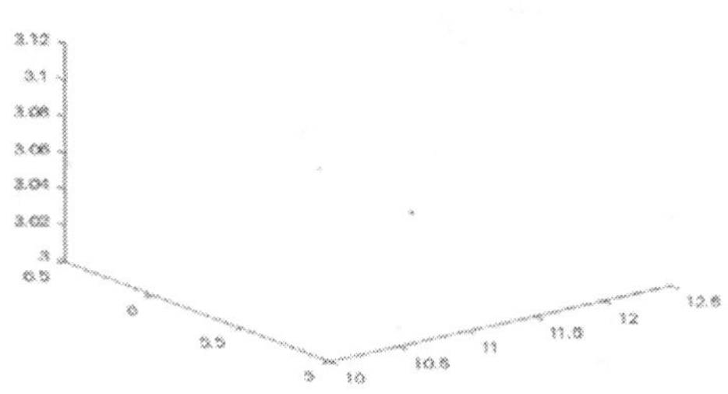

Figure 10 – The starting point of the robot (red dot) and the aim (green dot) at time t=0 in3 dimensional space

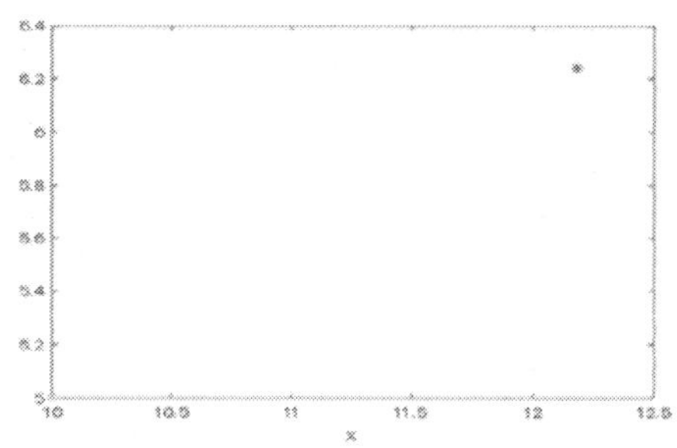

Figure 11 – Initial point of robot (red star) and the aim (green star located on extreme left bottom) in the horizontal (x,y) plane at time t=0

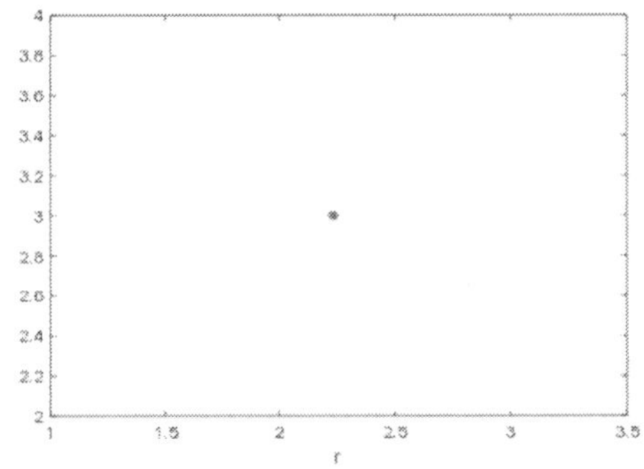

Figure 12 – Starting point of the robot (red star) and the aim (not visible at time t=0 due to the constraints provided) in the vertical (r,z) plane at time t=0

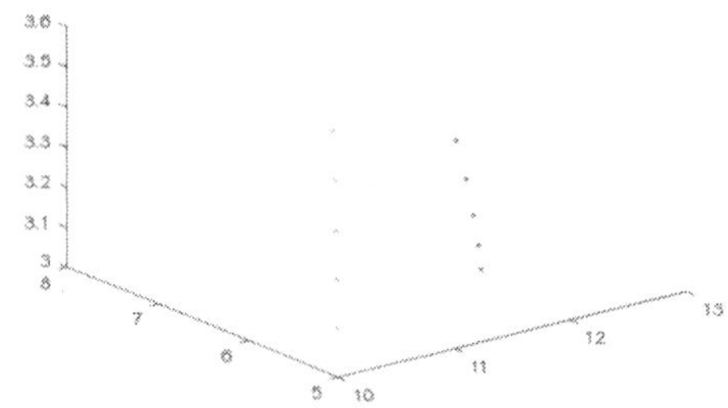

Figure 13 – The position of the robot and the aim after 5 steps in the 3 dimensional space (we can see that the robot is gradually approaching the target)

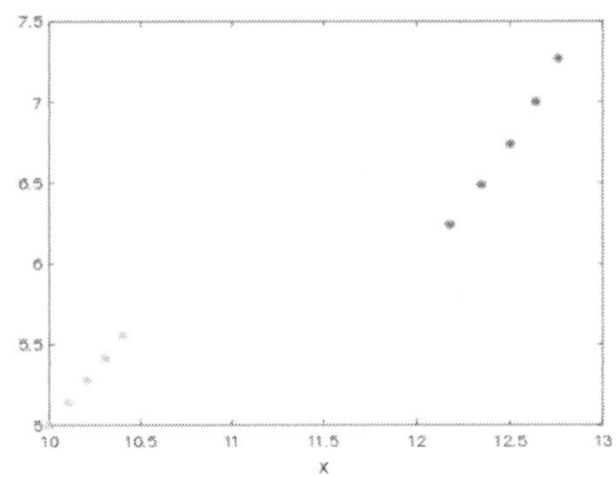

Figure 14 – The position of the robot and the target after 5 steps in the horizontal (x,y) plane

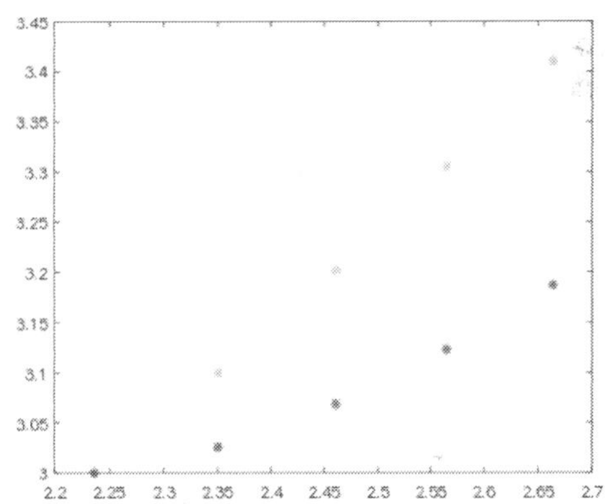

Figure 15 - The position of the robot (red star) and the aim (green star) after 5 steps in the vertical (r,z) plane

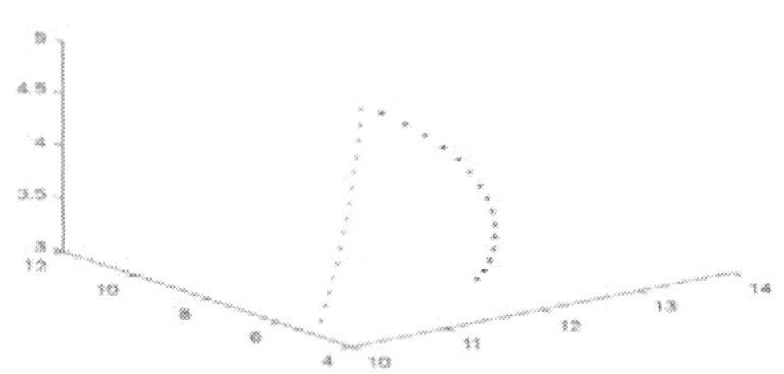

Figure 16- The position of the robot and the target after 15 steps in the 3 dimensional space (we can see that the robot has now reached the target)

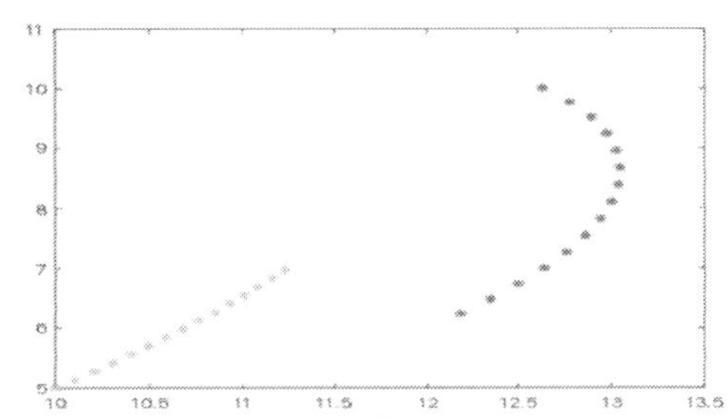

Figure 17 – The position of the robot (red star) and the target (green star) after 15 steps in the horizontal (x,y) plane

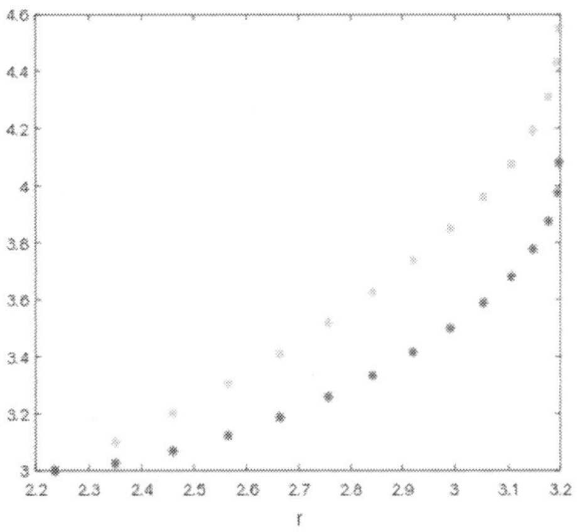

Figure 18 - The position of the robot (red star) and the target (green star) after 15 steps in the vertical (r,z) plane

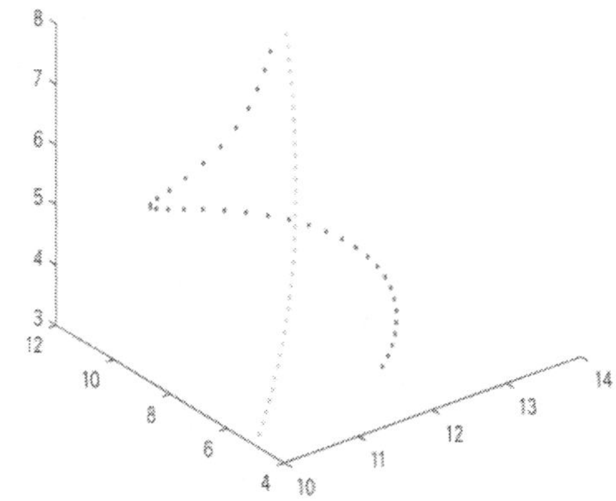

Figure 19 - The position of the robot and the target after 35 steps in the 3 dimensional space (we can see that the robot has once reached the target, bypassed it and is reached it again)

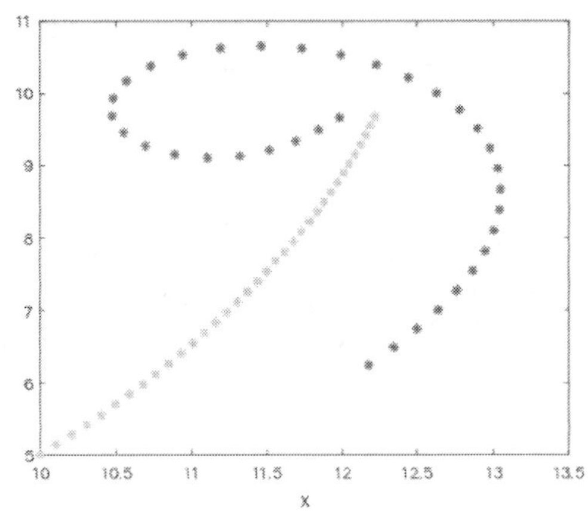

Figure 20 – The position of the robot (red star) and the target (green star) after 35 steps in the horizontal (x,y) plane

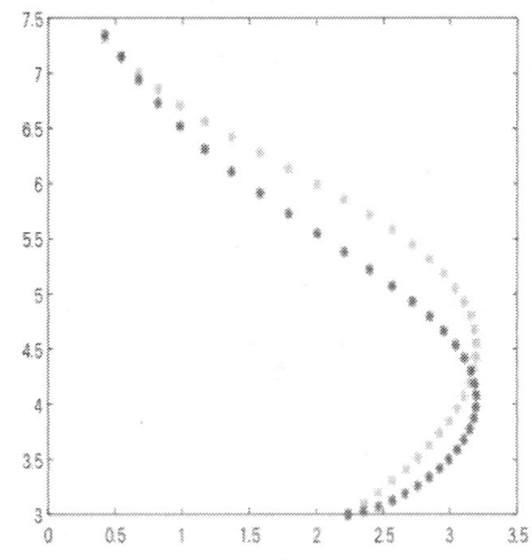

Figure 21 – The position of the robot (red star) and the target (green star) after 35 steps in the vertical (r,z) plane

IOP Conf. Series: Materials Science and Engineering **995** (2020) 012019 doi:10.1088/1757-899X/995/1/012019

4.2 Simulation results for obstacle avoidance mode

It is assumed that the robot deviates in the horizontal plane only towards a waypoint A. The values assumed for input variables are c1= Π/4, xg0 =13, xo=3, yg0 = 7, y0 = 5, zg0 =4, z0 = 4, xa = 3, ya = 4.

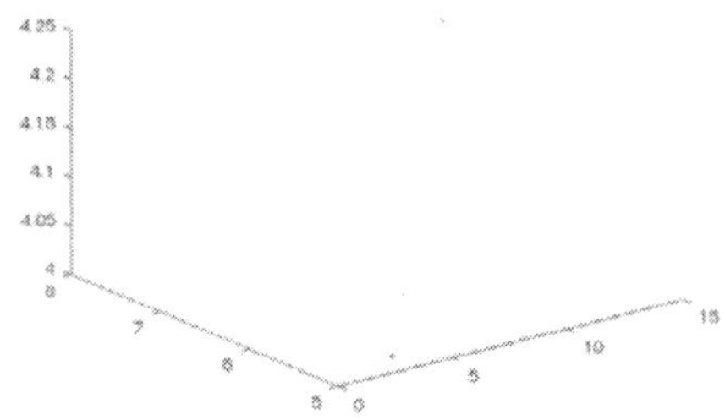

Figure 22 - The starting point of the robot (red dot) and the aim (green dot) at t=0 in the 3 dimensional space

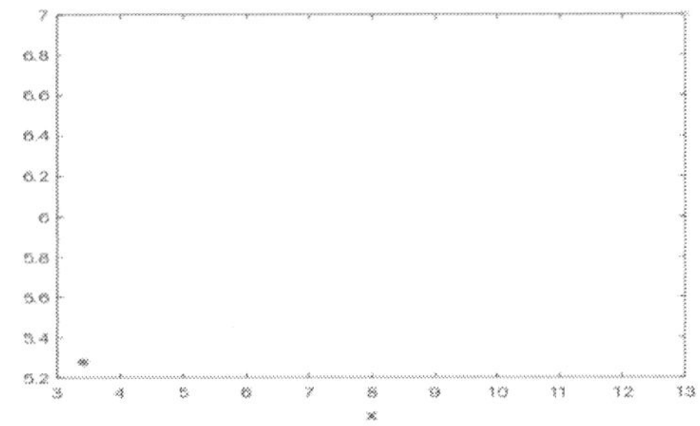

Figure 23 – The starting point of the robot (red star) and the aim (green star located at top right corner) at time t=0 in the horizontal (x,y) plane

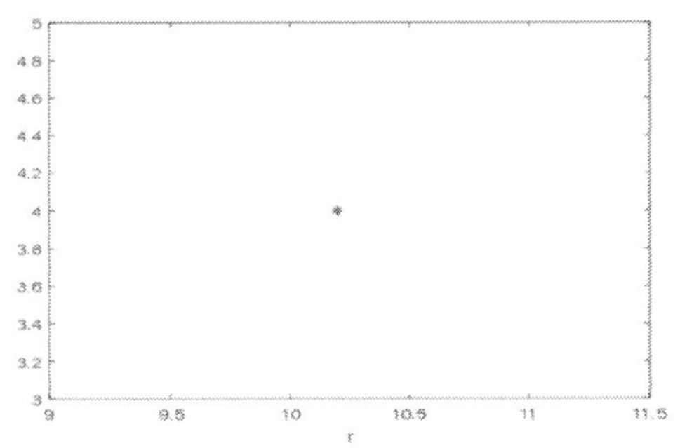

Figure 24 –The starting point of the robot (red star) and aim at time t=0 in the vertical (r,z) plane

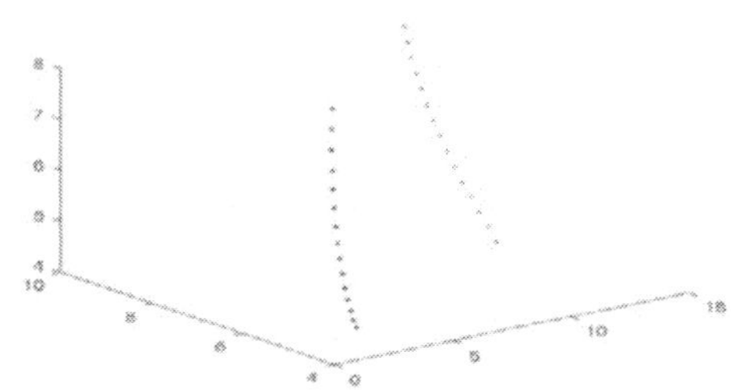

Figure 25- The position of the robot and the aim after 15 steps in the 3 dimensional Space (we can see that the robot is advancing towards the moving target)

IOP Conf. Series: Materials Science and Engineering **995** (2020) 012019 doi:10.1088/1757-899X/995/1/012019

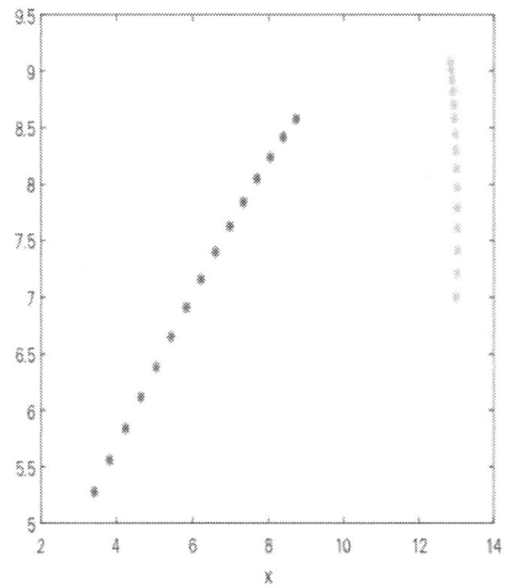

Figure 26 – The position of the robot and the aim after 15 steps in the horizontal (x,y) plane

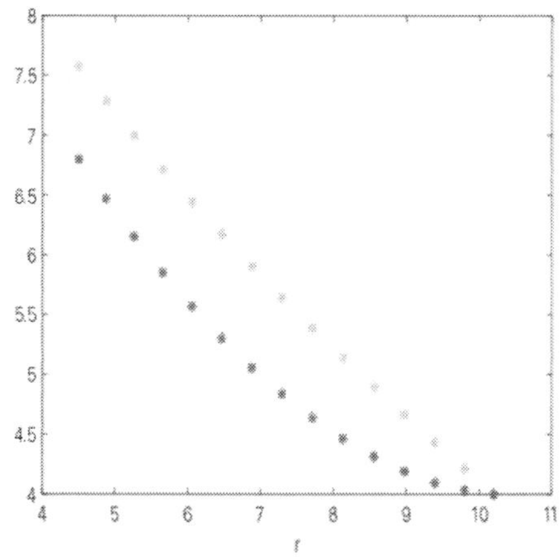

Figure 27- The position of the robot (red star) and the aim (green star) after 15 steps in the vertical (r,z) plane

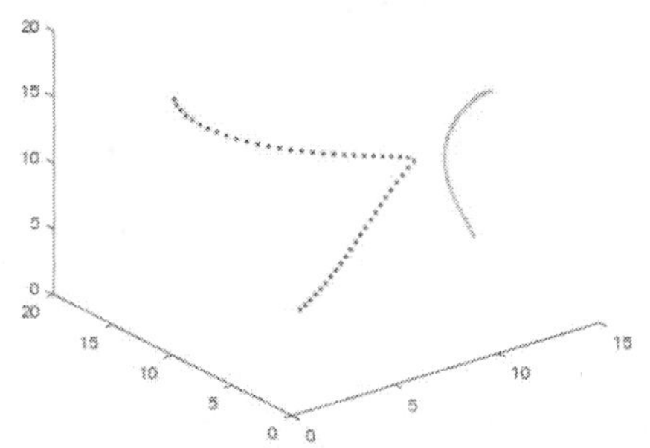

Figure 28 - The position of the robot and the aim after 50 steps in the 3 dimensional space (we can clearly see that the robot has deviated from its original course and is approaching the target gradually by following the alternate path)

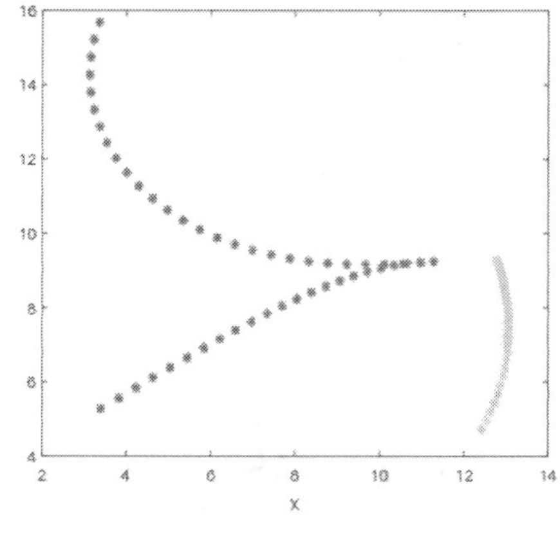

Figure 29 – The position of the robot (red star) and the aim(green star) after 50 steps in the horizontal (x,y) plane

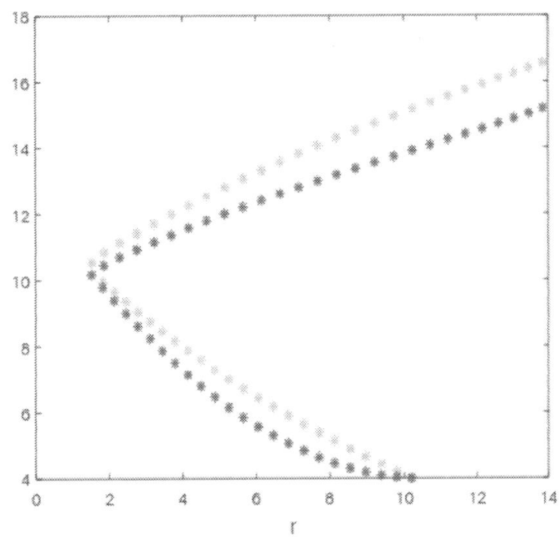

Figure 30 - The position of the robot (red star) and the aim (green star) after 50 steps in the vertical (r,z) plane

5 CONCLUSION

Project uses kinematics based algorithms for planning the path and navigation for Unmanned Aerial Vehicles. Kinematics derivations are elucidated considering the movement of a UAV and the units entailed in those equations are obtained and recounted. The plane decomposition practice is used to lessen the difficulty in planning the path in a 3D space by transforming it into 2D space. Both course of the flight or flight path angles and heading angle of the robot being ascertained by the line of sight angles. The Control laws are being obtained for the flight path and heading angles, this encompasses the navigation specifications such as proportionality and deviation factors. These navigation criteria are availed with differing degrees to reach the adequate trajectory control, incorporating path smoothness and obstacle avoidance. The obstacle avoidance means comprises the collision cone method. The notion of waypoints towards which the robot drifts in a variant path is also noted. Also the dynamic curbs hindering the movement of the robot and the aim in the 3D space are also contemplated. Finally, MATLAB code is availed for the simulation in two places: 1) Surveillance mode 2) Obstacle avoidance mode. Simulation shows the beneficiary effects of the above stated methods. Kinematics based algorithm is a high level control algorithm and it is very effectual than other algorithms like (k-means, kalman filter) and it is helpful in many military applications, weather forecasting, aerial photography by using surveillance mode.

References

[1] Ashraf, S., Aggarwal, P., Damacharla, P., Wang, H., Javaid, A. Y., &Devabhaktuni, V. (2018). A low-cost solution for unmanned aerial vehicle navigation in a global positioning system–denied environment. International Journal of Distributed Sensor Networks.

[2] Wierzbicki D. (2018). Multi-Camera Imaging System for UAV Photogrammetry. Sensors (Basel, Switzerland), Vol 18 No 8 p. 2433.

[3] Ahsan, Mansoor& Akhtar, Suhail&Mazhar, Farrukh& Khalid, Mudassir. (2012). An Algorithm for

Autonomous Aerial Navigation using MATLAB® Mapping Tool Box.
[4] Rhudy, M. B., Gu, Y., Chao, H., & Gross, J. N. (2015). Unmanned aerial vehicle navigation using wide-field optical flow and inertial sensors. Journal of Robotics, 2015.
[5] Cerón, A., Prieto, F., &Mejias, L. (2019). Vision-Based Path Finding Strategy of Unmanned Aerial Vehicles for Electrical Infrastructure Purpose. In Path Planning for Autonomous Vehicles-Ensuring Reliable Driverless Navigation and Control Maneuver. IntechOpen.
[6] Savkin, A. V., &Teimoori, H. (2010). Bearings-only guidance of a unicycle-like vehicle following a moving target with a smaller minimum turning radius. IEEE Transactions on Automatic Control, vol 55 No 10 pp 2390-2395.
[7] Kanellakis, C., & Nikolakopoulos, G. (2017). Survey on computer vision for UAVs: Current developments and trends. Journal of Intelligent & Robotic Systems, Vol 87 No 1) pp 141-168.
[8] Ramos, L. (2013). Unmanned Aerial Vehicle (Doctoral dissertation, Florida International University).
[9] Orsag, M., Korpela, C., & Oh, P. (2013). Modeling and control of MM-UAV: Mobile manipulating unmanned aerial vehicle. Journal of Intelligent & Robotic Systems, Vol 69 No 1-4 pp 227-240.
[10] Guido, G., Gallelli, V., Rogano, D., & Vitale, A. (2016). Evaluating the accuracy of vehicle tracking data obtained from Unmanned Aerial Vehicles. International journal of transportation science and technology, Vol 5 No 3 pp 136-151.
[11] Wu, Y., Sui, Y., & Wang, G. (2017). Vision-based real-time aerial object localization and tracking for UAV sensing system. IEEE Access, Vol 5 pp 23969-23978.
[12] Chopade, K., & GURU, S. (2015). A survey: Unmanned aerial vehicle for road detection and tracking. International Journal of Innovative Research in Computer and Communication Engineering, Vol 3 No 10 pp 10761-10766.
[13] P. Om Prakash, T. Surya, R. M. Rajkumar, M. Joseph Steffin, 2017, Navigation of Unmanned Aerial Vehicles in GPS Denied Region Using Vision Based Obstacle Detection, International journal of Engineering Research and Technology Vol 5 p 09.
[14] Wilhelm, J. P., & Clem, G. (2019). Vector Field UAV Guidance for Path Following and Obstacle Avoidance with Minimal Deviation. Journal of Guidance, Control, and Dynamics, Vol 42 No 8 pp 1848-1856.
[15] Zhai, R., Zhou, Z., Zhang, W., Sang, S., & Li, P. (2014). Control and navigation system for a fixed-wing unmanned aerial vehicle. Aip Advances, Vol 4 No 3 pp 031306.
[16] Wang, T., Wang, C., Liang, J., Chen, Y., & Zhang, Y. (2013). Vision-aided inertial navigation for small unmanned aerial vehicles in GPS-denied environments. International Journal of Advanced Robotic Systems, Vol 10 No 6 p 276.

Automation and performance analysis of tea filter bags packing process using delta configured robot

S Vigneshwaran[1], P Nagarajan [2], A Mohamed Ismail [3], K Yuvaraj [4]

[1,2,3,4] –Assistant Professor, Department of Mechatronics, Bannari Institute of Technology, Sathyamangalam –638401, India

vigneshvcetmdu@gmail.com

Abstract: A delta robot is a type of parallel robot that consists of three arms connected to universal joints at the base. Industries take advantage of these robots in packaging and pick and place operations due to its high accuracy and fastness. This three degrees of freedom robot is used in the process that automates the tea filter bags packing process through pick and place operation. Electrostatic gripper has been used as the end effectors. The tea filter bags are allowed to go through the conveyor and this delta configured robot is used for pick and place operation. This robot reduces the time complexity of the laborers and it also helps in increasing productivity by reducing the cost. Thus the tedious process becomes simpler through the delta configuration. Comparative analysis of the performance of robot with manual process is also shown in the results.

Keywords: Tea bags, Delta configuration, Electrostatic charges.

1. Introduction

A delta robot is a type of parallel robot that consists of three arms connected to universal joints at the base. The key design feature is the use of parallelograms in the arms which maintains the orientation of the end effectors. The delta robot allows for very high speed and high accelerations for very high speed and high accelerations. Having all the arms connected together to the end effector increases the robot stiffness but reduces its working volume. These delta robots are known for their fastness and high accuracy. Industries that take advantage of the high speed characteristic of delta robots are the packaging industry, pharmaceutical and medical industry .It can also be used for surgery because of its stiffness. Other applications may include high precision assembly operations. Today delta robots are well established in the automation industry. Unlike the larger articulated arms, delta robots are often kept as a stock item at many manufacturers such as yaskawa, Motoman and ABB. Such kinds of robots perform pick and place operations quickly and repeatedly too. The delta configured robot is also known as pickpac robot since it can pick one or more objects and pack or tray them. In this project the delta robot is used for the packing purpose. The tea filter bags are allowed to go through the conveyor and a delta configured robot is used to pick and place the tea filter bags process. Our delta robot has three degrees of freedom. Thus the tedious packing process becomes simpler through the delta configuration.

2. Literature survey

In [1], the paper deals with the handling materials and gives a brief idea about the vacuum gripper and electrostatic gripper. In [2], author discussed about a gripper that handles textile materials with the tactile sensors, camera modules, and it will be a type of two fingers which grasps the textile material. This gripper has been completely done for its flexibility. A complete case study of electro adhesion principle of handling flexible objects familiarly textile type objects are processed in a simple way through the electrostatic gripper

IOP Conf. Series: Materials Science and Engineering **995** (2020) 012020 doi:10.1088/1757-899X/995/1/012020

in [3]. In [4], the paper completely explains about the direct and inverse kinematics of the geometric analysis of the manipulators and end effectors orientation. From the literature survey of the above mentioned papers and more, we analyzed various ways available for handling the flat and non-rigid objects like the tea filter bags that we are trying to handle during this automation process and came to the conclusion about using the Electro adhesive gripper which comes under Adhesive gripping principle [5].

3. Problem identification

Durga Garments, a small scale textile industry having more than 10 number of employees are doing the process of cutting and packing of tea filter bags. 1 pack consists of 100 pieces. The employees are taking around 6 hours of work for cutting and another 6 hours of work for packing. The calculations are given below,

No of Employees = 10.
Each Employee Contribution approx. = 12- 20 packs.
Total amount of pieces per day = 20,000 pieces.
Total amount of pieces per month = 4, 80,000 approx.

Since it is taking a huge time for the packing process rather than cutting process, we have come up with a solution. We have developed a delta configured robot can reduce the time into half of the cutting process.

4. Robot design and fabrication

Clavel's delta robot design as shown in figure 1 is the best suited robot design structure according to their popularity in delta configured robots found in industries up to date. This delta robot has 4-degrees-of-freedom (DOF), 3-dof for XYZ interpretation, in addition to a fourth inward leg to control a solitary rotational opportunity toward the end-effector stage (about the hub opposite to the stage).This design serves us as the base design for our robot structure for the kinematics analysis.

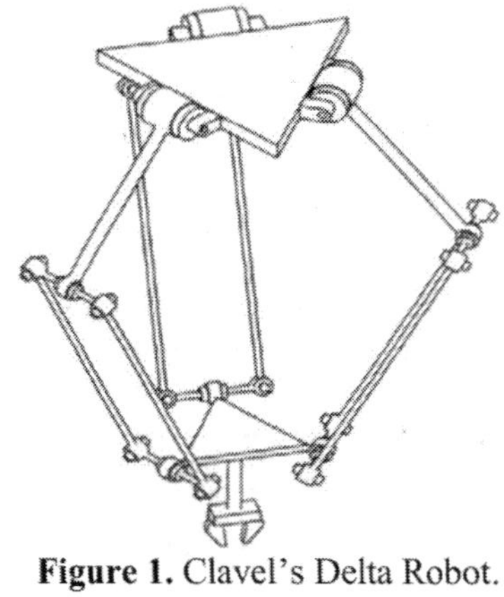

Figure 1. Clavel's Delta Robot.

The design objectives include low cost production, sustainability and maintenance, modularization, simplicity and order. The standard design is the initial process in any of the projects, the design includes the proper dimension of forearms, biceps, end effectors, motor coupler and frame of the robot. All the above said calculations and the analysis are considered while concluding with a standard design ready to be fabricated, purchased and assembled. The CAD model of the proposed setup is shown in figure 2.

Figure 2. CAD Design of the fabricated Delta Robot.

4.1 Fabrication

Dimensions for special parts include the design of biceps, forearms and end effector for the delta robot in this automation. The biceps and forearms are used to lift the objects or used to move the end effector from place to place with some variations in the joint angles of the revolute joints designed in between the actuators and biceps of the robot. The actuators are grounded to the base of the robot as shown in the figure 3. These actuators are housed in very precisely calculated ways such that they work according to the previously discussed Clavel's delta robot design.

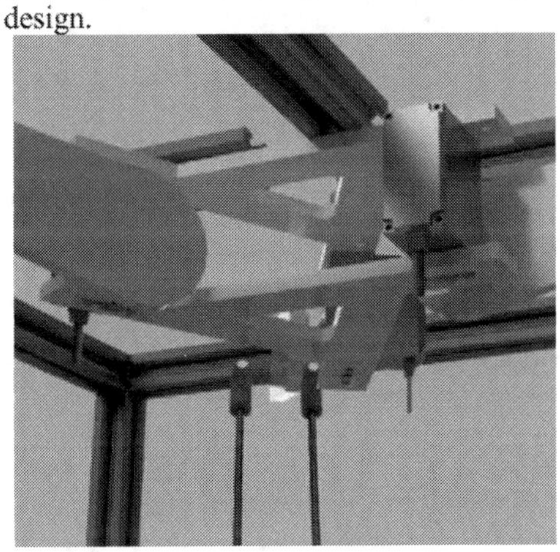

Figure 3. Motors attached to the Base.

Other special parts made for this robot are the horizontal and vertical pins used to form the required joint movements in between the biceps and the forearms and also in the joints between the end effector and the forearms. These pins are assembled in such a way that they provide revolute joints which are mutually perpendicular to each other as an alternative to the action provided by a spherical fisheye rod end intended to be connected at ends of the forearms first as shown in the figures 4 and 5.

IOP Conf. Series: Materials Science and Engineering **995** (2020) 012020 doi:10.1088/1757-899X/995/1/012020

Figure 4. Joint in biceps.

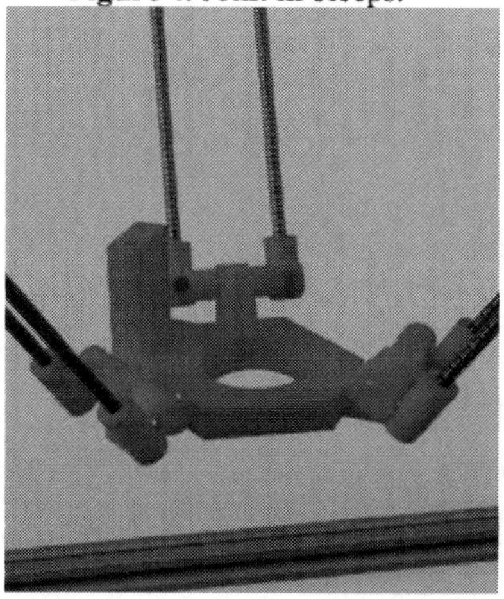

Figure 5. Joint in end effector.

4.2 Control and calibration

The control of each arm in the robot is one of the main goals in the project of pick and place the tea filter bags. The control flow of the robotic arm is as follows the figure 6

When the desired trajectory for the end effector is requested by the user or the mode at which the machine is running the controller generates the motor angles for each joint in the robot and passes these signals to the servo motors via PWM waves from the controller. These PWM signals are received by the driver modules in each servo motor, which takes the corresponding steps incremented or decremented according to the feedback signal from the position sensor attached to each joint.

The overall movement of these joints contributes the end effector being moved to the desired position.

IOP Conf. Series: Materials Science and Engineering **995** (2020) 012020 doi:10.1088/1757-899X/995/1/012020

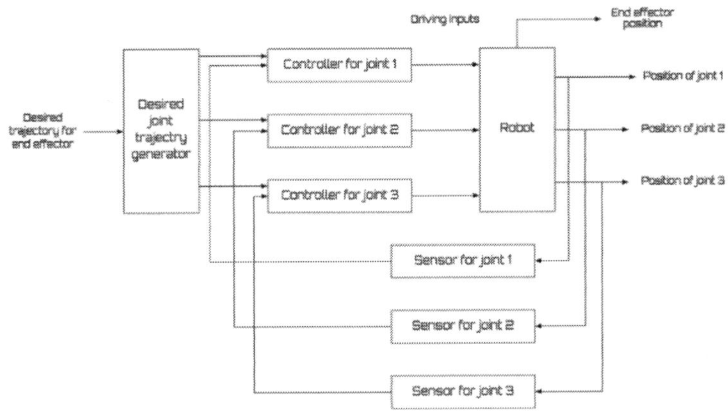

Figure 6. Control architecture of robot arm.

4.3 Position Calibration

When the full robotic setup is complete and ready for user, it requires to provide the positional coordinates as the input. Therefore, the appropriate position of the end effector becomes the input to the algorithm that controls the robot, which then performs the inverse kinematic calculation and finds three angles that control the servo motor that places the robot's arm in the correct position as requested by the user.

To perform these steps in the real world first the robot arms initial position needs to be calibrated. The calibration process involves manual calibration of the robot arm as the first step. As the first step the robot shoulder connected to the servo motor is removed. The robot's home location is then called through the interface on the command line. The disconnected arms are now precisely attached to the servo motor, so that the end effector lays neutral in its previously well-designed location at proper height. After the robot's manual calibration is finished in the real world then the joint movement machine checking is performed by calling a test function in the command line interface to ensure the proper movement of each joint up to its full extent.

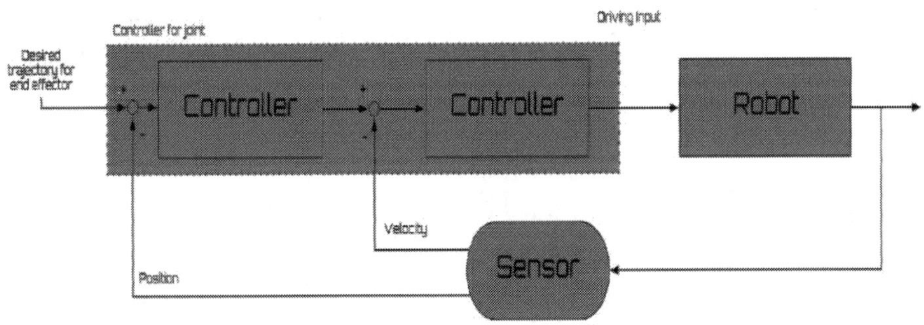

Figure 7. Flow Chart of Position velocity feedback system.

5. Experimental verification

Experimental trials were run with robots performing the task of picking and placing tea filter bags in the correct orientation for particular times versus manual pick and place of the same amount of tea filter bags in the same orientation. Initially the orientation of tea filter bags varied but showed a much less time period for the task completion with the manual attempts. After the orientation error is addressed with correction procedure, the task came close to timing with the manual attempt of counting and orienting the tea filter bags in required count and position. Each test was measured by means of a stopwatch to record the completion speed from the moment the operator pressed the start button on the controller until the last tea filter bag was placed.

Each pack was inspected for accuracy of placement. Each set of pack contains 50 tea filter bags, so each would receive an accuracy score out of 50 for correctly placed tea filter bag. The experimental data collected form running those trails are as follows in the Table 1

Table 1. Performance analysis of the developed Delta Robot.

Performance	Correctly packed packets	Time Taken(min)	Accuracy (%)
Trial 1	23	5.20	46
Trial 2	41	4.25	82
Trial 3	39	3.45	78
Trial 4	48	3.19	96
Trial 5	49	3.20	98
Trial 6	50	3.23	100
Trial 7	47	3.40	94
Trial 8	49	3.67	98

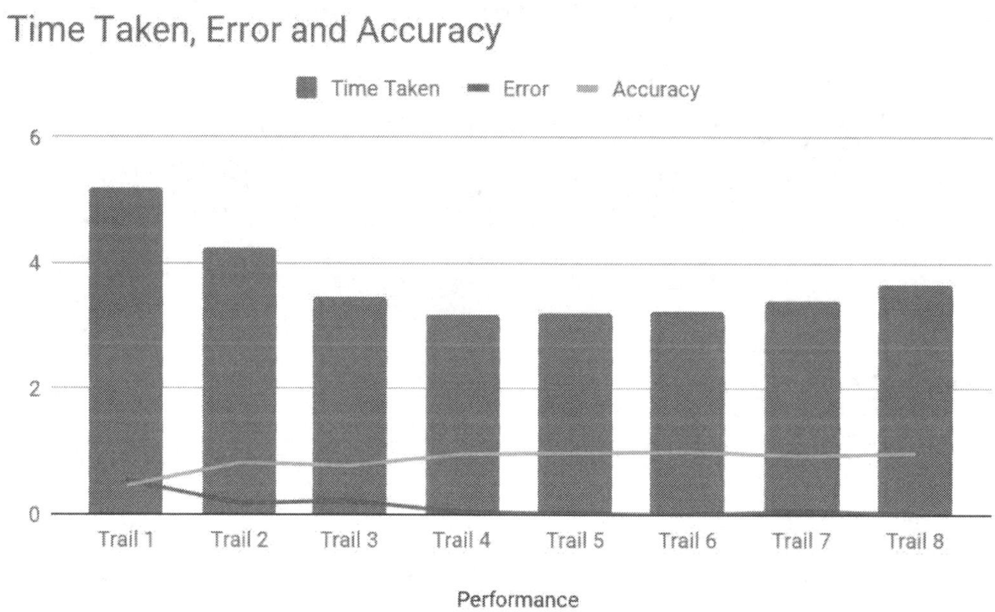

Figure 8. Performance Chart of the Robot.

IOP Conf. Series: Materials Science and Engineering **995** (2020) 012020 doi:10.1088/1757-899X/995/1/012020

6. Results and discussion

The experimental result suggests that the real time performance of the placement coordinates differ from actual position vs electronic coordinates calculated by the algorithm. The problem is with the slight backlash in the servo motor's gearbox. After those first two trials the variation in the X-axis is manually adjusted with reference to the error value in the program using the Visual Delta 3 Controller software. Then the following trails showed accuracy nearer to the expected result. The initial trial showed much time taken as a result of electro adhesive gripper lag in charging to discharge period. This is then fine-tuned to acquire much lesser time taken for task completion.

The cost estimate for this project was around twelve thousand rupees as attached in the Bill of Material without the cost of labor for the entire project. Though the initial investment seems high for the task to be considered but the overall long-term cost including the labor charge and the time taken for the task completion seems reduced with the involvement of this project.

7. Conclusion

The pick and place Delta Robot system proved to be not an accurate process as expected but definitely improved packing process efficiency in terms of fast task completion. From the figure 8, it is observed that even though the robot had a placement accuracy off by as much as 4-9%, it showed a strong repeatability. On a good note, it was accurate enough to put into the process with the current accuracy which doesn't matter in the packing process where count is perfect all time along the packing process.

Most importantly this system can easily differentiate the size of the product with ease making it more efficient for packing process. This proved to be more useful in the quick packing of tea filter bags.

References

[1]. Donfatto Paul Glick and Aaron Parness 2017 An Electrostatic Gripper for Picking an Object Proceeding of International Conference on Intelligent Robots and Systems pp 1172-1179

[2]. Erzincanli and J M Sharp 1997 Development of a non-contact end effector for robotic handling of non-rigid materials Cambridge University Press , Vol 15 No 3 pp 331-335

[3]. Hawkes E. W., Jiang H., and. Cutkosky M. R 2015 Three-dimensional dynamic surface grasping with dry adhesion The International Journal of Robotics Research

[4]. Kalantari A, Mahajan K, Ruffatto D, and Spenko M 2015 Autonomous perching and take-off on vertical walls for a quadrotor micro air vehicle in Proceedings of the IEEE International Conference on Robotics and Automation.

[5]. Parassidis Fahantidis N Petridis V Doulgeri Z Petrou L and Hasapis G 1995 A robotic system for handling textile and non-rigid flat materials Computers in Industry Vol 26 No 3 pp 303-313

[6]. Ethan W. Schaler Donald Ruffatto Paul Glick Victor White Aaron Parness 2017 An Electrostatic Gripper for Flexible Objects Proceeding of International Conference on Intelligent Robots and Systems pp 1172-1179

[7]. Shin, K.G. and Mckay, N.D. 1984 Open Loop Minimum Time Control of Mechanical Manipulations and its Applications Proceedings of American Control Conference pp. 1231-1236.

[8]. Thuy-Hong-Loan Le Michal Jilich, Alberto Landini Matteo Zoppi, Dimiter Zlatanov and Rezia Molfino 2013 On the Development of a Specialized Flexible Gripper for Garment Handling Journal of Automation and Control Engineering Vol. 1 No. 3 pp 255-259

[9]. P.M. Taylor, A.J. Wilkinson, G.E. Taylor, M.B. Gunner, G.S. Palmer 1990 Automated Fabric Handling Problems and Techniques Proceeding of IEEE International Conference on Systems Engineering pp 367-370

[10]. P.J Zsombor-Murray 2004 Descriptive geometric kinematic analysis of clavel's delta robot Centre of Intelligent Machines, McGill University

IOP Conf. Series: Materials Science and Engineering **995** (2020) 012021 doi:10.1088/1757-899X/995/1/012021

Design and Fabrication of Step Feeder automation Machine for the Automobile Steering Ball joints

G M Oorappan[1], A Jeevanandham[2], R Sivabalakrishnan[3], S. Sudhahar[4]

1, 3 Assistant Professor, Bannari Amman Institute of Technology, Department of Mechatronics Engineering, Erode, Tamilnadu, India.
2 Professor, Sri Krishna College of Engineering and Technology, Department of Electrical and Electronics Engineering, Coimbatore, Tamilnadu, India.
4 Assistant Professor, Bannari Amman Institute of Technology, Department of Electronics and instrumentation Engineering, Erode, Tamilnadu, India.

oorappan@bitsathy.ac.in, jeevanandhama@skcet.ac.in, sivabalakrishnan@bitsathy.ac.in, sudhahars@bitsathy.ac.in

Abstract. Automation feeding machine helps to replace the manual feeding of the components to achieve the production rate. Even so, all industrialists have been trying to get 100% efficiency from every machine and to give more production. The production rate of the machine is depended on certain factors like feeding, accuracy, Manpower, etc. However, feeding is the most important factor, because the production will be fast only while the component is federal quick. This paper is providing an automation solution of feed ball joints for an automotive industry when the vehicle gathers the steering assembly. There were many feeding mechanisms available. Still, the ball joint is such a large component and should need a certain interval to feed it correctly as required. In this machine, the hoppers to be a trapezoidal tanks stores the sorted of long components. The two-step feeders consist of cylinders, which make the movement of steps. The steps that are used the EN-8 materials are used to make the feeder assembly, and those are connected with the cylinders that are used to construct the up and down movements, when the cylinder makes the forward and retracts motion. The steps scoop up the components and feed them to the additive chute. The linear chute transfers the properly oriented components to the singling unit feeds the components to the machine for making further operations.

Keywords: Hopper, Step feeder, Linear Chute, Singling unit, PLC.

1. Introduction

Auto feeders are the easy way to feed and use the components without any time lag. Those are low cost when compared to robots and gantry systems. From starting to end of the machine, all the works are important. Initially, the design plays a major role in any machine. It is only way to interact with company mates and analyse the problem and how to found the solution should be discussed among the team of the company. However, the solution is simple that some companies will provide it Hence this machine provides certain solution of the problem will occur during the production process[1]. Same as like, this solution named STEP FEEDER AUTOMATION (which feeds ball joints) for an automobile company.

The features of step feeder are:

- Pneumatic operated and PLC control.
- 4-inch touch screen for monitoring and controlling.
- Easy Maintenance and friendly design.
- Easy to use and interface.

Machine involve in various sections, which are designing, manufacturing and assembly, programming, wiring and finally testing[2]. Design involves (every single and every part to be carried) whole design setup. To design the feeder SOLIDWORKS software is used. SOLIDWORKS is a conceptualization-designing tool for speedily creating many industrial design concepts with unique tools for design and manipulating.

This software is only to design the mechanical components and assembly section. PLC involved another part in the machine for proper controlling. This will be very useful for understanding of control wiring. When problems occurred, a new person can solve the problem easily using this PLC[3]. For this, the work involves in addition selecting a PLC (Programmable Logic Controller), calculating the SMPS (Switch Mode Power Supply) rating and MCB (Miniature Circuit Breaker), Control panel board design, programming part. This is the replacement of manual operation with increase in safety purposes.

2. Methodology

The alternating vibrations create a motion that guidance the raw components to proceed up the steps on the inside curved surface of the feeder product machine. The downward step feeder path has been custom-made fiduciary, so it allows the components in the prearranged orientation to make it to the next process[4]. The damaged or unwanted components get back from the machine because good components will reach the proper aligned position only after the several short-circuit trips in the feeder. The energy of vibration propelling the components to go into the spiral motion more than it leaves to the downward, indeed the components of the motion defied the gravity.

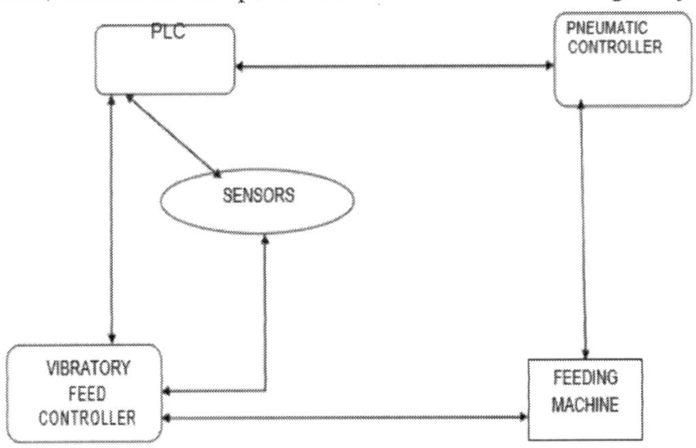

Figure 1. Block diagram of step feeder automation

The above block diagram Figure 1. shows that the process flow of step feeder, where the methodology says about the working process. First, the programmed control unit controls the whole process PLC (The Programmable Logic Controller) unit controls the feeding machine. In addition, it controls the vibratory controller, sensor, pneumatics and other functional units. Since this is a closed-loop control system, it will be having a feed after for each operation these feedbacks will help to the conventional method of using power and flexible way of feeding.

Surface mounted technology (SMT) manufactures produced great no of components more than 40,000 per hour to make this many large components[5]. Conventional XY-YZ assembly machine and improved feeding mechanism by reducing the feeding distance the step feeder is involved in feeding lager components such as vacuum tubes, etc.

3. Design and construction

The design is important part of any machine because if the design part invokes any failure the whole machine will drop down. Then it have to start the work/process from starting point. Hence, thus leads to unwanted work had to done and waste of money, labour and material. In this machine, various components are to designed and some samples are provided here[6].

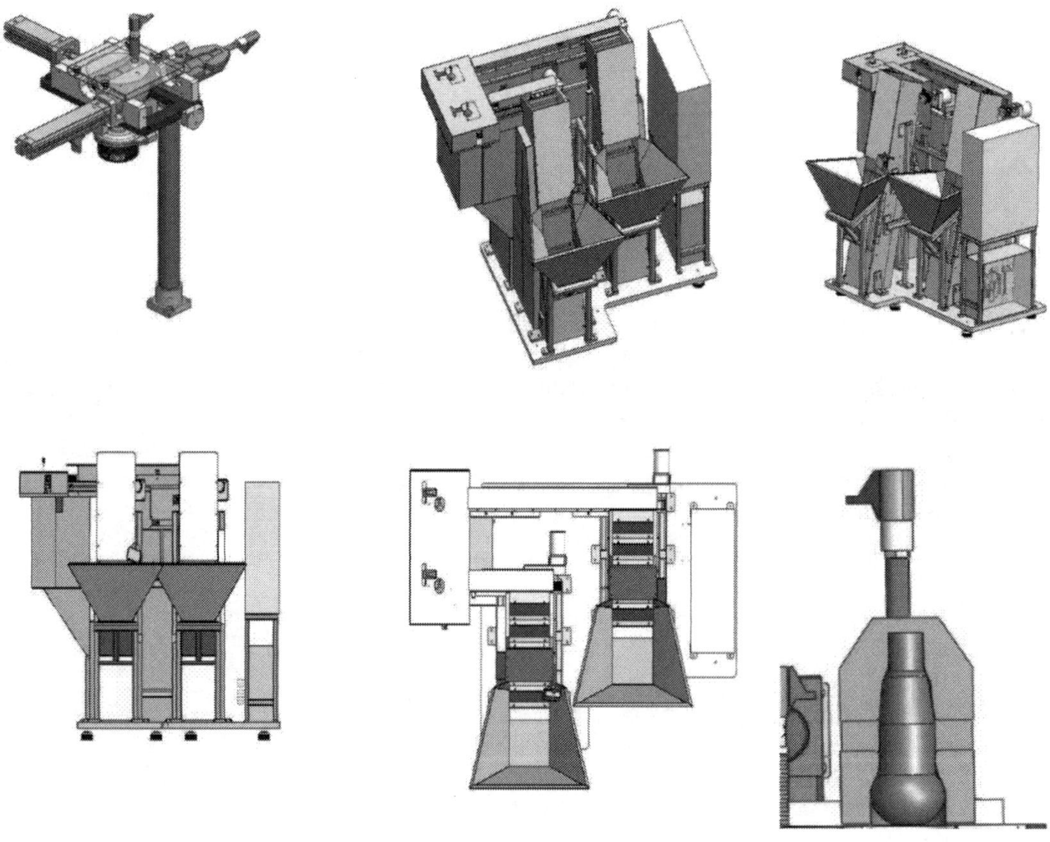

Figure 2. Post- design of step feeder machine for ball joint

At first, the design was made as in the Figure 3. but due to some of the problems raised like cost, space convenience, it has to change the design. So the design was made as per the picture in the Figure 2.

This step feeder is the compact ignoble noise alternative to a bowl feeding machine. In many occurrences, the components can be tooled without the aid or air such a way that the Hooper constructed in the step feeder machine. A very low filling height can be loaded manually. The components are collected from the Hooper by elevating plates; pre-sorted and feed to the top feeding technique is used to reach the desired transfer height. This system is manageable, low-cost, absorbs less power, and it be an easily operable so that the user can freely use this machine to feeding the steering ball joint in the automobile industry.

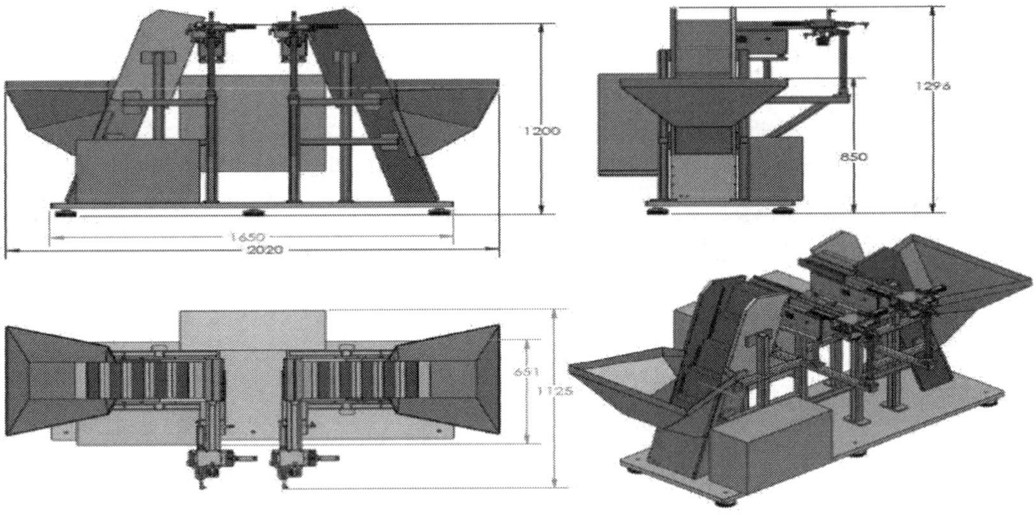

Figure 3. Pre- design of step feeder machine for ball joint

Mechanical components are made of CI which is used for working variable mechanical properties, mechanical fixtures and mechanical advantages most of the mechanical components Figure 4. shows are made by machine team only few complex design are manufactured in special purpose machining centres.[7]

Mechanical components are made of CI, which is used for working variable properties, fixtures and advantages. Most of the components are constructed from the machine team only few complex designs are manufactured in special-purpose machining centres.

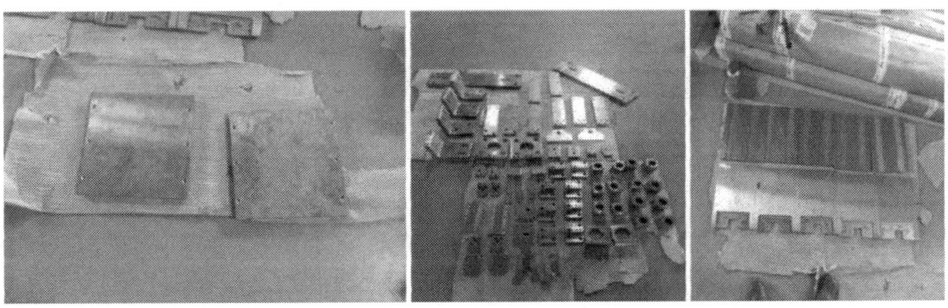

Figure 4. Constructed components for step feeder machine

Linear cylinders play major role for step feeding and for singling purpose. Machine has two types of cylinders one is "15 25", the other is "60 110." .The bigger one is used for feeding purpose. Small one is for singling. The rotary positioning pneumatic drive is used for positioning the ball joints before the singling cylinders are activated. This positioning drive is capable of higher accuracy as a servo drive does. The Filter, Regulator and Lubricator (FRL) unit is used to clean and lubricate the air inlet to the cylinders. This FRL unit is capable of regulating 50mpa of an air inlet. The valve terminal series is a combined unit of 5/2 valve. This valve terminal used in this machine consisted of series of 10 5/2 valves[8].

PLC is one of the main electric devices used to control the whole machine set up. It has two couples of 17in 17out PLC to control the unit. An eight channel relay system is capable of switching lines to control the unit as per the PLC program command is given. There will be six relay channels is used in this setup. The fuse ac and dc are used to save the sensors from higher frequency since the machine works at 460-volt supply. Table.1 shows the important devices of step feeder and their specification details.

Table 1. Important devices of step feeder and their specification details

S.No.	Important Device of Step Feeder	Specification	Size of the Components
1	Bowl	Capacity in litre	5
		Discharge height in cm	17
		Width of track in cm	18
		Track pitch (Spiral Distance in cm)	8
		Bowl weight in kg	9
		Total Discharge height in cm	82
		Fixing	Central
2	Level Controller	Inductive Sensor diameter in cm	1.2
		Operating voltage	10-30 VDC or 90 – 250VAC
		Capacity of the sensor	130mA or 200mA
		Total cable length in cm	150
3	Drive Units	Drive unit height in cm	6.5
		Drive unit diameter in cm	6
		Pitch between mounting/No. of bores	40/2
		Drive unit weight in kg	8
		Rating in Amps	4
		Length of connection cable in m	4
		Vibration frequency	100Hz
		Protective casing	Steel
4	Base Plate	Plate Diameter in cm	21.8
		Position of control box mounting in cm	12.5
		Fixing hole centre in cm	14
		Plate height in cm	2.3
5	Top plate	Plate Diameter in cm	40
		Plate height in cm	53.5

The fuses are categorized as AC & DC for typical sensors. The variable frequency drive is used to drive the whole units' power supply and controls the sensor power supply too. A switched-mode power supply (SMPS) is used to convert the AC to DC in a way to control the power level of supply from the main supply. The vibratory coil will provide vibration for the linear shoot to move the components from step to singling unit. It has 110-volts for this coil. The terminal blocks are used to connect the sensors, PLC, relay channel, SMPS, etc.

3.1 Determination of the bowl feeder force:

The assembly of the bowl feeder contains two main parts, a conveying element with a trapezoidal(T) form and a supporting cylindrical(C) element. Both parts are linked by springs and they are uniformly spread around their rotational axes y_T and y_S. Three flat spring packages are used here. Three flat spring packages are used here. Each package was declined by an angle β_{TC} to the axis plane. It had its stiffness, and the total values of all springs packages were interpreted as k_{TC}. The total damping coefficient was specified by valueb_{TC}.

The conveyancing element was described by its mass mT and by its moment of inertia JTy around rotational axis y. The supporting element had mass mC, the moment of inertia JCy, and it was placed on springs with total stiffness $kCRx$ and $kCRy$, and with total damping coefficients $bCRx$ and $bCRy$. The entire mechanism was settled with a rigid connection to the frame.

The movement of the bowl feeder Mechanism can be characterized with help of four coordinates, two translations yN , yS and two rotations φNy, φSy. The relative movements of both parts are defined:

$$y_{TC} = y_T + y_C \quad (1)$$

and

$$\varphi_{TCy} = \varphi_{Ty} - \varphi_{Cy} \quad (2)$$

In these aims, four differential equations specifying the movement of the bowl feeder system can be written:

$$m_T\ddot{y}_T + F_{bTCy} + F_{kTCy} + F_{VM}\cos\beta_{TC} = F_{TC} \quad (3)$$

$$J_y\ddot{\varphi}_{Ty} + M_{bTCy} + M_{kTCy} - r_{VM}F_{VM}\sin\beta_{TC} = 0 \quad (4)$$

$$m_C\ddot{y}_C + F_{CRy}\ddot{y}_C - F_{bTCy} + F_{CRy}\dot{y}_C - F_{kTCy} - F_{VM}\cos\beta_{TC} = -F_{TCy} \quad (5)$$

$$J_{Cy}\ddot{\varphi}_{Cy} + b_{\varphi CRy}\dot{\varphi}_{Cy} - M_{bTCy} + k_{\varphi CRy}\varphi_{Cy} - M_{kTCy} + r_{VM}F_{VM}\sin\beta_{TC} = 0 \quad (6)$$

Where force FVM is a connection force between the conveying part and the supporting part, which is transferred by the joint mechanism of both parts. Force $FTCy$ is the exciting force between the conveying part and supporting part.

4. Working principle

Auto loader is to load the components to the grinding machine. The machined components are drawn from it and finally load to container box. The detailed operation of an auto loader is involved through the gravity feeder, escapement valves, and conveyor. Initially, the components are loaded in the gravity feeder only[9]. The name itself indicates that, it feed the components to the conveyor with help of gravity. According to Newton's gravity laws, the components with certain mass always attracted towards the ground. The gravity feeder is at the angle of 40°, through the sliding force one by one the components will come to next position. In the gravity feeder, the guide strips are provided to avoid external problems and easy adjustment for varying lengths of components loading conditions[3].

To control the components to be fed on conveyor, escapement valves are used. Escapement valves are acts as a blocker of whole remaining components. In auto loader, machine consists of two escapements. These two escapements are operated with cylinder performances, which are controlled by PLC.

Auto loader consists of three conveyors are in-feed conveyor, Out-feed conveyor and External conveyor.

All these conveyors are used to transfer the component from one place to another[10]. In-feed and out-feed conveyor is of round belt conveyor, uses an induction motor for running with help of pulley & the External conveyor is the flat belt type. Figure 5. Shows the finalized manufacturing machine of steo feeder automation of ball joint. The already loaded component in the gravity feeder is fed to the conveyor through, the escapements' sequence the sequence as follows.

Escapement 1-up → Escapement 1-down [all components in the gravity feeder stepped down to one-step].

Escapement 2-up→ Escapement 2-down [a single row of aligned components fed to the conveyor].

The above-mentioned sequence will operate with help of proximity inductive flange type sensor. After the components are in the conveyor crossed the sensor, the next sequence starts. The proximity sensor is also placed in out-feed to verify all the components arrived or not.

Singling unit is a functional mechanical unit where the ball joint is positioned and pushed to final location of the feeder, this is done by two singling cylinders and a rotary cylinder.

Initially, the testing components are arranged on the gravity feeder in a proper manner. In this machine, there was the operation is controlled by two modes, which are manual and auto.

The operation type is selected by switch the specific mode. If the operation is in manual mode, the parts moment of the machine is controlled by the human with the help of HMI screen. If the operation is in auto-mode means to click cycle start option only on the HMI screen. The escapement 1 and escapement 2 are used to allow the testing components to the conveyor by the alternative up and down moments of the escapement cylinders.

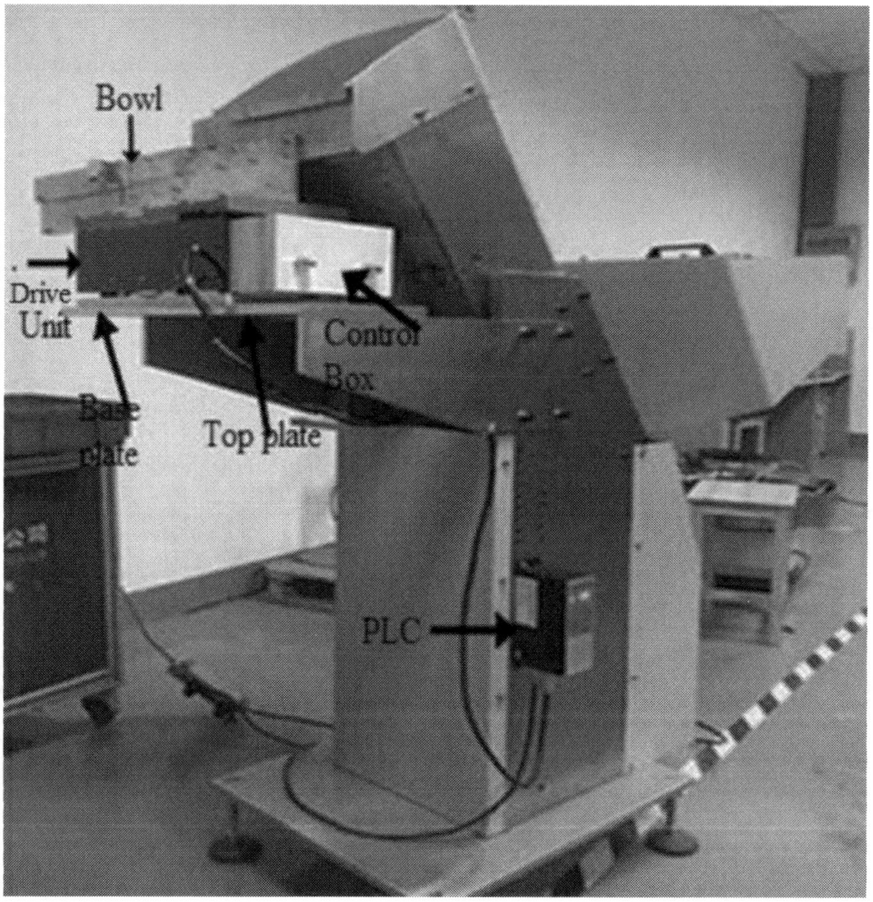

Figure 5. Finalized manufacturing machine of step feeder automation

Apart from designing, assembly plays a crucial role in working of machine. If mechanical components must be fitted according to design. The error may happen at machining or in design, the assembly section finds the problem occurred and also to discovery a suitable solution for the problem. This problems and solutions are discussed later.

In this machine, three components are used. Those vary according to the diameter namely 10 mm,12.5mm & 25mm. For each diameter size the guide way to be varied. Actually, the auto loads' component to the grinding machine for this purpose guide way is provided and also for better accuracy. The guide consisted of the strip provided with slot.

Spacer in the guide way is used to align the centre of in-feed and out-feed with grinding machine spacer in escapement cylinders is used to hold the components according to size as mentioned.

Escapement is to be changed according to component size. This is provided because if escapement is same for all means can damage the component.

5. Result and Discussion

This paper presented problems occurred during the testing of an auto loader. From all problems mentioned below are identified with the help of repeating testing and evaluation of the research manner. When the components fed from a row to the conveyor through in the sideway escapement, single component or more than one turns and fell into the conveyor in wrong direction. The space between gravity feeder and conveyor is varied (about 4mm distance) when the adjusted the conveyor, the problem vanished. In a row, there will be four components, but only three components will fall down to the conveyor, and one stuck in gravity feeder causing escapement error in HMI. After this problem, identified a bur in the edge of gravity feeder. This problem will also occurred if distance between the slide-way strips is uneven. Some component contains oil, adhesion problem existed. The components should not be grouped together while feeding.

For same position, the length of the round conveyor uneven belt helped to escape the components from the row. Rework took place. For many components, the rework is done [for example, Guide-way of 12.5 mm, escapement spacer of 25 mm, nylon strip (hole misaligned). Torque is not enough to pull the loaded components in conveyor, stacking of three or four components in the conveyor itself. Increase the length of in-feed conveyor.

This paper is give the solutions for the all researched problems of the machine after the continuous testing and at last machine is completed and successfully finalized.

6. Conclusion

In this paper, a conceptual foundation for automatic step feeding machine for ball joint has been presented. The framework introduces feeding aspects that cover how production is increases in automobile industry, and how the feeding process can be conducted for steering ball joint mechanism. The demonstrated work is expected to raise the reactiveness of automobile production companies, in terms of reducing the development time of production systems and Manpower.

The essential focus of the paper is to express how automatic step feeding mechanism can be sequence's ball joint, for which a sensing-based matchmaking procedure is used to conduct the identification of the steering wheel. Through the designs provided in this paper, the technological approach is rowed to be a satisfied approach to the modelling ball joint feeding mechanism, since it equips the modelling of constraints between abilities and proposal's continuous incorporation of steering wheel mechanism. Future work will be concentrated on stretching the capacity model for different parts feeding, together with implementation of the matchmaking procedure. Furthermore, since the modeling approaches only been used to design part feeding resources, further research has to be highly organized to corroborate the innovative approaches against other types of resources.

References

[1] Najmi S, Karimi A, Shadmani M and Saeed Z, A new three-dimensional dynamic machine and experimental validation for motion of a part in a vibratory bowl feeder. Mechanism and Machine Theory, 2020. 143.

[2] Buzzoni M. Battarra M, Mucchi E and Dalpiaz G, Motion analysis of a linear vibratory feeder: Dynamic machining and experimental verification. Mechanism and Machine Theory, 2017. 114: p. 98-110.

[3] Ghosh A, G.-N. Wang and J. Lee, A novel automata and neural network based fault diagnosis system for PLC controlled manufacturing systems. Computers & Industrial Engineering, 2020. 139.

IOP Conf. Series: Materials Science and Engineering **995** (2020) 012021 doi:10.1088/1757-899X/995/1/012021

[4] Sadizade B, Araee A, Nadimi S and Farshi V, Plateau honing of a diesel engine cylinder with special topography and reasonable machining time. Tribology International, 2020. 146.
[5] Kim N, H. Kim, and W. Lee, Hardware-in-the-loop simulation for estimation of position control performance of machine tool feed drive. Precision Engineering, 2019. 60: p. 587-593.
[6] Yong Y, Qi-Xin Z, Wei-Min A and Quan-Sheng J, Dynamic Characteristic Optimization of Ball Screw Feed Drive in Machine Tool Based on Modal Extraction of State Space Machine. IEEE Access, 2019. 7: p. 55524-55542.
[7] Hansson M N, Jarvenpaa E, Siltala N and Madsen O, Machineling Capabilities for Functional Configuration of Part Feeding Equipment. Procedia Manufacturing, 2017. 11: p. 2051-2060.
[8] Minglani D, Sharma A, Pandey H, Dayal R, Joshi B and Subramaniam S, A review of granular flow in screw feeders and conveyors. Powder Technology, 2020. 366: p. 369-381.
[9] Jia P, Y. Rong, and Y. Huang, Condition monitoring of the feed drive system of a machine tool based on long-term operational modal analysis. International Journal of Machine Tools and Manufacture, 2019. 146.
[10] Dakshin P. and S. Khurana, Frustum CAM (FC) mechanism: Achieving differentiating feed in continuous production machines of non-similar parts. Materials Today: Proceedings, 2020. 28: p. 920-924.

Smart Real-Time PV Surveillance Network with IOT

P Maniraj[1], A Lathika[2], C M Ravina[3], E Pradeebha[4], M Suresh[5], S Sundar[6]

1 Assistant Professor, M.Kumarasamy College of Engineering, Department of Electrical and Electronics Engineering, Thalavapalayam, Karur , Tamilnadu, India.
2,3,4 UG Student, M.Kumarasamy College of Engineering, Department of Electrical and Electronics Engineering, Thalavapalayam, Karur , Tamilnadu, India.
5 Assistant Professor, Kongu Engineering College, Department of Electrical and Electronics Engineering, Perundurai, Erode, Tamilnadu, India.
6 Assistant Professor, Bannari Amman Institute of Technology, Department of Electrical and Electronics Engineering, Sathyamangalam, Tamilnadu, India.

maniraj.angu@gmail.com, lathikaajay397@gmail.com, ravinaamurugesan@gmail.com , mailtopradeebha@gmail.com , msureshcit@gmail.com , sundars@bitsathy.ac.in

Abstract. The photovoltaic panel's network must stay in its extreme state to handle continuous energy efficiently. In the event of weather change, reliable energy yield changes. So, constant maintenance and monitoring will be needed. In this paper, we have projected an innovative real-time monitoring system to monitor the environmental conditions that are affecting the performance of panel such as fault, dust coverage or shading surfaces. Through this monitoring device, the deterioration of photovoltaic panel can also be defined and it also gives the information about the causes of current fault. An android app cayenne is interfaced with the system that will show the information about the temperature, rain and dust condition.

Keywords: IOT, Arduino UNO, Sensors, LCD Display.

1. Introduction

Rapidly rising needs for energy and global environmental issues involve the use of renewable energies. Solar panels are of the greatest future projection among renewable energy-generating technologies. Compared with the previous four decades, the involvement of solar Photovoltaic energy enlarged unusually. Solar PV nowadays represents the third-largest renewable energy source after wind and hydro.

Because of the huge electricity demand, India nowadays faces many power cuts. While a large number of renewable resources such as tidal, wind, geothermal, natural biofuel, and traditional biomass are present. But solar energy alone has enormous benefits. The rapid growth of society would increase the consumption of electricity. The only source for feeding such a mass population will be renewable energy.

After China, the USA, and Russia, India ranks fourth in energy consumption. India's consumption of electricity is 70%, whereas for 30 % of the population there is no electricity [1]. With the aid of the IoT, this issue can be solved.

IOP Conf. Series: Materials Science and Engineering **995** (2020) 012022 doi:10.1088/1757-899X/995/1/012022

Owing to the interconnection of different technologies, real-time data visualization, machine intelligence, product sensors and embedded devices the idea of the Internet of Things has evolved. Embedded systems, wireless sensor networks, monitoring systems, robotics (including home and building controls) enable the IoT and a lot more.

An IoT ecosystem involves smart web-enabled devices that store, communicate and use embedded processors, sensors and communication hardware to acquire records they receive from their environments. IoT systems share sensor data collected by linking it from an IoT access point or other edge devices in which data is either transmitted to the server to be analysed or localised. These devices also interact with related devices and rely on each other's data. A great deal of analysis is done by computers without human intervention, while people can connect with it.

The internet of things provides organizations several advantages which enable them to:

- Track product development as a whole
- Allowing customer service better
- Save money and time
- Increases productivity for employees
- Build-in and evolve business models
- Make better choices
- Plus sales production

This paper consists of certain fragments: Division II describes the analysis of literature in this field. Division III presents the current effort and proposed method which has been done and Section IV encapsulates the outcome.

2. EXISTING MODEL

A low-cost server uses ESP32 with physical system power for present collection of photovoltaic statistics. A data collection program for solar thermal parks. Data were gathered from PV park devices, dimension systems in the parking area (climate stations), and information bases information on the web. In terms of IoT (Internet of Things) spreads, there is a massive rise in uses in the smart city, smart parking, and so on. IoT reduces humanoid involvement by implementing system-to-system interaction, which is used to ease the analysis of multiple device modules. For this purpose the aforementioned surveillance system uses internet to link to remote services and provide a cloud platform for data collection to ensure the reliability and accuracy of the expected model. This makes the online monitoring and recording of ecological and knowledge conditions of the PV panel in real-time. It will also make the potential changes to the system smoother. For this cause, researchers have built further techniques in modern years to help gather and analyse PV data for the initial identification of faults. Built a control unit for synchronised monitoring of four PV screens. For the identification of faults the mean power of the PV boards and the voltage descent limit are used. Premeditated a mathematical method to track photovoltaic conditions and detect the hotspots. They follow the aspects (Cp and Rp) in the PV panel's equivalent method and use these limit projections to forecast when the PV board is part of the details or under hot-spot conditions.

3. PROPOSED MODEL

The system of solar cells is flattering a general choice as an energy source. To use sustainable electricity proficiently, the PV panel system must endure in the best possible circumstance. This device consists of a sensor such as a temperature sensor, a dust detector and a rain sensor. The temperature sensor is used to sense the temperature value and the dust detector is used to detect dust intense. The rain sensor is used to measure the rain value and all the parameters are displayed through LCD. The values will be uploaded to IoT.

IOP Conf. Series: Materials Science and Engineering **995** (2020) 012022 doi:10.1088/1757-899X/995/1/012022

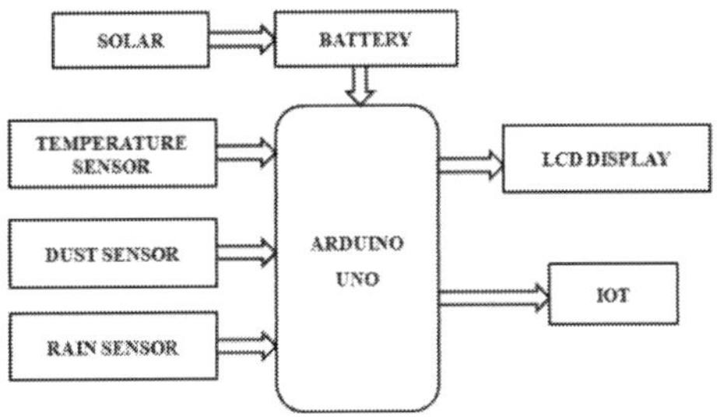

Figure 1: Proposed system smart real-time PV surveillance network with IoT

A. Arduino UNO

In Arduino UNO has 14 digital pins and is used as input or output with 6 analogue pins.As the specified operating situation, the rising pin can supply or receive 20 milliamperes, and also have an internal pull-up resistor (default detached) of 20-50 k ohm.

Indeed, certain pins have different functions:

• Serial / UART: Receiving (RX) TTL serial data. Pin number 0 and transmitting (TX) TTL serial data pin number. These terminals are linked to the respective ATmega8U2 USB-to - TTL Serial Chip pins.

• Peripheral interrupt pins 2 and 3. These pins can be designed to initiate a low-value interruption, an increasing or decreasing power, or a change in value.

• PWM pins: Pulse Width Modulation pins are 3, 5, 6, 9, 10, and 11 It will arrange for 8-bit PWM performance through analog Write() feature.

• SPI pins: Serial Peripheral Interface pins are 10 (SS), 11 (MOSI), 12 (MISO), and 13 (SCK).

These pins provision Serial communication through the SPI library.

Figure 2: Arduino UNO

B. Solar panel

Figure 3: Solar panel

The photovoltaic cells are assembled in solar panel and assembled on structural support. The solar unit may be used as a part of a broader PV network to produce and provide energy for business, domestic purposes under STC. Each unit is rated with its DC output power, which typically ranges from 100 to 320 watts. A small amount of electricity will be produced by a single solar module, so most of the systems have several modules. A PV system usually involves a panel or series of solar panels, an inverter and a battery and/or solar sensor and wiring link.

C. Battery

The specification of battery used in our system is given below.

Input power: 120V AC 50/60Hz 0.4A Max.

Battery voltage rating not less than 5.5V

Rating output: 12V DC 1.5A

Battery type: lead-acid battery

Maximum charge voltage: 14.4V

Maintenance charge voltage: 13.2V~14.0V

Operating environment:-10~40 ° C

Figure 4: Battery

D. Temperature Sensor

The humidity sensor detects, measures and regularly reviews the relative humidity levels. The sensor tests both the humidity and the temperature. Relative humidity expressed as a percent, is the ratio of actual humidity in the air to the maximum amount of humidity at which the temperature can be maintained. The warm air can hold the more water content in it. So the humidity levels changes with temperature variations.

Humidity sensor measures the relative moisture of the local areas in which they are situated. It calculates the temperature and relative humidity in the air and describes relative humidity as a proportion of the ratio of humidity to the average level that can be sustained in the mid-air at the actual atmospheric condition. As the air gets warmer, more water vapor is kept, so the humidity level changes with the atmospheric pressure.

Figure 5: Temperature sensor

E. Rain Sensor

Figure 6: Rain sensor

It provides a digital module and a programmable logic controller that "collects" droplets. As water drops are gathered on the power cord, parallel resistance paths are created which are measured by an op-amp. Lesser the resistance or more water is collected; the output voltage produced will be lesser.

F. Dust Sensor

Dust Sensor is a basic dust detection kit which can detect small particles greater than 0.8μm in diameter, much like the tobacco smoke. The analog signal sensor output is constant with dust intensity. Air Quality is an abstract concept which typically involves measures of many indicators: dust rates – calculated in μg / m3, with various scale particles such as Volatile Organic Compound Gasses (VOCs) – formaldehyde, benzene, and more – calculated in ppm (parts per million) of CO2 – measured in ppm.

Figure 7: Dust sensor

Sharp's GP2Y1010AU0F is an air feature sensor or optical pollution tracker. This Dust sensor is used to detect powder molecules. The IR emitting diode and the phototransistor are diagonal

IOP Conf. Series: Materials Science and Engineering **995** (2020) 012022 doi:10.1088/1757-899X/995/1/012022

directions organized in this device to sense the light reflection of dust in the air. This is particularly efficient in identifying extremely small elements such as tobacco burn and is widely used in air filter systems.

4. Output

Figure 8: Hardware Setup

The data's from various sensors are given to controller. Based on the data's, the output voltage produced is displaced in LCD. Through Cayenne mobile application the real time data can be viewed by using mobile.

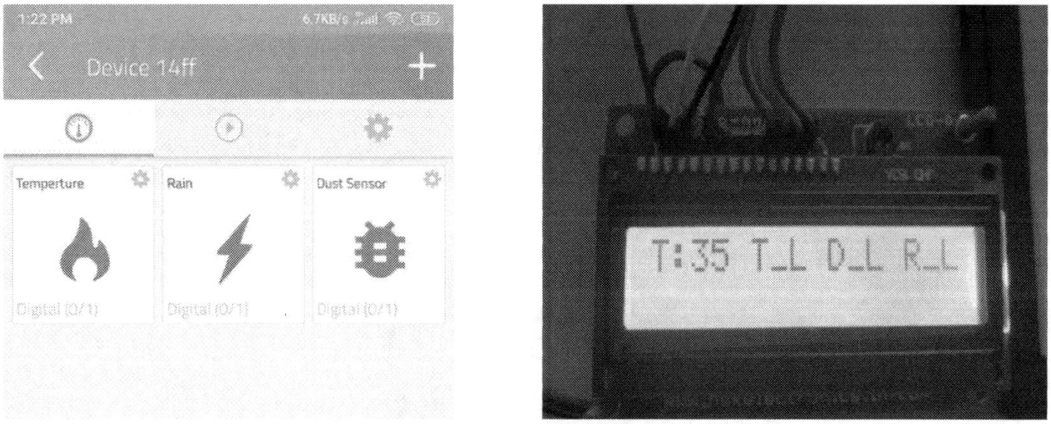

Figure 9: Output representation

The picture (figure 9) shows the various values of solar panel during normal condition. The values which are shown are lower than the ideal values that need to be given manually. The output shows the temperature, dust and rain are in normal value.

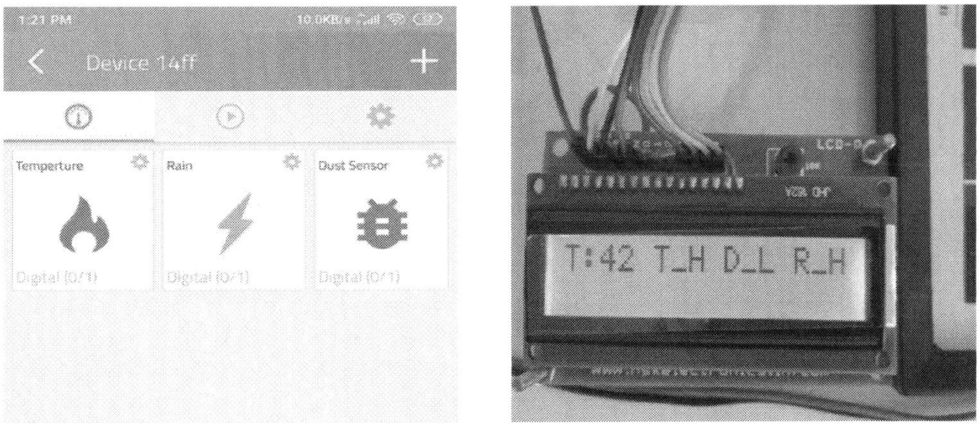

Figure 10: Output representation 1

The figure 10 shows the output,when the temperature is high and raining condition. The values are shown in LCD and also represented in mobile app.

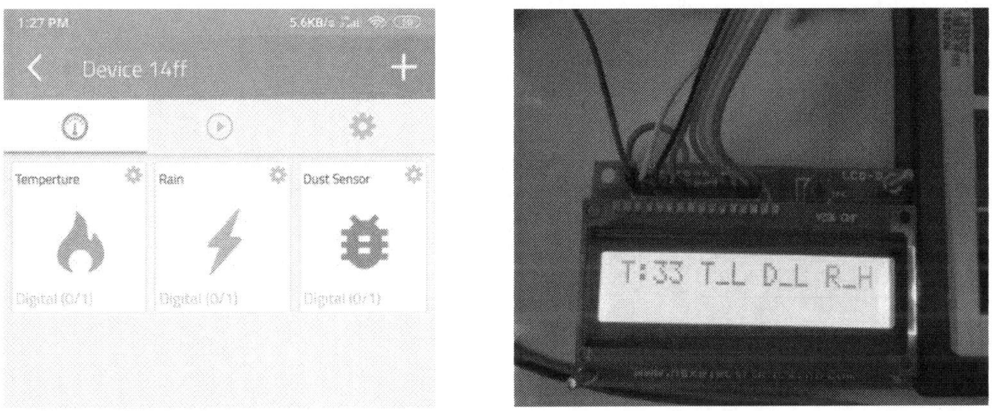

Figure 11: Output representation 2

The figure 11 shows the output,when there is a raining condition. The values are shown in LCD and also represented in mobile app.

5. Conclusion

Through this monitoring system, any individual PV panels that require maintenance can be identified. The reason for low output voltage (temperature or rain or dust) can be easily identified. The temperature, rain and dust data can be viewed through mobile. So, user can monitor the performance of solar panel at any place.

References

[1] Kumar 2015 Economic assessment of photovoltaic energy production prospects in India Procedia Earth Planetary Sci., vol. 11, pp. 425436

[2] Adhya S Saha D Das A Jana J and Saha H 2016 An IOT based smart solar photovoltaic remote monitoring and control unit 2nd International Conference on Controle Instrumentation Energy & Communication (CIEC).

[3] Benedek J Sebestyén T T and Bartók B 2018 Evaluation of renewable energy sources in peripheral areas and renewable energy-based rural development Renew. Sustain. Energy Rev vol. 90 no. 1 pp. 516535
[4] Mahdavinejad M S Rezvan M Barekatain M Adibi P Barnaghi P and Sheth A P 2018 Machine learning for Internet of Things data analysis A survey Digit. Commun. Netw vol. 4 no. 3 pp. 161175.
[5] Sethi P and Sarangi S R 2017 Internet of Things Architectures protocols and applications J. Elect. Comput. Eng vol. 2017 Art. no. 9324035.
[6] Ray P P 2016 A survey of IoT cloud platforms Future Comput. Inform. J vol. 1 nos. 12, pp. 3546.
[7] Natsheh M and Albarbar 2011 A Photovoltaic model with MPP tracker for standalone / grid-connected applications in Proc. IET Int. Conf. Renew. Power Gener. (RPG) Edinburgh U.K pp. 16.
[8] Natsheh M 2016 Power generation of solar PV systems in Palestine Appl. Sol. Energy vol. 52 no. 3 pp. 193196.
[9] Arribas L Cano L Cruz I Mata M and Llobet E 2010 PV wind hybrid system performance: A new approach and a case study Renew. Energy vol. 35 no. 1 pp. 128137.
[10] Omer S A Wilson R and Riffat S B 2003 Monitoring results of two examples of building-integrated PV (BIPV) systems in the UK Renew. Energy vol. 28 no. 9 pp. 13871399.
[11] Pietruszko M and Gradzki M 2003 Performance of a grid-connected small PV system in Poland Appl. Energy vol. 74 nos. 12 pp. 177184.

CONTROL AND STABILIZATION OF PITCH AXIS ANGLE IN 3 DOF TITLTROTOR

Mr.R.Dhanasekar[1], Ms.M.Kalaiyarasi[2], Mr.S.Arunjayakar[3], Mr.M.Raghunath[4], Mr.S.Raghavendra prabhu[5]

1,2, 3Assistant Professor, Bannari Amman Institute of Technology, Department of Electronics and Instrumentation Engineering, Erode, Tamilnadu, India.
4,5 Assistant Professor, Department of Mechatronics, Bannari Amman Institute of Technology, India.
dhanasekar1693@gmail.com , kalaiyarasim@bitsathy.ac.in, arunjayakars@bitsathy.ac.in, ragueee19@gmail.com, raghavendraprabhu@bitsathy.ac.in

ABSTRACT - Tiltrotor system is used in army for security purpose and mission attack. This system is affect due to environment conditions such as wind flow, rain and low atmospheric pressure, the system becomes unbalanced in working conditions. In recent days new control technique called fractional order PID have a conspicuous growth in the field of controller design for controlling the various linear and nonlinear system applications. The fractional order PID controller used in tiltrotor model, improves the system performance compared with other controllers. The proposed work deals with the Pitch axis control of tiltrotor system using FOPID controller and also performance has been compared with PID controller, with different tuning methods. The FOPID controller provide more flexibility in designing a controller than using PID controller due to selection of tuning parameters. FOPID and PID controller has been designed for pitch axis control, the performance characteristics of the system has been compared and analyzed by using MATLAB simulation.

Keywords : Tiltrotor – FOPID - Pitch axis - PID Controller.

I. INTRODUCTION

1.1 BENCH TOP TILTROTOR MODEL

Tiltrotor framework is made out of the base, utilized equalization, adjusting squares, propellers and some different parts. Equalization post to base as its support, and the pitching. Propeller and the equalization blocks were introduced at the two finishes of a parity bar. The propeller rotational lift, turning an equalization bar around the support so pitching moves, utilizing two Propeller speed contrast, turning a parity bar along the support to do rotational development. Parity the two shafts introduced encoder, used to quantify the turn hub, pitch hub point, in the two propeller associating pole introduced an encoder, which is utilized to gauge upset hub edge as appeared in figure 1.1.. Two propellers utilizing brushless DC engines, give the catalyst to the propeller. By modifying the equalization bar introduced in the side of the parity squares to diminish propeller engine yield. All electrical signs to and from the body are communicated by means of slip ring in this way wiping out the chance of tangled wires and decreasing the measure of contact and stacking about the moving pivot. Arrangement of the exploratory direction on the intention is to advise clients how to plan a regulator, to control the helicopter as per the ideal point and speed of development.

IOP Conf. Series: Materials Science and Engineering **995** (2020) 012023 doi:10.1088/1757-899X/995/1/012023

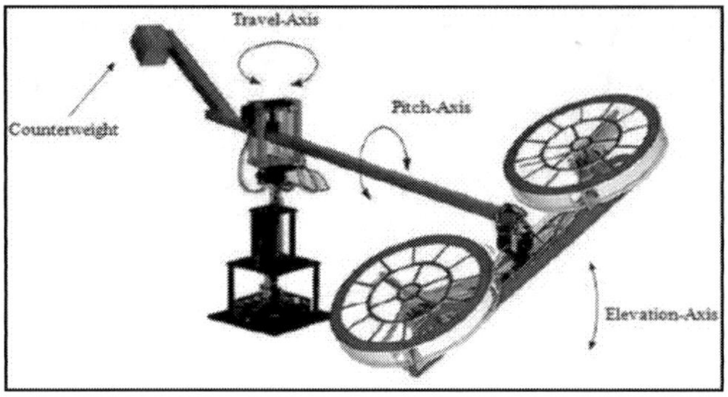

Figure 1.1 Bench Top Tiltrotor Model

1.2 PID CONTROLLER

Proportional integral derivative controller (PID controller) is a control loop feedback mechanism (controller) commonly used in industrial control systems. A PID controller continuously calculates an error value as the difference between a measured process variable and a desired set point. The PID control scheme is named after its three correcting terms, whose sum constitutes the manipulated variable (MV). PID controllers, when used alone, can give poor performance when the PID loop gains must be reduced so that the control system does not overshoot, oscillate or hunt about the control set point value. They also have difficulties in the presence of non-linearity, may trade-off regulation versus response time, do not react to changing process behavior (say, the process changes after it has warmed up), and have lag in responding to large disturbances.

1.3 FOPID CONTROLLER

A fragmentary request PID regulator is spoken to as 〖PI〗 ^ʎ D^μ. It permits us to alter subordinate and indispensable request notwithstanding the corresponding, essential and subsidiary constants where the estimations of ʎ and μ lie somewhere in the range of 0 and 1.FOPID regulator improves the quality and power of the framework than the ordinary PID regulator. This gives additional opportunity to work regarding two handles as,

- Order of differentiation
- Order of integration

This likewise gives greater adaptability to better the dynamical properties of the control framework. The fragmentary request regulator delights great heartiness. The power of partial regulators gets more featured within the sight of a non-linear actuator.

II. MATHEMATICAL MODELING FOR TILTROTOR SYSTEM

It is made out of the base, utilized equalization, adjusting squares, propellers and some different segments. Equalization presents on base as its support, and the pitching. Propeller and the equalization blocks were introduced at the two closures of a parity bar. The propeller rotational lift, turning a parity bar around the support so pitching moves, utilizing two propeller speed contrast, turning an equalization bar along the support to do rotational development. Parity the two posts introduced encoder, used to gauge the pivot hub, pitch hub point, in the two propeller interfacing bar introduced an encoder, which is utilized to quantify toppled hub edge. Two propellers, utilizing brushless DC engines, give the driving force to the propeller. By, changing the parity pole introduced in the side of the parity squares to diminish propeller engine yield.

IOP Conf. Series: Materials Science and Engineering **995** (2020) 012023 doi:10.1088/1757-899X/995/1/012023

2.1 PITCH AXIS

Expecting the move pivot zero, at that point the pitching hub force by two propeller engine lift the F1 and F2 as appeared in figure 2.1. Accordingly the pitch propeller pivot all out lift Fh=F1+F2. At the point when the lift Fh is more prominent than the gravity G helicopter rise. Rather the helicopter dropped. The pitch hub model for tiltrotor is indicated Fig.2. Presently, expecting zero move hub, the differential condition is,

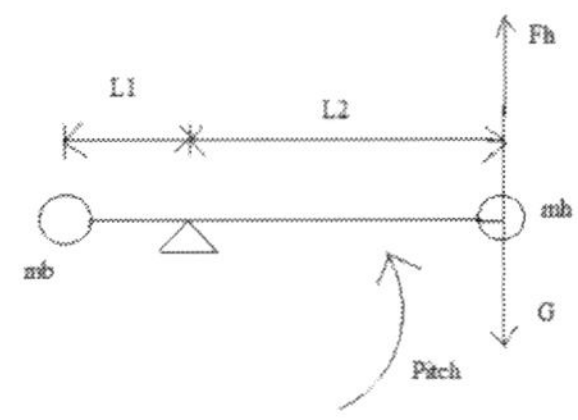

Figure. 2.1 .Schematic Diagram for Pitch control

$$J_p\ddot{\theta} = M_m diff(udiff) + Mg\theta + M_r cor\theta \qquad \text{---------------------- (1)}$$

Gravitational force and Coriolis force should be zero in motion state,

$$J_p\ddot{\theta} = M_m diff(udiff) \qquad \text{----------------------- (2)}$$

Where,

- J_p is the moment of inertia
- $M_m diff$ is the momentums caused by the motor
- $udiff$ is the difference between front motor and back motor
- $M_r cor$ is the Coriolis force

$$J_p\ddot{\theta}h = W.udiff \qquad \text{---------------------- (3)}$$

S.No	Tuning Method	K_p	K_i	K_d	ʎ	µ
1	Z-N	0.392	1.177	0.72	0.6	0.5
2	T-L	0.556	0.416	0.29	0.6	0.5

Laplace transform apply in equation (3)

$$\frac{d\theta h(s)}{d\theta w(s)} = \frac{udiff}{J_p s^2} \qquad \text{-------------------------- (4)}$$

By substituting the below values on the pitch axis transfer function in equation (4),

U diff=1.5N;j_p=0.03kg/m^2 At 45^0 motion position,

$$\frac{d\theta 1h(s)}{d\theta 1w(s)} = \frac{50}{s^2} \qquad \text{------------------------- (5)}$$

U diff=2.5N;j_p=0.03kg/m^2: At 70^0 motion position,

$$\frac{d\theta 2h(s)}{d\theta 2w(s)} = \frac{83.3}{s^2} \qquad \text{------------------------- (6)}$$

III CONTROLLER DESIGN FOR TILTROTOR MODEL

3.1 DESIGN OF CONVENTIONAL PID CONTROLLER

The design of the conventional PID controller is in need the exact values of the controller tuning parameters K_p, K_i, K_d in accordance with the system transfer function.

Tuning methods	Tuning parameters		
	K_p	**K_i**	**K_d**
Z-N	0.392	1.1 77	4.722
T-L	0.556	0.4 16	5.295

3.2 DESIGN OF FOPID CONTROLLER

The design of FOPID controller is in finding the five controller tuning parameters K_p, K_i, K_d ,λ and μ.

IV RESULTS AND DISCUSSION

The system model which consist of controller and the feedback loop for control the input flow in order to reduce the error by adjusting the tuning parameters using Ziegler–Nichols and Tyreus luyben tunning.

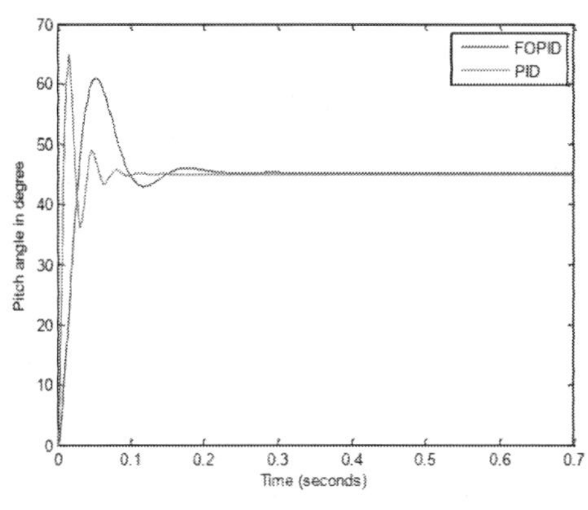

Figure 4.1 Closed Loop Response of Pitch Axis

IOP Conf. Series: Materials Science and Engineering **995** (2020) 012023 doi:10.1088/1757-899X/995/1/012023

Table.1.Closed Loop of PID Controller for Pitch Axis

Over shoot (%)	Rise time(seconds)	Settling time(seconds)
60	0.75	2.56

From the Figure.4.1 it is observed that angle for set point of 45 degrees, the PID controller and FOPID controller responses with settling time of 2.56 seconds. The Table.1 shows the closed loop of PID controller and FOPID controller for pitch axis. Also while comparing with open loop system, the PID and FOPID controller obtained the desired set point.

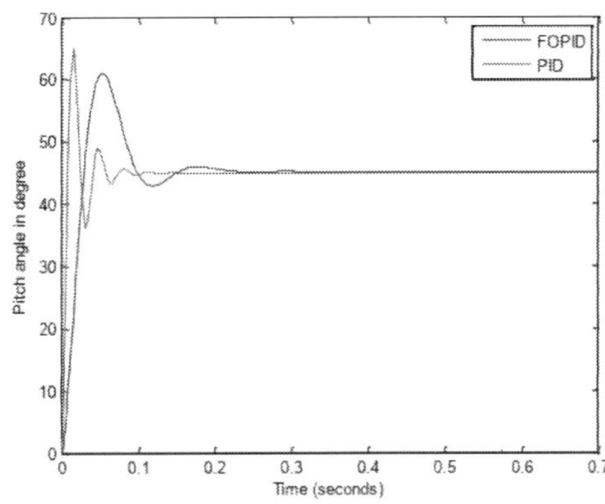

Figure 4.2. Response of Closed Loop for Pitch Axis

Table.2.Closed Loop with Disturbances in Pitch Axis

Overshoot (%)	Rise time(seconds)	Settling time(seconds)
60	0.75	2.56

From the Figure.4.2 it is observed that angle for set point of 45degrees, the PID controller and FOPID controller responses with settling time of 2.56 seconds. The Table.2 shows the closed loop of PID controller and FOPID controller for pitch axis. Also while comparing with open loop system, the PID and FOPID controller obtained the desired set point.

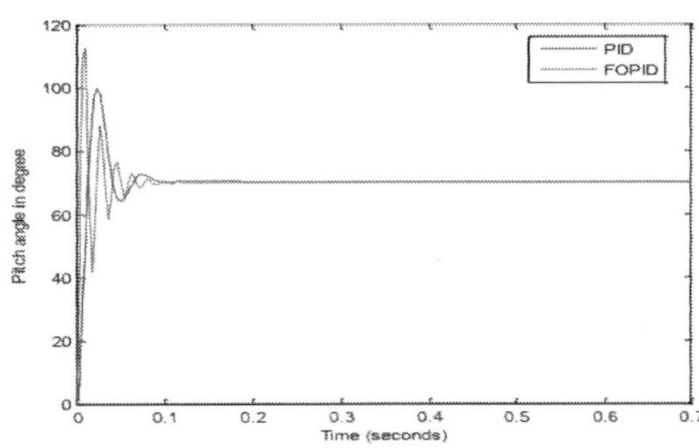

Figure.4.3. Response for closed loop of PID and FOPID Controller

Table.3. Closed Loop of PID and FOPID Controller for Pitch axis

Overshoot (%)	Rise time (sec)	Settling time (seconds)
40	0.5	39.4

From this Figure.4.3 it is observed that angle for set point of 70 degrees, FOPID and PID controller response are settling time of 39.4 seconds and over shoot of-. The Table.3 shows the closed loop of FOPID and PID controller for pitch axis. While FOPID controller is compared with PID controller, the FOPID controller overshoot is reduced.

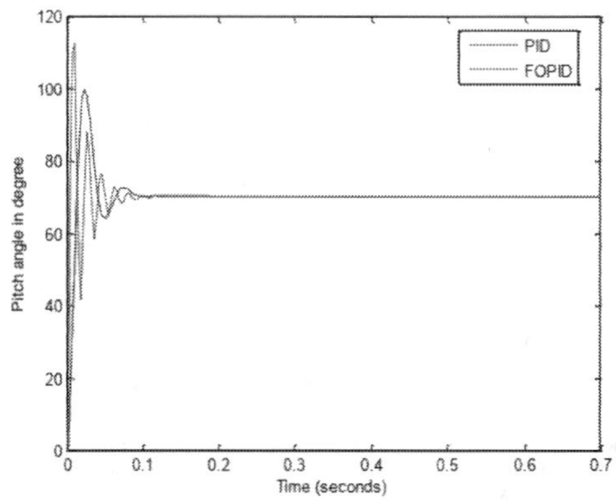

Fig.4.4 Response for Closed Loop with Disturbances in Tiltrotor Pitch Axis

Table.4. Closed Loop with Disturbances in Tiltrotor Pitch Axis

Overshoot (%)	Rise time(seconds)	Settling time(seconds)
40	0.2	2.56

From the Figure.4.4 it is observed that angle for set point of 70 degrees, the PID controller and FOPID controller responses with settling time of 2.56 seconds. The Table.4 shows the closed loop of PID and FOPID controller for pitch axis.

V. CONCLUSION AND FUTURE SCOPE

The tiltrotor bench top model is used to analyze with PID and FOPID controller. Mathematical model of pitch axis is derived and analyzed using different tuning methods. The control of pitch angle for 3 DOF tiltrotor can be done using some other advanced controller to improve the accuarcy.

REFERENCES

1. 3-DOF Helicopter-Laboratory Guide,(2012), Quanser 3-DOF Helicopter Laboratory Guide book, Quanser Inc.

2. Arbad NK (2014),"3DOF Model Helicopter with Hybrid Control", TELKOMNIKA Indonesian Journal of Electrical Engineering, Vol. 12,pp. 3863-3872.

3. Bo Zheng (2011) "Robust Attitude Regulation of 3DOF Helicopter Benchmark", IEEE, Vol. 5,pp.63-71.

IOP Conf. Series: Materials Science and Engineering **995** (2020) 012023 doi:10.1088/1757-899X/995/1/012023

4. Ge.Jinlai (2010),"Fuzzy Self-Tuning PID Controller in the 3-DOF Helicopter Experimental System", Information and Control. Vol.39, pp. 342-347.

5. He J, Wang Q,(1998),"PI\PID Controller Tuning via LQR Approach" .Proceeding of the 37th IEEE Conference on Decision & Control, pp. 1177-1182.

6. Mansor H (2010),"Design of QFT Controller for A Bench Top Helicopter System Model", International Journal of Simulation System, Science & Technology, Vol .11,pp. 8-16.

7. Technical Manual (2011), "3DOF Helicopter Systems", GOOGOL TECH

8. T.Kiefer (2004),Modeling and Flatness Based Control of a 3DOF Helicopter Laboratory Experiment

9. Yue Xin-cheng(2007),"Steady State Error Tracking Control of 3DOF Experimental Helicopter System", Journal of System Simulation, Vol.19, pp.4279-4283.

10. Zheng Fang (2012), "Robust Adaptive Integral Back stepping Control of a 3-DOF Helicopter," Int. Journal of Advanced Robotic Systems, Vol. 9, pp.980-992.

IOP Conf. Series: Materials Science and Engineering **995** (2020) 012024 doi:10.1088/1757-899X/995/1/012024

Performance Analysis of Various Edge Detection Techniques in X-Ray Imaging

R Sathish Kumar[1], R Karthikamani[2]

1 Assistant Professor, Department of Mathematics, Sri Ramakrishna Mission Vidyalaya College of Arts and Science, Coimbatore, Tamilnadu, India.
2 Assistant Professor, Department of Electronics and Instrumentation, Bannari Amman Institute of Technology, Sathyamangalam, Erode, Tamilnadu, India.

rsksrmv@gmail.com, karthikamani@bitsathy.ac.in

Abstract. Edge detection is an image processing method used for discover the limitations of objects within the image. It works by sensing incoherence in illumination. Edge detection is used for image segmentation and data extraction in areas such as image processing, computer vision and machine vision. In these four techniques such as Sobel Operator, Robert Operator, Prewitt Operator, and Canny Operator are used for detection of edge detection in which the canny operator is the most effective manner of finding the edge detection using X-rays. The key aim of this task is to sense human minor leg bone fracture from X-Ray imageries. The suggested structure has two steps namely pre-processing segmentation and fracture detection. Canny method produces very effective image. This paper compares the various edge detection techniques for detecting the bone fracture of lower leg bone and finds effective. The performance of the four segmentation techniques are compared based on the execution time and accuracy. The execution speed of Canny operator is 32.5 seconds and accuracy is 88.7%.

Keywords: X-ray, Sobel Operator, Robert Operator, Prewitt Operator, and Canny Operator.

Introduction

Image processing became a part of this world in every field like medical, etc. It also developed in medical applications to provide analysis of human body. Many digital images has been taken from the patients like X-ray, Computerized Tomography (CT), Ultra Sonography (US), Magnetic Resonance Imaging (MRI). These images are helpful for the diagnosis for physicians and research in medical. It is the very easiest way to know the problem of patient's body parts and location. Human body consists of 206 bones[1]. The bone fracture is usually detected using X-ray. It is commonly problem in human beings and classified in various ways. They are per prosthetic fracture: weakness of bone, traumatic facture: due to sustained trauma caused by road accident, fall; pathologic fracture: due to disease, weak the bones caused by osteoporosis[2], This paper analyse the natural leg bone and fractured leg bone. Edge detection algorithm is used in this paper to detect fracture of leg by using MATLAB software. Classifications of edges based on behaviour are: edge structure, edge orientation and noise environment. To perform edge detection, gradient method and laplacian method are used. Gradient method will detect the edge of image in maximum and minimum at first derivative. It will specify the threshold and finds magnitude of greater intensity in first derivative. Other method is laplacian method, which finds edge in zero crossing at second derivative. It has one dimensional ramp shape. Common operators used in gradient method are Robert operator, Sobel operator, Prewitt operator, Canny operator and others. These operators are helpful to analyse crack in bone by medical images[3].

1. Block diagram

The figure 1 shows the system flow diagram that has been implemented to get results of fractured bone and normal bone of leg using different algorithms. It is done using MATLAB software for better results. Input image is pre-processed using required filter to remove noise and image segmentation is applied using edge detection operators. Therefore final result is obtained using different operators of non-fractured bone and fractured bone.

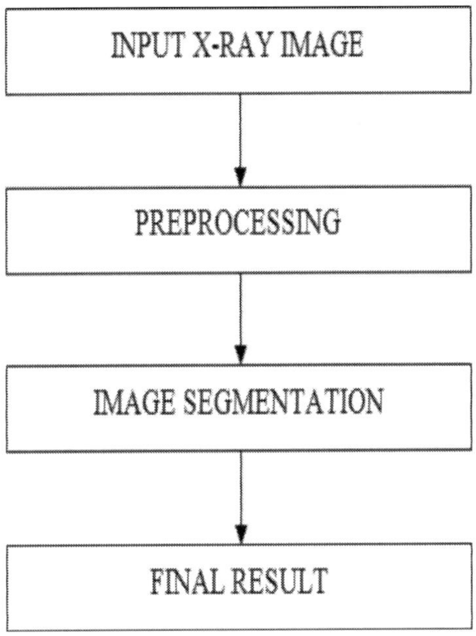

Figure 1. System Flow Diagram

2. EDGE DETECTION TECHNIQUES

2.1. SOBEL OPEARATOR

This operator has pair of 3*3 matrix convolution and rotated by 90. Based on pixels, it will responds in maximum to edges in vertical and horizontal directions. In each orientation of input image produces separate measurements from each kernals. This will helpful to combine each orientation in magnitude. Device the Sobel method for edges detection, which is created on a 3 by 3 array that is stirred over the main image[4]

-1	0	1
-2	0	2
-1	0	1

1	2	1
0	0	0
-1	-2	-1

Gradient magnitude is given by:

$$|G| = \sqrt{G_x^2 + G_y^2} \quad (1)$$

Approximate magnitude is,

$$|G| = |G_x| + |G_y| \tag{2}$$

Angle orientation relative to pixel grid is given by,

$$\theta = arc\tan\left({G_y}/{G_x} \right) \tag{3}$$

2.2. ROBERT OPERATOR

Robert operator is a two dimensional spatial gradient image measurement. It is simple and compute easy. This operator has pair of 2×2 matrix convolution kernel. Any of one kernel is rotated in 90. Based on pixels, it will responds in maximum to edges in vertical and horizontal directions at 45. In each orientation of input image produces separate measurements from each kernals. This will helpful to combine each orientation that is in magnitude[5].

Angle orientation relative to pixel grid is given by,

$$\theta = arc\tan\left({G_y}/{G_x} \right) - {3\pi}/{4} \tag{4}$$

2.3. Prewitt Operator

It is used to detect horizontal and vertical orientation of an image and developed by Judith M. S. Prewitt. This operator has pair of 3*3 matrix convolution and rotated by 90. Based on pixels, it will responds in maximum to edges in vertical and horizontal directions. In each orientation of input image produces separate measurements from each kernals. This will helpful to combine each orientation in magnitude [6]. The masks are as shown as

-1	0	1
-1	0	1
-1	0	1

-1	-1	-1
0	0	0
1	1	1

Gradient magnitude is given by

$$|G| = \sqrt{G_x^2 + G_y^2} \tag{5}$$

2.4. CANNY OPERATOR

This operator was developed by John F. Canny. The noise is identified in images for better analysis. Therefore, canny operator is efficient compared to other operators [7].

1. The canny algorithm shows good detection of real edges in images.
2. The edges detected by this operator appear close to the real image and there is no overlapping of edges.
3. The canny operator does not shows false edges.

The canny algorithm uses five steps to detect the edges:

1. The images are smoothened by removing noise.
2. The edges are found by marking the maximum gradient of the image.
3. Local maxima alone are marked as edges of the input image.
4. Thresholding values are used to determine the potential edges of image.

IOP Conf. Series: Materials Science and Engineering **995** (2020) 012024 doi:10.1088/1757-899X/995/1/012024

5. Only strong edges that are connected to a very certain point are marked as final edges.

3. SOFTWARE DESCRIPTION

3.1 MATLAB

The full form of MATLAB is MATrix LABoratory developed by MathWorks. It is a proprietary programming language and multi-paradigm computing environment. MATLAB allows algorithms to implement, data and functions plotting, programming interfaces with different languages like C++, C, Java, Python and user interfaces. It has a additional packages, Simulink, graphical simulations and used for engineering, science and economics. Therefore matlab code or simulation can be done to run the program or models for various applications. The file extensions using MATLAB are .m, .mat, .mlx, .rpt. Simulink extension are .mdl, .slx, .mdlp. MATLAB became important for application like deep learning, computer vision, signal processing, robotics, control system, image processing, etc.[10]

Features of MATLAB are:

- High level language for numerical computations, application development and visualization.
- Provides interact with environment for designing, problem solving and iterative exploration.
- Mathematical functions and graphics.
- Improve code quality maintain and performance.

Using this software the project is developed for detecting bone in human body from X-ray image as input in SIMULINK.

4. Implementation

The figure 2.1show the Simulink model for fractured leg. Various blocks have been used to detect input images.

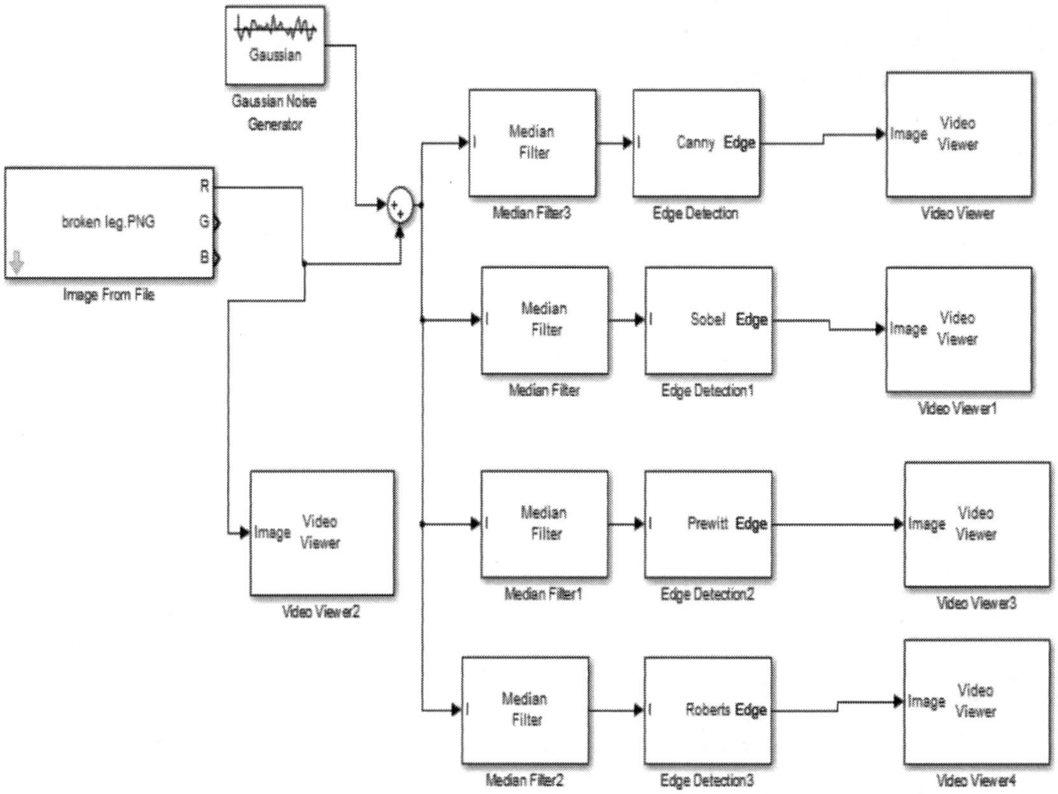

Figure 2. Simulink model for various edge detector operators of fractured leg

4.1. Working

The input x-ray image is taken from the hospitals or from internet. The image is uploaded in the MATLAB software. Then the image is pre-processed. The enhancement of input x-ray image is done in this stage, so that output image is obtained with better performance. Noise removal, image enhancement and colour enhancement are the main procedures. The unwanted pixels which affect the quality of image are called noise. Different types of noises are Poisson, Gaussian, salt and pepper, etc. the common type of noise found in x-ray image is Gaussian noise. It is very good to remove noise in image. After removal of noise the brightness and colour of the input image is enhanced to obtain the desired output image [8]. To reduce the processing time the image is converted to grey scale image. The region in which the red, green and blue elements have similar intensity is called 'grey' colour region. It contains grey scale values. The input image may contain some defects such as low contrast. These defects have huge impact on the contrast of an image.

The important step to analyse the image and read the data from the image is Image segmentation. Image segmentation is the processes of breaking down the picture to segments or pixels. It is used to identify the edges in an image. The different methods to find edges of an image are:

1. Based on discontinuities in intensity, finding boundaries.
2. By using pixel property thresholds are obtained.

3. Regions are found directly.

The problem is considered by the image segmentation technique. There is a region based method, which is found by continuity. Depend on some rules, image is divided into sub regions, like same grey scale level and pixel in one region. This technique depends on patterns of intensity values with nearby pixels. Based on anatomical and functional roles the clusters are identified.

The simplest way of image segmentation is by finding the threshold. Regions are classified based on range values by using Thresholding techniques. Thresholding is stated as the transformation of input image to an output image.

These methods are called as Edge or Boundary based methods. Edge detection is most commonly used technique. Various edge detection techniques used here are Sobel, Robert, Prewitt and Canny. These techniques are discussed above in chapter 3. Median filter is a non-linear filter used for smoothing of image by removing noise in an image. Image segmentation is done by analysing the changes in image intensity [9].

The output from the edge detector is viewed through video viewer. The video viewer 2 is used to view the gray scale image of the input image. The video viewer shows the output of the canny edge detection. The video viewer 1 shows the output of the Sobel edge detection. The video viewer 3 shows the output of the Prewitt edge detection. The video viewer 4 shows the output of the Robert edge detection [11].

5. Result

The results of normal leg and fractured leg images of various edge detection techniques are compared below.

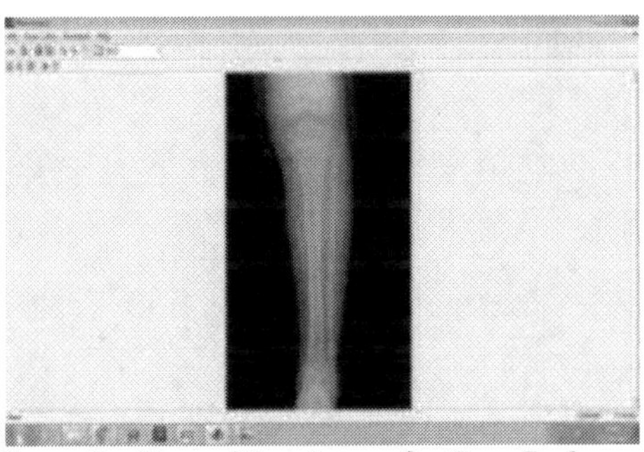

Figure 3.a Normal Leg Image for Grey Scale

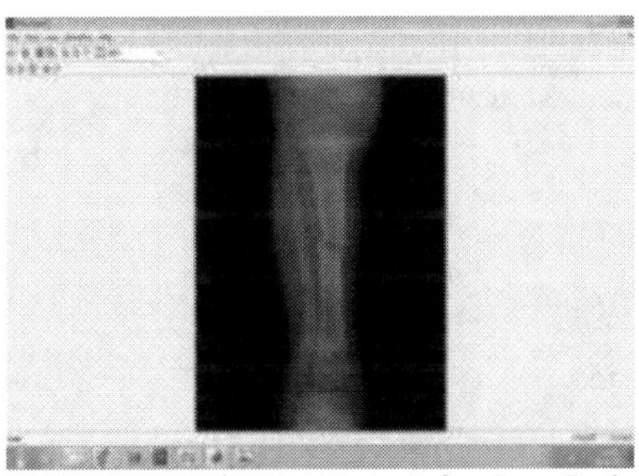

Figure 3.b Broken Leg Image for Grey Scale

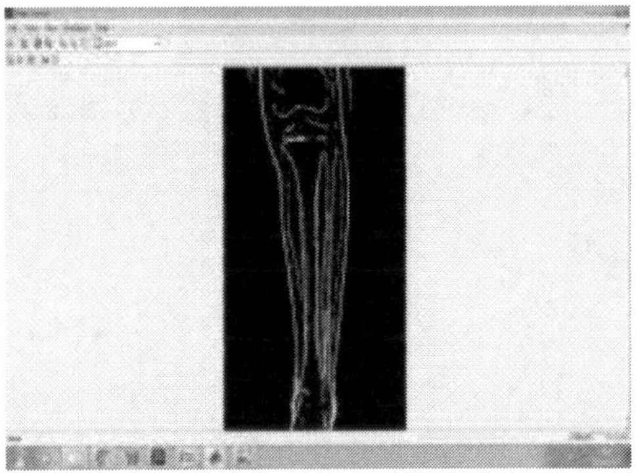

Figure 4.a Normal Leg Image for Sobel

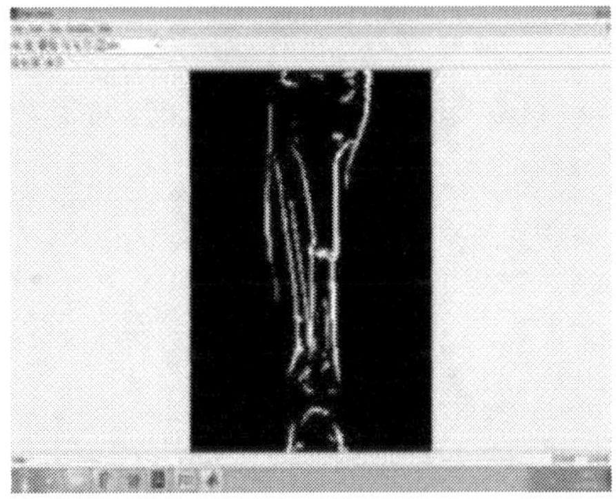

Figure 4.b Broken Leg Image for Sobel

IOP Conf. Series: Materials Science and Engineering **995** (2020) 012024 doi:10.1088/1757-899X/995/1/012024

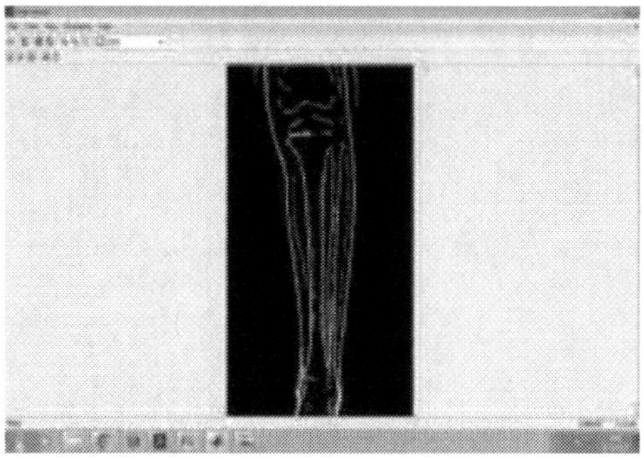

Figure 5.a Normal Leg Image for Prewitt

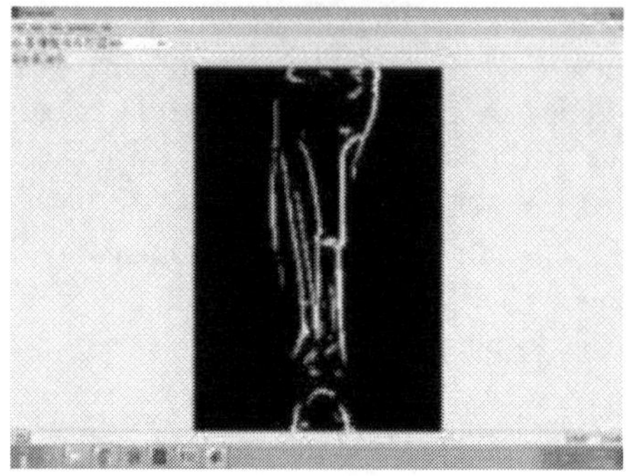

Figure 5.b Broken Leg Image for Prewitt

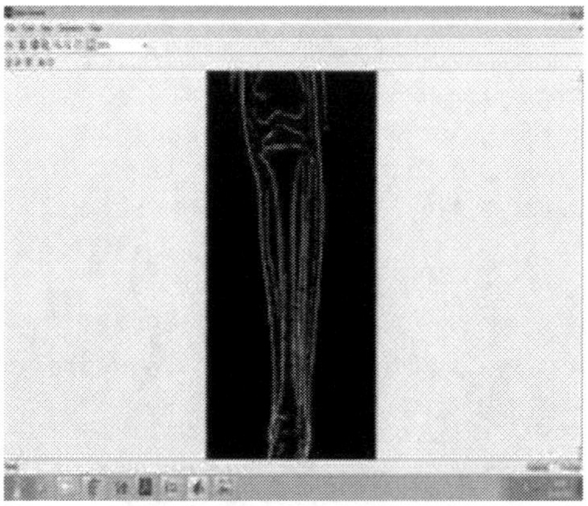

Figure 6.a Normal Leg Image for Robert

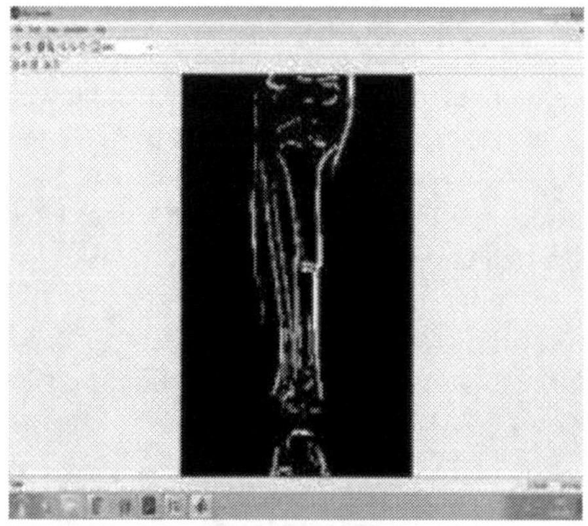

Figure 6.b Broken Leg Image for Robert

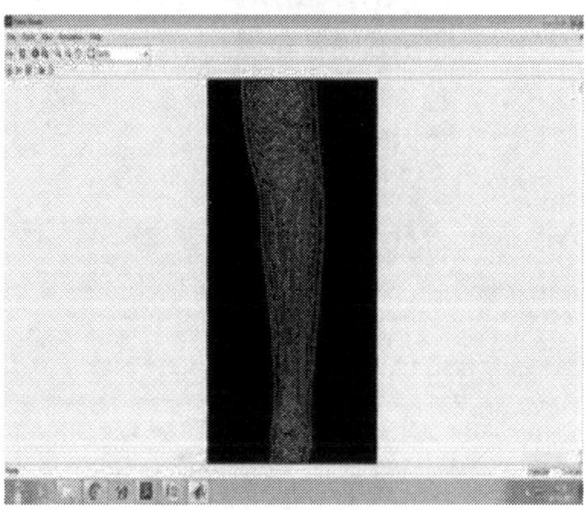

Figure 7.a Normal Leg Image for Canny

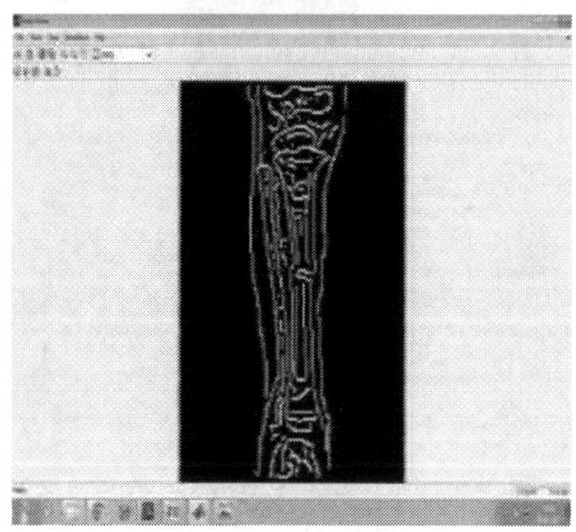

Figure 7.b Broken Leg Image for Canny

The above figures shows the image results of different edge detection techniques using MATLAB. The figure 3.a and 3.b shows the greyscale image of normal leg and broken leg respectively. The figure 4.a and 4.b shows the Sobel image of normal leg and broken leg respectively. The figure 5.a and 5.b shows the Prewitt image of normal leg and broken leg respectively. The figure 6.a and 6.b shows the Robert image of normal leg and broken leg respectively. The figure 7.a and 7.b shows the Canny image of normal leg and broken leg respectively.

Table 1. Analysis of Accuracy Measures of edge detection Techniques

Algorithm	Accuracy in %
Sobel	88.7
Prewitt	78.71
Robert	76.19
Canny	74.85

Table 2. Execution Speed of edge detection Techniques

Algorithm	Time in Seconds.
Sobel	34.9
Prewitt	35.26
Robert	36.1
Canny	32.5

From the Table 1 & 2 shows the analysis of accuracy and execution speed for four edge detection techniques. This paper compares the performance characteristics of canny algorithm with other edge detectors such as Sobel, Robert, Prewitt, that are applied to the x-ray image of lower leg bone. Thus canny operator provides better edge detection for bone structure.

7. Conclusion and Future Work

Edge detection is most important stage in pre-processing to obtain an information from the input image. It help to reduce the amount of data that need to be processed and as well as minimize the time for detecting the boundaries. This paper presents the image segmentation technique to detect the bone fracture. The fully automatic detection of fractures is important but also a difficult problem. Here in the four edge detection operators performance is analysed, and compared, which in turn discovered that the Canny operator attains an accuracy of 88.7 %, whereas the accuracy of Sobel, Prewitt, Robert are 78.71%, 76.19%, 74.85% respectively. The edge detection by canny Operator has improved performance over other operators because canny has better noise immunity than other techniques. The future work of this project are ankle fractures, ribcage analysis, spinal cord imaging, analysis of MRI images of brain, tumour detection, edge detection in tissues of plant and animals, and other medical imaging applications may be considered.

Reference

[1] Rakesh M. R (2017), 'Design and Simulation of Matlab/Simulink Model for Edge Detection Techniques in Image Segmentation', International Journal of Advanced Research in Electrical, Electronics and Instrumentation Engineering, Volume 2, Issue12.

[2] Irfan Khatik (2017), 'A Study of Various Bone Fracture Detection Techniques',

International Journal Of Engineering And Computer Science, Volume 6, Issue 5, Page No.21418-21423.
[3] Aung Soe Khaing, Hla Myo Tun, San Myint (2016), 'Detecting Leg Bone Fracture In X-Ray Images', International Journal Of Scientific & Technology Research, Volume 5, Issue06.
[4] Tamanna Sahoo, Professor Sandipan Pine- Design and Simulation of SOBEL Edge Detection using MATLAB Simulink, International Research Journal of Engineering and Technology (IRJET) Volume: 03 Issue: 05, May-2016.
[5] Anphy Jose, Deepa Merlin Dixon K, Naiji Joseph & Silpa George E, Ms. Anjitha V, 'Performance Study of Edge Detection Operators', International Conference on Embedded Systems - (ICES 2014).
[6] Abdel Rady A. S, Elaraby A. E. A, Hassan Badry Mohamed A. El-Owny (2014), 'New Algorithm For Edge Detection in Medical Images Based on Minimum Cross Entropy Thresholding', IJCSI International Journal of Computer Science Issues, Volume 11, Issue 2, No-1.
[7] Dr.Sana khudayer Jadwa (2016). 'Canny Edge Detection Method for Medical Image Retrieval', International Journal of Scientific Engineering and Applied Science (IJSEAS) – Volume-2, Issue-8.
[8] Susmitha , Ishani Mishra, Divya Sharma, Parul Wadhwa , Lipsa Dash - Implementation of Canny's Edge Detection Technique for Real World Images, International Journal of Engineering Trends and Technology (IJETT) – Volume 48 Number 4 June 2017.
[9] Swathika, Anandhanarayanan, Baskaran and Govindaraj- Radius Bone Fracture Detection Using Morphological Gradient Based Image Segmentation Technique. International Journal of Computer Science and Information Technologies, Vol. 6 (2), 2015.
[10] Rakesh M. R- Design and Simulation of Matlab/Simulink Model for Edge Detection Techniques in Image Segmentation, International Journal of Advanced Research in Electrical, Electronics and Instrumentation Engineering, Volume 2, Issue 12.,2017.
[11] Dr.Sana khudayer Jadwa- Canny Edge Detection Method for Medical Image Retrieval, International Journal of Scientific Engineering and Applied Science (IJSEAS) – Volume-2, Issue-8, 2016.

IOP Conf. Series: Materials Science and Engineering **995** (2020) 012025 doi:10.1088/1757-899X/995/1/012025

Efficiency assessment and improvement measures for thermal power plant in India

N Santhosh[1], M Srinivasan[2], M Chandrasekaran[3], E Sarangapani[2]

[1]Department of Mechatronics Engineering
[2]Department of Electrical and Electronics Engineering
[3]Department of Mechanical Engineering
Bannari Amman Institute of Technology, Sathyamangalam 638401, Tamil Nadu, India.

santhoshmech10@gmail.com, srinivasanm@bitsathy.ac.in, chandrasekaran@bitsathy.ac.in, sarangapani@bitsathy.ac.in

Abstract. Usage of conventional energy resources has high limitations and constrains now a days. The thermal efficiency of the equipments in the modern thermal based power plant is mandatory for effective power generation with low production cost and pollution. The systems and equipments in the thermal power plant should be operated with optimum efficiency. An assessment of these system performance and comparing it with the design value is necessary for its improvement. This paper presents the findings of the observations and trials carried out on thermal power plant energy equipments. The main objective of this work is to assess the performance of various thermal power plant systems in respect of efficiency and determining the main factors which contribute to the efficacy of the power plant by indirect method. Further, suitable measures are recommended to enhance the energy performance with cost benefits. While making the measurements, care is taken to maintain the accuracy and wherever any discrepancy is found, the readings were again taken to ensure the best possible results.

Keywords: Conventional energy, Thermal power plant, Indirect method

IOP Conf. Series: Materials Science and Engineering **995** (2020) 012025 doi:10.1088/1757-899X/995/1/012025

1. Introduction

Thermal power plants play the major role in electricity production in India. Performance assessment and consecutive improvement measure can lead to large contribution in monetary terms, which can be further exploited for capacity addition [1]. In the past period of time the power plant efficiency saw a surge from around 5-45%. The growing cognizance about the greenhouse gases further encourages the efficiency enhancement in the thermal power plants [2]. The efficiency of the power plant gradually deteriorates over a period of time. Retrofits can significantly increase the performance of older power plants. The maximum efficiency can be attained by regular monitoring and control of the sources where the energy loss takes place by appropriate methods [3]. The various losses in the components of the thermal power plant are assessed to identify the process irreversibility [4]. Researchers assessed the avoidable and unavoidable energy losses for investigating the cost associated with compressor, turbine, heat exchanger and combustion chambers [5]. Researchers proposed different energy resources to integrate with the coal based thermal power plant to enhance the efficiency such as utilizing the ocean thermal energy [6], solar power to pre-heat the feed water before the economizer of the boiler [7] and biomass for co-firing operations [8]. Research is done in various components of the coal based power plant such as boiler, turbine, condenser, etc. Among the various components of the thermal plant, the boiler plays a vital equipment and also it plays a significant role in the total efficiency of the power plant. So, the boiler is to be operated with optimal efficiency. Efficiency testing of the boiler is crucial and any deviations from the design value (norms) are easily identified and rectified [9]. Different work is proposed by researchers to assess and improve the efficiency of the boiler [10-11]. The turbine is another vital part of the thermal power plant and to enhance the efficiency different researcher proposed different techniques. An Artificial Neural Network (ANN) has been applied to control the important variables of turbine such as pressure, temperature, speed and humidity for effective operation of the turbine [12]. The condenser is yet another important system in the thermal power plant where the heat input to the condenser is just ejected into the atmosphere through the water cycle, which makes it the most inefficient part of the plant and research has been done to replace the coolant water to refrigerant which gets vaporized in the condenser section and used as a prime mover for a secondary turbine generator to produce additional electricity [13]. Before investigating different methods for energy enhancement of the power plants, certain considerations should be taken into account to minimize the energy loss in the basic stage. These energy saving precautions can avoid energy loss to a greater extent with very low investment. In this paper the performance assessment of thermal power plant in India, which has a production capacity of 1100 MW is considered and the assessment is done by indirect method where the heat losses are subtracted from 100 to assess its operating efficiency and energy savings opportunities. The performance and efficiency of the power plant is evaluated which then leads to the system optimization, simple modification and renovation of equipment. The general complications related with design, process and fuel inputs are examined and suitable solutions are recommended to improve the efficiency.

To assess the efficiency of thermal plant the various sections of the plant is considered such as boiler system, turbine system, unit auxiliaries, ash handling plant, pumping system, chiller, cooling tower, air compressor, lighting system, motor load and thermal insulation. Before the assessment of the various systems, the overall efficiency of the plant considered for a period of a year where the maximum efficiency is found to be 37.95% and the minimum efficiency is found to be 37.64% as shown in Fig. 1.

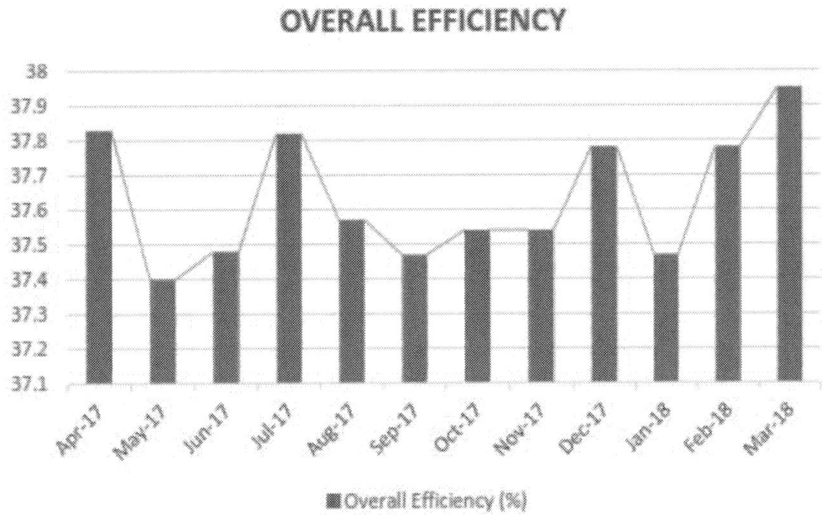

Figure 1 Overall efficiency

2. Efficiency assessment and recommendation of boiler system

During the assessment, the boiler is operated at 1485 tonnes per hour (TPH). The boiler efficiency trial is conducted as per the running load and the readings of all the parameters were recorded for a period of 3 hrs. Efficiency study involves in quantification of all the losses inherent to the boiler system and accordingly operating efficiency is calculated. During the trials, the key parameters, namely unit load, coal flow, coal analysis total air flow, combustibles in bottom ash and fly ash, flue gas analysis were recorded. The sample of fuel and ash is sent to the proximate analysis to the in-house chemical laboratory. During the trials, all relevant parameters, namely, coal, air, flue gas, water and steam were collected and efficiency assessment is carried out. The various losses of the boiler and its percentage in the overall boiler loss is shown in the table 1. The overall efficiency of the boiler is 88.25 and the remaining percentage is encountered as losses.

Table 1 Boiler losses and its percentage

S No	Description	Calculation	Energy loss in percentage
1	Unburnt carbon losses	Total combustibles (Combustables in fly ash + Combustables in bottom ash) x Calorific value of coal	0.82
2	Heat loss due to dry flue gas	[24 x Dry flue gas x (flue gas temperature(FGT) – combustion air temperature (CAT))] / Higher heat value (HHV) of fuel	6.05
3	Heat loss due to moisture in fuel	M x [584 + C_p x (T_f-T_a)] x 100 / GCV M - kg of moisture in 1kg of fuel C_p - Specific heat of superheated steam (0.45 J/K/kg)°C 584 is the latent heat (J/kg) corresponding to the partial pressure of water vapour	0.71

		T_f - Flue gas temperature T_a - Ambient air Temperature GCV – Gross calorific value	
4	Heat loss due to burning of hydrogen	[900 x (h_g – h_f)] / HHV h_g (Specfic enthalpy of saturated vapor) = 1055 + (0.467 x FGT) h_f (Specfic enthalpy of saturated liquid) = CAT - 32	3.72
5	Heat loss due to moisture in air	AAS x humidity Factor x Cp x (Tf-Ta) x 100 / GCV	0.13
6	Heat loss due to sensible heat of bottom ash	Total ash collected per kg of Fuel Burnt x GCV of bottom Ash x 100 / GCV of fuel	0.38
7	Heat loss due to sensible heat of fly ash	Total ash collected per kg of Fuel Burnt x GCV of Fly Ash x 100 / GCV of Fuel	0.12
8	Heat loss due to radiation and convection	Obtained from the standard chart available from the American Boiler Manufacturers Association (ABMA)	0.24

The various losses in the boiler such as heat loss due to dry flue gas, heat loss due to burning of hydrogen in fuel, heat loss due to moisture in fuel, heat loss due to moisture in the air, unburnt carbon loss, heat loss due to sensible heat of bottom ash and heat loss due to radiation and convection are calculated.

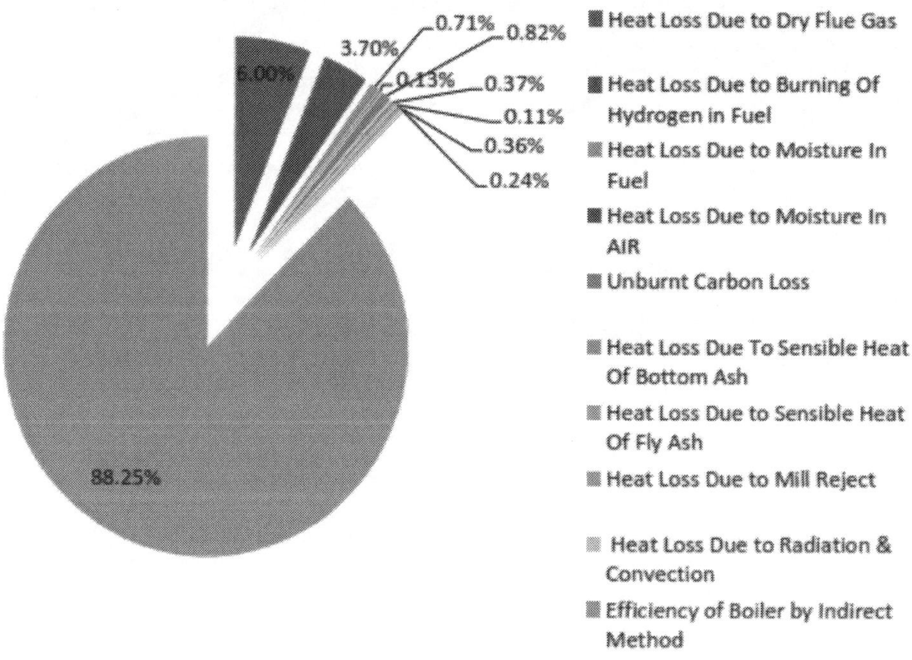

Figure 2 Boiler heat balance

It is observed from the figure 2 that the dry flue gas loss is a majority of all losses and this can be controlled by reducing the combustion air. This loss can also be controlled by keeping air leakage or ingress across ash handling plant (AHP) to a significant extent and ascertaining that the air supplied by the force draft fan (FDF) is fully utilized for combustion. Second major loss is the heat loss due to water evaporation due to hydrogen in the fuel, this loss is purely depends on the fuel composition and have no relevance to the operations of the boiler. Third major loss is heat loss due to moisture in the fuel and mainly incurred due to usage of high moisture content in the fuel. All though the coal quality is good and moisture present in fuel also is less compared to design coal values. All the other losses are inevitable and incurs due to various factors. It has been observed that the boiler efficiency is 88.25% as against the design efficiency of 89.02%.

The flue gas analysis is carried out at various locations in the flue gas path to arrive at the composition and temperature and in-leakage profile across the respective locations.

3. Efficiency assessment of turbine system

Steam turbine with a surface type condenser is used in the power plant and to assess the performance of the turbine, the various components are evaluated such as condenser, turbine heat rate and high pressure (HP)/Intermediate pressure (IP) of cylinder efficiency. The Table 2 shows the efficiency calculations of various parameters of turbine system.

Table 2 Efficiency calculation of turbine system parameters

S No.	Description	Efficiency calculation
1	Condenser Performance	$Qc = m_w \times C_p \times \Delta_t$ Qc - Heat rejected in condenser through circulating cooling water (kCal/h) m_w - water flow rate (kg/h) Δ_t - Average CW temperature rise ($T_{OUT} - T_{IN}$) T_{OUT} – CW outlet temperature (^{0}C) T_{IN} - CW inlet temperature (^{0}C)
2	Turbine Heat Rate	$W_s \times c \times \Delta_t$ W_s – Steam flow c – Specific heat capacity Δ_t – Temperature difference
3	HP and IP cylinder efficiency	Actual enthalpy drop / Isentropic enthalpy drop

The condenser performance is found to be 53% and to calculate the turbine heat rate the relevant parameters of the steam turbine (Main steam Pressure, Temperature, Flow, Cold reheat pressure, Temperature, Hot reheat pressure, Load etc.) were recorded for a period of 3 hours and the turbine heat rate are found to be 1944.5 kCal/kWh. The condenser is a water-cooled type with natural draft cooling tower. The design inlet water temperature of the condenser is 36 °C, at a turbine exhaust steam pressure of 0.107 kg/cm^{2}a, saturated steam temperature of 46.75°C at Turbine Continuous Maximum Rating (TMCR) of 1496.8 TPH.

It is found that the terminal temperature difference (TTD) of water cooled condenser (WCC) (steam saturation temperature - water outlet temperature) in as-run conditions is at 7.97°C at 30.7 °C inlet cooling water temperature, 39.9 °C outlet temperature, condenser vacuum of 0.113 kg/cm^{2}a against the design value of 0.75°C at 36 °C inlet CW temperature, 46 °C outlet cooled water temperature, condenser vacuum of 0.107 kg/cm^{2}a. This occurrence of high TTD and lower vacuum in spite of lower than design cooling water inlet temperature could be due to, condenser

IOP Conf. Series: Materials Science and Engineering **995** (2020) 012025 doi:10.1088/1757-899X/995/1/012025

leakages, reduced heat transfer effectiveness due to scale and fouled heat transfer area and drop in ejector performance. It is envisaged that a TTD of 5 ^{o}C should be targeted for, in day to day operations, which could enhance the unit output by around 1.66 MW. It is recommended to carry out helium leak test on condenser and rectify any leakages. Further, improving the heat transfer effectiveness by ensuring clean and unscaled/unfouled heat transfer area of chemical descaling treatment to remove the existing deposits. Incorporate the online condenser cleaning system is preferred.

4. Efficiency assessment of unit auxiliaries

Performance assessment of key plant auxiliaries is conducted and the auxiliaries include primary air (PA) fans, forced draft (FD) fans, induced draft (ID) fans, mills and pumps. The operating efficiencies of the PA, FD and ID fans were found by measuring various parameters such as flow, static head at the suction and discharge and their respective motor power consumption. The overall efficiency of the PA and FD fans is around 52 % and ID fan is 58%. The coal milling system is a critical area relating to plant operation and utmost care is accorded to their upkeep and operation. Measurements were undertaken to assess the performance of the mills and associated subsystem, this includes electrical measurements (kW, PF, Amps, and V) of operating mills. It is observed that the specific energy consumption of mill is in the range of 9.38 - 12.28 (kWh/Ton). To assess the performance of the pump, the electrical as well as flow parameters has been looked into and the overall efficiency is found to be 75.62 %.

The combined efficiency of the fans is lower at 52% as against 76% design value. The reason behind the poor performance of the fans are due to part loading and inefficient damper control for flow modulation. The pressure drop across the damper is a direct energy loss. This loss can be eliminated by the adoption of Variable Frequency Drive (VFD). VFD installation would also aid flow modulation without much reduction in head required, at the same time this does not create any artificial drop unlike damper.

Three condensate extraction pumps (CEP's) with the rated capacity of 900kW are installed and normally two CEP's are operated. The total power consumption of the two CEP's is 1279 KW. The operating discharge pressure is 28kg/cm^2, whereas the de-aerator pressure is only 9.55 kg/cm^2 indicating a pressure drop of more than 13 kg/cm^2 in the control valve. It is suggested to remove one operating CEP in one stage. Even after removal, the expected pressure developed by CEP is 22 kg/cm^2, which is sufficient to pump feed water to the deaerator. The estimated savings would be 16.6%, worth in annual energy savings of 16 lakhs kWh.

The cooling water pumps are designed to deliver a flow rate of 19500 m^3/hr at 23.5 m head and design pump efficiency is 84%. Presently pumps are operated at the efficiency from 49 to 71%, which is low compared to design and energy savings can be obtained by applying corrocoating for pumps to improve the pump efficiency. The expected efficiency improvement in pumps after corrocoating is to be 3%.

5. Efficiency assessment of pumping system

The performance of the water pumps is assessed by the measurement of flow, head and motor power. The pumps such as booster pumps, raw water pumps, fuel oil pumps are considered for performance assessment. The overall efficiency of booster pumps is found to be 67%, raw water pump efficiency is 68% and fuel oil pump is found to be 44%. The booster pumps are designed to deliver a flow rate of 1273 m^3/h at 120m head, but operating at an average flow rate of 1442 m^3/h at head 102 m. The higher flow rate measure Vs designed flow is due to less operating, which leads to shift in the performance curve of the pumps away from break even point (BEP) thus pumps are operating at a lesser efficiency against the design. Significant savings can be obtained by replacing pumps with energy efficient pumps. In order to avoid energy loss due to inefficiency, two pumps with a design head of 110 m and flow rate of about 1275 m^3/ h is proposed by considering a safety factor.

6. Efficiency assessment of chiller

Air conditioning is used to provide comfort and desired working conditions for the control rooms and office space, etc. The efficiency of chiller is determined by the relation M x C_p x ($T_1 - T_2$) / 3024; where, M- Water flow rate m^3/hr, C_p - Specific heat of water, i.e 1 kCal/kg ^oC, T_1 - Water inlet temperature, ^oC, T_2 - Water outlet temperature, ^oC and the overall efficiency is found to be 25%. There are three no's of vapor compression refrigeration (VCR) chillers, each of 160 tons of refrigeration (TR) are installed and out of 3 chillers two is operated. Performance study revealed that both the chillers are running in part load. Operating TR of the chiller is 43 TR (25%) and 76 TR (50%). The specific energy consumption of chillers alone is 0.70 and 0.89 (kW/TR). For each chiller one chilled water and condenser pump is running. There is a possibility to switch off one chiller and run only one chiller to further reduce chiller statistical process control (SPC) and overall SPC of the system. After switching off one chiller the total TR load will be shifted to one chiller which is 119 TR (75%) so still there is 25% further loading can be done.

7. Efficiency assessment of cooling tower

Different parameters measurement is conducted to asses the performance of cooling tower such as the cooling water flow, wet and dry bulb temperature and cooling water pump performance. The efficiency of cooling tower is calculated by [(Range)/(Range+Approach)] x100, where, Range (^0C) = Cooling water (CW) inlet temperature for cooling tower (CT) - CW outlet temperature from CT and Approach (^0C) = CW outlet temperature from CT - Wet bulb temperature. The effectiveness of the cooling tower is 76.89%, which is a satisfactory value.

8. Efficiency assessment of air compressor

The plant has separate air compressors for plant air system and instrument air system. The instrument air compressed system is used for control systems, solenoid valve operations, etc. and the service air compressed system is used for atomization, cleaning, maintenance, general service applications, etc. The main plant compressed air system comprises of two instrument air compressors operate parallel in tandem, between an operating discharge pressure window of 7.1 kg/cm^2 to 8.1kg/cm^2, (ie., the air compressor cuts in at 7.1 kg/cm^2 and cuts out at 8.1 kg/cm^2). The instrument air from the compressor at 7 kg/cm^2 is finally used at all end-users at less than 4 kg/cm^2 through FRL (filter - regulator - lubricator) side control. It is recommended to rationalize the compressor cut-in and cut-out pressure settings from existing 7.1 kg/cm^2 cut-in and 8.1 kg/cm^2 cut-out to 6.0 kg/cm^2 cut-in and 7.0 kg/cm^2 cut-out and thus avoid downstream throttling. The mean generating pressure reduction will be 1.10 kg/cm^2.

9. Efficiency assessment of lighting system

Extensive study of lighting is conducted during the course of study in the following areas such as control rooms, workplaces, boiler area, turbine area, pump houses and chiller area. The lighting fixtures installed in the plant are FTLs (Fluorescent Lamp), MVs (mercury-vapor lamp), SVs (Special Vehicle Lights) and LEDs (Light Emitting Diode) for indoor and outdoor lighting. FTLs and MVs fixtures have low efficacy. Already plant has replaced few of the lamps to LEDs so suggested to replace the remaining lamps to LEDs to achieve energy savings. There are about 4074 no's of 36 W FTL. It is recommended to replace these 36W FTL's with LED tube lights in a phased manner.

10. Efficiency assessment of motor load

From the motor load survey, it is seen that most of the motors are drawing power between 50 to 90 % of their rated capacity, which is a good sign of good maintenance practices.

11. Efficiency assessment of thermal insulation

The infrared thermal scanning of equipment having high surface temperature areas is carried out on hot surfaces including boiler, steam turbine, and steam pipe lines to identify hot spots/heat loss areas towards insulation review/upgrades. The thermography scanning is carried by infrared thermal imager. All the hot zone temperatures greater than 65 ^{o}C were noted down for further analysis. The heat losses were calculated only for those areas, where the temperature is higher than recommended norm. The formula used for heat loss estimation is given as $Q = (10 + (T_s - T_a) / 20) \times (T_s - T_a)$ in kCal / m^2/ hr where, T_s = Surface temperature in ^{o}C,T_a = Ambient temperature in ^{o}C, Heat loss = Q x Area of the surface (m^2) and the heat loss are found to be 70,563kCal/hr. It is recommended to improve the insulation level in the heat loss areas in order to minimize steam and subsequent fuel savings.

12. Conclusion

The overall efficiency of the thermal power plant is calculated by multiplying the thermal efficiency with the electrical efficiency. The key step to enhance the efficiency of the thermal power plant is the detailed study of the plant and then the efficiency calculation. The efficiency calculation by indirect method is the best way to account all the losses in the thermal power plant. From the detail study the following conclusions can be made:

- The dry flue gas loss is a majority of all losses in the boiler and this can be controlled by decreasing the dry gas. This is achieved by optimizing the amount of oxygen in the flue gas. Quality burners and accurate combustion are necessary for good results.
- In the condenser section it is recommended to carry out helium leak test and rectify any leakages. Further, the surface should be unsealed/unfouled to remove the deposits and achieve the effective energy transfer.
- Corrocoting and optimal design of pumps such as providing the accurate clearance inside the pump can improve the efficiency and also significant savings can be obtained by replacing pumps with energy efficient pumps.
- Optimal utilization of chiller plant can improve energy savings to a greater extent.
- It is recommended to replace these FTL's with LED tube lights in a phased manner to optimize the energy utilization.

13. Reference

1. Naveen S Seema S Kavita C Efficiency assessment and benchmarking of thermal power plants in India Energy Policy 2012 40:159-176
2. Chao F Rahul A Kristin J Truls G Thermal efficiency of coal-fired power plants: From theoretical to practical assessments Energy conversion and management 2015 105: 530-544.
3. Richard J Increasing the Efficiency of Existing coal fired power plants Congressional research service 2013.

4. Kamate S Gangavati P Exergy Analysis of Cogeneration Power Plants in Sugar Industries Applied Thermal Engineering 2009 29:1187- 1194.
5. Kotas T Exergy Criteria of Performance for Thermal Plant: Second of Two Papers on Exergy Techniques in Thermal Plant Analysis. International Journal of Heat and Fluid Flow 1980 2(4): 147-163.
6. Rodrigo S Julio V Thermal power plant efficiency enhancement with Ocean Thermal Energy Conversion Applied thermal engineering 2014 62(1): 105-112.
7. Hong H Peng S Zhao Y Liu Q Jin H A typical solar-coal hybrid power plant in China 2014 Energy Procedia 49 1777 – 1783.
8. Emmanouil K Panagiotis G Michalis A Ioannis V Emmanuel K Co - firing of biomass with coal in thermal power plants: technology schemes, impacts, and future perspectives Energy and Environment 2013.
9. Ashokkumar A Improvement of Boiler Efficiency in Thermal Power Plants. Middle-East Journal of Scientific Research 2012 12 (12): 1675-1677.
10. Bhatt M S Effect of air ingress on the energy performance of coal fired thermal power plants Energy Conversion and Management 2006.
11. Sumeet M Vijay M Boiler Efficiency Measurement and Energy Saving Steam Distribution System – A Case Study Journal for Research 2017.
12. Hosham S Khalid F Raheel J Enhancement of Performance for Steam Turbine in Thermal Power Plants Using Artificial Neural Network and Electric Circuit Design Applied Computational Intelligence and Soft Computing 2018.
13. Nitish K Mohanta D Enakshi K Dulal H Jaya M Efficiency enhancement of thermal power plants using refrigerant-R600a in condenser section International Conference on Innovations in Green Energy and Healthcare Technologies (IGEHT) 2017.

Simulation of Tic-Tac-Toe Game using LabVIEW

R Deepa[1*], M Janani[2] ,Ankita Singh[2]

1, Assistant Professor, Bannari Amman Institute of Technology, Department of Electronics and Instrumentation Engineering, Erode, Tamilnadu, India.
2 UG Student, Bannari Amman Institute of Technology, Department of Electronics and Instrumentation Engineering, Erode, Tamilnadu, India

rangarajdeepa@gmail.com, janani.ei19@bitsathy.ac.in, ankitasingh.ei19@bitsathy.ac.in

Abstract. Tic-Tac-Toe game is a two player game, in which a circle (O) or a cross (X) will fill the square block with three rows and three columns. There will be switching between the players in the game where each player will be provided a chance to make their move. If one of the two players have marked the same markers horizontally, vertically or diagonally then that particular player would awarded points. LabVIEW is the platform in which the above game is simulated. Tic Tac Toe is a two-player game which can be played using the LabView software. In this program the select function is used for the player to select the marker and the view feature in the program is to place the marker selected by the player in the desired position of that player. After each player makes their move the software will update and scan while it continues for the conditions of the game. Overall, this tic-tac-toe operates with the features of the programme and can be used without security problems.

Keywords: LabVIEW, Tic-tac-toe, Players

Introduction

Games offer a real wellspring for daily entertainment [1]. In addition, in optimising human physical and emotional health, games are successful. In addition to physical sports, people also play video games day after day. These games are not quite the same as those of video games because they do not involve a lot of physical exercise, emotionally and enthusiastically. The persistent problems that game developers are facing is that, getting games to react to a client of a game. And it's a nice game that does not test the gaming ability of the client and the game can be played within for a short period of time that will not hold the client's time. It is possible to apply this thought to any kind of game available today. When the rival individual players do not comprehend or get on, tabletop games are terrible all over. Developers are continually searching for new ways to make an additional computer game exciting and challenging for the client, with the latest PCs constantly propelling. Tic-Tac-Toe is a two-player game, where the square (3 x 3) is filled with symbols cross (X) or a circle (O). The game will switch between the two players, giving both player a chance to play. The software would indicate the player has won, regardless of whether X or O has won, whether one of the player's marker is filled in a line, vertical or diagonal line. The game Tic-Tac-Toe is popular among all of the day's gatherings. Tic Tac Toe is a user friendly game that make it perfect as an instructional tool for illustrating the principles of good sportsmanship. The game is the work of a wonderful mind. It includes looking ahead and trying to make sense of what the person playing against you is going to do immediately.

1.Literature Survey

The review of few research articles on the game Tic- tac-toe summarises the findings below. Tic Tac-Toe is an interesting, yet straightforward table game. Specialists have employed numerous approaches to deal with learning the game Tic Tac-Toe.

Fok and Ong (1996) uses the falsified neural organisation-based systems to play it for scale [2]. Citrenbaum and Yakowitz investigate games that share a few parallels with Spasm Tac-Toe [3]. Numerous product use of Spasm Tac game was accounted for and as of late it became accessible for advanced mobile phones, for example, the one for Android, and the other for Apple iPhone condition. In general, the Spasm tac-toe round is also a pencil, a paper game played by two individuals who go to place their pieces on a 3 times 3 network with the objective of being the main player to fill a straight, upright or slanted line with their markers.

Consider that possibility, as opposed to just one person. Playing against each other, does one play against a group of nine players, each capable of one cell in the multiple 3 networks? The group needs to organise its players in this novel way of playing the game, which behave autonomously based on their restricted data. Soedarmadji (2006) offers a response that can be generalised to the situation where two of these groups are playing against each other and other table games [4]. The arrangement essentially uses a decentralised structure which appears to interfere with the arrangement from the outset.

Stephen Mann presented the structure of an equivalent computerised circuit that estimates the neural network (NN) to determine the location of Tic Tac-Toe [5]. FPGA's are designed to upgrade custom computerised systems by actually designing ways in between the logical gadget entrances. Using an FPGA allows for restoration of the NN system with no money-related costs. The developer believes NN usage has superior execution over custom programming usage, as it utilises the inherent equal structure of the NN. Shahzeb Siddiqui *et al.* (2002) is running another NN play, expanding the game by adding two extra columns, two extra lines, and was reached [6]. Pinaki Chakraborty (2009) officially characterise the Spasm Tac-Toe game and develops fabricated knowledge-based strategies, which are equivalent to play later [7]. Leaw and Cheong (2010) plays out a mild quantization of the old spasm tac-toe style round, by enabling the superposition of old-style movements [8]. Edward (1996) uses an optically guided entryway display (OPGA) to perform a straightforward round of Spasm consideration toe to demonstrate the effectiveness of electrooptical circuits, and what's more, logical capacities in a solitary gadget [9]. Advances are also made on the nature of steering methods and the reasoning for replication. Alauddin (2013) promotes the use of the equipment for canny Spasm Tac toy. Use of Graphical LCD (GLCD) microcontroller and touch screen. The microcontroller gets GLCD player move (shown as X) and use informative calculation to investigate the action and find the best counter step [10]. The microcontroller shows the counter continuing as circle (O) towards the screen. The calculation selects the winner when the Spasm Tac play rule allows the game. The system is implemented using modestly accessible electronic segments off the rack, and tested to function rapidly and skilfully.

2. Proposed Work

In this work, LabVIEW is the software in which the 3x3 tic-tac-toe game is to be simulated. The proposed system is designed, so the two players can use LabVIEW programming to play tac-toe game. The display works and an option to switch between the chosen marker of the player is provided by the program. After each move of the player the game is restarted and finds the status of the game and the process repeats itself.

IOP Conf. Series: Materials Science and Engineering **995** (2020) 012026 doi:10.1088/1757-899X/995/1/012026

3. SYSTEM FEATURES:

The computer implementation has more features when compared to the traditional method of playing this game using pen and paper.

Let us see what makes computer implementation of playing Tic-Tac-Toe game different:

a. LabView software made this game user friendly
b. User is allowed to play the game any number of times without disturbance
c. The marker X/O can be chosen according to the user's desire
d. The game was created as a detailed Expert framework
e. The probability of this game is three which can be winning the game, draw the game or losing the game
f. This game is suitable for all age group people.

4. SOFTWARE REQUIREMENTS:

Two players can play the game. The game is initiated with three steps. 1. Start Game, 2. Make a Move, 3. End Game and the flow chart of the game is shown in figure 1, and the flow chart of the tic-tac-toe game for the proposed methodology is shown in figure 2.

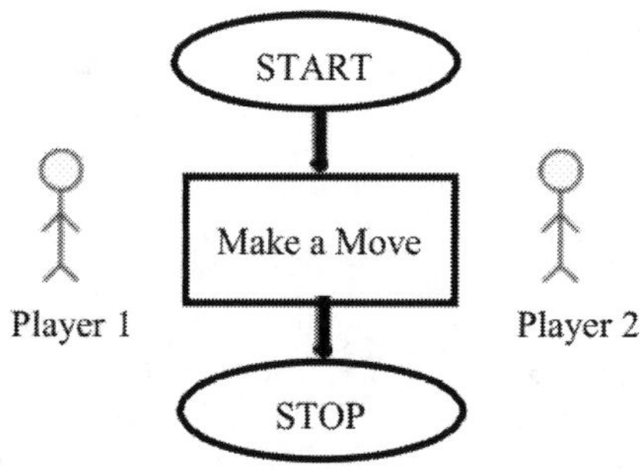

Figure 1 Flow Chart of the game.

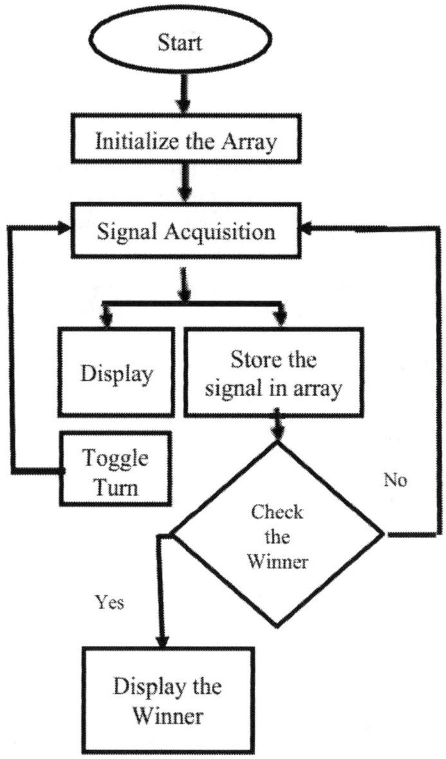

Figure 2 Flow Chart of the tic-tac-toe game

Rule for Tic-Tac-Toe:

- Tic tac toe is a two player game, let us assume the players as player 1 and player 2
- The two players choose their marker(which cane be 'X' or 'O')
- Player 1 begins his/her turn by putting his/her marker on one of the nine boxes ccording to his desire.
- The two players then make their turns again
- If any player is likely to have three of his/her markers in a straight, vertical or slant line dominates the match The aim of this design is to allow the players to play the tic tac toe usign the LabVIEW software. The inclusions of this software is the showcase function and the select function in which the player will be allowed to position his marker and at the same time switch between the markers such that each player will be given a chance for playing.The interface of the tic tac toe game in LabVIEW is in figure 1 in which one of the two players has won as that particular players marke(O) is position in the first column in all three rows.

Winning Criteria:

In a tic-tac - toe game, there are 8 chances to win the game. The choices are to horizontally acquire the three combinations, same as vertical, acquire two combinations diagonally and it is shown in figure3. Consequently, the best way to work out how to decide the winner is to verify the 8 combinations.

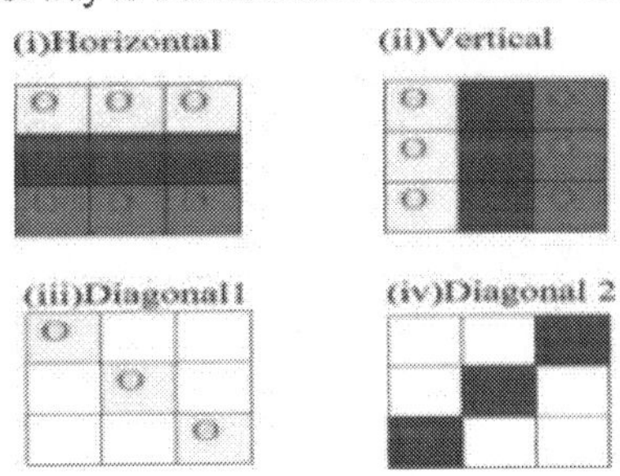

Figure 3 Winning Criteria

The aim of this design is to allow the players to play the tic tac toe usign the LabVIEW software. [7]The inclusions of this software is the showcase function and the select function in which the player will be allowed to position his marker and at the same time switch between the markers such that each player will be given a chance for playing.The interface of the tic tac toe game in LabVIEW is exposed in figure 4 in which one of the two players has won as that particular players marks(O) is positioned in the first column in all three rows.

IOP Conf. Series: Materials Science and Engineering **995** (2020) 012026 doi:10.1088/1757-899X/995/1/012026

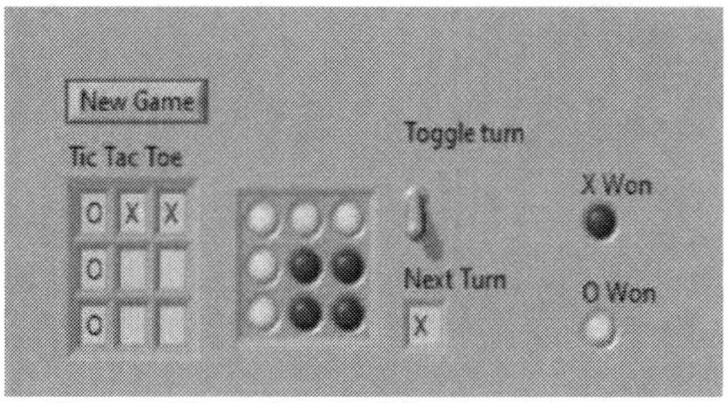

Figure 4 Front Panel of the tic-tac-toe game

The criteria for the tic tac toe game tha is designed in LabVIEW software are:

- Store the value of the marker(X or O) in an array.
- Switch between the players
- When clicked on 'New game' the previous moves are deletes and a new game is being started
- Check the value of the marker entered by the which has been stored in the array with the condition for victory that has been stored earlier

5. SYSTEM IMPLEMENTATION:

Use of programming is the step in the programming construction measure in which an executable programming framework is developed. At this point, where the conceptual plan becomes an operating arrangement. LabVIEW– (Laboratory Virtual Instrument Engineering Workbench) is provided by NI (National Instruments). LabVIEW is graphical programming that takes into account the operation of the instrument, the securing of information and the preparation of the information procured.

LabVIEW offers the adaptability of an outstanding programming language without the intricacy of customary criteria for evolution. Some of LabVIEW 's benefits include

- Understandability is elementary
- Full Functionality
- Requirements for optimised I / O
- No valid justification to compose programme line code
- Makes less of an optimal programming opportunity
- Cost and labour is the least needed

LabVIEW are called Virtual Instruments (VI) applications. Stress which controls equivalent sources of data, points equivalent yields.

VI is composed of three main components:

- Front Panel–The way customers interact with main VI.
- Block Diagram-the software control code.
- Icon / Connector-Way of VI to interfacing with different sub Vis or main VIs

For obtaining the client's image and show the images, a couple of cluster capacities have been developed, one for showing and the other to procure because LabVIEW does not have the capability to examine and display at the same time. They were set to 2 dimensional clusters once both clusters had been created.

- Creation of selection array: In order to obtain the markers from the player and is play them two,2-dimensional array functions are created for obtaining and displaying the markers respectively.

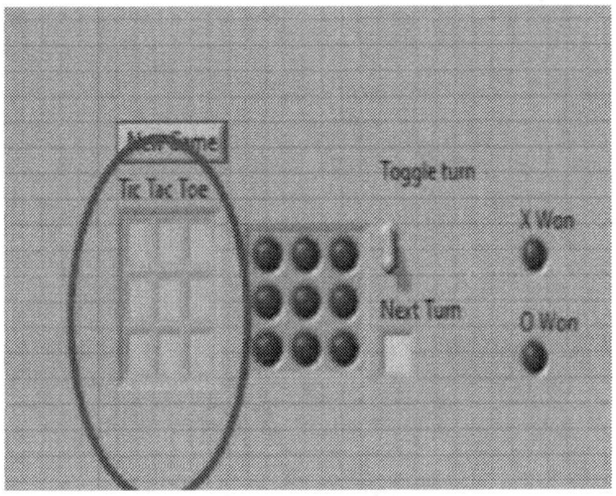

Figure 5 Creation of Array

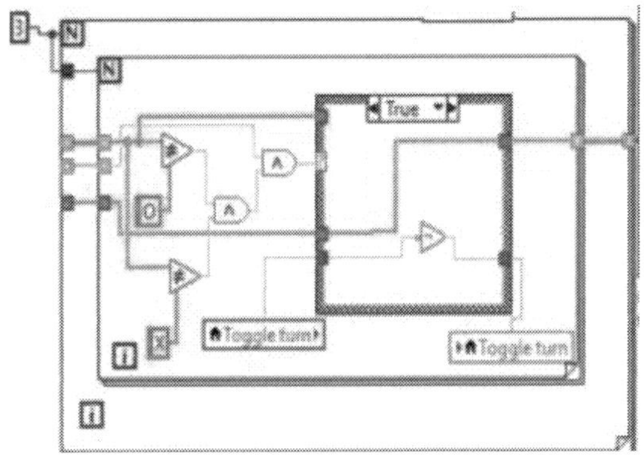

Figure 6 Creation of FOR Loop to represent an Array

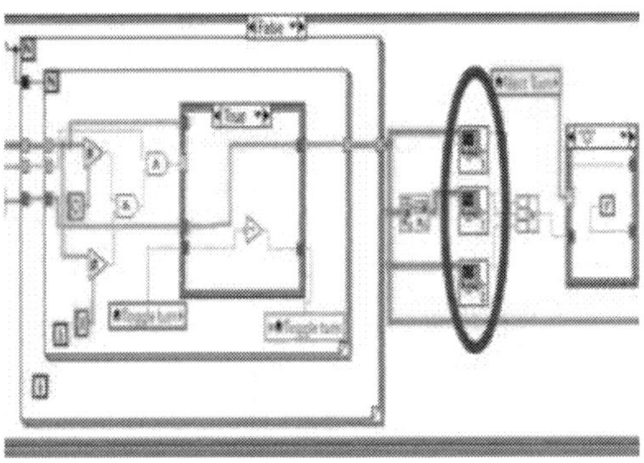

Figure 7 Creation of Sub VI

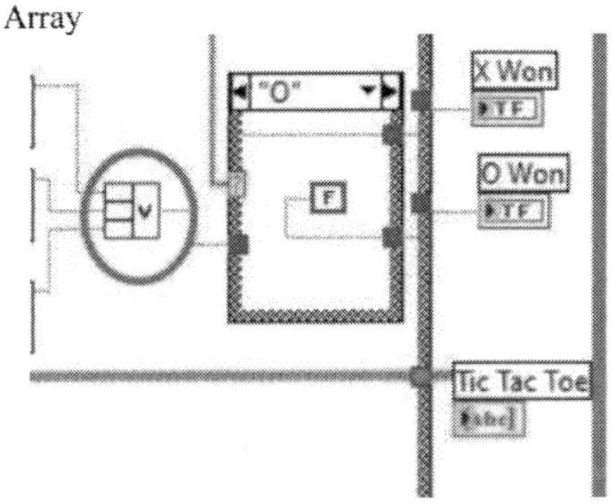

Figure 8 Combining 3 rules

Figure 5 shows the creation of the array in the front panel and the same represeneted in the block diagram using the FOR loop, outer FOR loop represents the row and the inner FOR loop represents the column for the tic-tac-toe game and it is shown in figure 6. This the 2 dimensional array in the which the value of the symbol is obtained and displayes.
Figure 7 shows the 3 SubVI which checks for three conditions i.e whether the same symbols are obtained vertically, horizontally and diagnolly.This is done using booleans.Whenever the player gives the input the value of the input is stored in an array anfd tranfered to sub vi which is then converted to binary digitsThe marker 'X' will stored as 1 whereas the symbol 'O' will be stored as 0. Figure 8 the encircled component is a 3 input OR gate which stores the players input values in terms of 1's and 0's.

The SubVI for the three criteria are given in the figure 9.

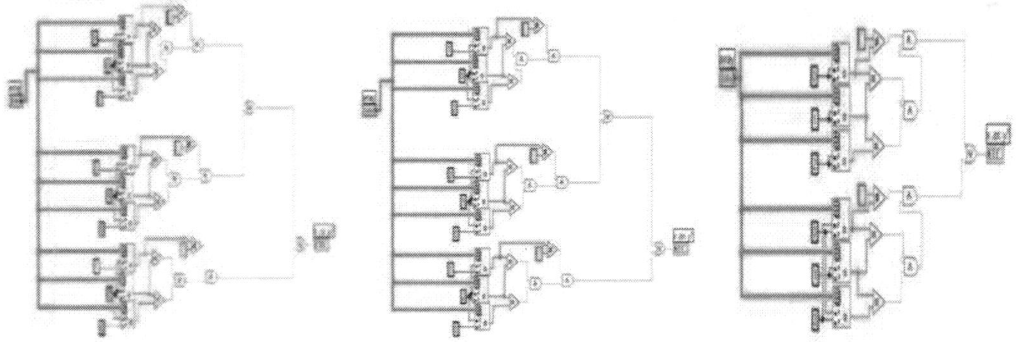

Figure 9 (i) Vertical Sub VI (ii) Horizontal Sub VI (iii) Diagonal Sub VI

The case structure is connected to two booleans for two different cases, if true it displays the winner, else start a new game.

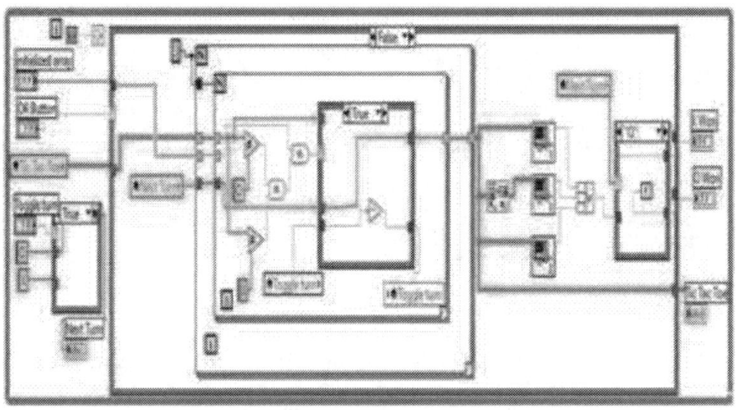

Figure 10 VI for tic tac toe game

The ouput of the sub VI is OR ed and given as the input to the case structure and it describes the true and the false statement and the Figure 10 describes the complete VI of the tic tac toe game.

6. SYSTEM TESTING:

The loop that uncovers blunders is called testing. When the source code has been made, as many blunders the software must be attempted to expose as may reasonably be anticipated before conveyance to the consumer. Programming testing may be expressed as approval period and validation of a Computer programme / application / item:

(A) further meet the requirements guiding its strategy, development

(B) true to form fills

(C) equivalent values can be executed

(D) satisfying the requirements of partners.

LabVIEW coding diagrams are called Virtual Instruments (VI). Different VIs are made for the spasm tac-toe framework; each VI speaks to one section in the company. Inspection is performed for each VI.

IOP Conf. Series: Materials Science and Engineering **995** (2020) 012026 doi:10.1088/1757-899X/995/1/012026

In LabVIEW the blunders are recognised while the VIs are created. VI won't work if mistakes do occur. This blunder should be corrected prior to the execution of a VI. Both VIs are tried independently and are triumphant in inspection. Numerous Virtual Instruments and sub Virtual Instruments are associated together in the system to create an absolute arrangement. In organised testing the coordination of all Virtual Instruments is annoyed and found to be successful. It is found that the bonds operate well. The framework is eventually tested for application testing to establish that the application fulfils the overall needs of all clients and that the system meets all user criteria as a whole and has been found to be successful.

7. RESULTS:

The game of tic-tac - toe is played with two players in LabVIEW. Figure 11 describes the overall view of the front panel window, it has the array of LED controls, the toggle switch is used to represent the turn of players. In figure 12, the player O has placed it horizontally and the X is waiting for the next turn. Similarly in figure 13, the player O has made it vertically, and the diagonally mapping is shown in figure 14, the player O is the winner. Similarly, the same process is repeated for the Player X as the winner by placing vertically, horizontally and diagonally.

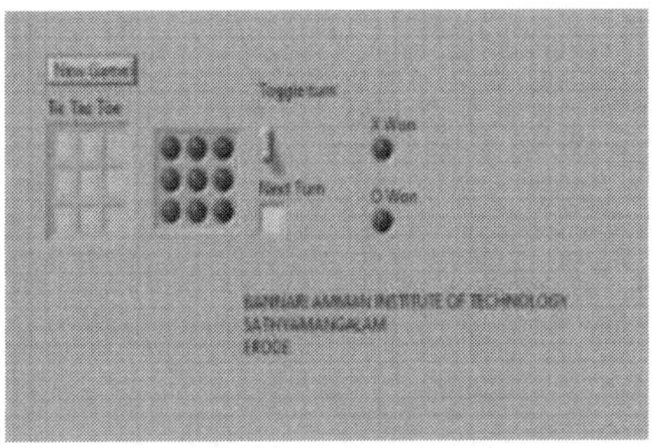

Figure 11 Start of the Game with player X and O

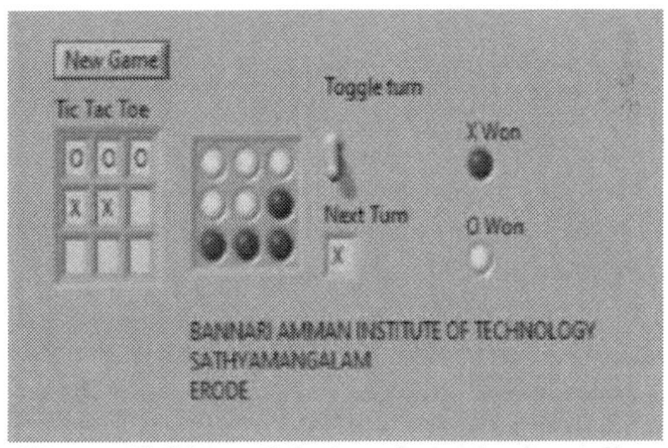

Figure 12 Player O is the winner, horizontally placed

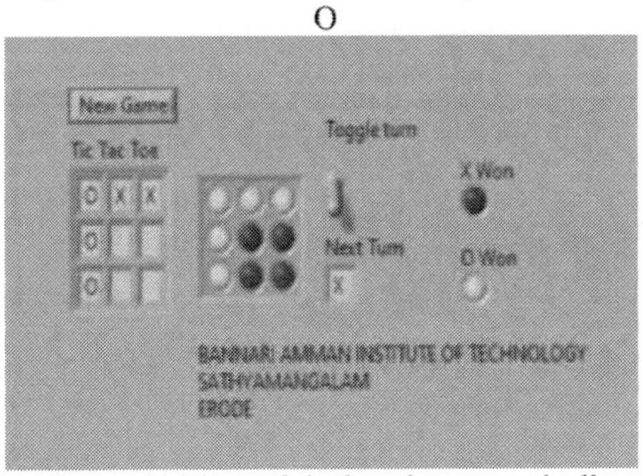

Figure 13 Player O is the winner, vertically placed

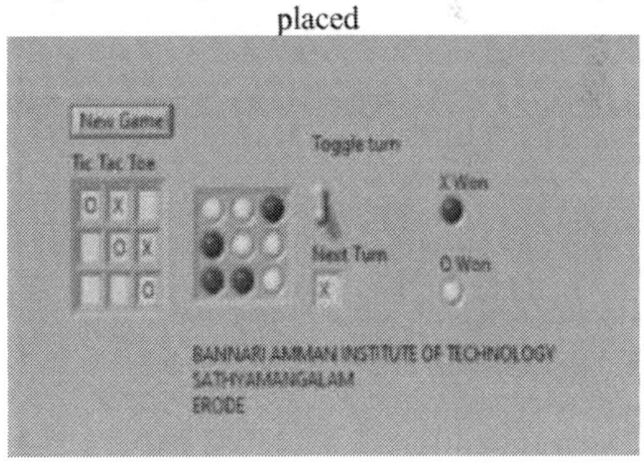

Figure 14 Player O is the winner, diagonally placed

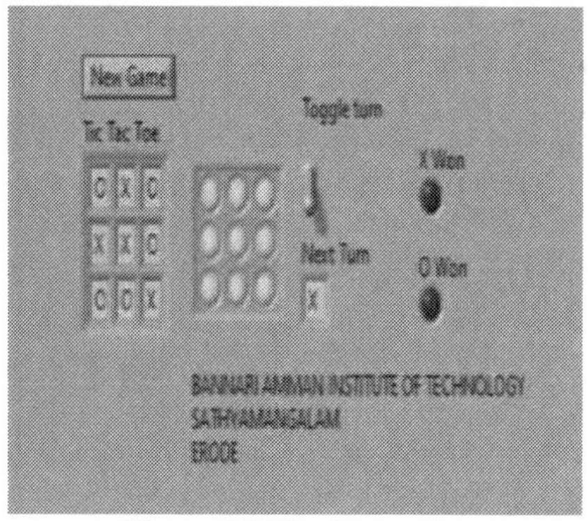

Figure 15 No player wins

The figure 15 shows where the player X and player O has not made the game successful, so it declared as game draw and new game is initiated again.

8. CONCLUSION

In this paper we provoked the tic-tac-toe game using the LabVIEW platform in an event driven GUI software. We made a 3x3 tic-tac-toe game using LabVIEW. The system is developed with the intention of allowing two players to play a round of tic-tac-toe using LabVIEW programming. The software will include a showcase job and find opportunity to position the image much like a switch between the images that allows each player to play the game. After each player makes their turn the programme will restart and search for the conditions of the game as it continues. The system typically works with no damages.

9. Acknowledgment

This work was supported by the Electronics and Instrumentation Engineering department that LabVIEW Software was provided by Bannari Amman Institute of Technology, Sathy.

10. References

[1] Tic-Tac-Toe game in wikipidia, http://en.wikipedia.org/ wiki /Tic-tac-toe
[2] S. C. Fok and E. K. Ong. A High School Project on Artificial Intelligence in Robotics, Artificial Intelligence in Engineering, Vol. 10, No. 1, 1996, pp. 61-70.
[3] R. L. Citrenbaum. Strategic Pattern Generation: A Solution Technique for a Class of Games, Pattern Recognition, Vol. 4, No. 3, 1972, pp. 317-329.
[4] Soedarmadji, Decentralized Decision Making in the Game of Tic-tac-toe, IEEE Symposium on Computational Intelligence and Games, May 2006.
[5] Stephen Mann and Matthew Netsch, "A parallel Embedded Neural Network for an Intelligent Turn-Based Engine".
[6] Shahzeb Siddiqui , Francis Mutuc and Nicholas Schmidt, " Designing a 5x5x5 Tic-Tac-Toe Game using a Neural Network with Backpropagation with a Twist", Evolutionary Computation, 2002.
[7] Pinaki Chakraborty, Artificial Intelligence Based Strategies to Play the Tic -Tac-Toe Game, Journal of Technology and Engineering Sciences, Vol 1, No. 1 January –June 2009.
[8] J N Leaw and S A Cheong, Strategic insights from playing quantum tic-tac-toe, Journal of Physics A: Mathematical and Theoretical Volume 43 Number 45, 2010.
[9] Edward P. Vogel, Tic tac toe game using an optically routed gate array, Proc. SPIE 2863, Current Developments in Optical Design and Engineering VI, 407, Nov 1, 1996.
[10] Alauddin Al-Omary, Machine- Human Tic-Tac game based on Microcontroller Technology, International Journal of Computer and Information Technology (ISSN: 2279 – 0764) Volume 02– Issue 05, September 2013.

IoT visualization of Smart Factory for Additive Manufacturing System (ISFAMS) with visual inspection and material handling processes

Sivabalakrishnan R[1], Kalaiarasan A[2], Ajithvishva M S[3], Hemsri M[4], G.M.Oorappan[5], Yasodharan R[6]

1, 5 Assistant Professor, Bannari Amman Institute of Technology, Department of Mechatronics, Erode, Tamil Nadu, India.
6 Assistant Professor, SNS College of Technology, Department of Mechatronics, Coimbatore, Tamil Nadu, India.
2, 3, 4 Student, Bannari Amman Institute of Technology, Department of EEE, Erode, Tamilnadu, India.

sivaeinfo@gmail.com, kalaiarasan.ee18@bitsathy.ac.in, ajithvishva.ee18@bitsathy.ac.in, hemsri.ee18@bitsathy.ac.in, oorappan@bitsathy.ac.in, r.yasodharan@gmail.com

Abstract. The challenges in a manufacturing system are lack of timely, accurate, and lack of information to featured product prediction, shop floor resources, product flow, product inspection, product status to customer, product delivery status and factory adaption for customized product. The proposed idea is to design IoT visualization based Smart Factory for Additive Manufacturing System (ISFAMS) that creates a way towards progressively from traditional automation to a fully connected mass customization and flexible cyber-physical system. The ISFAMS utilize a consistent stream of information from associated tasks and creating frameworks to learn and adjust factory productions to new requests from the customer. The system utilizes the Industrial Controller to control the operation of individual systems and sequence of product flow in the Smart Factory setup. The wireless sensor network acquires real-time manufacturing information and information is stored, accessed and visualized using cloud computing. The vision system and automated platform enable the inspection of products shape and dimensions based on the machine learning approach and to transfer the product from section to section and separate the product for packaging section. This digitization of manufacturing system increases flexibility, reliability, smart sensing and control, resource wastage, easy access to manufacturing information and logistics management.

Keywords: Smart Factory, Control Panel, Internet of Things (IoT), Industrial Automation, SCARA Robot, Vision System, Additive Manufacturing, 3D Printing Technology.

IOP Conf. Series: Materials Science and Engineering **995** (2020) 012027 doi:10.1088/1757-899X/995/1/012027

1. Introduction

As one of the essential things in Industry 4.0, the smart factory utilizes the recent advanced technology such as the Internet of Things (IoT), Additive Manufacturing, Cloud computing, and Mass Customization for cyber-physical systems and automated machines to real-time monitor of physical process, information storage, visualization through virtual system and optimal decision making to improve the production and quality [1]. The individual systems in the manufacturing area are able to interact and cooperate with each other within a smart factory to perform optimal decision making. IoT is an interconnection of computing devices with the manufacturing system for sending and receiving the product manufacturing information [2].

In the smart factory, the resources, processes, and products can be characterized by cyber- physical systems (CPS). The CPS is similar to the IoT technology which gives the higher coordination of process between the physical system and computational system of manufacturing unit. The digitization of manufacturing process is integrated with the smart factory that infers the real-time data, and statistics can be analyzed based on the information stored using cloud computing [3]. The IoT device communicates with all other devices to exchange of information remotely for monitoring and controlling the process. The cloud computing technology provides a platform to access the stored information about the manufacturing process and products for monitoring, analyzing and gives feedback to the smart factory through API of IoT services [4].

The automated machines involved in the smart factory to carry out the process like additive manufacturing, material handling (Pick and Place), material transfer using conveyors, product assembly, visual inspection (size, shape, and defects), product separation, packaging, and logistics management [6]. These sequences can be tracked using wireless sensor network during each process and send the data to the IoT enabled computing device. The information about the inspection can be monitored and segregate the data's like shape, dimensions, type of defects stored in cloud computing for ease of access by the authorized person (Both management and customer based on authorization level) [7]. The remote control is possible using the SCADA system interfaced PLC and IoT devices. The PLC performs the controlling operation by scanning the status of input terminal based on the ladder logic program stored in the memory.

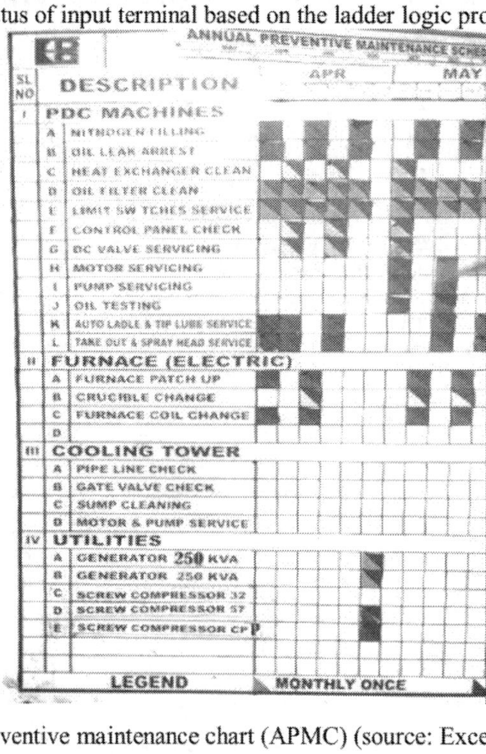

Figure 1: Annual preventive maintenance chart (APMC) (source: Excel die casting industry)

HMI display is provided in a smart factory for operator control and visualizes the machine data. The advanced development of automation provides automatic control of every process in a smart factory. The challenges in a smart manufacturing system are facing a lack of information about the individual process in each section of the smart factory set up and product flow to the management and status of the product at each stage in the manufacturing unit to the customer [8]. The top- can visually monitor the workers and machines involved in each process using CCTV camera and product flow at the instant of time. It is impossible to track the manufacturing operations such as timely process, machine operating periods, idle time and product status in each stage of the manufacturing process by using image processing or pattern recognition approach [10]. The survey on the industries like Die Casting Industry, Automation industry and automotive manufacturing industry, Chennai focuses the demand for the implementation of IIoT system for monitoring the machine parameter and creates an alert based on the annual preventive maintenance schedule of the conventional machine in their industry during industry visit for consultancy project shown in figure 1 and figure 2.

Figure 2: Die casting machines for parameter monitoring in control panel

The proposed system creates a solution for on-demand need of industries like internet based monitoring of production process, inventory management, material usage, and inspection result tracking and product status to the customer. Our proposed system consists of hardware and software contributions. In hardware, any manufacturing industries require four major sections like material storage, product fabrications, material handling, assembling, inspection and packaging. Similarly our ISFAMS consists of six sections such as Feeding section, 3D printing section, Conveyer, Assembling section, visual inspection section and packaging section. Along with this, our solution includes data visualization system integrated with manufacturing sections to track process, material position, product status and statistics. Our system interlinks the top level management to access complete status of manufacturing process and products.

2. Experimental setup

The conceptual design depicts that the customer can order the product with required specification details like variant, shape, dimensions through mobile or web application shown in Figure 3. The customer order and tracking the product status can be stored in a database through cloud computing using IoT device and API. The Information is accessed, and the manufacturing process is planned by the data analysis application in a distributed computer system. The 3D printer starts printing the parts (toys, home appliance material, health care products, etc.) based on customer data.

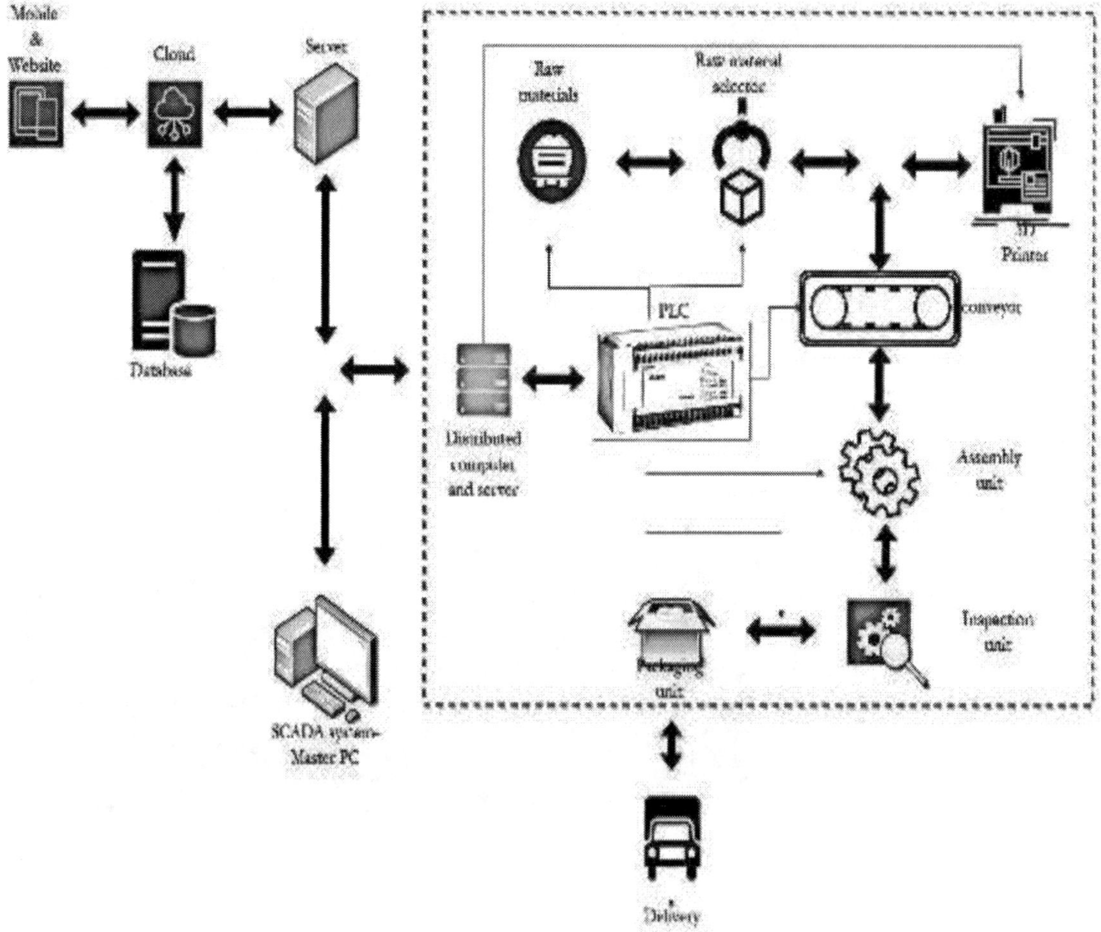

Figure 3: Conceptual Design

After the part feeder will feed the part to the conveyor, then printed parts get transferred by the single arm to the conveyor. When the conveyor receives part 1 & part 2 and starts moving the part to the inspection section after the proximity sensor senses the availability of both parts. The conveyor stops when a part reaches the assembled section and parts are assembled by the SCARA arm for making it as a single product. The assembled product is further moved to the inspection section for identifying the product variant, shape, dimensions and defects. The finished product is packed by packaging section with barcode and the entire sequence is controlled by PLC. The corresponding information about product status, inspection data, manufacturing details, idle time, and maintenance report is stored and accessed by IIoT device integrated with this manufacturing system.

Progress of the system structure

On viewing our progress towards project development based on the priority of hardware system of any manufacturing industry, we have developed

i) Feeding section

ii) 3D Printing section

iii) Conveyor section with semi-rotary arm actuation at the beginning of conveyor

Feeding section

Feeder unit has stack of readymade part. When entire system is switched ON through HMI or IoT dashboard, part is ejected from the stackholder using pneumatic cylinder. The speed of cylinder actuation can be controlled by the flow control valve and meter in meter out circuit. PLC gives signal to solenoid valve to pick the part from feeder unit and place it in conveyor using 1800 pneumatic operated rotary actuator.

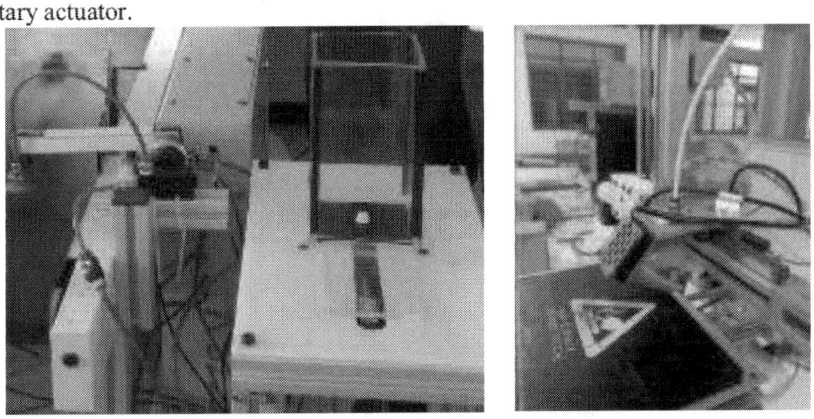

Figure 4: a. Feeder section, b. 3D printer

3D printer section

3D printer consists of stepper motor, microcontroller, driver, linear guide rails, display, extruder and heat bed. CAD design can be converted into .stl file using REPTIER host. The 3D printer will starts printing the 3D part variant after firmware gets order information from the client dashboard. Once part is manufactured, sensor triggers the PLC input to start the pneumatic actuator for placing 3D part in the conveyor shown figure 4.

Conveyor section with semi-rotary arm actuation

Conveyor has developed to transfer product from one section to another section for every pulse frequency input from PLC to servo amplifier. The servomotor has run the conveyor PVC belt length of 1.5 meters and servomotor can be attached with two non-identical timing pulleys (Ratio of 1:2). The conveyor can move the product weight of nearly 500 ~700 gms.

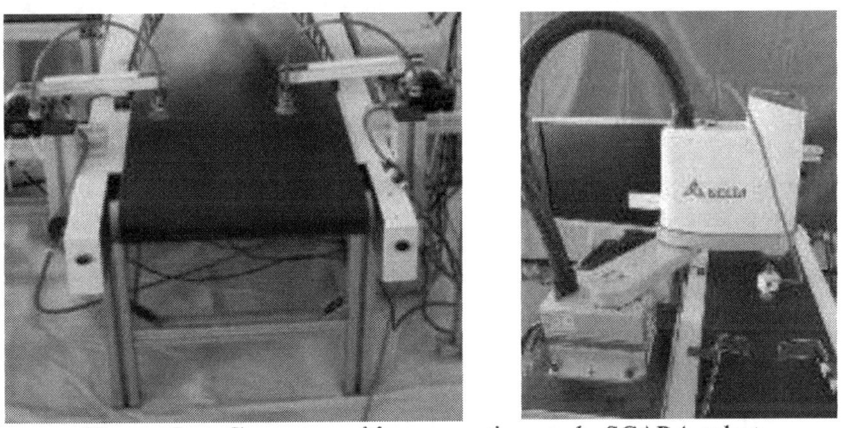

Figure 5: a. Conveyor with pneumatic arm, b. SCARA robot

IOP Conf. Series: Materials Science and Engineering **995** (2020) 012027 doi:10.1088/1757-899X/995/1/012027

At the beginning of the conveyor, two semi-rotary arm actuation mechanisms are attached for the pick and place of product from one section to another. The arm can be actuated using 1800 rotary pneumatic actuator with controlled speed using flow control valve. The material can be grasped using suction cup and it will maintain the direction throughout the rotation shown in figure 5.

SCARA Robot for assembly

Conveyor transfers both the parts to assembly section. Photoelectric sensor detects the part that reaches the assembly section. Then PLC triggers the I/Os of four axis SCARA Robot that starts assemble the parts. SCARA I/O sends the signal to PLC input which indicates the completion of assembling process. Conveyor sends the assembled product to the inspection sections.

Vision inspection system

Once the product reached inspection section, the visions system ensure the correctness of shape and dimension as well as quality of the product using image processing techniques. If a product does not have any defect, it will go automatically to packing process. If a product has defect in it, then the vision system I/O will send the information about the defect product and registered in admin dashboard to improve product quality shown in figure 6.

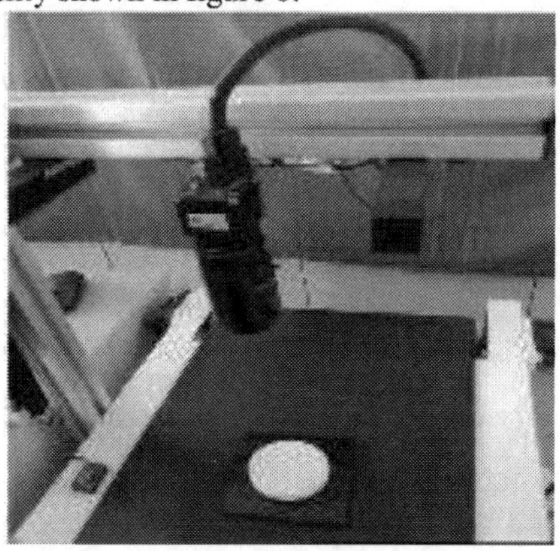

Figure 6: vision inspection system

3. Methodology

The methodology describes the flow of data or information tracked from the each section of the proposed system through Ethernet switch and updated in the cloud database by acquiring the data from the corresponding address specified in the Delta PLC shown in figure 7. The Client and admin Dashboard has been created using PHP and Database information retrieval and storage using MYSQL. The client should log-in to his account and place an order for required product. In addition to this, client dashboard has the provision for client to track the status of his ordered product. Admin can control the smart factory machineries through net from anywhere. Admin can also track the product flow, raw material usage, product status, and inspection data, and machine operation, factory conditions like temperature, delivery status, defect products and product quality with respect to time. These data's can be analysed to improve production rate, product quality and reduce the wastage of materials. This dashboard implementation helps the management to get real insights from shop floor, machineries and manufacturing process and also provides the production control through plant head. The utilization of resources and waste management is possible by the collecting and analysing the IoT data.

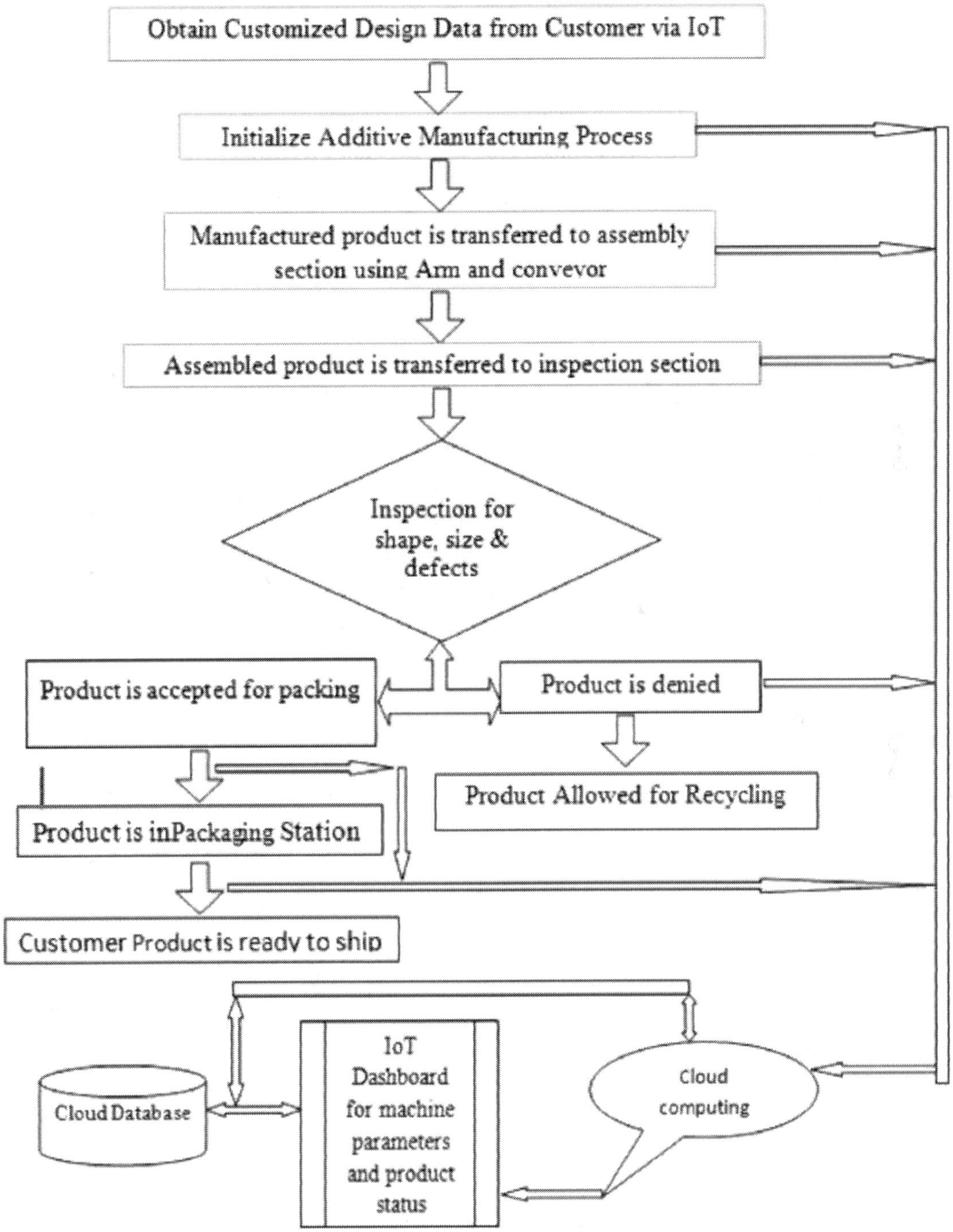

Figure 7: Conceptual flow of control algorithm

IOP Conf. Series: Materials Science and Engineering **995** (2020) 012027 doi:10.1088/1757-899X/995/1/012027

4. Result and discussion

The proposed system was designed and fully automated by the programmable logic controller which takes care of the control towards each system in the corresponding section shown in figure 8. It will track the flow of material in the conveyor by using the sensor. The vision data was captured and identified the faulty product and information is sent to the IoT dashboard and the user can diagnosis the fault and rectifies the system malfunction by analysing the previous process data of manufacturing system. The customer can order the product form the remote location and monitor their product build status and expected completion.

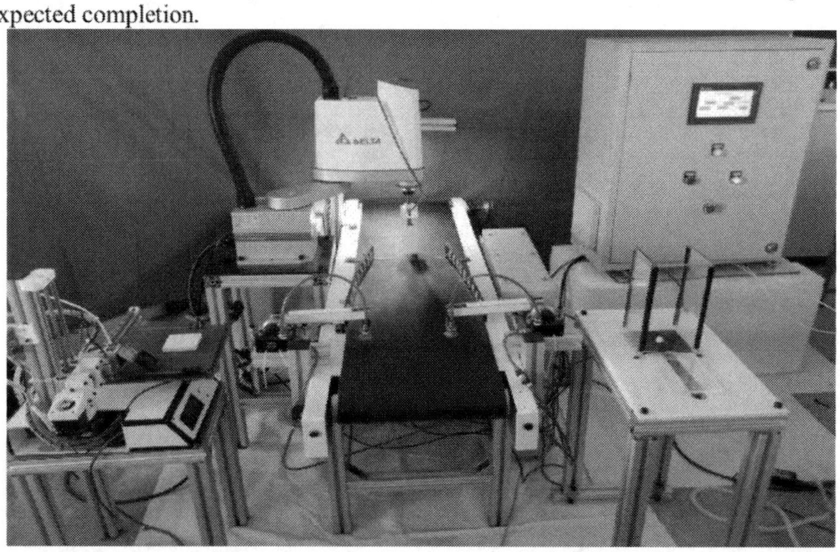

Figure 8: Smart factory layout for additive manufacturing system

The customer can pre-plan their work according to the timely information getting from the ISFAMS. The supervisor is eliminated by implementing the proposed idea in every manufacturing system. IoT dashboard has individually built for client and management. The data visualization was unique for the each dashboard based on their hierarchy level in industry and customer visualization is limited to the product status.

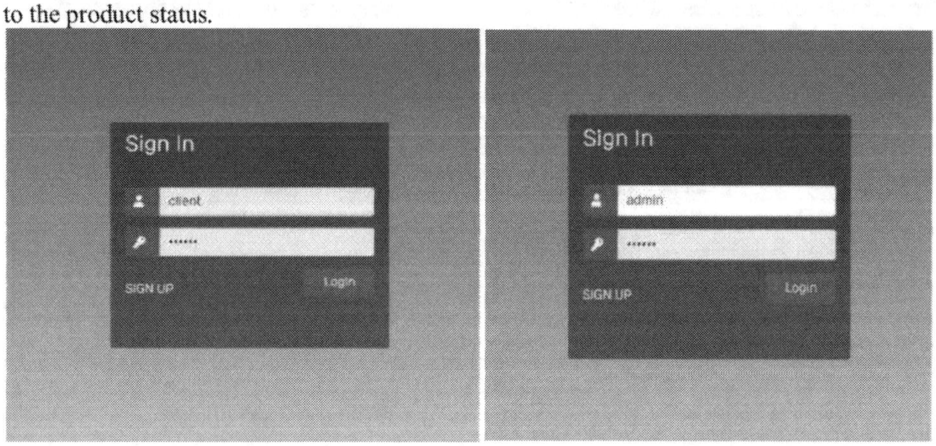

Figure 9: Admin and client IoT dashboard

IOP Conf. Series: Materials Science and Engineering **995** (2020) 012027 doi:10.1088/1757-899X/995/1/012027

The IoT admin dashboard shown figure 10 and 11 gives all the information about the production status like resource available, material wastage, number of product finished based on time history, faulty products and its reason, number of pending orders, products demands, priority of products. Based on this information, the management can be focused towards the planning of business to the market trends, customer needs, product demand and IoT interconnected industry for predicting market strategies. The IoT visualization helps the maintenance engineer to get updated for predictive maintenance schedule alert like replacement of oil, diagnosis of machineries, and calibration of 3D printing, optimal duration of operation and numbers.

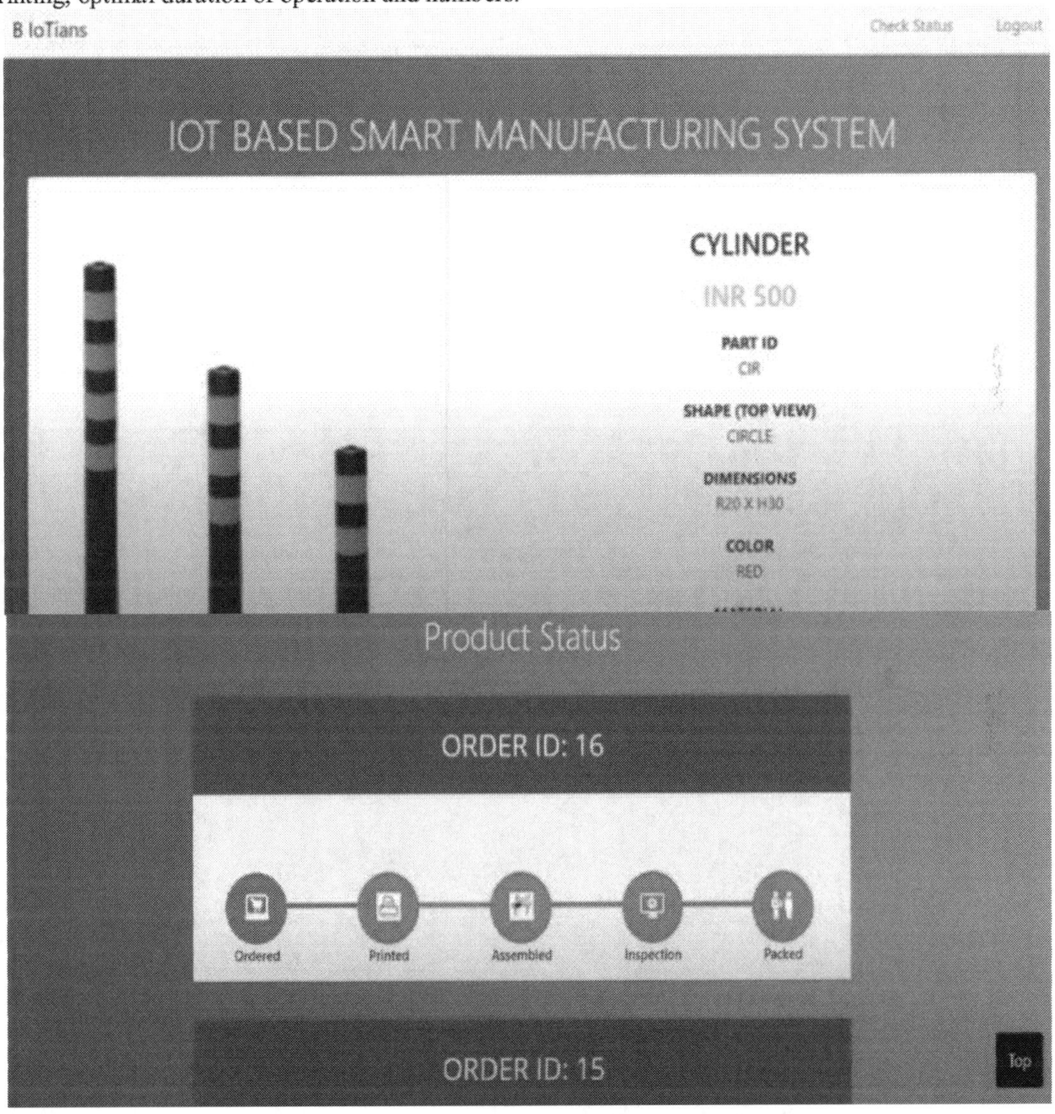

Figure 10: Client visualization for ISFAMS

The priority is assigned for the newly order product based on analyzing the previous data of the same product variant on the basis of most maximum satisfactory condition to the benefits of concern with less span of time and material availability. The priority assigning algorithm will assign weight to each ordered product by comparing the past data with higher possibility of delivering product with short span of time. The sample set of ten products and its production dada analyzed in the given table 1 and priority for each product was assigned in the table. This priority data has to utilized by our IoT application and create weight for newly ordered product.

PR. ID.	Order in nos.	Delivered in nos.	Processing time in hours	Material availability	Due period in Days	Required number of products	Priority assign
PRD001.	20	20	2	Less	10	20	9
PRD002.	40	20	3	Enough	15	25	4
PRD003.	30	25	2	Enough	20	25	3
PRD004.	25	25	3.5	Enough	21	30	5
PRD005.	40	30	1.5	Enough	30	45	2
PRD006.	20	20	1.2	Enough	10	50	1
PRD007.	15	15	2.5	Enough	25	10	7
PRD008.	40	20	3	Enough	20	20	6
PRD009.	40	35	2	Less	20	25	8
PRD0010.	35	35	5	Less	10	30	10

Table 1 Priority assignment for the ordered product

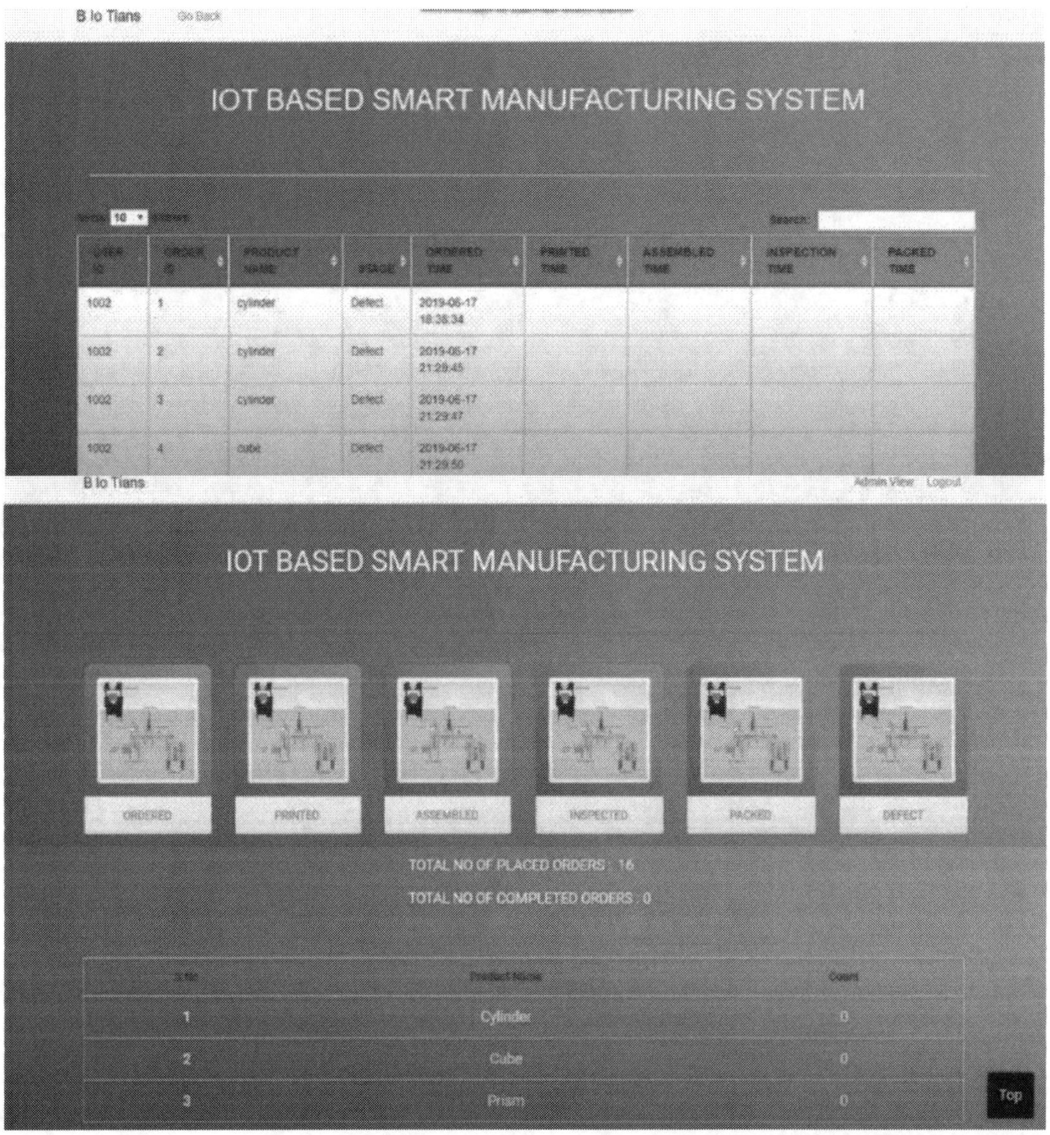

Figure 11: Admin visualization for ISFAMS

5. Conclusion

The ISFAMS creates novel layout for additive manufacturing system and its enables process and data visualization to the management and products status updates to the remote customer. The system proposes the idea of accessing the machine, process data from Data register in Delta PLC by specifying its address. The retrieved data is to be uploaded in cloud for monitoring and visualization and future analytics purpose to determine the product demand, market strategies, resource utilization, waste material management and predictive maintenance schedule monitoring and alert based on implementation of IIoT technologies. The data based been visualized in the both admin and client dashboard prepared using PhP and python integration. The current status and material flow as indicated in the admin dashboard with pictorial representation for the monitoring purpose of top level management. The future scope to the system is to implement and interconnecting of multiple IIoT industries to predict the market demand and enables integrated product manufacturing.

Acknowledgement

This research was supported by the Mechatronics Department that was provided Bannari Amman Institute of Technology, Erode and was recognized in the Delta Cup and Mitsubishi Electric Cup 2020.

References

[1] Kumar, A., 2018. Methods and materials for smart manufacturing: additive manufacturing, internet of things, flexible sensors and soft robotics. Manufacturing Letters, 15, pp.122-125.

[2] Zhong, R.Y., Xu, X. and Wang, L., 2017. IoT-enabled smart factory visibility and traceability using laser-scanners. Procedia Manufacturing, 10, pp.1-14.

[3] Haleem, A. and Javaid, M., 2019. Additive manufacturing applications in industry 4.0: a review. Journal of Industrial Integration and Management, 4(04), p.1930001.

[4] Wang, S., Wan, J., Li, D. and Zhang, C., 2016. Implementing smart factory of industrie 4.0: an outlook. International journal of distributed sensor networks, 12(1), p.3159805.

[5] Chen, B., Wan, J., Shu, L., Li, P., Mukherjee, M. and Yin, B., 2017. Smart factory of industry 4.0: Key technologies, application case, and challenges. IEEE Access, 6, pp.6505-6519.

[6] Bahrin, M.A.K., Othman, M.F., Azli, N.H.N. and Talib, M.F., 2016. Industry 4.0: A review on industrial automation and robotic. Jurnal Teknologi, 78(6-13).

[7] Dilberoglu, U.M., Gharehpapagh, B., Yaman, U. and Dolen, M., 2017. The role of additive manufacturing in the era of industry 4.0. Procedia Manufacturing, 11, pp.545-554.

[8] Kang, H.S., Do Noh, S., Son, J.Y., Kim, H., Park, J.H. and Lee, J.Y., 2018. The FaaS system using additive manufacturing for personalized production. Rapid Prototyping Journal.

[9] Mehrpouya, M., Dehghanghadikolaei, A., Fotovvati, B., Vosooghnia, A., Emamian, S.S. and Gisario, A., 2019. The potential of additive manufacturing in the smart factory industrial 4.0: A review. Applied Sciences, 9(18), p.3865.

[10] Rong, W., Vanan, G.T. and Phillips, M., 2016, September. The internet of things (IoT) and transformation of the smart factory. In 2016 International Electronics Symposium (IES) (pp. 399-402). IEEE.

[11] Mabkhot, M.M., Al-Ahmari, A.M., Salah, B. and Alkhalefah, H., 2018. Requirements of the smart factory system: a survey and perspective. Machines, 6(2), p.23.

[12] Chen, B., Wan, J., Celesti, A., Li, D., Abbas, H. and Zhang, Q., 2018. Edge computing in IoT-based manufacturing. IEEE Communications Magazine, 56(9), pp.103-109.

[13] Kang, H.S., Lee, J.Y., Choi, S., Kim, H., Park, J.H., Son, J.Y., Kim, B.H. and Do Noh, S., 2016. Smart manufacturing: Past research, present findings, and future directions. International journal of precision engineering and manufacturing-green technology, 3(1), pp.111-128.

[14] Osterrieder, P., Budde, L. and Friedli, T., 2020. The smart factory as a key construct of industry 4.0: A systematic literature review. International Journal of Production Economics, 221, p.107476.

[15] Sivabalakrishnan, R., Dineshkumar, M., Sharon, P.B., Naveenkumar, P. and Vignesh, S., 2019. Design and fabrication of 3D printed Acrylonitrile Butadiene Styrene (ABS) dam automation structure with integrated flood monitoring system using data analysis and computation techniques, Springer LNME (No. 2215). EasyChair.

[16] Yasodharan, R., Sivabalakrishnan, R. and Devendran, P., 2015. Trusted routing with an efficient

certificate revocation for mobile Ad Hoc Network. Internaional journal of innovative Science, Engineering and Technology, ISSN, pp.2348-7698.

Classification of Benign or Malignant Tumor Using Machine Learning

MKalaiyarasi[1], RDhanasekar[1], S Sakthiya Ram[1], P.Vaishnavi[1]
1Assistant Professor, Department of Electronics and Instrumentation, Bannari Amman Institute of Technology, Sathyamangalam, Erode 638 401,

kalaiyarasime@gmail.com, dhanaekar@bitsathy.ac.in, sakthiyaram@bitsathy.ac.in, vaishnavi@bitsathy.ac.in

Abstract. The abnormal growth of the cell in the human body may leads to tumor. There are almost 100 types of cancer will affect the different parts of the human. The affected human may have symptoms like lump, abnormal bleeding, prolonged cough, weight loss etc... And depends on the part where the tumor formed. Breast cancer is one of the types among the 100 types of cancer and it is most commonly found on the female than the male. Breast cancer is a disease that caused due to the over growth of the cell in the breast area and they started to form lump over the breast. The people who are suffering from the breast cancer will have many emotional side effects and they are supped to undergo lot of pain in their day to day life. The most important problem for the health care people is earlier prediction. So, it can be rectified by employing machine learning algorithms in the prediction stage. In this paper the classification algorithms are used to classify whether the tumor is benign or malignant. The supervised learning algorithms of machine learning such as logistic regression, Support vector machine and K Nearest neighbour algorithm are usually used to analyse the tumour detection. Stacking ensemble method used in order to combine the entire three algorithms is proposed and the performance of algorithm is compared with the logistic regression, Support vector machine and K Nearest neighbour algorithm in order to get an efficient model for the classification.

Keywords: Tumor detection, Machine learning.

IOP Conf. Series: Materials Science and Engineering **995** (2020) 012028 doi:10.1088/1757-899X/995/1/012028

1. Introduction

For teaching machine to handle data more effectively, the machine learning comes to picture. Use machine learning technique, in-case after analysing data, if the pattern interpretation and information extraction are not possible (1). The machine learning worth increases as the dataset quantity gets more. Many applications like process industries, medicine, defence and banking uses machine learning for extracting significant data and information. Many algorithms are available to make the machine to learn themself (2).Breast cancer is a dangerous disease that leads to death and this type of cancer mostly affect only women in the entire world (3). For reducing the death rate and control the spreading of cancer to the other parts of the body. Image processing technique is used to detect the cancer at starting stage (4).Breast cancer may spread over the other organs of the body through blood or lymph system and it may cause other tumors to grow. The type of Treatment procedures that need to be followed will be varying according to the type of the cancer and its stage. For every individual patient the treatment procedure includes at least one of the following methods like surgery, chemotherapy, and radiation therapy. Breast cancer may occur in two different categories like malignant and benign tumors and it is the difficult task for the physicians to classify the type of tumor. The malignant and benign tumor is shown in the figure 1. So, the health care domain requires some reliable methodologies to predict the tumor cell and its type. The automation in the tumor identification can be achieved by employing machine learning algorithm for the detection purpose. The earlier detection of the disease will help the doctors to cure the disease so; it may reduce the cause of death and can save many patients life.

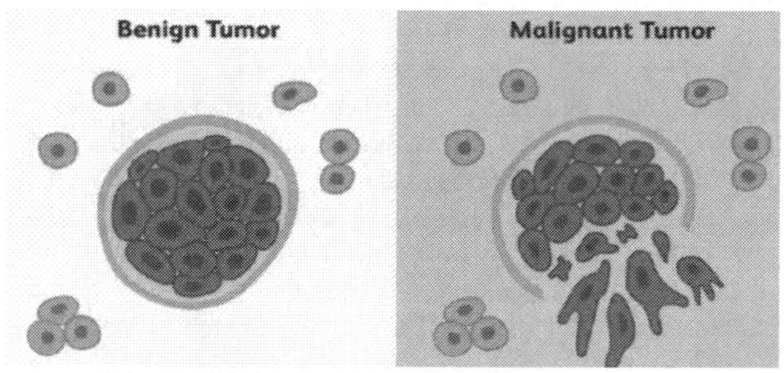

Figure 1. The benign and malignant tumor

Machine learning is a technique where, the model will label from the given dataset according to the type of leaning preferred. It builds the mathematical model based on the provided sample data. It can be mainly used for prediction of the label, classify the label and form a cluster according to their input parameters. In order to classify the tumor from the malignancies, the supervise algorithm can be used to classify the required part that may be considered as tumor. The classification will come under the type of supervised learning. In this study, the dataset is taken from Kaggle website and the three different supervised algorithm is used to classify the dataset and the output of each algorithm is compared with the proposed method by the metrics value.

The main objective of machine learning is classification of data. The classification may be anything either binary or multiple. The application of binary classifications like Decision tree, Boosted Tree, KNN, SVM, Logistic Regression and Random Forest. Is needed in many fields These popular ML algorithms have been applied to many fields, such as recognition of hand pose [5],face recognition (6),text separation (7), emotion identification (8), object detection at multi view and point (9), prediction of molecular hybridization (10), imageand graphic (11), (12), machine vision (13), stability estimation of hard rock pillar (14), packet classification (15), data mining (16), Chess Games (17), remote control vehicles (18) and classification of MRI sequence images (19).

IOP Conf. Series: Materials Science and Engineering **995** (2020) 012028 doi:10.1088/1757-899X/995/1/012028

1.1 Data

The dataset for tumor detection is taken from Kaggle website. The dataset contains radius, texture, perimeter, area, smoothness, compactness, concavity, concave points, symmetry and fractal dimension of each cell nucleus. The mean, standard error and largest value of the above mentioned features were computed and it results in 30 features. The target contains tumor types like benign and malignant. The features are really very large, it leads to usage of more memory and execution time. So the important features were extracted based on the correlation. By using the heatmap plot the correlation between the features and target was found. The correlation plot is given in the figure 2. From the plot the less correlated features like symmetry standard error, smoothness standard error, texture standard error, fractional dimensional mean and standard error were found and dropped.

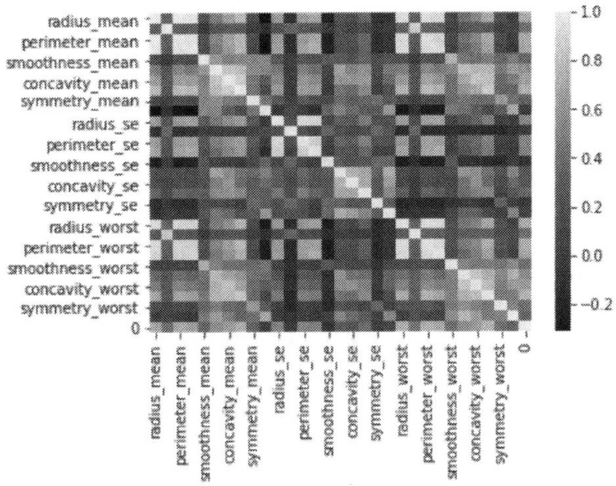

Figure 2. Correlation between the features and tumor

1.2Logistic Regression

For classification, one of the famous algorithm is Logistic Regression. It is used for both binary classification like '0' or '1', 'Yes' or 'No' and multiple classifications. The base of this algorithm is linear regression [20]. The Sigmoidal function is used for modelling the data which eliminate the drawbacks of linear regression classification. The sigmoidal function is given in the equation 1.

$$g(z) = \frac{1}{1+e^{-z}} \quad (1)$$

The threshold value is very important for classification. The setting of threshold value changes based on the application. Precision and recall plays a vital role to decide the threshold value.

The feature [input] matrix is denoted as

$$X = \begin{bmatrix} 1 & \cdots & x_{1j} \\ \vdots & \ddots & \vdots \\ 1 & \cdots & x_{ij} \end{bmatrix} \quad (2)$$

Where in x_{ip},

n→ith observation

j→jth feature

Recall Linear regression

$$h(x_i) = \beta_0 + \beta_1 x_{11} + \beta_2 x_{12} + \cdots + \beta_p x_{np} (3)$$

$$(x_i) = \beta^T x_i \quad (4)$$

Where,

$h(x_i)$→ Classified target for nth observation

β→ Regression coefficient vector/matrix

$$\beta=\begin{bmatrix}\beta_0\\ \beta_1\\ \vdots\\ \beta_p\end{bmatrix} \quad (5)$$

By applying sigmoidal function in linear regression

$$h(x_i) = g(\beta^T x_i)\frac{1}{e^{-\beta^T x_i}} \quad (6)$$

The sigmoidal graph is shown below

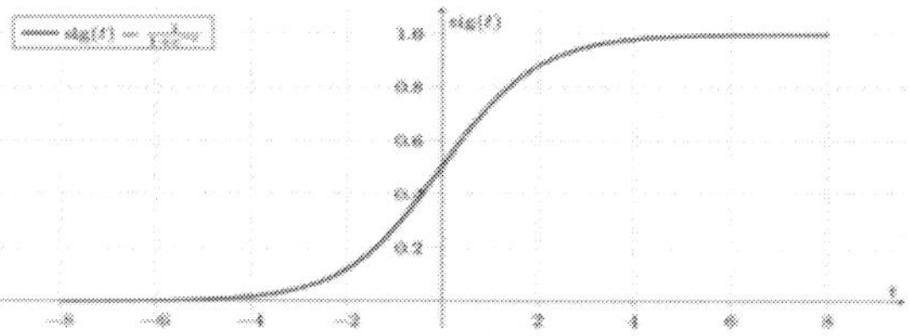

Figure 2. Sigmoidal Graph

From the graph it is infer that

$$g(z) = 0\ atz \to -\infty \quad (7)$$

$$g(z) = 1\ atz \to \infty \quad (8)$$

After describing conditional probabilities for 2 targets

$$P(y_i = 1|x_i;\beta) = h(x_i) \quad (9)$$

$$P(y_i = 0|x_i;\beta) = h(x_i) \quad (10)$$

The equation 1 and 2 can be combined as

$$P(y_i|x_i;\beta) = \left(h(x_i)\right)^{y_i}\left(1 - h(x_i)\right)^{1-y_i} \quad (11)$$

Likelihood parameters of β

$$L(\beta) = \prod_{i=1}^{n} P(y_i|x_i;\beta) \quad (12)$$

Apply log on both side

$$log\left(L(\beta)\right) = log\left(\prod_{i=1}^{n} P(y_i|x_i;\beta)\right) \quad (13)$$

$$log\left(L(\beta)\right) = \sum_{i=1}^{n} y_i log\left(h(x_i)\right) + (1 - y_i)log\left(1 - h(x_i)\right) \quad (14)$$

Cost Function of logistic Regression

$$J(\beta) = \sum_{i=1}^{n} -y_i log\left(h(x_i)\right) - (1 - y_i)log\left(1 - h(x_i)\right) \quad (15)$$

Gradient descent for minimizing the cost function

$$\frac{\partial J(\beta)}{\partial \beta_j} = (h(x) - y)x_j \quad (16)$$

Where,
y→ Actual response
$h(x)$→Predicted response

For getting minimum $J(\beta)$

$$\beta_j = \beta_j - \alpha \sum_{i=1}^{n}(h(x_i) - y_i)x_{ij} \quad (17)$$

Where,
α→ Learning rate
The β value updated until the error minimized.

By using the Logistic Regression Algorithm, the classification of cancer will be done by using the dataset after data pre-processing techniques. Then the confusion matrix of Logistic Regression is given in the figure 4. This figure gives the details of True Positive, True Negative, False Positive and False Negative of cancer detection using Logistic Regression.

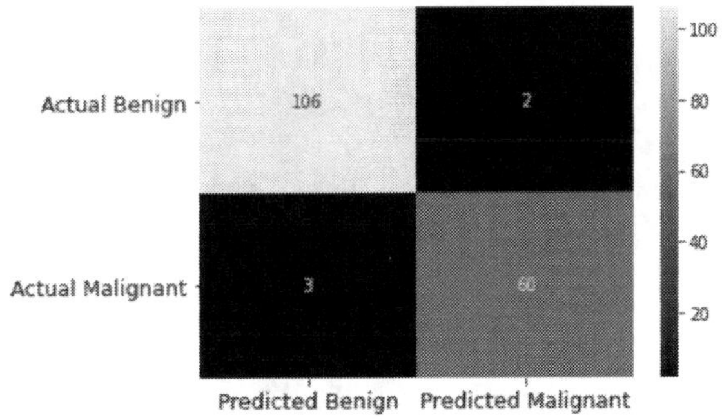

Figure4. Confusion matrix of Logistic Regression

1.3Support Vector Machine

The support vector machine comes under supervised learning. It is suitable for both prediction and classification. The target data plotted in n-dimensional space, n is nothing but the number of features in the dataset. After plotting the hyper-plane has to draw. In 1992, the SVM which was first projected by Vapnik et al. It depends on VC dimension concept. Now, the SVM algorithm has certain constraints. The performance of SVM purely based on kernel selection. Still the perfect selection of kernel for some problems are not found. The prevailing SVM algorithm focussed on penalty coefficient [25]

Rules for hyper-plane:

1. Hyper-plane should segregate 2 classes very well.
2. The distance between the hyper-plane and the nearest data in the class is called margin. The hyper-plane should have maximum margin for 2 classes.
3. Different kernel tricks are available. Based on the non-linearity of data distribution, the kernels are selected.

The details of margin, hyper planes of Support Vector Machine classification is shown in the figure 5

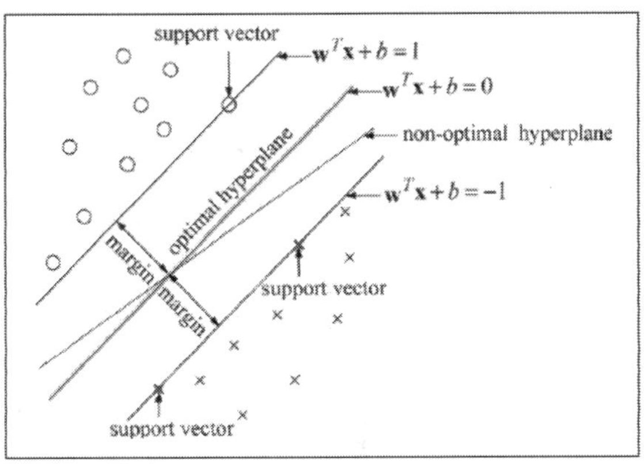

Figure5. Support Vector machine working [24]

The hyperplane line is used to segregate the dataset into two classes. The hyper plane is a straight line. The equation of straight line is

$y = a * x + b$ (18)

The equation can be reframed as

$a * x + b - y = 0$ (19)

The compact form of the equation

$W.X + b = 0$ (20)

Where

$X = (x, y)$ (21)

$W = (a, -1)$ (22)

The SVM classifies data by the given format

$y_i = 1$, If $w.x_i + b \geq 1$ (23)

$y_i = 0$, If $w.x_i - b \leq 1$ (24)

Margin is the distance between hyperplane and the support vector. The margin will be represented as

$M = \frac{1}{|w|}$ (25)

For optimal classification the margin should be maximum

$max\frac{1}{||w||}$ or $\min ||w||$ (26)

For better result l2 optimization is used

$min\frac{||w||^2}{2}$ (27)

Sometimes very few class 'A' object falls in class 'B'. This is called slack variable. If tries to make the slack variable zero the margin will reduce. So regularization parameter 'C' is introduced to maintain the trade-off between the margin width and slack variable.

$min(w, b)\left\{\frac{||w||^2}{2} + C\sum_i \xi_i\right\}, y_i(w.x + b) \geq -1 - \xi_i$ (28)

For non-linear distribution of data, kernel tricks is used for classification. The kernel just map the non-linear data into high dimensional space for easier classification.

Different Kernels:

Linear Kernel: $K(x_i, y_i) = x_i x_j$

Polynomial Kernel: $K(x_i, x_j) = (x_i.x_j + c)^d$

RBF Kernel: $K(x_i, x_j) = exp(-\gamma||x_i - x_j||^2)$

Cost Function of SVM

$J(\beta) = \sum_{i=1}^{n} y_i max(0,1 - \beta^T x) + (1 - y_i)max(0,1 - \beta^T x)$ (29)

By using the SVM Algorithm, the classification of cancer will be done by using the dataset after data pre-processing techniques. Then the confusion matrix of SVM is given in the Figure 6. The Figure 6 gives the details of True Positive, True Negative, False Positive and False Negative of cancer detection using SVM.

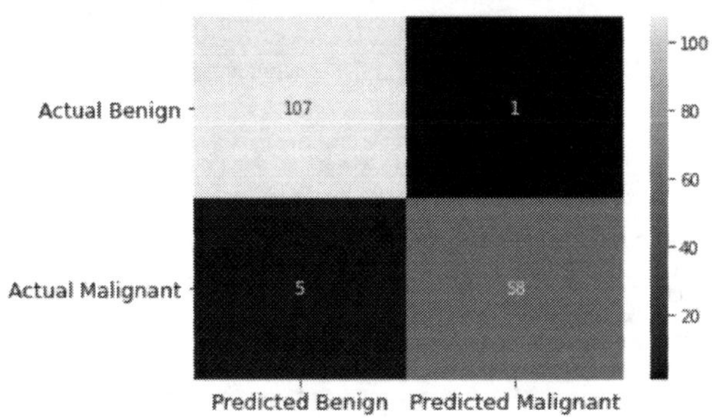

Figure 6. Confusion matrix of SVM

1.4K-Nearest Neighbor

The KNN algorithm is also a supervised algorithm. It is used for both classification and prediction problems. The KNN classify the data based on the majority of nearest data to the tested data [21]. The KNN was first proposed by Cover and Hart in 1968, it is a type of classification algorithm. KNN is usually called lazy learning method. The explicit learning phase and training process are not available in KNN [26]. The K value chosen in order to reduce training error and validation error. If K is small the training error will be zero but the validation error will become high. If K is high both errors will be high. So the K has to be chosen in-order to minimise the two errors. K-Nearest neighbor for classification is shown in figure 7. The K value is nothing but the number of neighbour data taken for classification. Always choose odd number as a K value.

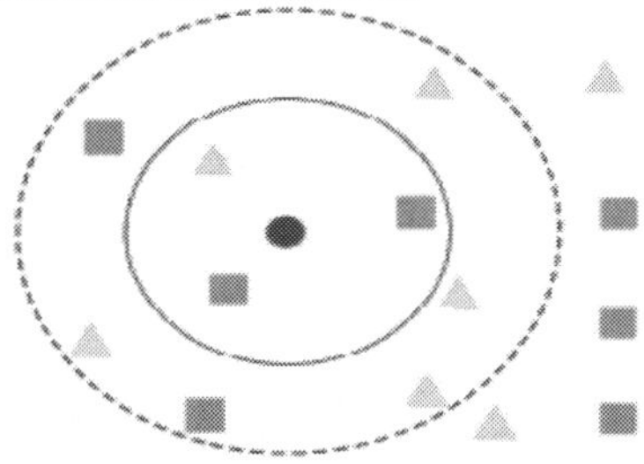

Figure 7. K-Nearest neighbor for classification [22]

The nearest neighbour is calculated by using the Euclidean distance

$$d(q,p) = \sqrt{(q_1 - p_1)^2 + (q_2 - p_2)^2 + \cdots (q_n - p_n)^2} \quad (30)$$

$$d(q,p) = \sqrt{\sum_{i=1}^{n}(q_i - p_i)^2} \quad (31)$$

The Euclidean distance is calculated for all target points in the dataset. Then based on the K values minimum distance of K data points is taken. Based on the target majority, the classification is done. The number of samples in the region is denoted by $k(y)$

The posterior probability $p(y/x)$ is given

$$P(y/x) = \frac{P(x/y)P(y)}{P(x)} = \frac{k[y]}{k} \quad (32)$$

The classification is done by using the following equation

$$g(x) = \begin{cases} 1, & k[y=1] \geq k[y=-1], \\ -1, & k[y=-1] \geq k[y=1] \end{cases} \quad (33)$$

By using the KNN Algorithm, the classification of cancer will be done by using the dataset after data pre-processing techniques. Then the confusion matrix of KNN is given in the Figure 8. The Figure 8 gives the details of True Positive, True Negative, False Positive and False Negative of cancer detection using KNN.

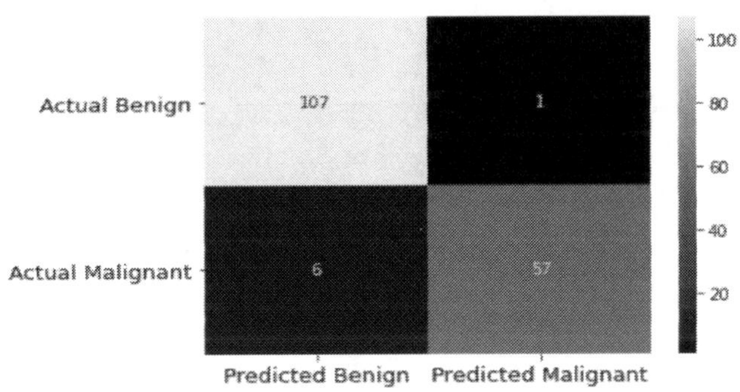

Figure 8. Confusion matrix of KNN

2. Proposed Method

Ensemble stacking technique is used for creating new model. It associate the choices from multiple models and create new model from those model for improving the performance of the tumour detection process [23]. The step involved in proposed method is given.

1) The training dataset divided into small parts. For each training dataset the Logistic regression model is fitted to 9/10 part of the set and test take place in 1/10 part of the set.
2) Then fit the logistic regression model to all the training dataset

3) By using this model, the classification is done for all the testing dataset.
4) Repeat the steps 2 to 4 for SVM and KNN model, it give some other classification results for training and testing dataset
5) The classification results from the training dataset are used as an input for building a novel model.
6) By using this model the final classification is made on classified testing dataset

By using the proposed method, the classification of cancer will be done by following the above procedure. Then the confusion matrix of proposed method is given in the Figure 9. The Figure 9 gives the details of True Positive, True Negative, False Positive and False Negative of cancer detection using proposed method.

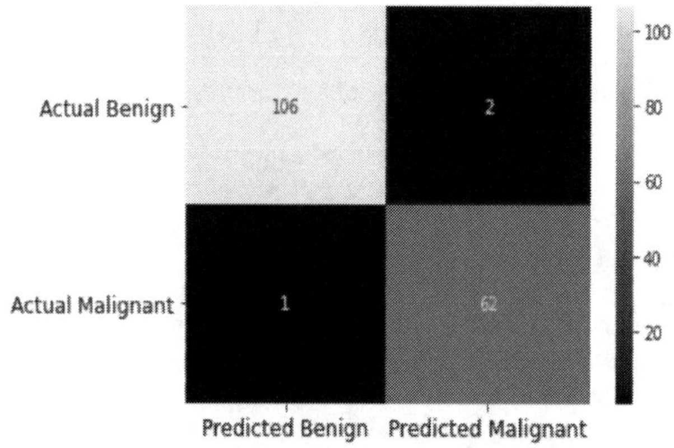

Figure 9. Confusion matrix of proposed method

3. Results and Discussion

The breast cancer is classified by using 3 conventional machine learning algorithm and it is compared with the proposed stacking ensemble method by using the metrics. The metrics used in this research are accuracy, precision and recall. The formula used to calculate those metrics are given in the below equation

$$Accuracy = \frac{TP + TN}{TP + TN + FP + FN} \tag{34}$$

$$Precision = \frac{TP}{TP + FP} \tag{35}$$

$$Recall = \frac{TP}{TP + FN} \tag{36}$$

The bar-graph is used for the comparison of proposed method with the conventional methods. The figure 10 displays different algorithms and its metrics value in percentage. From the below graph it is clear that the proposed method give better accuracy and recall. It shows that the proposed method gives more true prediction and less false prediction values than the conventional methods like SVM, KNN and Logistic Regression.

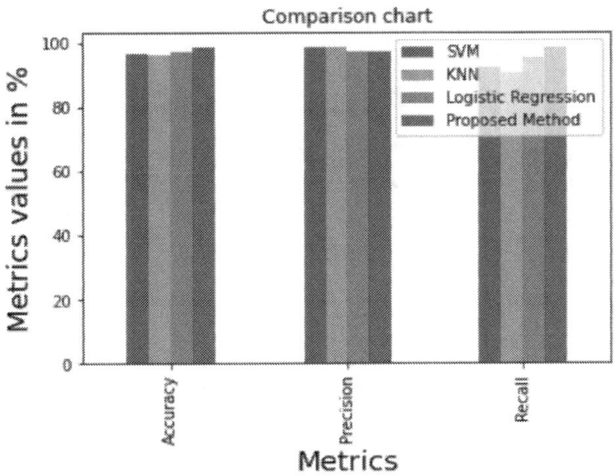

Figure 10. Performance comparison of Algorithm

Table 1. Performance comparison in numeric value

Metrics	SVM	KNN	Logistic Regression	Proposed Method
Accuracy	96.49	95.90	97.07	98.24
Precision	98.30	98.27	96.77	96.87
Recall	92.06	90.47	95.23	98.41

This table 1 evidently displays the value of three metrics like accuracy, precision and recall. From this table the conclusion is made that the proposed algorithm gives better result than the conventional algorithms.

4. Conclusion

The breast cancer dataset are taken from the kaggle website. The data are split into 7:3 ratio for training and testing. The more important features are taken based on the correlation matrix. The three conventional algorithms like SVM, KNN and logistic regression are applied to find the classification model. Then the proposed method also used to predict the classification model. After designing the model, based on the metrics the more efficient model for classification is found. The proposed method gives better satisfactory result for tumour classification whether the given data is benign or malignant. In future for getting more than 99% metrics, the optimization techniques are planned to use.

Reference

[1] M. Bowles, ‘Machine Learning in Python: Essential Techniques for Predictive Analytics’, John Wiley & Sons Inc., ISBN: 978-1-118- 96174-2
[2] W. Richert, L. P. Coelho, ‘Building Machine Learning Systems with Python’, Packt Publishing Ltd., ISBN 978-1-78216-140-0
[3] J. Ferlay, I. Soerjomataram, R. Dikshit et al., ‘Cancer incidence and mortality worldwide: sources, methods and major patterns in GLOBOCAN 2012’, International Journal of Cancer,

vol. 136, no. 5, pp. 359–389, 2014.
[4] J. G. Elmore, C. K. Wells, C. H. Lee, D. H. Howard, and A. R. Feinstein, 'Variability in radiologists' interpretations of mammograms', New England Journal of Medicine, vol. 331, no. 22, pp. 1493–1499, 1994.
[5] T. Parag and A. Elgammal, 'Unsupervised learning of boosted tree classifier using graph cuts for hand pose recognition', Proc. Brit. Mach. Vis. Conf., pp. 1259-1268, Sep. 2013.
[6] X. Peng, S. Setlur, V. Govindaraju and S. Ramachandrula, 'Using a boosted tree classifier for text segmentation in hand-annotated documents', Pattern Recognit. Lett., vol. 33, no. 7, pp. 943-950, 2012.
[7] Demirkir and B. Sankur, 'Face detection using boosted tree classifier stages', Proc. IEEE 12th Signal Process. Commun. Appl. Conf., pp. 575-578, Apr. 2004.
[8] M. Day, 'Emotion recognition with boosted tree classifiers', Proc. 15th ACM Int. Conf. Multimodal Interact., pp. 531-534, 2013.
[9] B. Wu and R. Nevatia, 'Cluster boosted tree classifier for multi-view multi-pose object detection', Proc. IEEE Int. Conf. Comput. Vis., pp. 1-8, Oct. 2007.
[10] W. Zhu, Y. Han, H. Wu, Y. Liu, X. Nan and Q. Zhou, 'Predicting the results of molecular specific hybridization using boosted tree algorithm', Concurrency Comput. Pract. Exper..
[11] C. Shen, Y. Chen, G. Yang and X. Guan, 'Toward hand-dominated activity recognition systems with wristband-interaction behavior analysis', IEEE Trans. Syst. Man Cybern. Syst..
[12] Bera and D. Bhattacharjee, 'Human identification using selected features from finger geometric profiles', IEEE Trans. Syst. Man Cybern. Syst..
[13] G. Yu, J. Yuan and Z. Liu, 'Unsupervised random forest indexing for fast action search', Proc. IEEE Comput. Vis. Pattern Recognit., pp. 865-872, Jun. 2011.
[14] J. Zhou, X. Li and H. S. Mitri, 'Comparative performance of six supervised learning methods for the development of models of hard rock pillar stability prediction', Natural Hazards, vol. 79, no. 1, pp. 291-316, 2015.
[15] J. R. Quinlan, 'Learning efficient classification procedures and their application to chess end games'in Mach. Learn, Berlin, Germany:Springer, pp. 462-482, 1983.
[16] X. Dong, M. Qian and R. Jiang, 'Packet classification based on the decision tree with information entropy', J. Supercomput., vol. 74, pp. 1-15, Jan. 2018.
[17] M. Jaworski, P. Duda and L. Rutkowski, 'New splitting criteria for decision trees in stationary data streams', IEEE Trans. Neural Netw. Learn. Syst., vol. 29, no. 6, pp. 2516-2529, Jun. 2018.
[18] M. McClelland and M. Campbell, 'Probabilistic modeling of anticipation in human controllers', IEEE Trans. Syst. Man Cybern. Syst., vol. 43, no. 4, pp. 886-900, Jul. 2013.
[19] Y. Zhang, D. Kwon and K. M. Pohl, 'Computing group cardinality constraint solutions for logistic regression problems', Med. Image Anal., vol. 35, pp. 58-69, Jan. 2017.
[20] Lei Liu, 'Research on Logistic Regression Algorithm of Breast cancer Diagnose data by Machine Learning', International Conference on Robots & Intelligent System (ICRIS) - e-ISBN: 978-1-5386-6580-0,2018
[21] Osisanwo F.Y , Akinsola J.E.T, 'Supervised Machine Learning Algorithms: Classification and Comparison', International Journal of Computer Trends and Technology (IJCTT) – e-ISSN: 2231-2803 Volume 48 Number ,2017
[22] Ayon Dey, 'Machine Learning Algorithms: A Review', (IJCSIT) International Journal of Computer Science and Information Technologies, Vol. 7 (3) , 2016
[23] Opitz, R. Maclin, 'Ensemble of machine learning algorithms using the stacked generalization approach to estimate the warfarin dose', Journal of Artificial Intelligence Research, 11, Pages 169- 198, 1999
[24] D. Meyer, 'Support Vector Machines - The Interface to libsvm in package e1071', August 2015
[25] X. Li, M. Fang, H. Wang and J.-J. Zhang, 'Supervised transfer kernel sparse coding for image classification', Pattern Recognit. Lett., vol. 68, no. 12, pp. 27-33, Dec. 2015.

[26] X. Shen, L. Niu, Z. Qi and Y. Tian, 'Support vector machine classifier with truncated pinball loss', Pattern Recognit., vol. 68, no. 8, pp. 199-210, Aug. 2017

Performance Evaluation of Domestic Refrigerator with Waste Heat Recovery System

P Thangavel[1], S.Mokesh[2], SA.Reavanth[3], K.Pragatheeswaran[4]

1 Associate Professor, Department of Mechanical Engineering, Kongu Engineering College, Perundurai-638060, Tamilnadui, India.

2, 3, 4 UG Students, Department of Mechanical Engineering, Kongu Engineering College, Perundurai-638060, Tamilnadui, India.

ptvelmech@gmail.com, mokesh998@gmail.com, reavanth18@gmail.com, pragateesh19@gmail.com

Abstract. The objective of the article is to study the performance of domestic refrigerator with waste heat recovery system. The usage of domestic refrigeration systems are increasing every day due to the requirement. During the operation of domestic refrigeration system, heat is rejected through the condensing coil to the atmosphere and leads to thermal pollution. Hence this heat can be recovered by suitable waste heat recovery system and employed for useful domestic applications. Hot chamber box has been fabricated and assembled with the domestic refrigerator. Experimental analysis has been carried out under air cooled condenser and hot chamber condenser with different loading conditions like no load, 1 litre, 3 litres, 5 litres, and 10 litres of water in the evaporator unit. COP is calculated for all the loading conditions and compared with air cooled condenser and hot chamber condenser. COP of all the loading conditions is maintained almost in the same range value. COP value is obtained in the range of 5 to 5.5 for the air cooled and hot chamber condensation modes. Hence the performance of domestic refrigerator is not changed due to the running of hot chamber box. Due to installation of hot chamber box waste heat is recovered from the condenser coil. The maximum heat recovery is arrived in the range of 31.15 to 33.66 kJ/kg.

Key words: COP, Domestic refrigerator, Hot chamber box, Waste heat recovery.

1.INTRODUCTION

Refrigeration is the science of producing and maintaining temperature below that of the atmosphere temperature. This indicates that the removing the heat from a substance through a refrigeration cycle. Maintaining the low temperature in a space using refrigeration system could be employed for preservation of food materials, vegetables, fruits, medicines, etc. Heat always travels from higher temperature to lower temperature region. With help of external sources, heat can travel from lower temperature to higher temperature. Refrigerators are working based on the statement of second law of thermodynamics. Most of the domestic refrigerators are belongs to Vapour Compression Refrigeration System (VCRS) and the main components of VCRS are compressor, condenser, expansion device and evaporator. The working fluid of the refrigeration system is termed as refrigerant which is defined as a substance that absorbs latent heat through vaporization and rejects the same during the condensation. During the condensation process, large amount of heat is released from the condenser coil of refrigerator to the atmospheric air. Hence the cooling effect is obtained in the evaporator of the VCRS. Heat rejected from the condenser coil can be recovered by a suitable waste heat recovery system. By retrofitting a waste heat recovery system, condenser heat can be recovered and can be utilised for various domestic applications like maintaining food items at warm condition. It is possible to improve the overall performance of the refrigeration system by utilizing the waste heat generated from the refrigerator. Development of such efficient methods can be employed for heating the food items which reduces the extra energy cost incurred to heat the food items separately.

Waste heat from the domestic refrigerator was utilized for water heating application which can be used for different applications in houses. Maximum temperature of 44.3°C for water is obtained from the refrigeration system with the domestic refrigerator capacity of 165 litres and overall performance of the refrigerator is also enhanced [1]. Domestic refrigerator with waste heat recovery system provides an increase in overall effectiveness of domestic refrigerator and provides energy saving. There is an improvement in COP of the domestic refrigerator and waste heat could be used for heating applications [2]. The use of waste heat recovery system in a domestic refrigerator is employed for water and air heating applications. Temperature gradient of 10°C between air inlet & outlet was achieved and water temperature has increased upto 46°C. COP of the domestic refrigerator is also enhanced without change of normal refrigeration effect [3]. Experimental study was executed on VCRS under different evaporator loads like 50,100,150,200 and 250 Watts. During the process, the heat rejected in the condenser was applied for sub cooling purpose of another refrigeration system. Results show that improvement in condenser heat rejection rate, refrigeration effect and COP of the system [4]. Condensing coil length plays an important role in the heat rejection rate from the condenser and variable lengths of condenser coil have executed in this study. Usage of condenser coil length about 9.75 m in domestic refrigerator offers 5.74% improvement in COP compared to the existing system. Net refrigeration effect is improved about 5.81% more than the existing system and the power consumption is about 5.48% less than the existing VCRS. Hence more heat rejection occurs in the condenser coil and improvement in heat rejection is about 4.85% more than the existing condenser [5].

Heat rejected from the condenser coil of 1 TR air-conditioning system was examined. Rejected heat was used for water heating applications and mass flow rate water was varied from 20 to 200 lph. At lower mass flow rate of water, the water temperature was achieved about 60°C. Cooling and heating effect are achieved by using air-conditioning system [6]. Waste heat recovery system has assembled with the refrigeration system to recovery the condenser heat and recovered heat was used for water heating application. Temperature gradient was obtained about 46°C [7]. Modified heat exchange has employed in air-conditioning system and the experimental results shows that the waste heat is recovered from the air conditioning system. Increase in COP is obtained about 13% from the analysis and thermal pollution is also reduced [8]. 165 litres capacity refrigerator has used and the condenser heat is recovered by the waste heat recovery system. Water temperature was reached about 50.6 °C in outside the cabin [9].

Water cooled condenser was assembled with the refrigeration system to receive the heat rejected from the condenser. Maximum water temperature is reached about 50°C in 3 hours of operation and average temperature was obtained about 40°C. Overall performance of the refrigeration system was enhanced without changing the normal cooling effect [10]. Water tank capacity of 100 litres has used to observe the heat rejected from the condenser of the refrigerator and the maximum temperature was achieved by 45°C under full load operation. Based on the quantity of water in the tank, the heating time has varied. This provides better performance of the refrigeration system and reduces the thermal degradation of the environment due to condenser heat [11]. Performance evaluation of vapour compression refrigeration system was completed under different loading conditions in the evaporator chamber [12].

In summary, VCRSs are widely used for domestic applications in present scenario. Hence the performance improvement of the refrigeration system should be concentrated by reducing the electricity consumption and also proper utilization of waste heat from the domestic refrigerator for different applications. Literature shows that most of the studies are carried out to improve the performance of the domestic refrigerator by utilizing the waste heat from the condenser coil effectively. Recovered waste heat could be used for numerous applications like heating of food items, cookies, hot water, etc. Experimental study has carried out on a domestic refrigerator with a waste heat recovery system and discussed the practical feasibility.

2. MATERIALS AND METHODOLOGY

The main components of VCRS are compressor, condenser, expansion valve and evaporator. Experimental arrangement has been arranged with two modes of condensation namely air cooled condenser and hot chamber condenser. Figure 1 represents the experimental arrangement of the domestic refrigerator with the proposed condensation modes. The capacity of the domestic refrigerator is 195 litres. Existing air cooled condenser (copper) is having the dimensions of 0.005m in diameter and 8.40m in length with equally spaced aluminium fins. For the same area of the existing air cooled condenser, with the help of copper tube of 0.008m diameter a hot chamber box condenser has fabricated and the hot chamber condenser is assembly with the domestic refrigerator at the top. Pressure and temperature measurements are observed at different state points like 1, 2, 3 and 4 by pressure gauge and thermocouples respectively. Thermocouples are also used to measure the temperature inside the hot chamber box. Gate valves are mounted between the air cooled condenser and hot chamber condenser in order to bypass the flow of refrigerant in different modes. Photographic view of the experimental arrangement is shown in Figure 2.

Flow diagram of the experimental system is shown in Figure 3. Vapour refrigerant at low temperature and pressure (1) enters into the compressor where it is compressed isentropically. Subsequently its pressure and temperature increased (2). Then the high-pressure vapor state refrigerant enters into the hot chamber box from compressor and the heat available in the refrigerant is used to heat the hot chamber box. After the hot chamber box, vapour state refrigerant enters into the air cooled condenser coil where it is condensed into the high-pressure liquid state refrigerant (3). High h pressure/low temperature liquid state refrigerant is expanded through the expansion valve (4) and allowed to flow inside the evaporator where the cooling effect is induced. Due to absorption of heat inside the evaporator chamber, the refrigerant is converted into low pressure/temperature vapour refrigerant (1) and the same cycle repeats.

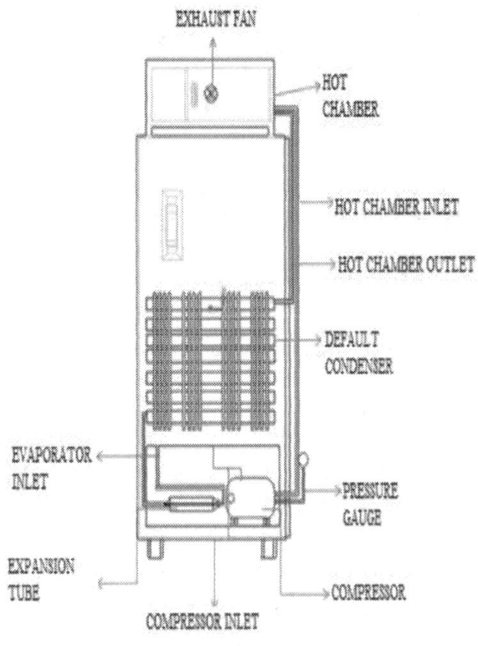

Figure.1. Experimental arrangement of domestic refrigerator with hot chamber box condenser

Figure.2. Photographic view of the experimental arrangement

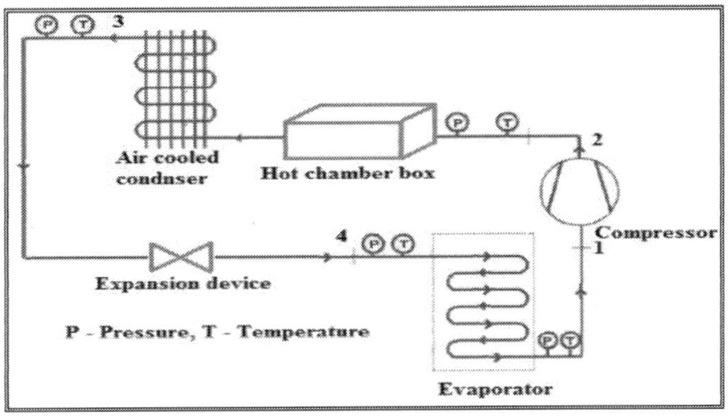

Figure 3 Flow diagram of the experimental system

A temperature controller unit is assembled with the experimental system and temperature is controlled in the hot chamber box by accepting signals from the temperature sensors through thermocouple. It compares the actual temperature which is produced in the hot chamber box with the desired control temperature or set point temperature. Based on the set point temperature limit, the exhaust fan will rotate automatically to drain the excess hot air which accumulated inside the hot chamber box. When the set point temperature is reduced inside the hot chamber box, the exhaust fan get stopped its operation and allows the hot chamber box to reach the required set point temperature. Temperature sensor is placed inside the hot chamber box and received input signals are given to the temperature controller. Measured temperature is indicated in digital temperature indicator and set point temperature is also arranged in the indicator. The photographic view of the temperature controller is shown in Figure 4.

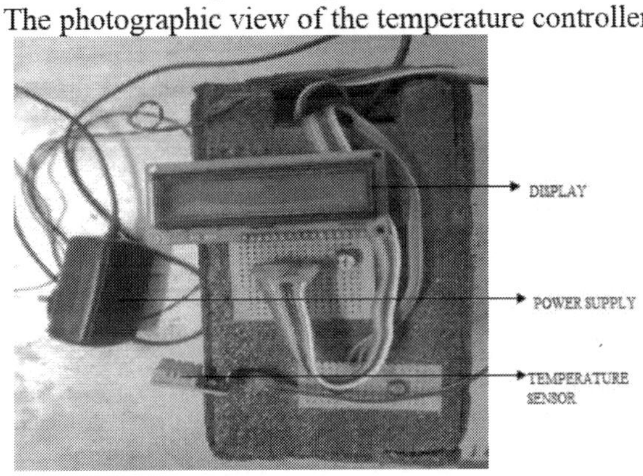

Figure 4 Photographic view of temperature controller

3. EXPERIMENTAL ANALYSIS

Experimental analysis is executed in two modes of operations like air cooled condensation and hot chamber box condensation. In order to run the experimental system in the above modes, valves are employed at different locations of the experimental arrangement. By opening and closing of these valves, the experimental analysis has executed in air cooled condensation and hot chamber box condensation. These valves provide the change of circulation of refrigerant at the desired modes and bypass the flow of the refrigerant to the two different condensers. First the refrigerant is circulated to flow through existing air cooled condenser and the experimental analysis is completed with different loading conditions like no load, 1 litre, 3 litres, 5 litres, and 10 litres of water in the evaporator. For each loading condition, experimental system has been run for the time period of 100 minutes and each 10 minutes interval temperature and pressure measurement has noted at different locations like compressor inlet and outlet, condenser outlet and evaporator inlet. This experimental observations are repeated for the air cooled condenser and hot chamber condensation modes. During these modes of operation, hot chamber temperature is also noted in the middle and bottom positions. The temperature controller controls the temperature inside the hot chamber box according to the set point temperature. R134a is charged in the

VCRS. The following empirical correlations are used to calculate the various output parameters like work input supplied, condenser heat rejection rate, refrigeration effect and COP.

Work input supplied (W_{in}) $= m_r(h_2 - h_1)$ (1)

Condenser heat rejection rate (Q_{rej}) $= m_r(h_3 - h_2)$ (2)

Refrigeration effect (RE) $= m_r(h_1 - h_4)$ (3)

$$COP = RE/W_{in} = (h_1 - h_4)/(h_2 - h_1) \quad (4)$$

4. RESULTS AND DISCUSSION

Enthalpy values are calculated at different states like compressor inlet (h_1) and outlet (h_2), condenser outlet (h_3) and evaporator inlet (h_4) using the empirical relations. Based on the enthalpy values, the various output parameters like work input supplied, condenser heat rejection rate, refrigeration effect and COP are calculated at different loading conditions. COP is the prime parameter for any refrigeration system to identify its performance. Hence, COP has been calculated for the different loading conditions in evaporator for air cooled condenser and hot chamber condenser. The variation in COP is plotted with respect to experimental time. 100 minutes has been taken for each experiment in air cooled condenser and hot chamber condenser modes. The variation in COP with respect to experimental time for air cooled condenser and hot chamber condenser is represented in Figure 5,6,7 8 and 9 for the different loading conditions like no load, 1 litre, 3 litres, 5 litres, and 10 litres respectively. The COP values are arrived about 5 to 5.5 for the both mode of condensations. Due to change in loading conditions in the refrigeration system, COP values are varied for the hot chamber condenser and air cooled condenser.

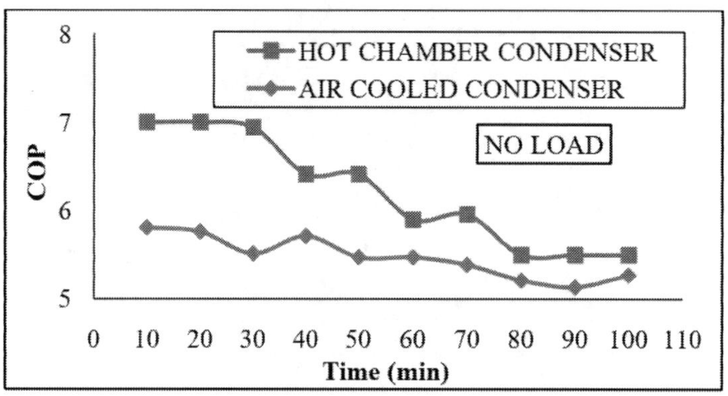

Figure 5 Time Vs COP at no load

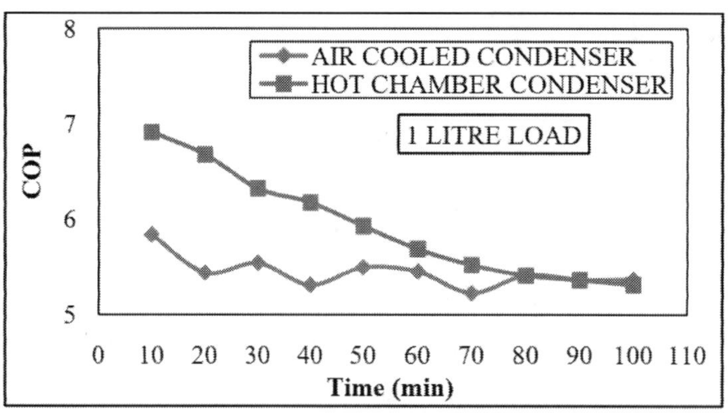

Figure 6 Time Vs COP at 1 litre load

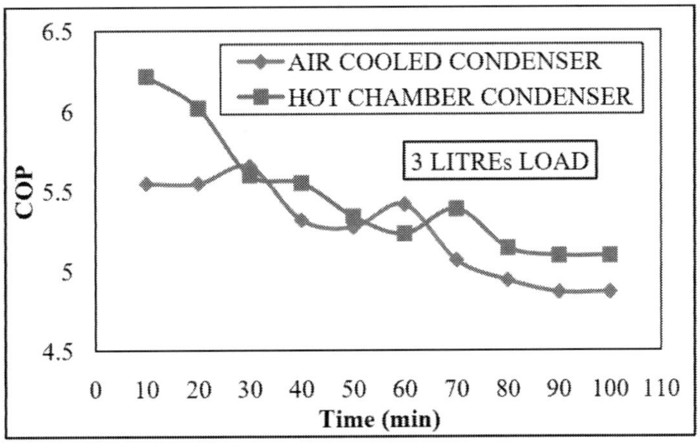

Figure 7 Time Vs COP at 3 litres load

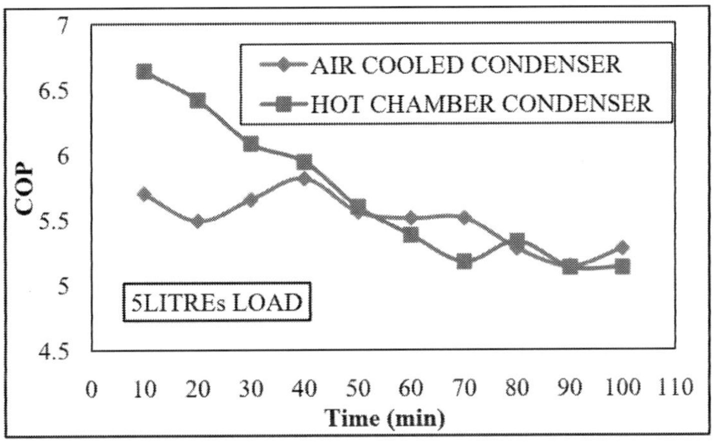

Figure 8 Time Vs COP at 5 litres load

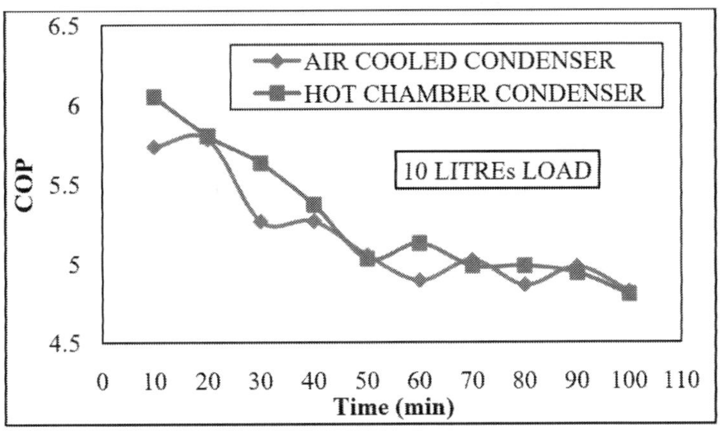

Figure 9 Time Vs COP at 10 litres load

Experimental study has been executed for the period of 100 minutes under each loading conditions. When the time increases, the required cooling effect is also progressed in a increased mode. Hence, COP of the refrigeration system is decreased with increase in time under air cooled condenser and hot chamber condenser operations. Figure 10 represents the variation of waste heat recovered in the hot chamber box at different loading conditions. Waste recovery rate is increased when the experimental time is increased for all the loading conditions. The maximum heat recovery is arrived in the range of 31.15 to 33.66 kJ/kg. Based on the different capacity of domestic refrigerators, hot chamber box can be designed and implementation of this concept in the domestic refrigerator increases the overall performance of the system. Cooling and heating effect are achieved by using the domestic refrigerator and holds economic feasibility for real time implementation.

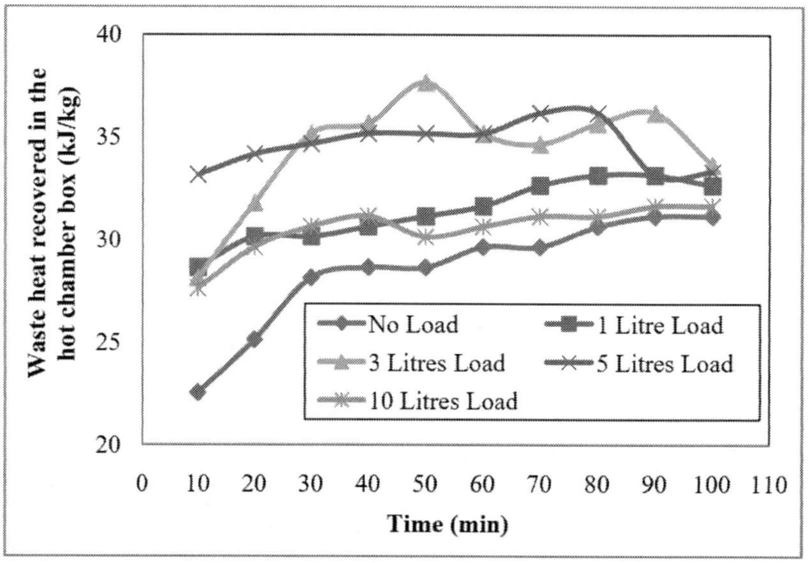

Figure 10 Waste heat recovered in the hot chamber box at different loading conditions

5. CONCLUSION

Experimental analysis has been performed on domestic refrigerator with waste heat recovery system and the following points are noted based on the results obtained.

- Domestic refrigerator with waste heat recovery system was experimentally investigated at various load conditions like no load, 1 litre, 3 litres, 5 litres, 10 litres and the performance parameters are calculated and compared on each load conditions.
- Waste heat recovery system performs well along with the domestic refrigerator. Hence the condenser heat can be trapped and applied for various domestic applications like maintaining food at warm condition.
- The performance of domestic refrigerator is indicated by the term COP. COP of all the loading conditions is maintained almost in the same range value. COP value is obtained between 5 to 5.5 for the air cooled and hot chamber condensation modes respectively.
- Hence the performance of domestic refrigerator is not changed due to the running of hot chamber box.
- Waste heat recovery rate is increased when the experimental time is increased for all the loading conditions. The maximum heat recovery is arrived in the range of 31.15 to 33.66 kJ/kg.

REFERENCES

[1] Soma A. Biswas and Sandeep M. Joshi, "Waste Heat Recovery from Domestic Refrigerator", International Journal of Current Engineering and Scientific Research, Vol. - 4, Iss. - 9, 2017.

[2] A. M. Vibhute and Avinash M. Patil, "Waste Heat Recovery in Domestic Refrigerator", International Engineering Research Journal, Special Issue, 2015.

[3] Lakshya Soni, Pawan Kumar and Rahul Goyal, "Waste Heat Recovery System from Domestic Refrigerator for Water and Air Heating", International Journal of Engineering Sciences & Research Technology, 5(7), 2016.

[4] S. Panneerselvam, P. Thangavel and S. Kannaki, "Performance Analysis of Vapour Compression Refrigeration System with Mechanical Subcooling", International Journal of Innovative Technology and Exploring Engineering, Vol. - 8, Iss.- 8, 2019.

[5] T. Mahaboob Basha, H.Ranganna and G. Maruthi Prasad Yadav. "Optimum Length of A Condenser for Domestic Vapor Compression Refrigeration System", International Journal of Science, Engineering and Technology Research, Vol. - 4, Iss.- 2, 2015.

[6] Suvarna Shivaji Ghadge and V.N. Kapatkar, "Heat Recovery Analysis of Window Air Conditioner", International Journal on Theoretical and Applied Research in Mechanical Engineering, Vol. - 4, Iss.- 1, 2015.

[7] Anshul Sarin, Aatish Goel, Tushar Sharma, Abhishek Misra and Shivam Tyagi, "Effective Utilization of Condenser Heat of a Refrigerator in a Heating Chamber", International Journal of Scientific and Research Publications, Vol. - 7, Iss.- 12, 2017.

[8] R.B. Lokapure and J.D. Joshi, "Waste Heat Recovery through Air Conditioning System", International Journal of Engineering Research and Development, Vol. - 5, Iss. - 3, 2012.

[9] S. B. Lokhande and S. B. Barve, "Design & Analysis of Waste Heat Recovery System for Domestic Refrigerator", International Journal of Modern Engineering Research, Vol. - 4, Iss. - 5, 2014.

[10] A. Anwaar, N. Prashant, S.B. Patil and G. Priyanka, "Design and Development of a Waste Heat Recovery Unit for Household Refrigerator System", International Research Journal of Engineering and Technology, Vol. - 6, Iss. – 11, 2019.

[11] Y. A. Patil and H. M. Dange, "Improving the Performance of Household Refrigerator by Recovering Heat from the Condenser, International Journal of Science and Research, Vol. - 2, Iss. - 6, 2013.

[12] P.Thangavel and P. Somasundaram, "Part Load Performance of Vapour Compression Refrigeration System with Hydrocarbon Refrigerants", Journal of Scientific & Industrial Research, Vol. – 72, 2013.

IOP Conf. Series: Materials Science and Engineering **995** (2020) 012030 doi:10.1088/1757-899X/995/1/012030

Performance Evaluation of Portable Mist Humidifier

M. Mahesh[1], P.Thangavel[2], K. Bhuvaneshwaran[3], V. Boopathi Raja[4], S. Dinaesh Krishna[5]

1, Assistant Professor, Department of Mechanical Engineering, Kongu Engineering College, Erode, Tamil Nadu, India.
2Assistant Professor, Department of Mechanical Engineering, Kongu Engineering College, Erode, Tamil Nadu, India.
3, 4, 5 Student, Bachelor of engineering in the Department of Mechanical Engineering, Kongu Engineering College, Erode, Tamil Nadu, India.

mahesh25391@gmail.com, ptvel@kongu.ac.in, bhuvaneshkrish007@gmail.com, boopathirajavbr3599@gmail.com, dinaesh1998sdk@gmail.com

Abstract. This paper deals with the performance evaluvation of portable mist humidifier, for producing a cooling effect in a specified area. A humidifier is an electrical device that increases the humidity of an environment. So the room temperature got reduced with the help of increasing the humidity. This devise provides a cooling effect at low cost without health hazards. The experimental arrangement made with the components of axial fan, copper tubes, hydraulic pump, nozzles, respectively. The misting nozzles are placed into the holes which are made in the copper tube and the copper tube is fitted over the frame on the axial fan. The hydraulic pump is kept in the water tank to circulate the water to the copper tube. When the water flows through the tube it emits water droplets with the help of misting nozzles. Due to the high-speed axial fan, the emitted water droplets are sprayed into the atmosphere as a mist. When there is a increase in the moisture content there will be decrease in room temperature. Due to this effect the room temperature has been decreased by 12% with the help of humidifier.

Key words: Cooling Effect, Copper Tube Humidifier, Humidifier, Humidification

1. Introduction

The process by which the moisture or water vapour or humidity is added to the air without changing its dry bulb temperature is called a humidification process. This process is not possible actually since the humidification is always accompanied by cooling or heating of the air. A humidifier is an electrical appliance device that increases humidity (moisture) in a single room or an entire building. In-home applications, humidifiers are commonly used to humidify a single room. For entire house or industrial applications, the capacity of the humidifier to be increased and provides humidification effect for the entire house. "Portable humidifier" is smaller in size and capacity which could be employed for small rooms. In this humidifier, water is supplied by manually in the tank on a periodic basis. Using appropriate mechanism water diffused into the air and space will be cooled. The various studies indicate that the comfortable value Relative Humidity (RH) for human beings is in the range of 35 to 55%. During the summer season, the atmospheric air is in dry condition and human comfort could not be maintained. Humidification of dry air is essential during the summer season. Based on the financial constrain, the portable mist humidifier unit can be used to create the required environmental condition.

Xiaong Huang et al. studied the performance characteristics like heat and mass transfer of corrugated plate spray humidification air coolers under wet working conditions. So that the humidification process can be done by a corrugated plate. Zhuangbofeng et al. studied that low level of indoor humidity, when human exposes to this will have respiratory symptoms or dryness (for skin, pharyngeal, eye etc.). It is closely related with air quality and indoor thermal comfort. This study showed that humidification improves the breathing quality of human beings.

Mostafasharqway et al. presented the cross-flow packed-bed humidifier. The experimental performance of this humidifier was studied with spraying hot water in a cross-flow arrangement over packing material where air flows through it. So that humidifier was made by simple water flow set up. J. Moureh et al. studied the performance of Refrigerated Display Cabinets (RDCs) and the use of mist flow. To improve the performance, fine water droplets were injected into the air curtain. So that the spraying of water droplets into the atmosphere would increase the humidity level.

Stuart Gaffeny et al. stated that humidification was concerned with the addition of water vapour to a gas. It can be measured as either relative or absolute. This study showed that higher relative humidity could be delivered by active humidification devices (including hot water baths). But were associated with potential hazards and higher cost. Pedro MagalhãesSobrinho et al. discussed about the factors that were directly connects with the energy consumption of a small air conditioning system. This study had been showed that the amount of energy needed for the humidifier.

In most of the literature, the cooling effect was done by refrigerants like HCFC and CFC which are having high ODP and GWP values, which results in Global warming. Low level of indoor humidity exposure may cause respiratory problems, so the usage of humidifier helps to increase the humidity level. A humidifier increases the air-side heat transfer coefficient. In a cross-flow arrangement, water is sprayed on the packing materials where the air flows.

2. Methodology

The problem in the current air coolers was identified by physical observation and experimentation. The literature related to portable mist humidifier were collected from different sources and observed the methodologies used. From the collected literature it has been analysed the gap and drawbacks found, nature of the model and cost involved. Conceptualization of the model to solve the problems found in the currently available methodologies by applying the engineering principles. Feasibility of the model checked by analysing the process parameter in the portable mist humidifier with the proposed mechanism and materials were selected.

Design calculations are carried out to meet up the process required for each component such cooling effect. Materials are purchased for each component according to the requirement and design. Fabrication of each component was done according to design and assembled to achieve the proposed mechanism. The model is operated to perform the cooling effect and results were noted and analysed the obtained results with the other equipment. The performance of the machine is analysed continuously to improve the quality of the work.

3. Materials and Specification

The following table contains the materials chosen for the making of experimental setup and its specification.

IOP Conf. Series: Materials Science and Engineering **995** (2020) 012030 doi:10.1088/1757-899X/995/1/012030

Table 1. Components description

Sl.No.	Parameters	Pictures	Range
1.	Axial fan		900 rpm
2.	Copper tube	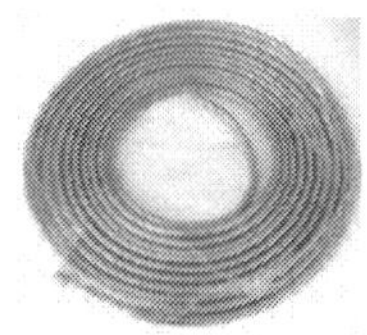	7mm-outer diameter, 5mm–inner diameter. 5ft length
3.	Mist Nozzle		0.2 mm (hole diameter)
4.	Hydraulic pump		1100 L/hr.

4. Fabiricated Model

Figure 1. Assembly of Portable Mist Humidifier

The assembly unit consists of an axial fan, copper tubes, misting nozzles, hydraulic pump, water tubes and the tank is shown in Figure 1. First, the copper tube is rounded as a coil and then the holes are made around the tubes. Then the misting nozzles are fitted into the holes made in the copper tube. Now the copper tubes are mounted over the frame of the axial fan. The sucking part of the pump is connected to the water tank and the delivery end of the pump is connected to the one end of the copper tube. The water from the copper tube is connected to the tank.

5. Working Process

The power supply was provided to the axial fan and pump for water circulation. Then the water is pumped at high pressure from the water tank to the copper tube. The water is passed through the copper tube and returned to the water tank and this cyclic operation is carried out during the entire process. When water flows through the copper tube, the water molecules are sprayed through the misting nozzles into the atmosphere. When the power supply started supplying the axial fan will provide high velocity of the air and the misting effect of the water particles, the diffusion of water molecules carried out into the atmosphere.

6. Design and Mathematical Calculation

To carry out a design Process, different software are available such as Auto CAD, Solid Works, CATIA V5R20, Creo, etc. The design of the proposed model is shown in Figure. This model is prepared by using CATIA software. This CATIS software was selected as it is most common software for solid modelling and more user-friendly. This setup consists of a table fan, Water tank of 600mmx600mm, the copper tube of diameter 12mm, Water pump, and Misting nozzle, respectively. The designed model is shown in Figure 2.

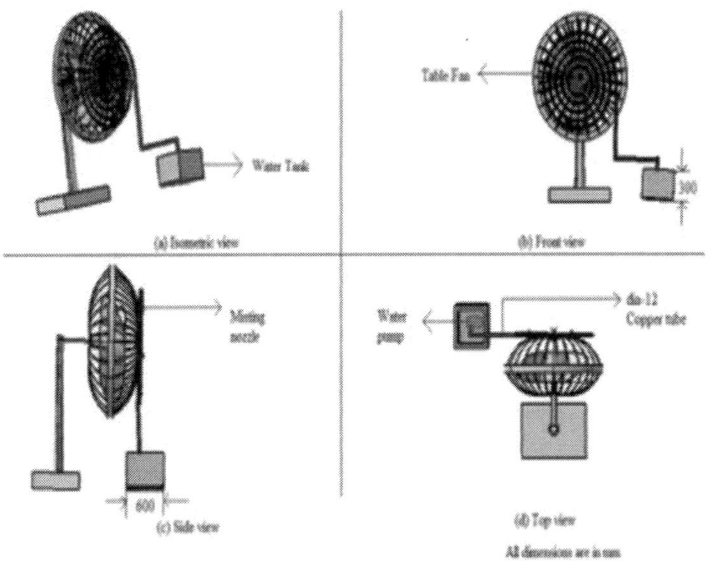

Figure 2. Design of Portable Mist Humidifier

6.1 Calculation

For normal room,

Length of the room (L) = 10 ft.
= 3.048 m.

Width of the room (W) = 8 ft.

= 2.438 m.

Height of the room (H) = 12 ft.

= 3.657 m.

Volume of the room,

$$V = L \times W \times H \quad (1)$$

$= 3.048 \times 2.438 \times 3.657$

$= 27.18\ m^3$.

$= 27\ m^3$ (approximately).

By using the dimensions, the volume of the room is calculated.

6.2 Cooling Effect in the room

$$Q = \frac{(C \times V \times E \times \Delta T)}{3600} \quad (2)$$

Where,

Q = cooling capacity of the room.
C = Air changes.
V = Volume of the room.
E = Energy per cubic meter of the air.
ΔT = Temperature difference.
= (Initial temperature – Final temperature).

For 15 minutes,

$$Q = \frac{(2 \times 27 \times 9.1 \times 10^3 \times (30-29.5))}{3600}$$

= 0.0682 kWh.

For 30 minutes,

$$Q = \frac{(2 \times 27 \times 9.1 \times 10^3 \times (30-28.75))}{3600}$$

= 0.1706 kWh.

$$\text{Cooling effect} = \frac{33-28.75}{33} \times 100$$

=12.87%

7. Result and Discussion

This model had been tested in the room conditions and the results are noted. The test results of the fabricated model are given in the form of the table 2.

Table 2. Performance analysis of humidifier

Sl.No.	Period of Time (minutes)	Temperature (Before humidifying) (in °C)	Temperature (After humidifying) (in °C)	Cooling Capacity (kWh)
1.	0-5	30	29.9	0.01365
2.	10	30	29.7	0.04095
3.	15	31	29.5	0.06825
4.	20	32	29.25	0.10237
5.	25	33	29.0	0.13650
6.	30	33	28.75	0.17062

Comparison of results with and without humidifier is shown in the figure 3. At the starting time of 5 minutes the room temperature is at 30°C. This study is done at day time in UPS room at kongu engineering college, so the temperature of the room was varying with the climatic condition and the usage of UPS. When the time increses there is a increase in the atmospheric temperature of the room in the normal condition. While using the fabricated humidifier continuously for 30 minutes the cooling capacity reaches 0.17062kWh.

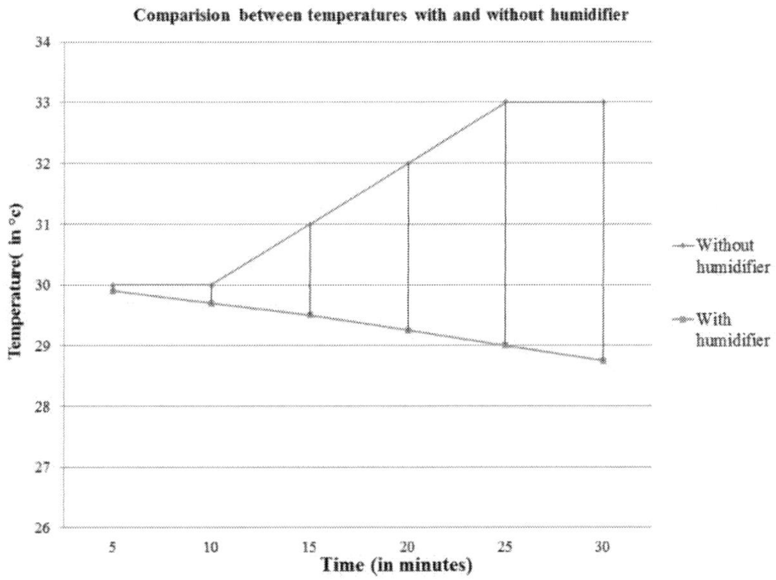

Figure 3. Performance graph for humidifier

8. Conclusion

Thus the developed portable mist humidifier can be used for cooling of a small area. The cooling effect produced in the humidifier is comparatively higher than most of the currently available machines. The hydraulic pump and the axial fan used in the model work flawlessly to produce a good cooling effect. The developed mist humidifier is simple and it can be used by anyone and it is non-toxic. The portable mist humidifier is cost-efficient and it is done at a very affordable price. Comparison of room conditions with and without humidifier is done and results are analyzed and it shows cooling effect. Using this humidifier, the cooling effect is increased by 12.87%.

Acknowledgment
This research was supported by the mechanical engineering department that was provided Kongu Engineering College, Erode

References

[1] Xiaoqinghuang, dongliang zhang, xuzhang spray humidification air coolers, volume 205, 2017, pages 1886-1892.
[2] Zhuangbo Feng, Xiaoqing Zhou, ShihanXu, Junwei Ding, Shi-JieCao, portable ultrasonic humidifier, Building and Environment, Volume 133, April 2018, Pages 62-72.
[3] Mostafa H. Sharqawy, Ibrahim Al-Shalawi, Mohamed A. Antar, Syed M. Zubair, cross-flow humidifier, Applied thermal engineering, volume 117, May 2017, pages 584-590.
[4] TingfenKe, Xin Huang, Xiang Ling, Heat and mass transfer in humidifiers, Applied thermal engineering, volume 156, 25 June 2019, pages 310-323.
[5] Stuart Gaffney, Andrew Dalton, Humidification devices, Anaesthesia and intensive care medicine,

volume 19, issue 8, August 2018, pages 397-400.
[6] Moureh J, Letang G, Palvadeau B, Boisson H, Mist flow process, International journal of refrigeration, Volume 32, issue 2, March 2009, pages 203-219.
[7] PedroMagalhãesSobrinho, Celso Eduardo Tuna, Humidity in air conditioning system, Experimental thermal and fluid sciences, Volume 49, September 2013, pages 152-159.
[8] Peng Xu, Xiaoli Ma, Xudong Zhao, Kevin Fancey, Applied energy, Volume 203, October 2017, pages 761-777.
[9] Samira Pourhedayat, instant running water cooler, Applied energy, volume 229, November 2018,pages 364-374
[10] M. Ghazikhani, I. Khazaee, S. Vahidifar, Energy analysis of humidifier, Energy and buildings, volume 124, July 2016, pages 129-140.
[11] W. Chen, S. Shang, B. Wang, X. Li, Y. Cao, and W. Shi, "Experimental study on effects of supply-air humidification on energy and emission performance of domestic gas boilers," Energy Build., vol. 209, p. 109726, 2020, doi: 10.1016/j.enbuild.2019.109726.
[12] E. Eder and M. Preißinger, "Experimental analysis of the humidification of air in bubble columns for thermal water treatment systems," Exp. Therm. Fluid Sci., vol. 115, no. November 2019, p. 110063, 2020, doi: 10.1016/j.expthermflusci.2020.110063.
[13] W. F. He and D. Han, "Performance Simulation of an Air-Heated Humidification Dehumidification Desalination System," IOP Conf. Ser. Earth Environ. Sci., vol. 73, no. 1, 2017, doi: 10.1088/1755-1315/73/1/012004.
[14] S. Rabczak, D. Proszak-Miąsik, and K. Nowak, "Energy consumption in humidification process," IOP Conf. Ser. Earth Environ. Sci., vol. 214, no. 1, 2019, doi: 10.1088/1755-1315/214/1/012129.
[15] R. Tu, J. Li, and Y. Hwang, "Fresh air humidification in winter using desiccant wheels for cold and dry climate regions: Optimization study of humidification processes," Int. J. Refrig., vol. 118, pp. 121–130, 2020, doi: 10.1016/j.ijrefrig.2020.04.009.

IOP Conf. Series: Materials Science and Engineering **995** (2020) 012031 doi:10.1088/1757-899X/995/1/012031

PERFORMANCE ANALYSIS OF COLD ENERGY STORAGE USING PHASE CHANGE MATERIAL

R.Karthikeyan1*, Dr.P.Thangavel2

1 Assistant Professor, Kongu Engineering College, Department of Mechanical Engineering, Perundurai, Erode,Tamilnadu, India.
2 Associate Professor, Kongu Engineering College, Department of Mechanical Engineering, Perundurai, Erode,Tamilnadu, India..

Karthi.mech@kongu.edu , ptvel.mech@kongu.edu

Abstract. Refrigeration is a method of extracting heat from low-temperature reservoir and shifting it to a high-temperature reservoir. The energy consumed by the refrigerator is increased day by day due to the changes in the standard of living and quick industrialization. In household appliances, the domestic refrigerator is the most energy consuming devices due their continuous operation. Efforts have been made in many countries to boost the COP of the refrigerators and reduce the energy consumption in the compressor. The above problems are rectified by using of Phase Change Material (PCM) in the domestic refrigerators. Due to the fact that over a large number of freezing / melting periods, PCM must be chemically and thermally stable. In the case of thermal energy storage in refrigerators, the PCM must be applicable and the choice of PCM in refrigeration systems is an important issue. The use of PCM in the evaporator segment minimizes compressor dome temperature fluctuations and provides stable conditions against changes in thermal loads. The main objective of the present research is to investigate the household refrigerator using Phase Change Material. The experimental results showed important impacts on the performance of the system, the on-off compressor cycle and the reduction in electricity consumption. The results showed that 0.69 % increase in COP, 1.8 % decrease in compressor work and 6.7 % increase in refrigerator Effect. The dome temperature is reduced to 3.91% when compared to without PCM in the evaporator.

Keywords: Phase Change Material (PCM), Refrigerating Effect, Co Efficient of Performance, Compressor work, Evaporator, Household Refrigerator, Vapour Compression Refrigeration System, Dome Temperature.

1. Introduction

Refrigeration is a method of extracting heat from and shifting it to a high-temperature reservoir from a low-temperature reservoir. The heat, magnetism, electricity and lasers are used to drive the heat transfer work. The applications of the refrigeration are refrigerators, air – conditioning, cryogenic applications and freezers. The air-conditioning systems are designed reversible to the heat pump that use the outlet temperature in the cooling process. Jonathan Cofre – Toledo (2003) et al. describes the use eutectic phase change material i.e., combination of ammonium chloride and PLUSICE E-10. They have used ammonium chloride at 19.5%. Their concentrations and thermal storage values are determined by phase shift, and their heat capacity. The density values are considered at the temperature of 20 ° C. The material for phase change is located parallel to the aluminium tubes. On either side of the evaporator, two rows of six tubes are mounted parallel to the evaporator tubes with self-locking nylon cables. In freezing food, the average temperature is decreased to 37.2% and 56.9% 19.5 wt. % aqueous solution of $NH4Cl$.

Subhanjan Bista (2008) et al. Explain the benefit of using phase change material in condenser compressor and evaporator sections to reduce the amount of power supplied. In this they have used both the organic, eutectic and inorganic phase change materials separately at different times and analyzed the power consumption. Organic phase change materials like n-Tetradecane, Paraffin (C14) and inorganic phase change materials like Hexadecane + tetradecane (2:3–0:1 by volume) and Octadecane + docosane and eutectic phase change materials like $LiClO3 \cdot 3H2O$ or water. By applying these different phase change materials show improvement in system performance and in reducing noise of the refrigerator. It also helps to keep the items stored in times of power outages. Yusufoglu (2009) et al. used two different kind of phase change material in two different models of refrigerator to optimize the compressor on/off time. They have increased the surface area of the condenser in the design to increase the effect of phase change material. They used distilled water in their first model of the refrigerator behind the evaporator. In the second refrigerator model the circulated cold air in the evaporator through tubes. The result shows 9.4% saving in energy by using 950 grams of phase change materials used in the evaporator. R. Elarem (2005) et al. used the ICE PCM are filled in the entire evaporator of twelve U-type tubes. The change in temperature range of 4 ° C is produced by using the above mentioned phase change material. The results show that 8 % increase in COP and 12 % decrease in the power consumption by using the new PCM in the refrigerator when compared to base type refrigerator.

Xi Guo (2008) et al. have prepared composites with expanded PCM of graphite stabilised paraffin and wood/polyethylene. The simple and efficient stable PCMs are formed by the base PCMs are mixed with porous material. The result indicated that the Thermal Energy Storage (TES) composites have the good thermal capacity when compared to the Refrigerant. D. Shikalgar Niyaj (2010) et al. explains the use of propane as refrigerant rather than R134a. The positive result shows that COP of the system was increased by incorporating the Phase Change Material (PCM) in the evaporator in addition to provide the cooling to the material. Depending upon the thermal load, the refrigerator was tested with Ethylene Glycol. The running time of the compressor per cycle was decreased and the co efficient of the performance was increased when R290 as a refrigerant. The selected refrigerant is equal to R134a and the torque of the compressor was reduced. The life time of the compressor was increased due to the reduction in torque of the compressor.

Kun Du (2012) et al. observed that the energy storage is maximum which leads to store more heat of fusion when compare to conventional energy system. The latent heat storage is the alternate method for the conventional structure. The phase change material is coated inside the evaporator backside. The compressor inlet and outlet temperature varied from -20 °C to 200 °C was measured by the application of the phase change material.

Suxin Qian (2014) et al. The performance of refrigerator was reviewed with and without PCM. In this refrigerator, the new design of refrigerated space which is positioned after the condenser. The selected PCM is N-Octadecane and blending temperature of 27.5 °C. The refrigerator COP was increased to 9.58% due to the lower temperature of condenser. The system performance was increased due to the modified design and refrigerant properties. Malik Muhammad Umair (2019) et al. analyzes the performance of the refrigerator and reported that the system performance was increased and leakage of the refrigerant was reduced due to the energy storage in the organic phase change material. The energy produced on or after the organic phase change material is maintaining the shape stability. The performance of the system was increased and work input to the compressor was reduced.

Kin Yuen Leong (2019) et al. use the nano-particles for the upgrading of energy storage capacity of the refrigerant. The phase change material having the capacity to release or store the latent heat that is used for maintaining the cooling. The effectiveness of the substance was lowered due to the lower energy storage capacity of the substance. The energy storage capacity of the substance was improved by the mixing the nano particles to the refrigerant. As a result , the effectiveness of the substance was increased.

Vapor-compression refrigeration is the cycle that used in the all refrigerators and air conditioning. The latent heat of vaporization is the one of the technique used in the domestic refrigerator which refrigerant flowing through tubes. It is often used in domestic refrigerators and vehicle air conditioning, large refrigerated or frozen food and meat storage plants, refrigerated trucks and railway vehicles. In this refrigeration cycle, the heat is removed from the low temperature source and rejected into a high temperature sink by the external power source. This is the application of second law of thermodynamics. The main problem behind the refrigerators is the energy consumed by the compressor was high and also lower performance. The above problem is rectified by the employ of phase change material in the refrigerator. The phase change material stores the energy when energy is excessive and release the energy when required.

The main objective of the research work is to improve the performance of the refrigeration system and reduce the energy consumption in the refrigerator using phase change material in the evaporator.

2. Methodology

The below figure 1 describes about the methodology of the work. The initial task in the project is to create a model of the refrigeration system. The literatures about the research done in the field of refrigeration are studied. The phase change material is selected based on the temperature ranges. The PCM is selected and then incorporated in the evaporator section and subjected to analysis. The results obtained through analysis are checked and verified. The report is prepared using the obtained results from the analysis and concluded.

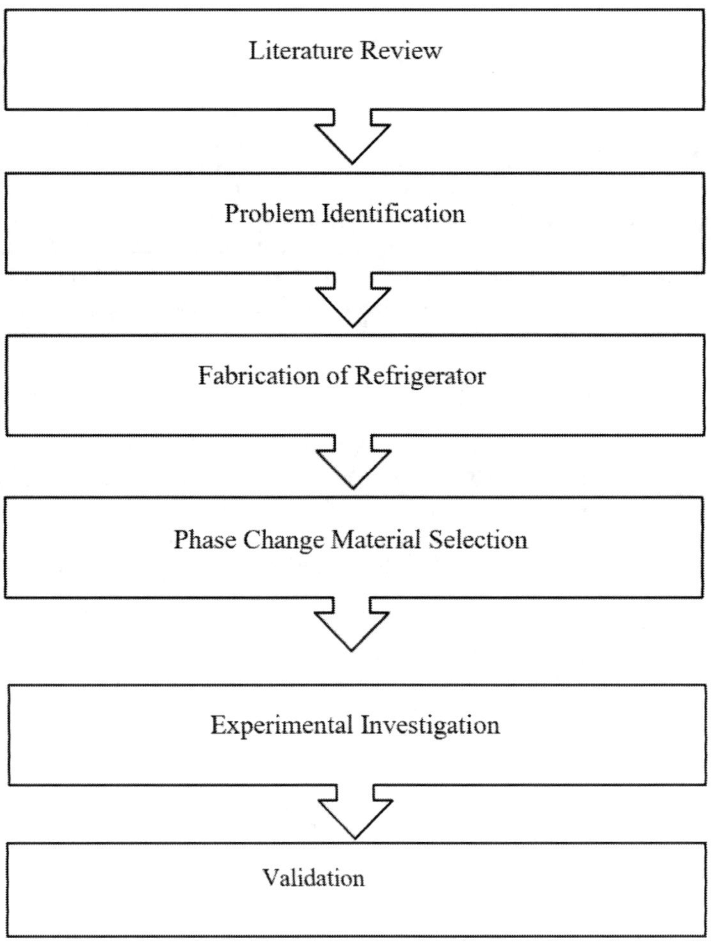

Figure 1 : Proposed methodology

IOP Conf. Series: Materials Science and Engineering **995** (2020) 012031 doi:10.1088/1757-899X/995/1/012031

2.1. *Experimental detail*

The figure 2 shows the experimental setup of the refrigeration system. The refrigeration system consists of a (1/5 HP) compressor, condenser, expansion valve, evaporator, plywood, aluminium paper and phase change material in liquid state. The components are fixed in their locations through welding at some places and riveting at other locations. The refrigerant used is R134a and it made to cycle inside the modal. The equipment needs external power supply for its function so external source is applied. The equipment is placed where temperature can be controlled and readings were taken at the same location.

Figure 2: Photographic view of the working model

IOP Conf. Series: Materials Science and Engineering **995** (2020) 012031 doi:10.1088/1757-899X/995/1/012031

2.2. *Experimental Procedure*

The experiment is connected to the power supply and the process begins with the compression of R134a refrigerant with the compressor, the refrigerant in the vapour phase gets compressed and the temperature of the refrigerant reaches super saturation temperature and also the pressure of the refrigerant increases. Then the refrigerant with high temperature and high pressure moves to the condenser and then the condensing of the refrigerant takes place. The condensation process decreases the temperature of the refrigerant and does not cause any change in pressure. The refrigerant with high pressure and low temperature moves inside the expansion valve and the pressure suddenly decreases and also the temperature because of the sudden change in volume. Now with low pressure and temperature the refrigerant enters into the evaporator and absorbs the heat inside and comes out of the evaporator. The refrigerant emits cold energy by absorbing heat energy in every cycle. The phase change material is placed inside the evaporator so during each cycle the phase change material absorbs the cold energy and solidifies and when the temperature is reached the compressor cut off takes place and the working stops. When the temperature of the evaporator increases the compressor is cut on and cycle continues and thecooling takes place. During the cut-off the phase change material emits the cold energy by absorbing the heat energy and liquefies. Because of the phase change material action the compressor cut-on time is increased. So when the compressor work is reduced so the power consumption is reduced.

2.3. *Selection of PCM*

The selected phase material is the savE® OM03 and is used in the refrigerator. The phase change material is organic material which stores the energy in the shape of suppressed temperature and release the power once necessary. The phase change material which store the heat without change in the material and compound properties of the substance. The freezing and melting temperature of the savE® - OM03 organic material are 3.0°C and 4.0 °C

The Table 1 shows the Properties of Organic Phase change materials.

Table 1. Properties of Phase Change Material

S.NO	Property	Value
1.	Melting Temp (°C)	4.0
2.	Freezing Temp (°C)	3.0
3.	Latent Heat (kJ/kg)	229
4.	Density of Liquid (kg/m3)	835
5.	Solid Density (kg/m3)	912
6.	Liquid Specific Heat (kJ/kgK)	1.91
7.	Solid Specific Heat (kJ/kgK)	1.76
8.	Thermal Conductivity of Liquid (W/mK)	0.146
9.	Thermal Conductivity of Solid (W/mK)	0.224
10.	Base Material	Organic

3. Results and Discussion

As per the analysis the phase change material is placed in the evaporator. The phase change material solidifies at -5 degree. The readings obtained in the systems with and without PCM are compared. The performance of the system is increased and power consumption is decreased by using PCM in the evaporator.

3.1. Effect on Compressor Cut-Off With and Without PCM

The Figure 3 shows the time versus compressor outlet temperature and of compressor cut–off cycle in the definite period was noted. The difference in the cut-off conditions show that the cut-off of the refrigerator model with PCM takes place before the cut-off of the model without PCM. The Cut off cycle with PCM is lesser than that of exclusive of PCM. As a consequence, the compressor mode is extended off. The cut-off cycle was reduced certain period of time.

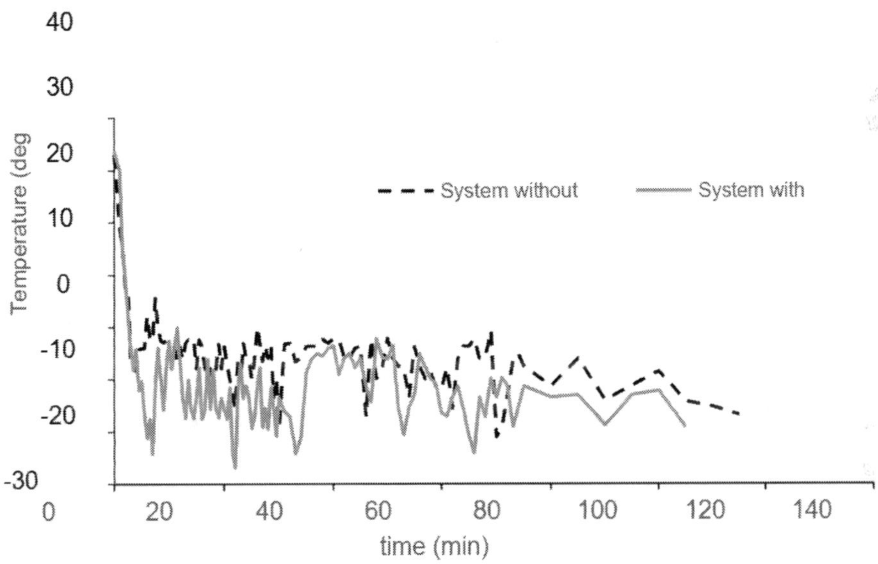

Figure 3: Effect of compressor cut off

3.2. *Effect on Co-Efficient of Performance With and Without PCM*

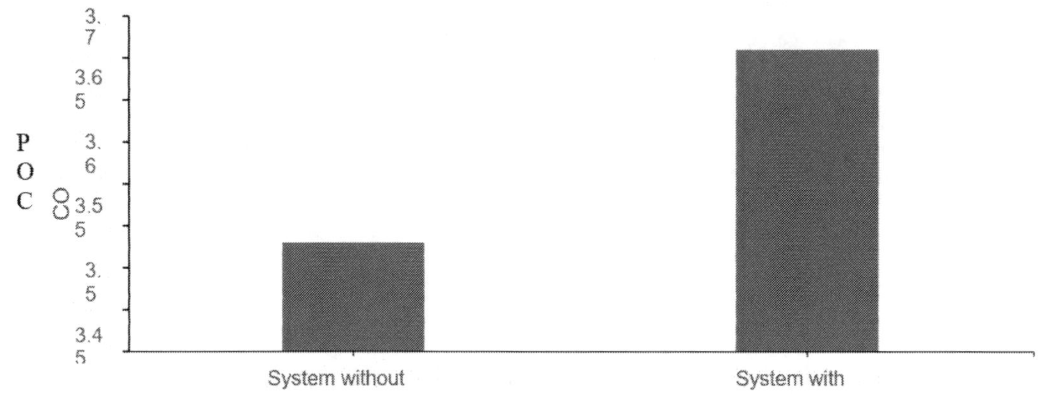

System without PCM   System with PCM

Figure 4: Effect on co-efficient of performance

Fig 4 shows the comparison between co-efficient of performance of the refrigerator with model PCM and without PCM. As shown in the figure the co-efficient of performance of the model with PCM is more when compared with the refrigerator model without PCM. This is because of the cold energy stored in the PCM during every cycle of the refrigerant flow that emits the cold energy stored during the cut- off time so the performance is increased. As the results showed using PCM, the COP is increased by 6.7% than the usual system without using PCM and based on the quantity of PCM , the COP value is increased.

3.3. *Effect on Refrigerator Effect with and without PCM*

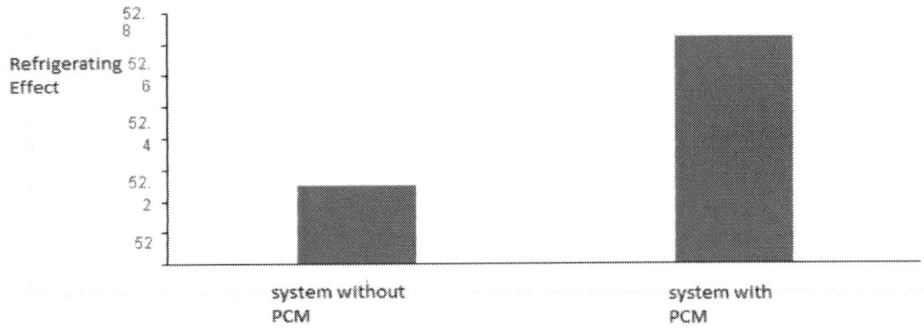

Figure 5: Effect on Refrigeration effect

The Figure 5 explains the comparison between the amount of refrigeration effect produced by the refrigerator model with and without PCM. The refrigerating effect and COP are directly proportional to each other. As seen in the figurethe refrigeration effect produced by the system with PCM is more when compared with the system without PCM. The reason behind this is the PCM stored in the evaporator section stores the cold energy and emits it back when there is a difference in the evaporator section. As the results showed, using PCM the refrigeration effect increased by 0.69% than the usual system which is modeled without using PCM.

3.4. *Effect on Compressor Work with and without PCM*

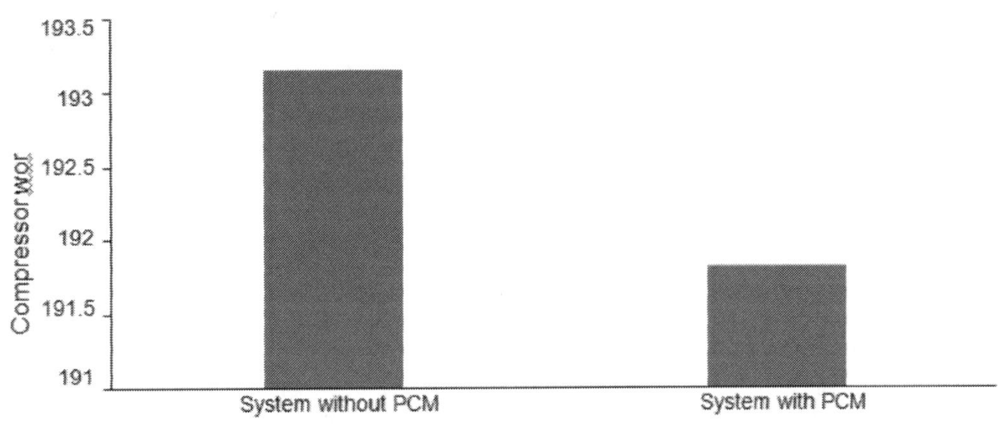

Figure 6: Effect on Compressor work

The Figure 6 explains the difference in compressor work takes during the cycles of refrigeration systems designed without PCM. The difference shows the system designed with PCM shows reduced in compressor work while the compressor without PCM is more. The reason for reduction of compressor work is the prolonged compressor cut off – on time. The compressor cut off –on time is reduced due to the placing of PCM in the evaporator. As the results showed using PCM, the compressor work decreased by 1.8% than the usual system.

3.5. *Effect on Dome Temperature with and without PCM*

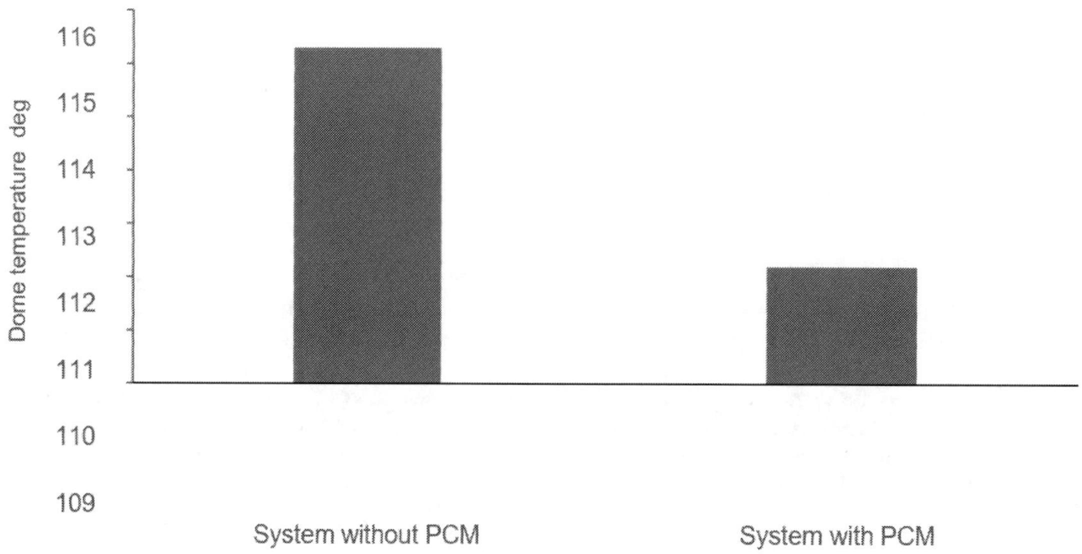

Figure 7: Effect on Dome temperature

The Figure 7 shows the comparison between the dome temperature produced during the compressor operation on both the systems with and without PCM. The interval between cut-off and cut-in condition the compressor work reduces due to the PCM in the evaporator. As a result, the dome temperature produced in the compressor system with PCM is 3.91 % less when compared with the system without PCM.

4. Conclusion

In household appliances, the domestic refrigerators are the most energy consuming devices. The energy consumed by the refrigerator and COP are the most challenging research in the refrigerators. The duration of the compressor cut off-on was long due to their lower energy storage capacity if the material. The above problem in the refrigerator is rectified by the bring into play of Phase Change Material (PCM) inside the refrigerator. The PCM is an organic material that is stored the energy when excess energy is produced and release the energy whenever required. The PCM of Liquid state to be made used to store the energy in the evaporator. The cold energy storage setup is analyzed a as per the plan. The thermal storage material be used to increase the COP, reduce the Energy Consumption and prolonged compressor cut off on time. The result shows that increase in COP by 6.7%, refrigeration effect by 0.69% and reduction in compressor work by 1.8%. The dome temperature in the compressor was reduced to 3.91 % when compared refrigerator without PCM.

References

[1] A. Abhat (1983),' Low temperature latent heat thermal energy storage heat storage materials'. (1983), pp. 313-332.
[2] Al-Abidi (2012),' The review of thermal energy storage for air conditioning systems renew. Sustain'. Energy Rev., 16 (8) (2012), pp. 5802-5819.
[3] Azzouz (2008), 'Performance enhancement of a household refrigerator by addition of latent heat storage'. Int. J. Refrigeration., 31 (5) (2008), pp. 892- 901.
[4] Azzouz (2009), 'The performance of household refrigerators with latent heat storage: an experimental investigation'. Int. J. Refrigeration". 32 (7) (2009), pp. 1634-1644.
[5] S. Naik (2010), 'Solar dryer with thermal energy storage systems for drying agricultural food products: a review. Renew. Sustain. Energy' Rev., 14 (8) (2010), pp. 2298-2314.
[6] Bansal (2003), 'Developing new test procedures for domestic refrigerators: harmonization issues and future r&d needs' – a review. Int. J. Refrigeration., 26 (7) (2003), pp. 735-748.
[7] Cabeza (2011). 'Materials used as {PCM} in thermal energy storage in buildings': a review. Renew. Sustain. Energy Rev., 15 (3) (2011), pp. 1675- 1695.
[8] Cheng W.L (2017), 'Analysis of energy saving performance for household refrigerator with thermal storage of condenser and evaporator'. Energy Conversion. Management 132 (2017), pp. 180-188.
[9] W.-L. Cheng (2011), 'A novel household refrigerator with shape-stabilized PCM (Phase Change Material) heat storage condensers': an experimental investigation, Energy 36 (2011) 5797–5804, https://doi. org/10.1016/j.energy.2011.08.050.
[10] M.I.H. Khan (2016), 'Conventional refrigeration systems using phase change material': a review, Int. J. Air-Cond. Refrig. 24 1630007, https://doi.org/10.1142/ S201013251630007X.
[11] V. Kumar (2016), 'Energy Saving Using Phase Change Material in Refrigerating System Energy Saving using Phase Change Material in Refrigerating system'.
[12] G. Cerri (2003), 'Identification of domestic refrigerator models including cool storage', in: Int. Congr. Refrig., Washington, DC.
[13] Neto I de Marchi (2009), 'Refrigerator COP with thermal storage', Appl. Therm. Eng. 29 2358–2364
[14] A. Abhilash (2015), 'Comparison of a conventional domestic refrigerator with a PCM encapsulated refrigerator' 4 582–7.
[15] G. Sonnenrein (2015), 'Copolymer-bound phase change materials for household refrigerating appliances': experimental investigation of power consumption, temperature distribution and demand side management potential.
[16] C. Veerakumar (2016), 'Phase change material based cold thermal energy storage: materials, techniques and applications' - a review, Int. J. Refrig. 67 (2016) 271–289.

Performance analysis on Solar Air Heater with Discrete Arc Shaped Rib Element in Absorber Plate

M.Makeshkumar[1], P.Thangavel[2]
[1] Assistant Professor, Department of Mechanical Engineering, Kongu Engineering College, Perundurai, Erode
[2] Associate Professor, Department of Mechanical Engineering, Kongu Engineering College, Perundurai, Erode

makesh.mech@kongu.ac.in , ptvel@kongu.ac.in

Abstract. Solar Air Heater (SAH) absorbs the irradiation and converts it into thermal energy, which is supplied to flowing fluid for the purpose of drying agricultural products, curing of industrial products, space heating applications and so on. An experimental examination has been conducted to assess the thermal performance of SAH having discrete arc shaped rib element in the absorber plate and analyzed against Flat Plate Solar Air Heater (FPSAH). Performance has been assessed by identifying the various parameters like useful heat gain, surface temperatures of collector plate, flow rate of working fluid, temperatures of air at inlet and exit. Investigation has been performed for flow rates of 0.028 kg/s and 0.045 kg/s. The results indicate that the absorber plate with artificial roughness element shows considerable increase in exit air temperature, heat transfer rate and thermal efficiency. Thermal efficiency of about 70 % is obtained for 0.045 kg/s, which is 13.1% more than that of FPSAH.

Keywords: Flat plate, Thermal resistance, Roughness geometry, Discrete ribs, Heat gain

1. Introduction

Renewable energy sources become more promising alternative for fossil fuels since it reduces the green house gas emissions. Among all the renewable energy, solar energy is voluminous, inexhaustible and clean energy source which has been used by humans for thousands of years to grow crops, dry farm products, to stay warm and so on[1-3]. Therefore, it is necessary and efficient to harness the solar energy for various industrial and agriculture applications. In recent days solar energy is used to generate electricity with the help of photovoltaic technology, heat water with the help of evacuated tubes, space heating with the help of air heaters.

SAH are the devices extensively utilized for low temperature operations due to its nature of heating the fluid to low and moderate temperatures. The development of viscous sub layer nearer to the plate lessens the heat transfer between collector plate and air, which leads to reduction in thermal gain of SAH [4]. The sub layer increases the thermal resistance, which can be eliminated by various techniques. Researchers investigated the thermal performance by providing various structures of protruded elements on collector plate [5]. The idea of providing dissimilar roughness geometries breaks the sub layer by creating disturbance in the flowing fluid nearer to the surface of the collector plate.

Various dissimilar structures and arrangements of roughness configurations helps in enhancing the heat transfer are transverse and inclined arrangements of broken and continuous ribs, chamfered ribs, wire mesh, dimple shaped protrusions, solid baffles, perforated baffles and so on[6-12]. Many researchers indicated that the usage of rib shaped roughness elements increase the performance of SAH than other forms of designs. In this work, the performance of SAH with newly designed rib element on the collector plate was examined for different mass flow rates and it was compared with Flat Plate Solar Air Heater (FPSAH).

2. Experimental arrangement

An experimental arrangement is constructed to test the thermal performance of newly designed absorber plate. The performance results obtained from new design is to be compared with the flat plate collector. The pictorial view of experimental arrangement is shown in Fig 1. It incorporates a collector box, absorber plate with ribs attached to it, plain glass cover at the top of collector box, centrifugal blower for forced convection, thermocouples for temperature measurement of air and absorber plate, anemometer for fluid flow measurement, collector frame to support the collector box and PVC pipe connections for fluid flow path. The dimensions of the collector box are 1600 mm x 600 mm x 50 mm and the length of test section is 1400 mm. Aluminium sheet of 1 mm thickness coated with black paint is adopted to fabricate the absorber plate. The aluminium material is chosen due to its high thermal conductivity compared to GI material. The height of the rib is 20 mm, the rib width is 40 mm, spacing between the ribs are 30 mm and the thickness of the rib is 2 mm.

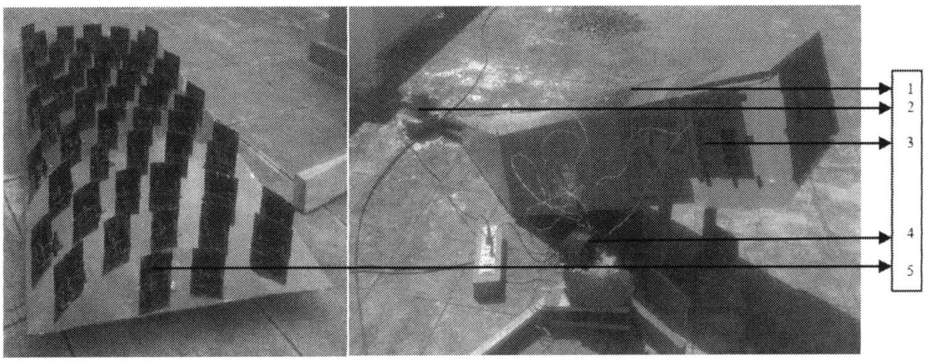

(a) (b)

Fig.1. (a) Photographic view of discrete arc shaped rib element in absorber plate, (b) Photographic view of experimental arrangement.

1 - Glass cover, 2 - Centrifugal blower, 3 – Wooden box, 4 – Digital temperature indicator, 5 – Absorber plate with arc shaped rib element

3. Thermal Performance Analysis

In the following section, the equations used for the analysis of SAH are discussed as below.
The useful gain in thermal energy absorbed by the working fluid (Q_u) when running across the collector plate is given by [13],

$$Q_u = m\, C_p\, (T_o - T_i) \tag{1}$$

where m is the mass flow rate of working fluid through the test section (kg/s), T_o is the temperature of working fluid at outlet (K), T_i is the temperature of working fluid at inlet (K) and C_p is the specific heat capacity of working fluid (J/kg K). The working fluid employed for heat absorption is air.
The heat energy harnessed by the collector plate due to irradiation incident on the projected area of collector plate is given by,

$$Q_a = I\, A_p \tag{2}$$

where I is the amount of radiation intensity incident upon the absorber plate(W/m^2) and A_p is the projected area of collector plate in test section(m^2).
The value of convective heat transfer coefficient between the collector plate and working fluid flowing over it (h) can be calculated by,

$$h = Q_u / A_p\, (T_p - T_f) \tag{3}$$

where T_p is the mean temperature of collector plate (K), T_f is the mean air temperature flowing across the absorber plate (K) and h is in the unit of W/m^2 K.
The value of T_p is calculated by taking mean value of temperatures at nine points of the absorber plate and it is given by,

$$T_p = (T_1 + T_2 + T_3 + \ldots + T_9)/9 \quad (4)$$

The value of T_f can be calculated by

$$T_f = (T_o + T_i)/2 \quad (5)$$

4. Results and discussion

The thermal behavior of newly designed SAH having discontinuous pattern of arc shaped rib element in absorber plate is investigated. The mass flow rates are chosen as 0.028 kg/s and 0.045 kg/s for better comparison with previous studies. The functioning of SAH with discrete rib elements have been studied and compared with the FPSAH. The experiment was tested in daytime during March 2020.

The amount of radiation falling on the absorber plate was captured with the aid of solar power meter. During start of the day, the irradiation was found to be 507 W/m^2 and then it gradually increases until noon followed by reduction in irradiation when the sun moving towards west. The maximum value of irradiation incident upon the upper layer of the collector plate was found to be 960 W/m^2. Thermocouples were utilized to determine the temperature of working fluid at entry and exit. Variance in temperature on hourly basis between the outlet and inlet air is seen in Fig.2. It is evidently understood that maximum temperature variance is noted between 12:00 to 14:00 hr. Maximum temperature difference identified for 0.028 kg/s is 18 K whereas 20 K for 0.045 kg/s. Many researchers identified that the maximum temperature difference observed in flat plate collectors are not more than 15 K. Therefore, it is observed that more than 5 K of temperature difference is obtained with rib shaped element on collector plate with higher flow rate. The temperature variance is low for lower flow rate, increases in flow rate increases the difference. The ribs provided on the surface of aluminium plate help air to collect more heat due to increase in surface area.

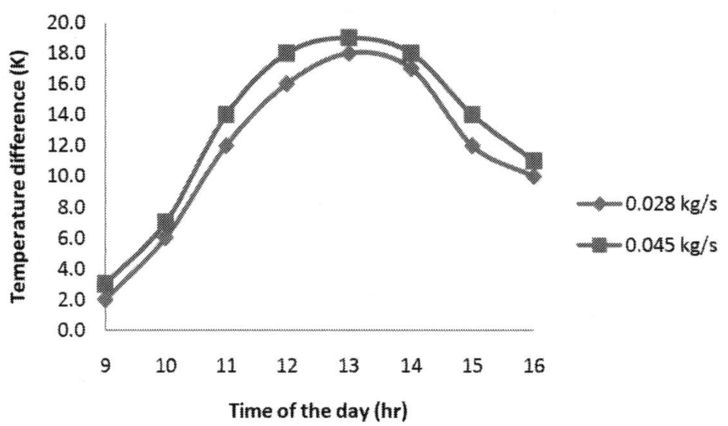

Fig. 2. Comparison of temperature difference for different flow rates

Fig. 3 manifests hourly variation of useful heat gained by the absorber plate for chosen mass flow rates. At lower rate of flow, the maximum thermal energy acquired by the air when passing across the collector plate is 506 W. Similarly at higher mass flow rate the maximum energy acquired by the air is 534 W. From above observations, it is evidently understood that increase in flow rate facilitates air to absorb more heat. In FPSAH, maximum useful thermal energy acquired by the air is found to be 420 W. Compared to flat plate collectors, more amount of heat is absorbed by newly designed air heater at higher mass flow rate.

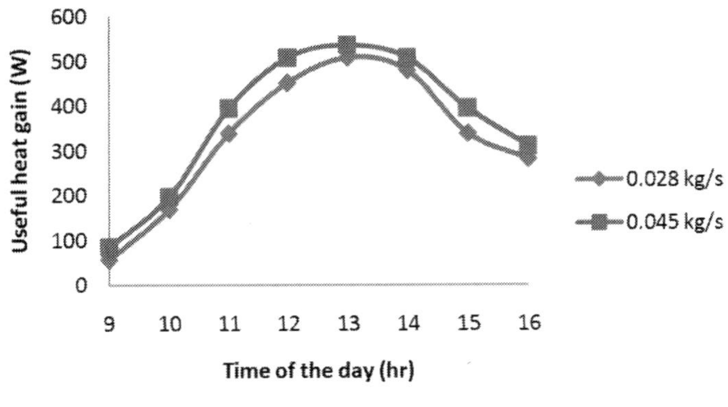

Fig.3. Variation of useful heat gain for different flow rates.

The variation of thermal efficiency obtained by newly designed SAH for dissimilar flow rates is shown in Fig. 4. The thermal efficiency gradually increases from morning to the afternoon session and then gradually decreases. The thermal efficiency of highest value is obtained for 0.028 kg/s flow rate is 66% whereas the maximum efficiency obtained for 0.045 kg/s is 70 %. Therefore, it is evidently understood that increase in efficiency is witnessed for higher mass flow rate than the FPSAH, since many researchers identified that the maximum efficiency identified by FPSAH was 45%.

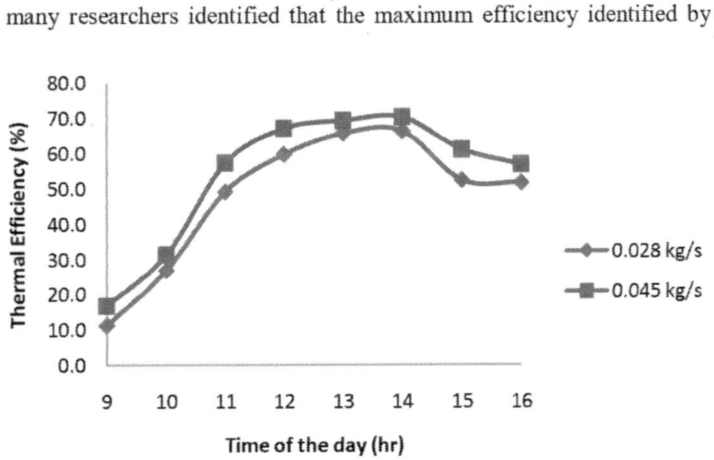

Fig. 4. Variation of useful heat gain for different flow rates.

The value of h between air and collector plate is calculated by convective heat transfer equation. The variation in value of h for dissimilar fluid flow rates is shown in Fig. 5. The value of h can be calculated for supply of air at 0.028 kg/s and 0.045 kg/s.

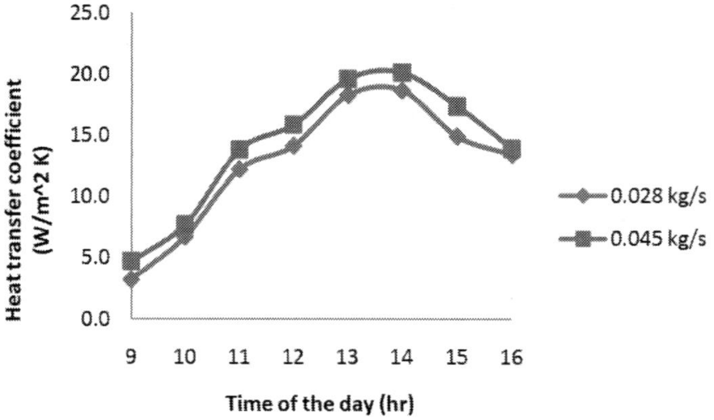

Fig. 5. Variation of heat transfer coefficient for different flow rates

Convective heat transfer coefficient of absorber plate with rib shaped roughness element has higher value, when it is evaluated against FPSAH. The maximum value is found to be 20 W/m^2K for 0.045 kg/s obtained at the time of 14:00 hr., whereas the value of h is 18 W/m^2K for the supply of air at 0.028 kg/s. The increase in value of h is due to increase of heat gain at higher mass flow rate.

5. Conclusion

Experimental examination has been executed to observe the heat exchange rate and thermal gain of SAH with roughness geometry on the surface of absorber plate. The thermal efficiency has been investigated for flow rates of 0.028 kg/s and 0.045 kg/s. Outlet temperature of air for arc shaped rib element is more for higher flow rate than lower flow rate. The maximum thermal efficiency obtained for 0.045 kg/s is 70% whereas 66% for 0.028 kg/s. From the above results, it is concluded that usage of roughness element increases the turbulence, thereby increases the value of h. Also increase in flow rate increases the thermal efficiency of the SAH. Therefore, it is suggested to use SAH with artificial roughness geometries on the surface of absorber plate with higher mass flow rate to have better thermal performance.

References

[1] Chen X Yang H Lu L Wang J Liu W 2011 Experimental studies on a ground coupled heat pump with solar thermal collectors for space heating Energy 36 pp 5292 - 5300.
[2] Sharma V K Colangelo A Spagna G 1993 Experimental performance of an indirect type solar fruit and vegetable dryer Energy Conversion and Management 34 pp293 - 308.
[3] Sreekumar A 2010 Techno-economic analysis of a roof integrated solar air heating system for drying fruit and vegetables Energy Conversion and Management 51 pp2230 - 2238.
[4] Varun Saini RP Singal SK 2007 A review on roughness geometry used in solar air heaters Solar Energy 81 pp1340 – 1350.
[5] A.E. Kabeel M.H. Hamed Z.M. Omara A.W. Kandeal 2017 Solar air heaters: Design configurations, improvement methods and applications – A detailed review Renewable Sustainable Energy Reviews 70 pp1189–1206.
[6] Chabeane F Moummi N Benramache S 2014 Experimental study of heat transfer and thermal performance with longitudinal fins of solar air heater Journal of Advanced Research 5 pp183–192.
[7] D. Jin S. Quan J. Zuo S. Xu 2019 Numerical investigation of heat transfer enhancement in a solar air heater roughened by multiple V-shaped ribs, Renewable Energy 134 pp78–88.

[8] A.P. Omojaro L.B.Y. Aldabbagh 2010 Experimental performance of single and double pass solar air heater with fins and steel wire mesh as absorber Applied Energy 87 pp3759–3765.

[9] A.S. Abdullah M.M. Abou Al-sood Z.M. Omara M.A. Bek A.E. Kabeel 2018 Performance evaluation of a new counter flow double pass solar air heater with turbulators Solar Energy 173 pp398–406.

[10] S. Singh 2020 Experimental and numerical investigations of a single and double pass porous serpentine wavy wire mesh packed bed solar air heater Renewable Energy 145 pp1361–1387.

[11] P.T. Saravanakumar D. Somasundaram M.M. Matheswaran 2019 Thermal and thermo-hydraulic analysis of arc shaped rib roughened solar air heater integrated with fins and baffles Solar Energy 180 pp360–371.

[12] Kumar A Bhagoria J L Sarviya R M 2009 Heat transfer and friction correlations for artificially roughened solar air heater duct with discrete W-shaped ribs Energy Conversion and Management 50 pp2106–2117.

[13] A.E. Kabeel M.H. Hamed Z.M. Omara A.W. Kandeal 2018 Influence of fin height on the performance of a glazed and bladed entrance, Single-pass solar air heater Solar Energy 162 pp410–419.

Design and implementation of low-cost electric bicycle with battery level indicator

Rajalashmi K[1,]*, Vignesh S[2], Chandru M[3], Arulshri K P[4]

[1, 3] Department of Electrical and Electronics Engineering, Bannari Amman Institute of Technology, Sathyamangalam – 638401, India
[2] Department of Mechatronics, Bannari Amman Institute of Technology, Sathyamangalam – 638401, India
[4] Department of Mechanical Engineering, Nehru Institute of Technology, Coimbatore – 641105, India

*rajalashmikannan@gmail.com

Abstract. In the recent days, carbon emission from the vehicles are found to increase the level of air pollution in a huge manner. In order to reduce the pollution due to transportation, electric vehicles are seen as an alternative. This approach paved way for developing an electric bicycle by incorporating 250 W BLDC hub motor, 295.6 WH lithium-ion battery and pedal assist system (PAS). The maximum speed of the bicycle is 25 km/h. The distance covered using PAS is 75 km. The maximum carrying capacity is 120 kg with a charging time of 2.5 to 3 hours. The pedal assisting system helps to share the torque between manual pedalling & electric motoring. This ensures that the battery is used in an efficient manner. The cost of this bicycle includes cycle assembly, battery, hub motor, controller circuit and wireless odometer. The total expenses involved in developing this prototype is nearly 38.57% lesser without profit when compared to other products available in the market. The pulling capacity of the cycle can be enhanced by increasing the battery capacity which would add additional cost based on capacity.

Keywords: Lithium-ion, Pedal assist system, Transportation, Battery level indicator, Hub motor

1. Introduction

As we move towards the decade's end, a ton's evolving. The purchaser notions to auto organizations' standpoint to vehicles in India, there's a ton occurring in this area. Indeed, the legislature is attempting to acquire electric vehicles to India and actualize it by 2032. However it would appear that India will require more opportunity to acclimate to this change. Electric vehicles and Hybrid vehicles are the discussion of town in view of decreased outflows, yet additionally in light of the fact that they give the proprietor a status in the public eye, that of one aware of nature, a green crusader of sorts.

E vehicle is very useful in our country, because in this way electric bicycle was designed and developed successfully [1-2]. Features of this vehicle are 250W BLDC hub motor, 295.6WH lithium-ion battery, pedal assist system (PAS), maximum speed 25 km/h, range by using the PAS is 75 km in single charging, maximum carrying capacity 120kg, and charging time 2.5 to 3 hours. The pedal assisting system helps to share the torque between manual pedaling & electric motoring. So that, battery is used in efficient manner [3-5]. The cost of this cycle includes the price of the cycle assembly, battery, hub motor, controller circuit, wireless odometer, etc. The total expenses involved in developing this prototype is Rs.21,500/- which is nearly 38.57 % lesser without profit when compared to other products available in the market.

IOP Conf. Series: Materials Science and Engineering **995** (2020) 012033 doi:10.1088/1757-899X/995/1/012033

The pulling capacity of the cycle can be enhanced by increasing the battery capacity which would add additional cost based on capacity. The decrease of air contamination should be possible by advancing the utilization of electric bicycles [6-8]. An electric bike also known as an eco-accommodating bi-cycle assimilated with electrically powered appliance is mainly preferred. There are huge number of e-bicycles reachable all over the globe in which only a few cases tend to assist the rider's ability in pedaling the bicycle [9-11].

There are enormous kind of e-bicycles which are legally named bikes or mopeds [12-14]. When we focus upon the activity and accreditation of the incredible bikes such as electric cruisers [15], there are several laws which tend to exclude them throughout the process. There is a clear statement through electric bike laws that the e-bicycles are categorized self-sufficiently.

2. Methodology

The e-bicycle has been developed by incorporating the BLDC hub motor through hall sensor feedback. The user interface provides an opportunity to handle throttle and PAS with the control of signal line, control line and power line. The appropriate power source is provided to the experimental unit. This paves way for development of e-bicycle with enhanced efficiency. Figure 1 shows the schematic sketch of the proposed e-bicycle.

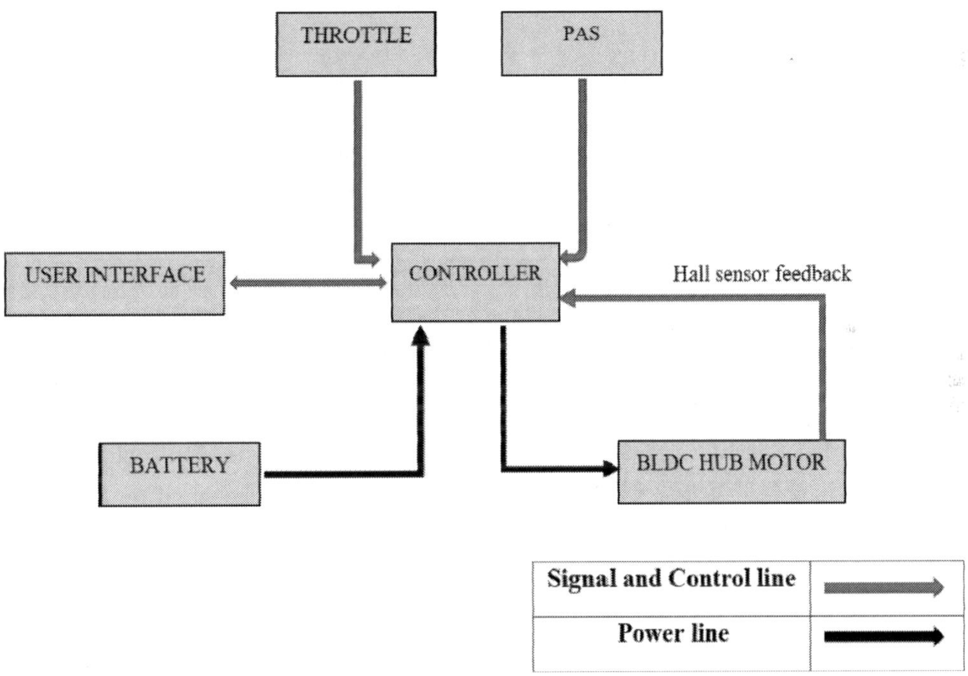

Figure 1. Schematic sketch of the proposed e-bicycle

The BLDC motor gets the position of the sensor and feed the sensor output to the controller. The controller gets the variable voltage from the throttle and drives the motor accordingly to the variable voltage with the help of the battery connected to the controller. The pedal assist sensor (PAS) helps the person to pedaling by driving the motor.

3. Design, development and implementation

3.1 *Calculation of power required to traction*

The mathematical steps involved in the proposed work to calculate the power required to traction through the basic concept is given in equation no. 1.

$$F = m*a \text{ (N)} \tag{1}$$

As we know, the value of mass 'm' is 150 kg. In this scenario, the maximum mass is assumed for carrying out the work. The value of acceleration 'a' is assumed as 6.94 m/10s. The force is calculated in equation no. 2. Here, the value of wheel diameter is 6". The torque is calculated using equation no. 3.

$$\text{Force} = 150 * 6.9410 = 104.1 \text{ N} \tag{2}$$

$$\text{Torque } T = r * F \text{ Nm } r = \text{radius} \tag{3}$$

Thus, the below mathematical steps will provide the power required for the wheel of bicycle. It is calculated using equation no. 4. The wheel radius is 262 = 13 inch = 0.3302 m.

$$\text{Torque, } T = 0.3302 * 104.1 = 34.373 \text{ Nm}$$

$$\text{Power, } P = T * n \text{ W}$$

$$\text{Power, } P = 34.373 * 6.94 \text{ cycle} \tag{4}$$

$$P = 238.548 \text{ W}$$

It is almost 250 W power required to the center of the wheel of bicycle.

3.2 Selection of motor

There are various types of motor used for Electric vehicles applications. But among them BLDC motor and DC motor are best suitable for low power applications in terms of cost, size and weight. Hence the comparison between BLDC motor and DC motor is mentioned in Table 1.

Table 1. BLDC and DC motor comparison

	BLDC Motor	DC Motor
Efficiency	85% - 90% Efficient	75% - 80% Efficient
Maintenance	Low maintenance	High maintenance
Noise	Noiseless	Comparatively noisy

From the comparison, the BLDC motor is efficient and suitable for the E-Cycle. Usually the BLDC motors have hall sensors which is used to detect the position of the poles inside the motor. The position of the poles is sent to the controller through digital signals to the motor controller. The controller sends the signals to energize the respective rotor poles.

3.3 Selection of battery

There are many type of batteries available in market. The most commonly used batteries used for automobiles are Lead acid batteries and Li-ion batteries. The comparison between those batteries are mentioned in Table 2.

Table 2. Lead acid and Li-Ion battery comparison

	Lead acid battery	Li-ion battery
Energy density	Low	High
Specific energy	Comparatively low	Comparatively high
Temperature sensitivity	Degrades significantly above 25°C	Degrades significantly above 45°C
Depth of discharge	Around 50%	Around 80%
Self-discharge	High	Very low
Size and weight	Large and Comparatively heavy	Compact and Comparatively light

From the above table the Li-Ion battery seems to be compact and power efficient. Hence Li-Ion batteries are used in this e-bicycle.

IOP Conf. Series: Materials Science and Engineering **995** (2020) 012033 doi:10.1088/1757-899X/995/1/012033

3.4 Technical specifications

The model that has been designed comprises of specifications as follows: 250 W BLDC hub motor, 295.6 Wh lithium-ion battery, maximum speed 25 km/h, maximum carrying capacity 120 kg, charging time 2.5 to 3 hours. The charger used is 42volt, 2A, range 75 km in a single charge. It is attached with pedal assisting system battery level indicator (displays voltage, current and power consumed).

3.5 Selection of controller

The BLDC motor controller is a gadget used to drive the motor. It acts as an intermediate between the motor and the battery. The BLDC motor actually works on AC voltage. The controller converts the DC voltage from the battery to the AC voltage with the help of power electronic switches. The variable voltage is received from the throttle and thus the motor speed is controlled. The various input and output are mentioned below.

- Electrical brake pin
- Throttle pin
- Motor pin
- Battery pin
- Hall effect pin

3.6 Influence of electrical brake pin

If the brake is applied in the vehicle, mechanically the wheel is stopped. But the motor is not electrically stopped. This may damage the coils of the motor. So the electrical brake pin is used. The brake pin is connected to the brake lever. So when brakes are applied automatically the supply to the motor is also stops. This protects the motor coil damage.

The throttle pin is used to get the variable voltage signals to vary the speed of the motor. Motor pin is connected to the motor to drive the motor accordingly to the throttle signal pulse. Battery pin is connected to the battery with which the controller operates. Hall-effect pin gets the signal pulse from the hall sensors in BLDC motor and runs the motor. The wattmeter is used to measure the power consumed by the cycle when operated. The wattmeter is connected across the battery and the motor controller.

Figure 2. Developed model of e-bicycle

The Figure 2 shows the electric bicycle which has been implemented with all the necessary specifications. The test-run of the electric bicycle also provided good user experience and extreme comfortness.

4. Conclusion

The electric bicycle is one of the emerging technologies and it has been launched by so many reputed firms. Due to the reduction in fossil fuels like petrol and diesel, etc., the usage of electric vehicles will get increased in the near future. The use of proposed electric bicycle would reduce the greenhouse gas emissions from automobiles in the atmosphere in a considerable manner. The tool is also simple to manage and it is not much expensive. There are many technical features, such as, odometer, electrical and mechanical braking system pedal assist, hall-effect sensor provided in this unit which can make the electrical cycle smoother compared to other two wheelers.

References

[1] Bigazzi Alexander and Kevin Wong 2020 Electric bicycle mode substitution for driving, public transit, conventional cycling, and walking *Transportation Research Part D: Transport and Environment* **85** 102412

[2] Bai Lu, NN Sze, P Liu and AG Haggart 2020 Effect of environmental awareness on electric bicycle users' mode choices *Transportation Research Part D: Transport and Environment* **82** 102320

[3] Zhiwei Chen, Yucong Hu, Jutint Li and Xing Wu 2020 Optimal deployment of electric bicycle sharing stations: model formulation and solution technique *Networks and Spatial Economics* **20(1)** 99-136

[4] Shang Tian-Tian, Xiao-Ming Miao, Di Lu and Xiao-hua Xin 2020 The evolutionary process of disruptive green innovation in China: evidence from the electric bicycle sector *Technology Analysis & Strategic Management* 1-13

[5] Johny Sachin 2020 Design and fabrication of foldable electric bicycle *Materials Today: Proceedings*

[6] Nematchoua ModesteKameni 2020 Evaluation of the potential of classic and electric bicycle commuting as an impetus for the transition towards environmentally sustainable cities: A case study of the university campuses in Liege *Belgium Renewable and Sustainable Energy Reviews* **119** 109544

[7] Li Ying 2020 Fire Risk of Electric Bicycle Based on Fuzzy Bayesian Network *Journal of Physics: Conference Series* **1578(1)**

[8] Atmika I 2020 Study of Mechanical Properties of Hemp Fiber Composites for Electric Bicycle Frames *Materials Science Forum* **1000**

[9] MacArthur J, Dill J and Person M 2014 Electric bikes in North America: Results of an online survey *Transp. Res. Rec. J. Transp. Res. Board* 123–130

[10] Pucher J, Peng Z R, Mittal N, Zhu Y and Korattyswaroopam N 2007 Urban transport trends and policies in China and India: Impacts of rapid economic growth *Transp. Rev.* **27** 379–410

[11] Fishman E and Cherry C 2016 E-bikes in the Mainstream: Reviewing a decade of research. *Transp. Rev.* **36** 72–91

[12] Peine A, Van Cooten V and Neven L 2017 Rejuvenating design: Bikes, batteries, and older adopters in the diffusion of e-bikes *Sci. Technol. Hum. Values* **42** 429–59

[13] Zhao Shuo 2020 Design of Resonant Wireless Charging Device for Public Electric Bicycle *IOP Conference Series: Materials Science and Engineering* **738(1)**

[14] Yuan Min, Linpeng Hou and Hui Jing 2020 Analysis of the Damage of Cyclists in Electric Bicycle-Sedan Angle Collision: Electric Bicycle-Sedan Angle Collision *International Journal of Ambient Computing and Intelligence* **11(1)** 99-114

[15] Hung Nguyen Ba and Ocktaeck Lim 2020 A review of history, development, design and research of electric bicycles *Applied Energy* **260** 114323

Fuzzy Logic based control system and Artificial Intelligence in Industrial Automation

R Deepa[1], Sri Rakshitha A K[2], Gopala Krishnan V[3]

1 Assistant Professor, Bannari Amman Institute of Technology, Department of Electronics and Instrumentation Engineering, Erode, Tamilnadu, India.
2, 3 UG student, Bannari Amman Institute of Technology, Department of Electronics and Instrumentation Engineering, Erode, Tamil Nadu, India.

rangarajdeepa@gmail.com, srirakshitha.ei19@bitsathy.ac.in, gopalakrishnan.ei19@bitsathy.ac.in

Abstract. Amidst the AI driven world, the Automation plays a major role in every sector. Usually the control systems in industries are manually controlled in unpredictable conditions. We are in need of a human to control a system. In this affair an AI trained machine acts as a human replica to drive it fully automated. The highly labor-intensive industries can be benefited by the use of Artificial Intelligence in the automation part, this reduces the human errors and safeguard them from risky areas. The fuzzy based PID controller is used here because it acts as Human-in-the-loop behind the whole process. This helps in tuning the PID controller to set a desired value which enhances the gains, adding to that it gives a high accuracy and brings up more flexibility in changing the rules. The result which is obtained is simulated in MATLAB with overshoot difference.

Keywords: Fuzzy Logic Control (FLC), Artificial Intelligence (AI), Proportional Integral Derivative (PID), Precision Drilling.

Introduction

Industrial Automation is a major factor to avoid human errors where a small issue leads to big accidents in industries. To make it as an unmanned one, AI takes responsibility in achieving this. Artificial intelligence is a human brain designed machine brain. It is trained by data, tests, predicts it and becomes efficient day by day by emphasizing it and taking decisions on its own. AI technology can be blend in the Oil and Gas sector, improvising the efficiency in operation. The vital accidents takes place in onshore gas station, a high tracking facility infused with AI is a must to ensure the safety of the employees and avoid such happenings. These are often observed by AI-powered cameras, robots, drones, etc. AI models are trained on boundless amount of production data that validates the accuracy of the analysis done and predicts spontaneously. The Fuzzy logic technology is now mainly used in control that includes transfer of technology as a transfer learning method, turbine startup with a high speed is controlled, optimization of turbine cycling is done and finally controls the power supply. Thus use of these technologies in real-time is going to be a biggest game changer in upcoming generation.

IOP Conf. Series: Materials Science and Engineering **995** (2020) 012034 doi:10.1088/1757-899X/995/1/012034

1. Fuzzy Logic Control

The fuzzy Logic control (FLC) behaves as human interpretation by understanding the tasks of a person which is trained to it in terms of data. The effectiveness of the work done by FLC stands for its accurate results.

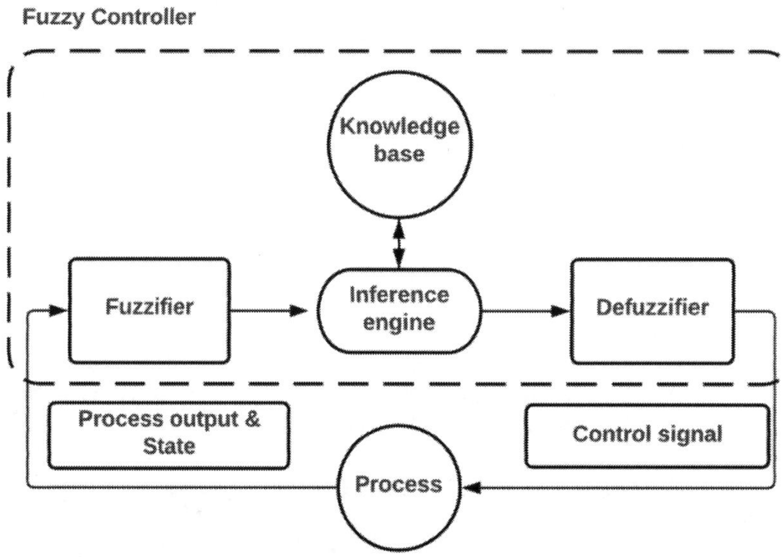

Figure 1. Structure of Fuzzy control

1.1. *Major components of FLC*

The following are the major components of Fuzzy Logic Control as shown in the above figure 1.

- ***Fuzzifier*** – Fuzzification is the method of converting the input values into fuzzy values to set labels.
- ***Knowledge Base*** – The input and output relationship is stored in this knowledge base.
- ***Inference Engine*** – Inference engine behave as a support to the knowledge base by deducing the new information based on fuzzy rules.
- ***Defuzzifier*** – Defuzzifier is known for its work of converting values from inference engine to a crisp output.

1.2. *Use of Fuzzy logic in PID controller*

In these modern days, technology reaches its peak in every sector, in this list Fuzzy logic stands for its broad potential of its application in industrial automation. These are all proven concepts in which engineers rely on. PID controller assumes itself in a linear behaviour where this simplification can be made in stable conditions, the operating point is pushed far away by strong disturbances from the operating point. In this circumstance, the linear approach does not run. The similar one can take place when we change the parameters. In such conditions the PID controller comes into action and it is replaced instead of old methods.

In the PID controller we have this tuning system both manually and automatically. These are not encouraged in certain conditions where we have to control a high pressure using PID.

Manual PID tuning: This type of tuning is said to be done by humans manually. They tune it to a value to obtain gains and the adjustments are made prior.

Auto PID tuning: It is also called 'Self-tuning'. Usually PID controllers are designed to simple down matters by selecting their own tuning ranges based on automated analysis. For formulating and auto tuning it we always want a mathematical procedure which is processed to an extent by input/output predictions.

Taking these two tuners into consideration, some complication occurs in uncertain conditions. To avoid this Fuzzy logic steps into PID for controlling and tuning automatically.

1.3. Implementing Fuzzy logics in PID tuning

The Mathematical application of fuzzy logic is applied here. Mainly the tuning is done with the fuzzy on applying its logics. This is done by Zeigler-Nichols control theory. The fuzzy control decides how the tuner should be tuned to gain appropriate values. For control application the fuzzy logics has always been a trial and error method. The change that occurs in the fuzzy tuned PID controller is far better than normally tuned PID controllers. The optimizations are simulated in the results below.

2. Artificial Intelligence in Control

Artificial intelligence is an upcoming technology in which data can be predicted and analysis is done on historical data. AI is useful to us in recent years, this was first tested by Alan Turing and he came up with the theory of Machine Thinking where the machine decides itself to think and act as a human. This AI is used in control systems in Industries.

2.1. AI in Oil and Gas industry

- ***Production Area***

AI gives hands in oil and gas industries for production to find out the area of efficiency to get profits and produce. AI stands for its standard of data analysis that it can train an enormous number of data in the production side as a raw one to refine the data to generate analytics.

- ***Reservoir Management***

Maintenance and the degree of optimization is high at many technicalities like seismic interpretation, management of reservoirs, geology and on the production side these come under oil reservoir management in industries. AI models can be trained for such technicalities in the field of surveillance etc. by their information.

- ***Communication:***

AI supports in bridging the gap between customers and the industry by designing the Chabot with a recommendation system in it and providing information, queries etc. to the customer to engage with.

- ***Continuous Monitoring:***

We have some safety measures always in industry where they continuously monitor the safety of the workers in oil fields, mines, plants etc here the AI cameras and bots come into action and alerts the person or the authority of the risk and take predictive measures to avoid such accidents.

- ***Industrial Automation:***

Certain works in industries are sensitive and to be done with utmost care of workers. In this circumstance, AI automates everything and does the task in a perfect way. The only thing is to train it and it will take up the rest. This will avoid many human errors.

These are the advantages of using AI in oil and gas industries. The very concerned part of application is feasibility. Drilling for production fuel and other products have been tough on off-shore has its done by the precision drilling. In this part the historical data of the drills and the mining are collected and trained with the model. This enhances in collecting real time data like the pressure under the ground, temperature till how the driller can withstand and finally seismic reports both in land and ocean.

3. Results

The simulation done in MATLAB software is given here. This is compared with the traditional PID controller and non-linear PID controller.

3.1. Fuzzy in PID

Two different trials are simulated here. The data is collected from the areas where the PID is applicable. In the First trial it gives a normal output by testing the model with turbines. This is trained by calculating

IOP Conf. Series: Materials Science and Engineering **995** (2020) 012034 doi:10.1088/1757-899X/995/1/012034

its speed in the run time. The first trial gives an output with limited benefit. So, while heading up to second trial, the data from the pressure steam of reduced one is taken to train the model. This time it causes a reduction of thirty percentages in the gain. The major requirement is to deduce it to get a crisp output. Hence the second trial acts better than first and gives a good benefit in low gains.

The simulation of the Fuzzy PID controllers are done for the trial which is obtained and the difference can be seen in the graph. The table 1 depicts the trial and error method and it is given below:

Table. 1 Trial and Error Method

Trials	Indexes	Relays	Fuzzy	Field PI	Fuzzy
1	Overshoot (%) Settling time Rise time(s) IAE	0.1081 139.6400 129.9700 2.6147e + 04	0.1556 139.5350 129.8650 2.1756e+04	0 146.1650 136.4100 4.6981e+05	0 141.3350 131.3900 1.6515e+05
2	Overshoot (%) Setting time Rise time(s) IAE	0.0361 139.9150 130.1950 3.4055e+04	0.1274 139.7200 129.9800 3.0209e+04	0 154.8950 139.2200 6.6299e+05	0 143.1050 132.8350 2.2842e+05

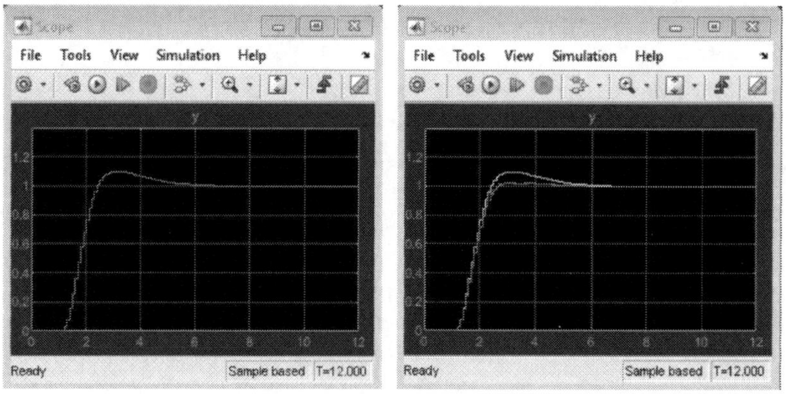

Figure 2. MATLAB Simulation

3.2. *Artificial Intelligence in industry*

Hence Artificial Intelligence Technology plays a major role in Oil and gas industries and other sectors by providing safety and monitor measures to the employees and a user friendly support system. AI powered predictive systems analyse huge data from sensors placed in some areas and predict it. AI helps in reducing production costs in order to widen the profit. AI cut production costs from 10% to 20%. This upstream could result in savings of capitals upto $100 billion to $1 trillion by 2025. This prevents cyber threats by its high security. In the areas like using big equipments, the precision drilling uses AI and contributes more on control.

4. Conclusion

Thus the paper forecasts, the future problems and analyses the data on industries using technology, which saves the mankind from certain risks. On a fuzzy approach the physical components like pressure, temperature and humidity can be controlled, monitored and maintained in a smooth way using this approach. The tuning method became easy and it is essential in using Fuzzy logic control. On the other hand, Artificial Intelligence with its algorithms from Machine learning, Deep Learning, Natural Language Processing, Neural networks etc. these entire stand for a big gain in the automation sector.

5. Reference

[1] Li W.-L. "Nonlinear uncertain turbine governor design based on adaptive backstepping method.", 31st IEEE Chinese Control Conf. (CCC), 2012, 744-749.
[2] Morris, A. "Adaptive digital control of a steam turbine.", Proc. Institution of Electrical Engineers. 123 (6) IEE, 1976.
[3] Bolek,et al. "Two-valve control of a large steam turbine.", Control Engineering Practice, 10 (4), 2002, 365-377.
[4] Fang, et al. "Nonlinear inversion-based output tracking control of a boiler-turbine unit.", J. Control Theory and Applications, 2005, 3 (4): 415-421.
[5] Wang, D., Wang, Z., Meng, L., Han, P. "Multi-model based IMC design for steam temperature system of thermal power plant.", Int. Conf. on Test and Meas. ICTM'09. 2009: 5-9.
[6] Kordestani, M., Khoshro, M.S., Mirzaee, A. "Predictive control of large steam turbines.", 9th Asian Control Conference ASCC 2013, 2013.
[7] Ding, H. “Multivariable self-tuning control of a boiler-turbine system.”, Int. Conf. Advances in power system control, operation and management, APSCOM-91, 1991: 498-502.
[8] Liu, X.-J., Lara-Rosano, F., Chan, C.W. "Neurofuzzy network modelling and control of steam pressure in 300MW steam-boiler system.", Engineering Applications of Artificial Intelligence, 2003.
[9] Industrial Automation and Process Control Book by Jon Stenerson.
[10] MODERN DATA ANALYTICS: Applied AI and Machine Learning for Oil and Gas by Tatyana Plaksina (Author).
[11]Hung Yuan, Bo Chen, Jin Lin. A PI-type Fuzzy Controller with Self-tuning Scaling Factors [J]. Fuzzy Sets and Systems, 2005, 93 (2):23 - 28.
[12]Caihua Lu, Yuewen Xu, and Weiming Yang. Permanent magnet linear synchronous motor feed system for fuzzy PID control [J]. Journal of Electrotechnics, 2007, 22 (9): 59-63.

Dynamic Clustering-Genetic Secure Energy Awareness Routing To Improve the Performance of Energy Efficient In IoT Cloud

E Karthika[1], S Mohanapriya[2]

[1] Assistant Professor, Dharmapuram
Gnanambikai Government Arts College for Women
Department of Computer Science, Mayiladuthurai, Tamil Nadu 609001, India
[2] HOD/BCA, Sona college of Arts and Science,
Department of Computer Application, Salem, Tamil Nadu 636005, India

karthikathendral@gmail.com, mohana_laks@yahoo.co.in

Abstract: The Internet of Things (IoT) is one of the emerging technologies that has attracted the attention of researchers in the field of education and industry. The idea behind the Internet of Things is that things and devices are operated on the Internet to achieve some common goals with humans. It serves as a platform for monitoring the collection process, controlling the cyber-global fire world, and collecting data and analyzing data using IoT sensor nodes. Power saving is important for battery-powered devices. The amount of previous work is small, although it has a large tip capacity, especially in terms of battery life and energy efficiency. This regulates the development of IoT a simple and energy efficient routing scheme for war sensor networks. The nodes require an increased lifespan of intelligently transmitted data communications. To conserve the energy of IoT nodes, the Dynamic Clustering-Genetic Secure Energy Awareness Routing Protocol (DC-GSEARP) utilizes clustering, cluster head selection and energy and efficient path calculation for efficient and real-time routing. To accomplish the efficient cluster head selection, to have used Dynamic Clustering algorithm. The proposed DC-GSEARP protocol achieve an improved that can be used to construct an optimal path for energy efficient data transmission for IoT sensor nodes. DC-GSEARP, which uses the integration of clustering supported by short path resolution to provide energy efficient and advanced routing capabilities for IoT, ensures entering a path with lower power consumption and enhanced QoS measurements. The results of the simulations show that the proposed approach can reduce the power consumption of sensor nodes, data packet losses and extend network life.

Keywords: Dynamic Clustering -Genetic Secured Energy Aware Routing Protocol (DC-GSEARP), Internet of Things (IoT), Cluster Head Selection and Energy-Efficient Path.

1. Introduction

IoT-based data security in the digital world has always been a major global concern for the risks and stakes associated with the region. It is becoming very sensitive to new technologies such as end-to-end communication with disparate frameworks such as IoT and sensors. Its purpose is to make human life more convenient, but at the same time started to take on new challenges of security in information. Impressive works can be found in the literature and are performed on protected data and information with various cryptographic techniques.

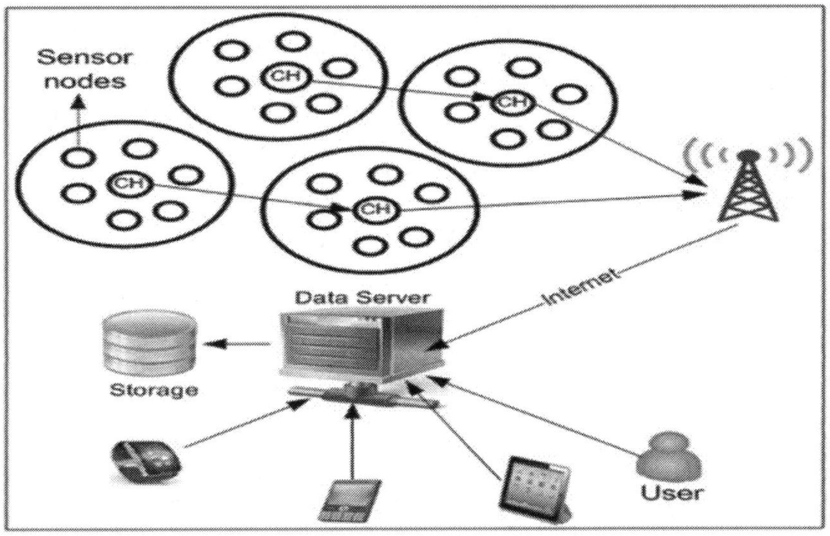

Figure 1. Cluster Based Energy Efficient Routing Schemes.

Therefore, we need a simple solution that can effectively save the energy of the equipment. Equal routing Distribute processing load across all devices in a network Energy efficient networks, especially one, including coordinated devices, are very important. This guarantees that the more devices you have, the longer you will be working. Conversely, if the processing load is due to routing, only a subset of the IoT nodes in the network will have a higher battery drain, but in Figure 1, this subset of equipment will have its energy at a faster rate is passed to consume an early break network. For the purposes of this article, the proposed routing protocols, in turn guarantee, solve this problem by increasing the time of network traffic and distributing routing efforts. The work of this protocol is described in detail in the following chapters. It should also be noted that the term refers to the IoT network of both node and device sensor nodes and is used interchangeably in text.

With this huge number, devices can collect personal information and also monitor lifestyles, activities and behaviors. The Internet of Things has created new types of applications and services that are beneficial to human life, but they can cause huge prices for privacy and security. Wireless sensor networks (WSNs) play an important role in the Internet of Things. It is easy to use and easy to deploy in harsh areas. These sensors monitor and recognize the surrounding environment. When a sensor monitors and reports an object, such as the appearance of a sensor node, a rare animal is sent a series of messages via multi-hop radio communication to the sink.

2. Related Work

Energy consumption is given priority over energy demand due to rising and rising energy prices, global warming and energy waste. [1] This is achieved by raising awareness about the Internet and energy consumption from the home model. This data is compatible with real-time cloud-based data analysis [2] using IoT (Internet of Things) for laboratory or auditorium tables and occupation levels. Energy consumption generates a warning if further inquiries and action schedules or occupation levels are not met [3]

Specifically, DBSCAN (Density-Based Spatial Clustering of Applications with Noise) per cluster is used to obtain the radius of each cluster region [4] at a distance between users within the outer boundaries of each cluster. However, there is an inherent contradiction between power consumption reliability and time: additional reliability usually leads to higher power consumption than reliance on traditional restoration [5].

Some customer satisfaction / feedback is analyzed as Quality of Experience (QoE) and Quality of Service (QoS) are not a good candidate for multimedia exchange as described in Emotions and User

Expectations [6]. An important factor. In Noma, however, the multiple transfer energy consumption that leads to simultaneous transmission (non-orthogonal multiple access) is also good, in fact, the power consumption of the circuits is an important device because it controls IoT [7] energy. Efficient use of energy resources requires Low Power Wide Area (LPWA) technology because it characterizes the ability to distribute low power connected devices on a large scale, at low cost and over a wide area [8].

Energy Harvesting (EH) considered IoT a tool for large-scale use of vast trees and completely independent spawning trees (CISTs).High performance EH technology [9], which provides a permanent network operating environment around, eliminating the need to change the power source frequently. Clustering algorithms play an important role in energy saving and energy controlled networks. [10] This allows you to distribute loads to the appropriate network for cluster heads, which reduces energy consumption and increases lifespan.

Predicts energy consumption in the short term, [11] providing the best way for communication between energy operators and consumers. Based on the margin device of the energy forecasting mechanism based on real-time energy management, the complexity of time saving is minimal and the main contribution is to monitor the servers of the best standardized technology choices via cloud-based public data. There is a new learning sequence [12] with an error rate.

In addition, the proposed method uses feedback from signs of full pre-heavy activity to predict each data packet [13]. Studies of heavy quality significantly reduce overall energy consumption once the simulated results are proposed. Furthermore, the method proposed for false alarms, such as irregular messages, is more sensitive than work [14].

As a single statistical device, we redone the numerical solution by obtaining the best closed solution with some useful properties based on the Karush-Kune Tucker (KKD) position which is unnecessary. In many demographic devices, EE (Energy Efficient) can propose an efficient and somewhat favorable method for solving the maximum problem [15]. Internet of Things task processing requests are provided by colleagues. Host Virtual Machine (VM) IoT and relays of the object representing those in the proposed P2B network. We are considering three cases to investigate the capabilities of power saving and work processing systems. The first display was the "relay only" mode, [16] which was used for task requests only and was processed by the relay.

At the same time, circuit power [17] is taken into account to reduce the overall power consumption of the network through integrated capacity control and time allocation. If multiple sources simultaneously send information to their respective locations via a [18] large array relay, multiple pairs of relay decoding and sharing are considered a network-dependent population energy saving design.

In particular, access is detected via random access distributed based on the probability of allocation detection caused by power consumption, data collision speed, and its recovery accuracy [19] by transmitting sensor data for each sensor node. Determine the contribution of. At the beginning of each super frame, the base station raises the schedule broadcast and contains the network panel information used to turn the IoT device type into a sleeping group [20]. In the Random Access (RA) function of LTE-based systems, the Timing Advance (TA) value is added to the device that integrates its development time synchronization. This letter uses the [21] TA value, which proposes a simple preferred Energy-Efficient Random Access (EERA) program for standard IoT networks.

3. Materials and Methods

Dynamic Clustering -Genetic Secured Energy Aware Routing Protocol based network clusters are made up of nodes with higher energies that are more complex in sending packets than nodes with lower energy levels. Nodes with high energy are called root nodes. The root node broadcasts a data packet containing information that the destination arrives through the root node. A node whose first high threshold energy level requires traditional energy to respond to packets on a node passing a cache entry and broadcasts an active receive packet. This guarantees that the identified root node knows the root of the node known as the leaf node. As a cluster composition, the head determines that it is important for the cluster to have a complete routing trench. The figure 2 represent a proposed method block diagram and route selection process.

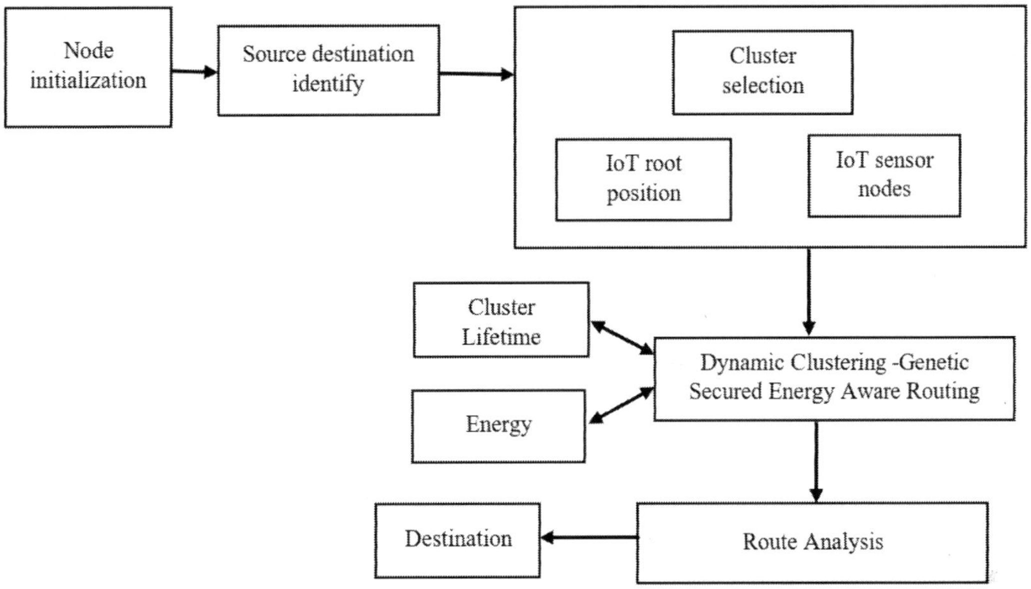

Figure 2. Proposed System Block Diagram.

The choice of spacing between the cluster head station and its adjacent nodes for its strength and fluidity is based on the extended innovation node search energy level. Node clusters with maximum power, minimum mileage with its adjacent nodes, and medium velocity deflection characteristic will be the selected head, if the next step and the cluster structure is advanced, the cluster member may be selected. It is based on masonry head properties.

3.1 Cluster formation

First, the basis of network topology is the development of a set of IoT sensor nodes. These nodes are collected with the help of dynamic clustering algorithms. A cluster is often defined as a group of very similar nodes that are located or co-occur. Clustering aids reduce energy consumption and improve data transmission. A detailed description of the dynamic clustering analysis algorithm is given below.

For each C_i //un-clustered node
Empty cluster to create (C_e)
Clustered multimedia node C_i
C_i *-the first member of cluster* C_e
For each C_i // un-clustered multimedia node //
$While\ (C_i < max\ generation)$
For every i
For every k
Update the function and light intensity
End for
End for

3.2 Cluster Head Selection

After the cluster is formed, the cluster heads are selected based on a cluster. The cluster head cluster is responsible for collecting messages from all population sensor nodes and sending them to the base station. Therefore, the cluster head cluster consumes more processing and energy, and is then involved in generating the described routing path. However, it drains the cluster head energy very quickly. The IoT node with the maximum remaining power in a given cluster will be the cluster head for subsequent rounds. This rotating routing effort of cluster head channels is distributed evenly between devices to increase the operating life of the network. Each device in the first round will have the same amount of power, so the cluster heads in this round will be randomly selected. In this dynamic cluster distance are calculate using an equation 1.

Run Dynamic Clustering algorithm to create cluster set C_s

$$\begin{aligned}
&For\ C_i \in C_s\ do\\
&For\ i \in C_k.nodes\ do\\
&If\ d.current_energy < i.current_energy\ then\\
&d = i;\\
&Set\ C_i\ Cluster_head = d;\\
&To\ Connecting\ all\ cluster\ heads\\
&Return\ routing\ path;\\
&While\ C_i - 1 > 1\ do\\
&Select\\
&C_k, C_{k+1} \in d_{i+1}; \qquad (1)\\
&End\ while\\
&For\ i = 1\ to\ P\ //denotes\ number\ of\ nodes\ per\ cluster\\
&If\ number\ of\ nodes\ C_k\ equals\ to\ P\\
&End\ for\\
&End\ if
\end{aligned}$$

3.3. Route discovery

When a node in IoT tries to send a data packet to a synchronous node with another uncertain path, it violently uses a path finding process to determine such a path. Route Requirement Network path detection is performed by flooding the network with RREQ (also known as query) packets. Each node receives the request and rebroadcasts them until they are in sync path or it must be in cache memory. Path RREP pocket responses and node responses returned to the actual source of the request. Route request and response packets are redirected to the source. A request to create a path to travel at present. A routing error packet is sent to the supported IoT node to clear the cache for each object in the path, including the broken connection path. If the path is still needed, the new path finding process must start from the source and there is no alternative to temporary storage.

3.3.1 Route maintenance

Route maintenance is a message packet along an outgoing or forward source route executed by each IoT node. Each node such as has the legal responsibility to ensure that the data packet was sent to the next one-hop neighbor along the starting node specified in the packet. The data packet retransmits until a confirmation is received (maximum number of attempts).

3.3.2 Energy Efficient Route Establishment

The network is formed by a "divide and rule" policy for sending node data packets to the destination. Tree structures and virtual skeletons are used for fusion within the network to ensure that data packets are sent to receivers with good cache hit rates and energy. Nodes that enter the network are associated with one of the specific roots. If its energy level is otherwise a root, higher than the identified root node, and it is a node of the IoT leaf node to be the root. The node with the highest energy in the virtual backbone domain is connected at one end to the other end to establish a path from the node within the IoT. The following sections detail the route / path discovery and routing / route maintenance algorithms.

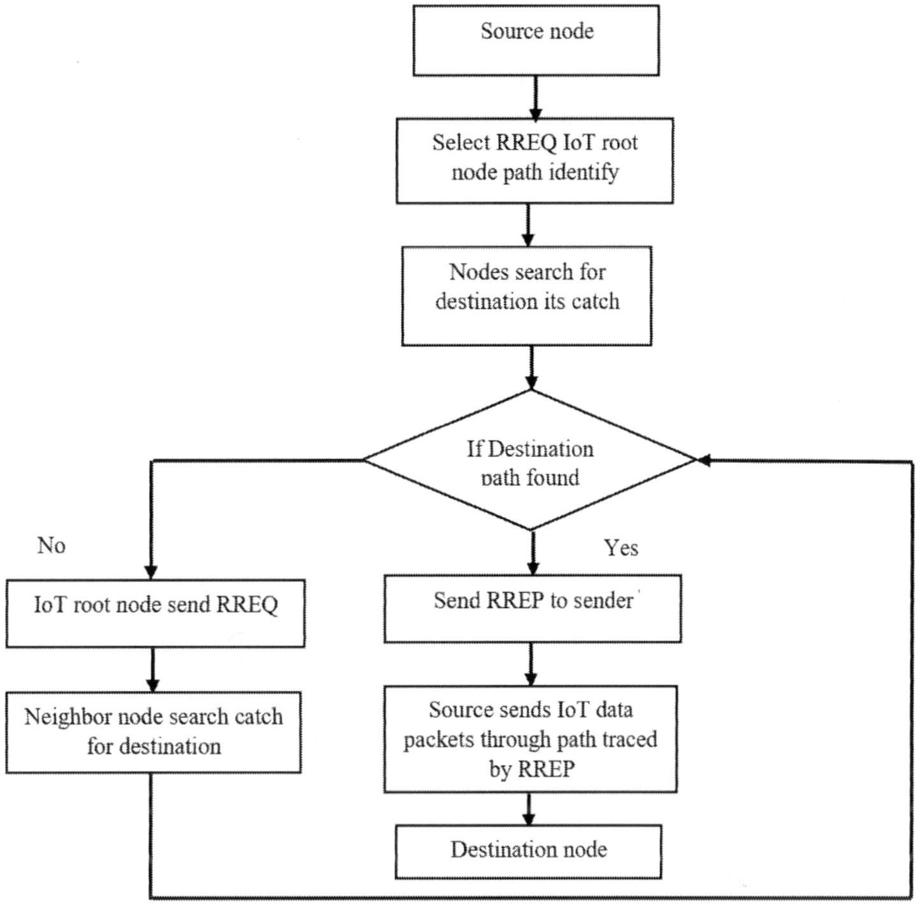

Figure 3. Analysis of Energy Efficient Route Establishment.

It is used only by normal IoT sensor nodes to sense responsibility for environmental data and data transmission, it flow diagram is shown in following figure 3. The advanced IoT node is responsible for receiving the data detected from the node in the component IoT, and the node's advanced energy will $(1 + h) * E_n$

The energy of the normal node is n $* E_n$and the energy of the altitude node is A $(1 + h) * E_n$as follows. Therefore, the total energy of the nodes in the network is expressed

$$Total\ Energy = n * E_n + A\,(1 + h) * E_n \qquad (2)$$

The Dynamic Clustering -Genetic Secured Energy Aware Routing Protocol ((DC-GSEARP)) broadcast path detection data packets and answers respond to the sink node entry in its cache or path response message. The developed protocol is that if the packet has the response of the path to the cache entry of the pocket, or if the pocket is sent that way, the path will be sent that way.

4. Result and Discussion

Network simulation (version 2), commonly referred to as NS2, and is simply a simulation device driven by the phenomenon that the dynamic properties of resource communication networks have proven to give a valuable result. Network functions (UDP, TCP, routing algorithms, etc.) and protocol emulation should be possible using wired and wireless NS2. Table 1 shows the simulated network model. The simulation parameters are evaluated through the proposed Dynamic Clustering - Genetic Secured Energy Aware Routing Protocol (DC-GSEARP). The network area consists of a maximum number of randomly deployed IoT nodes, and it is assumed that the maximum transmission distance of each IoT sensor is 100 meters.

Table 1. Simulation Parameters.

Parameters	Value
Data rate	2 Mbps
Application Type	Constant bit rate (CBR
CBR interval	1.0 (second)
Simulation Time	350 s
Simulator	NS-2
Data Speed	20 m/s
Total energy	500joule

Table 1 describes the resources needed in this proposed system protocol. This section describes the comparison of existing Density-Based Spatial Clustering of Applications with Noise (DBSCAN), Karush-Kuhn-Tucker (KKT), Completely Independent Spanning Trees (CISTs) and the proposed Dynamic Clustering -Genetic Secured Energy Aware Routing Protocol (DC-GSEARP). This experimental result describes the analysis of parameters are analysis of throughput, energy efficient, route availability analysis, life-span of the network and packet delivery ratio.

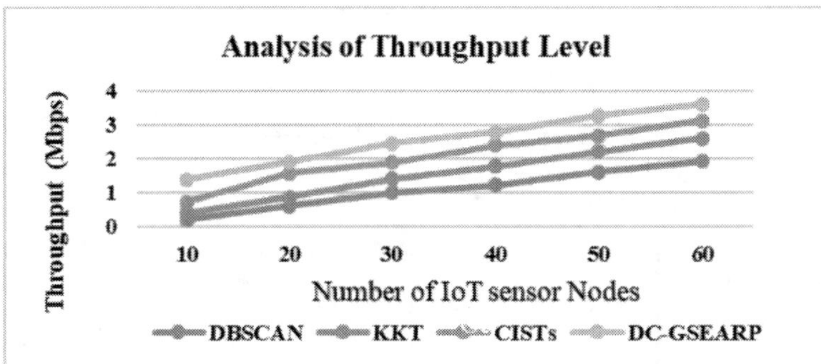

Figure 4. Analysis of Throughput Level.

IoT can determine the good performance of the sensor node by measuring the transfer success on a regular basis. A successful distribution plan in which the deep-rooted path was suddenly broken and without loss. DC-GSEARP ensures that the rehearsal team maintains the right path through a rule-based system with improved performance.

$$Throughput = \frac{Packets\ Received\ (n) * Packet\ size}{200} \quad (3)$$

The throughput evaluation in the figure 4 shows the performance of the DC-GSEARP under variations in the availability of nodes at regular period of intervals shows the supreme performance of the DC-GSEARP with 3.59 Mbps and its heightened throughput compared to the previous methods (DBSCAN with 1.9 Mbps, KKT with 2.6 Mbps, and CISTs with 3.1 Mbps).

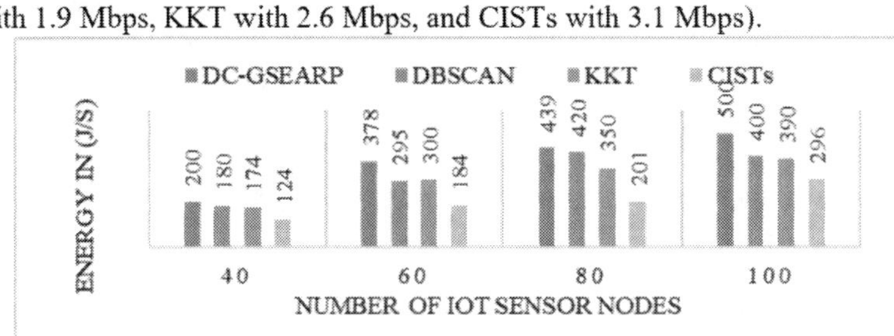

Figure 5. Energy efficient Level Analysis.

The first IoT sensor node or IoT node in the set of power means of the network cycle times runs out. The network life is determined by the remaining power supply network, which can simply be

IOP Conf. Series: Materials Science and Engineering **995** (2020) 012035 doi:10.1088/1757-899X/995/1/012035

defined. The DC-GSEARP protocol and existing method DBSCAN, KKT, CISTs comparison of energy consumption shown in figure 5.

$$Energy\,Efficiency = \frac{Energy\,Density_{(Discharge)}}{Energy\,Density_{(Charge)}} * 100 \qquad (4)$$

In the path analysis calculate total number of IoT node, number of available path it's divided by total number of connection.

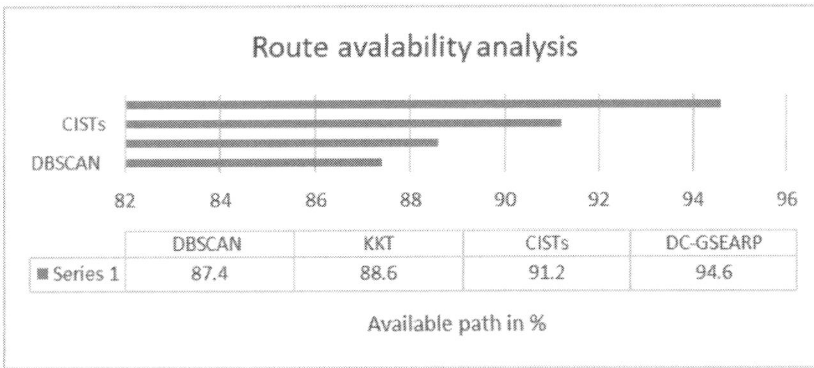

Figure 6. Route Availability Analysis.

An important advantage of using path transmission is inherent path diversity (ie, the loss process is expected to operate independently on different paths). To achieve this goal, energy efficient routing is an effective method as shown by network data that can be shared via multiple paths to reduce network congestion. Comparison of the proposal shown in figure 6 with existing methods. The existing methods DBSCAN in 67%, KKT in 71%, CISTs in 86%, and the proposed method DC-GSEARP with 95%.

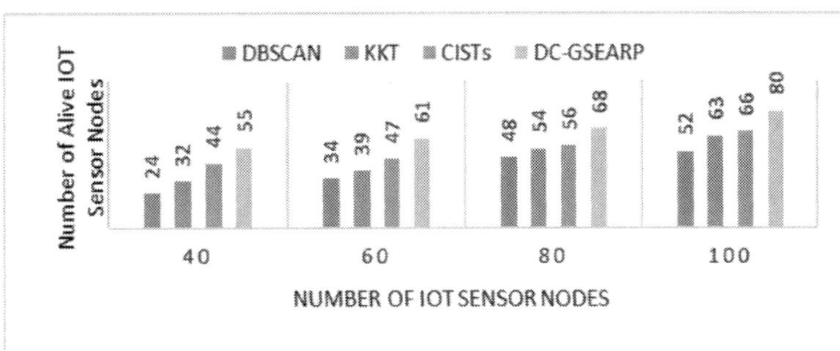

Figure 7. Life-Span of the IoT Sensor Node.

Network life is determined using the direct IoT nodes available on the network. The batteries of the electric IoT sensor nodes are clustered and fully designed to avoid unnecessary use of the power of these sensor nodes and to extend the life of the system in a well-organized manner through proper head selection. In this comparison of proposed method and existing method is show in figure 7.

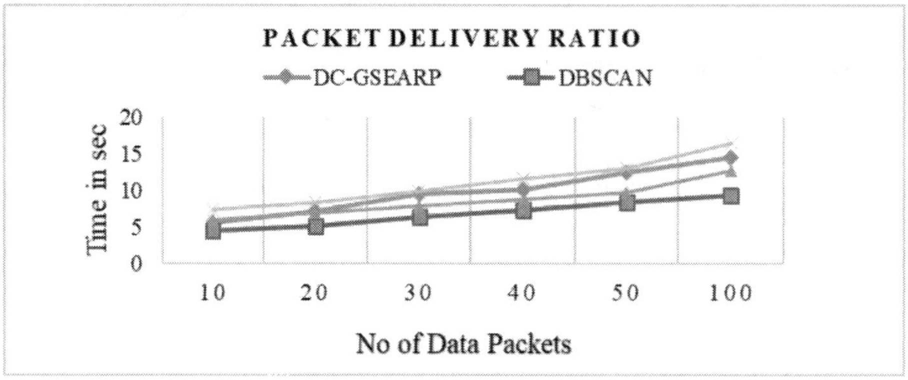

Figure 8. Packet Delivery Ratio.

Packet Delivery Rate (PDR) to assess the quality of the network. It defines the ratio between the packets received by the destination and the data packets generated by the source. It can be obtained by using the awk script, which produces a trace file and the result.

PDR = *Received packets/ Generated packets * 100* (5)

The proposed method DC-GSEARP and existing method DBSCAN, KKT, CISTs comparison of packet delivery ratio shown in figure 8. The analysis result proposed method 9.4 sec of lower time-based packet delivery provide compare to existing method DBSCAN has 14.5 sec, KKT with 12.76 sec, and CISTs has provided a 16.45 sec.

5. Conclusion

Today's IoT networks work effectively in many applications. For this reason, the IoT of the various sensors must be selected at the nodes required to create cluster heads for data transfer. Energy is the basic source of wireless sensor networks. The gateway used to establish the connection is a simple IoT based on DC-GSEARP, which avoids the waste generated by head and cluster mode selection and can adopt a properly designed DC-GSEARP. It is a cluster in which the use of DC-GSEARP can greatly improve the energy efficiency of the network. Energy is an important factor in improving online life. The formal configuration of the Internet in the network of things provides a long life for cluster routing. Consumption, as compared to the cluster-based routing technology (ESCHS, Fuzzy Logic, ARSH-FATI), significantly reduces the energy of the sensor node, another type of routing. The proposed DC-GSEARP based analysis of throughput level is 3.59 mbps, energy efficient level analysis is 500 joules, route availability analysis level is with 95%, life-span of the IoT sensor node level is 80 alive of sensor nodes ,packet delivery ratio is 9.4 sec.

Reference

1. Sindhanaiselvi, D., Keerthikaa, S., & Shanmuganantham, T. (2018). Energy Efficient Modern Home Using Smart Boards for IoT Energy Automation. 2018 International Conference on Circuits and Systems in Digital Enterprise Technology (ICCSDET). doi:10.1109/iccsdet.2018.8821228.
2. Aditya, J., Nagateja, T., & Vaddi, R. (2017). Tunneling Field Effect Transistors for Energy Efficient Logic, Sensor Interface and 3D IC Circuits for IoT Platforms. 2017 IEEE International Symposium on Nanoelectronic and Information Systems (iNIS). doi:10.1109/inis.2017.28.
3. Mahato, D. K., Yadav, S., Saxena, G. J., Pundir, A., & Mukherjee, R. (2018). Image Processing and IoT Based Innovative Energy Conservation Technique. 2018 4th International Conference on Computational Intelligence & Communication Technology (CICT).
4. Zhu, R., Zhang, X., Liu, X., Shu, W., Mao, T., & Jalaian, B. (2015). ERDT: Energy-Efficient Reliable Decision Transmission for Intelligent Cooperative Spectrum Sensing in Industrial IoT. IEEE Access, 3, 2366–2378. doi:10.1109/access.2015.2501644.
5. Wonseok Lee, Bang Chul Jung , And Howon Lee (2020) DeCoNet: Density Clustering-Based Base Station Control for Energy-Efficient Cellular IoT Networks,pp - (120881-91).

6. Al-Kadhim, H. M., & Al-Raweshidy, H. S. (2019). Energy Efficient and Reliable Transport of Data in Cloud-Based IoT. IEEE Access, 7, pp – (64641–50). doi:10.1109/access.2019.2917387.
7. Ali Hassan Sodhro, Sandeep Pirbhulal, Luo Zongwei, Khan Muhammad, Noman Zahid (2020) Towards 6G Architecture for Energy Efficient Communication in IoT-Enabled Smart Automation Systems, IEEE Internet of Things Journal, pp - (1-8).
8. Wu, Q., Chen, W., Ng, D. W. K., & Schober, R. (2018). Spectral and Energy-Efficient Wireless Powered IoT Networks: NOMA or TDMA? IEEE Transactions on Vehicular Technology, 67(7), pp – (6663–7). doi:10.1109/tvt.2018.2799947.
9. Popli, S., Jha, R. K., & Jain, S. (2018). A Survey on Energy Efficient Narrowband Internet of things (NBIoT): Architecture, Application and Challenges. IEEE Access, 1–1. doi:10.1109/access.2018.2881533.
10. Nguyen, T. D., Khan, J. Y., & Ngo, D. T. (2018). A Distributed Energy-Harvesting-Aware Routing Algorithm for Heterogeneous IoT Networks. IEEE Transactions on Green Communications and Networking, 1–1. doi:10.1109/tgcn.2018.2839593.
11. Pan, J., Jain, R., Paul, S., Vu, T., Saifullah, A., & Sha, M. (2015). An Internet of Things Framework for Smart Energy in Buildings: Designs, Prototype, and Experiments. IEEE Internet of Things Journal, 2(6), pp – (527–7).
12. Behera, T. M., Mohapatra, S. K., Samal, U. C., Khan, M. S., Daneshmand, M., & Gandomi, A. H. (2019). Residual Energy Based Cluster-head Selection in WSNs for IoT Application. IEEE Internet of Things Journal, 1–1. doi:10.1109/jiot.2019.2897119.
13. Han, T., Muhammad, K., Hussain, T., Lloret, J., & Baik, S. W. (2020). An Efficient Deep Learning Framework for Intelligent Energy Management in IoT Networks. IEEE Internet of Things Journal, 1–1. doi:10.1109/jiot.2020.3013306.
14. Z. Shi, N. Yao, and F. Zhang (2017) Scheduling feasibility of energy management in micro-grids based on significant moment analysis in Cyber-Physical Systems: Elsevier, 2017, pp. 431-9.
15. Wang, B., Sun, Y., & Xu, X. (2020). A Scalable and Energy-efficient Anomaly Detection Scheme in Wireless SDN-based mMTC Networks for IoT. IEEE Internet of Things Journal, 1–1. doi:10.1109/jiot.2020.3011521.
16. Huang, Y., Liu, M., & Liu, Y. (2018). Energy-Efficient SWIPT in IoT Distributed Antenna Systems. IEEE Internet of Things Journal, 5(4), pp – (2646–6).
17. Al-Azez, Z. T., Lawey, A. Q., El-Gorashi, T. E. H., & Elmirghani, J. M. H. (2019). Energy Efficient IoT Virtualization Framework With Peer to Peer Networking and Processing. IEEE Access, 7, pp – (50697–9). doi:10.1109/access.2019.2911117.
18. Yang, Z., Xu, W., Pan, Y., Pan, C., & Chen, M. (2018). Energy Efficient Resource Allocation in Machine-to-Machine Communications With Multiple Access and Energy Harvesting for IoT. IEEE Internet of Things Journal, 5(1), pp – (229–5).
19. Lv, T., Lin, Z., Huang, P., & Zeng, J. (2018). Optimization of the Energy-Efficient Relay-Based Massive IoT Network. IEEE Internet of Things Journal, 5(4), pp – (3043–58). doi:10.1109/jiot.2018.2829827.
20. Sun, P., Tian, Z., Wang, Z., & Wang, Z. (2018). PRSS: A Prejudiced Random Sensing Strategy for Energy-Efficient Information Collection in the Internet of Things. IEEE Internet of Things Journal, 1–1. doi:10.1109/jiot.2018.2873756.
21. Al-Janabi, T. A., & Al-Raweshidy, H. S. (2019). An Energy Efficient Hybrid MAC Protocol with Dynamic Sleep-Based Scheduling for High Density IoT Networks. IEEE Internet of Things Journal, 1–1. doi:10.1109/jiot.2019.2905952.

Anti Fungal Studies of Tea Leaf mediated Silver Nano Colloid

S. Heera[1*], I. Dhanya[1*], LiyaMariya Varghese[1], Malathy Krishna[1] and Reny Renji[1]

1*Department of Physics, Catholicate college, Pathanamthitta, Kerala, India 689 6452

heeraammu95@gmail.com, dhanya.iv@gmail.com

Abstract. Silver nano colloids are prepared using Tea leaf as reducing and stabilizing agent. Samples with different concentrations of the reducing agents are irradiated with gamma rays of various doses. Transmission Electron Microscope images of the samples show agglomerated clusters of nano particles before irradiation while the images of after irradiation show chain free individual nano particles with reduced size and spherical morphology. The Selected Area Electron Diffraction pattern confirms the crystal nature and face centred cubic structure of nano silver. UV-Visible spectroscopy gives absorption peaks in the blue region. Gamma irradiation results in the reduction in peak intensity without any change in wavelength. A statistical approach is done on treating the gamma irradiated colloidal sample with the as prepared silver nano colloid for measuring the affected area of fungal infection on a potato-agar medium. It is found that the fungal growth is declined rapidly with the introduction of gamma irradiated silver nano colloid than the as prepared colloid.

Keywords: silver nano colloid, Tea powder, gamma irradiation, spherical morphology, antifungal studies.

1. Introduction

Silver nanoparticles are noble metal nano particle most often used in the fields of green chemistry and nano technology branches due to their high surface to volume ratio [1]. Silver nanoparticle is an essential component in electronic industries due to semiconducting properties and fast switching mechanism [2]. Its unique properties make it nearly impossible to substitute and are used in a wide range of applications in consumer products like antifungal creams, edible packages and in medicines [3]. Solubility and stability of silver nano particle with polymers and carbon compounds are excellent. Silver can be tuned in research for improving the physico-chemical properties [4]. In Biomedical research, silver is highly used as fluorescent nano particle for drug delivery applications [5, 6]. Silver nano particle usually prepared from plant extract shows fluorescent quenching effect [7, 8]. Antimicrobial properties of silver nanoparticles and silver ions have been known for last few centuries [9]. Silver nanomaterials have also been shown to have antifungal properties. Most important findings were related to toxic effects towards Candida albicans species [10]. It was shown that colloidal silver nanoparticles in very low concentrations may have substantial antifungal impact in vitro. Even now the mechanisms of antibacterial nano silver action are not completely understood. It is supposed that silver inhibits the function of proteins and DNA replication [11]. Another theory is that silver causes damage to the bacterial membrane by incorporating itself into-membrane structure and interacting with the major building elements [12]. This structural damage causes change in membrane permeability to water and various ions, which leads to bacterial death.

Aim of this work is to enhance the antifungal property of silver nano particle by irradiation treatment. The attempt is also made to reduce the agglomeration and to maintain the average particle size in nano regime by suitable irradiation of highly energetic rays like gamma ray.

IOP Conf. Series: Materials Science and Engineering **995** (2020) 012036 doi:10.1088/1757-899X/995/1/012036

2. Experimental Details

Silver nitrate powder (Sigma Aldrich Co. USA: purity-99.9999999%) is used as the source material for our study. In order to prepare Silver nitrate solution, weigh out 10 mg silver nitrate powder using electronic weighing balance and is dissolved with 150ml de-ionized water using a magnetic stirrer. 10 grams of tea leaves from the Kannandevan hills, Kerala are cleaned well using tap water several times and then with double distilled water. Particular leaf extract is taken by boiling it with 100ml deionized water. Extract is double filtered using Whatman NA 125mm Transfer Paper. Different concentration like 2ml, 4ml, 6ml, 8ml and 10ml of filtered tea-leaf extract is added to each beaker containing 30ml AgNO3 solution. Magnetic stirring is provided for uniform concentration for 10 minutes while adding the extract. Keep the solution for 2-3 days under constant room temperature to get colloidal silver nano particle. The colour of the solution changes from colourless to yellowish brown.

The as prepared silver nano particles are exposed to Co-60 irradiator- model no. LDI 2000-BRIT. It is having an irradiation rate at a maximum order of 2 gray/minute with an attenuation of 50%. The samples are irradiated at a dose rate of 1 Gray/min. Varying dosages like 4Gray; 6Gray and 8Gray are given to silver nano colloids by controlling the exposure time. A visible colour change from dark red to pale yellow is observed for the colloidal samples.

Structural and morphological properties are analysed through Transmission Electron Microscope (TEM) Model JEOL JEM 2100 operated at 200KV and Selected Area Electron Diffractogram (SAED) is recorded. Optical absorption spectra are performed in a wavelength ranging from 300 to 900 nm using a double beam UV- Visible spectrophotometer, JASCO-V550 with deionized water as the reference. HRTEM images are further analysed using Image J software to get the average particle size and lattice spacing.

Antifungal activity of gamma irradiated biosynthesized silver nano particle is performed using Dextrose Agar media according to well diffusion method. Antifungal activity was assessed against Rhizopus Stolonifer. 30ml of gamma irradiated silver nano particle is added in wells of 1.5 mm diameter wounds in Petri plates with potato- dextrose –Agar (PDA). A colony of Stock cultures from Rhizopus Stolonifer already prepared and maintained in PDA slants were transferred to the medium. Zone of inhibition is continuously examined using a meter ruler. A histogram is plotted based on the statistical analysis for a period of one week. During the period the specimen were well maintained at constant temperature of 200C.

3. Result and Discussions

Green synthesis of silver NPs is done with tea-leaf extract as reducing agent. The appearance of yellowish-red colour of the samples suggests the formation of nano silver. As the concentration of the extract increases the nano silver colloidal solution changes from orange to dark red in colour. This may be due to the production of large amount of silver nano particles in colloidal suspensions. Silver nano colloids with varying concentrations of reducing agents undergo gamma irradiation at different doses like 4 gray, 6 gray and 8 gray. The Figure 1 shows the colloidal samples with gamma irradiation.

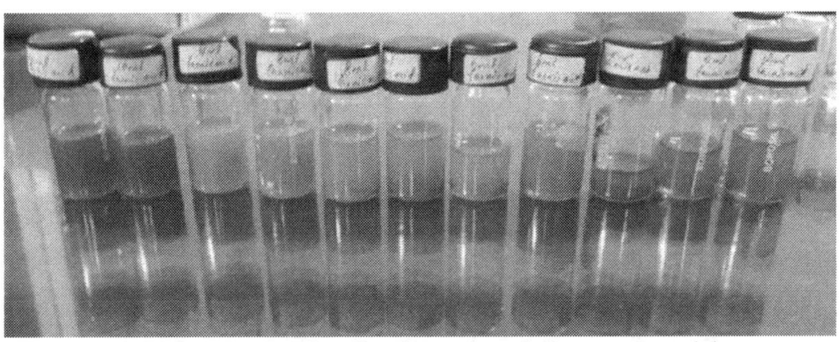

Figure 1: Gamma irradiated samples of silver nano particles

IOP Conf. Series: Materials Science and Engineering **995** (2020) 012036 doi:10.1088/1757-899X/995/1/012036

The morphology, size and lattice spacing of synthesized silver nanoparticle is determined by TEM, HRTEM and SAED pattern. Figure 2 shows TEM, HRTEM and SAED image of silver nanoparticle under 20 nm and 2 nm (inset: HRTEM) resolution after irradiation. It reveals the lattice spacing between adjacent fringes is 0.82 nm. SAED analysis can index with the help of C-Spot software treatment and found to have FCC lattice with orientation of peaks along (111), (220) and (311).

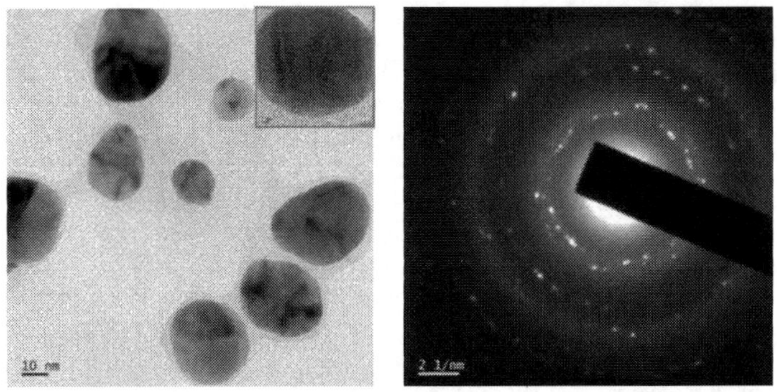

Figure 2: TEM, HRTEM and SAED pattern of Gamma irradiated samples of silver nano particles

The UV-Visible absorption diagram of gamma irradiated nano silver colloid and Tauc plot is shown in Figure 3. From the figure, it is found that the intensity of absorption of gamma irradiated sample becomes 1.35 and from Tauc plot the band gap energy is found to be 2.11±0.02 eV. It may be due to the colour change from dark brown to pale yellow.

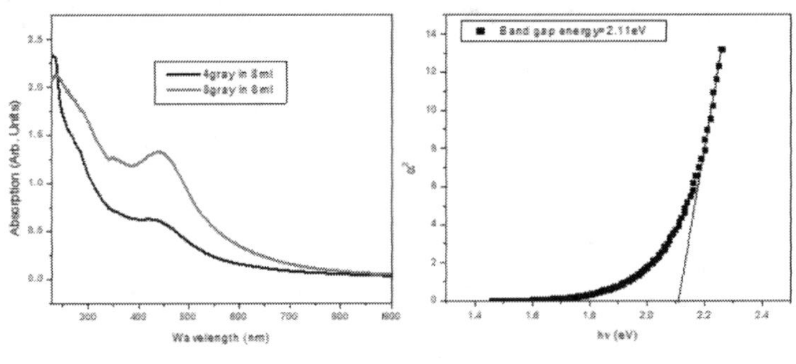

Figure 3: Optical absorption spectra and Tauc plot for Gamma irradiated silver nano particle

Antifungal studies are done on the Potato Agar solution by making three wounds of 1-1.5 mm. The control is taken as tea-leaf extract media only and is inserted in one wound. The other two is filled with silver nano particles before irradiation and the last with gamma irradiated silver nano particles. The study is repeated for 4 Gray and 8 Gray samples of 2ml, 4ml and 8ml tea-leaf extract added samples. The best inhibition rate is obtained for 8gray-8ml tea leaf extract mediated sample. The affected area of fungal growth is monitored in a continuous duration of 24 hrs and is continued for one week. The diameter of affected area is measured using a meter ruler and histogram is plotted. Figure 4 represents the fungal affected areas on different media and the histogram of timely response.

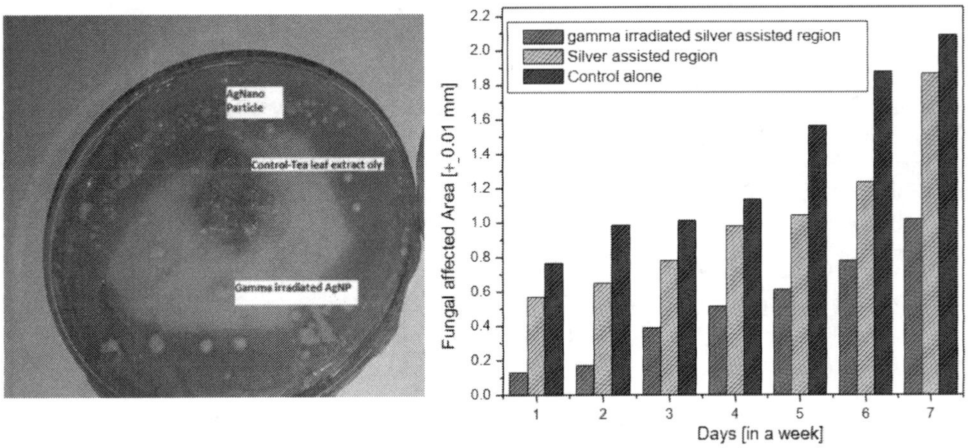

Figure 4: Antifungal activity and histogram for Gamma irradiated silver nano particle

4. Conclusion

In this work, green synthesis of silver nanoparticles is prepared through tea-leaf extract of different concentrations. Gamma irradiation of various dosages are done on samples and are found to have a reduction in yellowish-brown colour in aqueous solution to pale yellow due to a reduction in average particle size to controlled nano regime. The characteristic colour is also an indication of excitation due to surface plasmon vibrations in silver nanoparticles. From TEM analysis of nano silver samples without irradiation, it is found that morphology is core-shell in nature with agglomerated nano particles. It got disappeared in gamma irradiated samples and individual nano particles are found with a reduction in average particle size. The uniform distribution of nano particle in gamma irradiation is due to the effective photo induced reduction of particle size by irradiation effect. The SAED pattern of silver nanoparticles with and without gamma irradiation show white ring like pattern. The ring-like diffraction pattern indicates that the particles are crystalline. The diffraction rings could be indexed on the basis of image software and are found of having FCC structure for the nano silver. Optical studies of bio synthesized silver peaks show almost similar broadness in variety of samples. Prepared nano silver is found to have a direct allowed transition with a wide band gap. Effect of gamma irradiation further reduces the peak intensity and broadness while the band gap energy is not much affected. From the antifungal studies of silver nano particles, the culture media alone shows a drastic growth in fungi for an observation period of one week while in the samples incorporated with gamma irradiated silver nano particles, the growth is much reduced.

Acknowledgement

Author(s) are thankful to Low-level radiation research laboratory, Division of BARC, Kollam, Kerala, India for providing the necessary gamma irradiation facility and STIC, CUSAT for providing us the TEM-SAED facility.

References

[1] Alsharach E, Alazzam S, Ahmed F, Arshi N, Al-Hindawi M and Sing G K 2017 Green synthesis of silver nano particle and their reduced graphene oxide nano composites as anti bacterial agents: a bio inspired approach Act. mett. Sini. (English letters) vol. 30 no. 1 pp. 45-52

[2] Zhasnakunov Z A, Satyvaldiev A C, Doolotkeldieva T, Omurzak E, Bobusheva S T, Kelgenbaeva Z and Abdullaeva Z 2018 Synthesis of biologically active silver and copper nano composites Mat. Res. Exp. vol. 5 no. 8 pp. 085404.

[3] Lin S K and Cheng W T 2020 Fabrication and characterization of colloidal silver nano particle via photo chemical synthesis Mat. Lett. vol. 261 no. 15 pp. 127077.

[4] Chang K, Jian X, jeong H M, Kwon Y, Lu Q and Cheng M J 2020 Improving CO_2 electrochemical

reduction to CO using space confinement between gold and silver nano particle J Phys. Chem. Lett. vol. 11, no. 5, pp. 1896-1902.

[5] Valsalam S, Agastian P, Arasu M V, Al-Dhabi N A, Kareem A, Ghilan M, Kaviyarasu K, Ravindran B, Chang S W and Arokiyaraj S 2019 Rapid biosynthesis and characterization of silver nanoparticles from the leaf extract of tropaeolummajus and its enhanced in-vitro antibacterial, antifungal, antioxidant and anticancer properties J. of Photochem. and Photobio B: Bio. vol. 191 pp. 65-74.

[6] Maghimaa M and AAlharbib S 2020 Green synthesis of silver nanoparticles from curcuma longa and coating on the cotton fabrics for antimicrobial applications and wound healing activity J. of Photochem. and Photobio B: Bio. vol. 204 pp. 111806.

[7] Hosny A M S, Kashef M T, Rasmy S A, Aboul-Magd D S and El-Bazza Z E 2017 Antimicrobial activity of silver nanoparticles synthesized using honey and gamma radiation against silver-resistant bacteria from wounds and burns Adv. in Natural Sciences: Nanoscience and Nanotechnology vol.8 No.4 pp.1-7.

[8] Bozkurt P A 2017 Sonochemical green synthesis of Ag/graphene nano composite Ultrasonics Sono chemistry, Vol. 35, Part A, pp. 397-404.

[9] Nasrollahzadeh M, Shafiei N, Eslamipanah M, Fakhri P, Jaleh B, Orooji Y and Varma R S 2020 Preparation of Au nanoparticles by Q switched laser ablation and their application in 4-nitrophenol reduction Clean Tech. and Env. Policy vol. 22 pp.1715–1724.

[10] Huang W, Yan M, Duan H, Bi Y, Cheng X, and Yu H 2020 Synergistic antifungal activity of green synthesized silver nanoparticles and epoxiconazole again stsetosphaeria turcica Journal of Nanomaterial Article ID 9535432.

[11] Li J, Zhu Z, Liu F, Zhu B, Ma Y, Yan J, Lin B, Ke G, Liu R, Zhou L Tu S and Yang C 216 DNA-mediated morphological control of silver nanoparticles, Small. Vol. 12 no. 39 pp.5449-5487.

[12] Qais F A , Shafiq A, Khan H M, Husain F M , Khan R A, Alenazi B, Alsalme A , and Ahmad I 2019 Antibacterial effect of silver nanoparticles synthesized using murrayakoenigii (L.) against multidrug-resistant pathogens Bioinorg. Chem. and Applied Physics B pp 1565-3633.

Autonomous Vehicle for Drug Delivery

K Nandakumar[1], S R Surya [2], K Karthick[3] ,N Raviraj[4] , B Cheran[5], V Poovizhi[6], S Gokul[7], C Nimisha[8]

[1] Assistant Professor, Department of Mechatronics Engineering, Bannari Amman Institute of Technology, Erode, Tamilnadu, India.
[2,6,7] Student, Department of Mechatronics Engineering, Bannari Amman Institute of Technology, Erode, Tamilnadu, India.
[3,4,5] Student, Department of Electrical and Electronics Engineering, Bannari Amman Institute of Technology, Erode, Tamilnadu, India.
[8] Student, Department of Electrical and Communication Engineering, Bannari Amman Institute of Technology, Erode, Tamilnadu, India.

nandakumar@bitsathy.ac.in, surya.mc16@bitsathy.ac.in

Abstract. Autonomous vehicles are intended to move in an environment without any human intervention. In order to determine the structure of the environment, we have to rely on sensors data to build the map and to localize the vehicle within its environment .This paper presents the autonomous vehicle built using 2D LIDAR and Robot Operating System(ROS).If the environment is to be explored is outdoor we can rely on GPS to track the location the autonomous vehicle, whereas in an indoor environment we cant do so. For building the map of the environment and for the vehicle's localisation simultaneous localisation and mapping (SLAM) is used. The LIDAR data is used to find out the corresponding distance of the obstacle from the robot. Based on this point cloud data map is generated denoting the obstacle, explored area, unexplored area and the location of the robot with respect to the map. Data from the encoder is used to navigate the robot. Path planning is done using A* algorithm to find out the shortest distance between the starting and the destination point. The experimental result helped us in comparing the performance of the autonomous mobile robot built using ultrasonic sensor with the autonomous mobile robot built using LIDAR.

Keywords: 2D LIDAR, Autonomous mobile robot, SLAM, ROS, A* algorithm

1. Introduction

As the future is moving towards automation, robotics has gained its importance and it is likely replacing humans. Robots perform human work with better repeatability and accuracy. So, we prefer employing robots to finish the desired work with better efficiency and to reduce time consumption. The autonomous robot built using ultrasonic sensor has wider angle of divergence. So, it is quite difficult to predict the exact location the obstacle, whereas in LIDAR the point of divergence is comparatively less, so the map generated with LIDAR data will give the accurate location of the obstacle with respect to the map. LIDAR throws out approximately 650-point cloud data for every successive revolution. The line by line execution of the python code may lead to the loss of certain LIDAR data. So we go with robot operating system (ROS) which does multiprocessing. The entire process is divided into subprocess called nodes which gets executed concurrently. Encoder attached to the wheels of the robot is used to calculate the distance travelled by the robot. Having a dedicated robot to perform some specific application avoids the need of a person to carry out that task.

2. Literature review

Genghang Zhuang et all conducted numerous experiments for localisation of autonomous vehicles using GPS and INS. But localising the autonomous robot without using devices like GPS for indoor application can be achieved using 2D LIDAR and MCL . Scanning of the point cloud data is done using laser and a prior map is generated using the scanned data. Once the bot moves from its position new map is generated using the scan points and this updating process is done with help of Monte Carlo Localisation (MCL) [1].

Ilya Afanasyev et all conducted a feasibility test using gazebo simulation approach. In order to study the characteristics and behaviour of the robot under different conditions, robot simulators have been developed to pre-analyse the end results. The experiment is conducted using PR2 robot built under SLAM in ROS environment[2].

John Kerr et all conducted a research to optimize the existing 16 ROS into a single ROS which would benefit their students. To understand the concept of ROS, working to normal operating system is studied so that the same can be implemented with ROS. The first test started with comparing the 16 ROS with criteria's like name of the developer, latest update date, number of tutorials available. At the end of first comparison 6 ROS were eliminated .To filter the 10 ROS, second stage of testing is carried out by setting some criteria and each ROS is tested based on the criteria's provided and scores were given out of 10.At the end of second stage of testing 7 ROS were eliminated. The final test is conducted with practical testing. The end results demonstrated that a student with a enough coding knowledge with no prerequisite knowledge in ROS was able to control a robot in a time span of 25 hours [3].

Hugh Durrant-Whyte et all proposed a SLAM technique that has been proposed over the past decades. SLAM technique is adopted to localize the mobile robots in an unknown environment. The environment can be indoor, outdoor, or underwater. Over the past decades the problem of SLAM has been unsolved. In the present the problems are over- thrown with the help of techniques like EKF-SLAM, Rao blackwellised filter[4].

Jie Hu et all proposed a A-star algorithm which is widely used in unity 3 games in finding shortest path between the nodes. The possible ways to reach the destined location is analysed. Then the weightage of each node is calculated, the node will comes up with the least value is considered as the shortest path to reach that destined location. The algorithm showed best results in finding the shortest path[5].

3. Existing methodology

One of the most commonly used range/distance finding sensors is ultrasonic sensor (HC-SR04). This sensor detects the distance of the objects by emitting high frequency ultrasonic waves and waits for the waves to get reflected back from the objects. Using the time taken for the sound wave to get reflected back, the distance can be computed. This sensor has a measurement range of 2cm- 400 cm and a measuring angle of 15 degree.

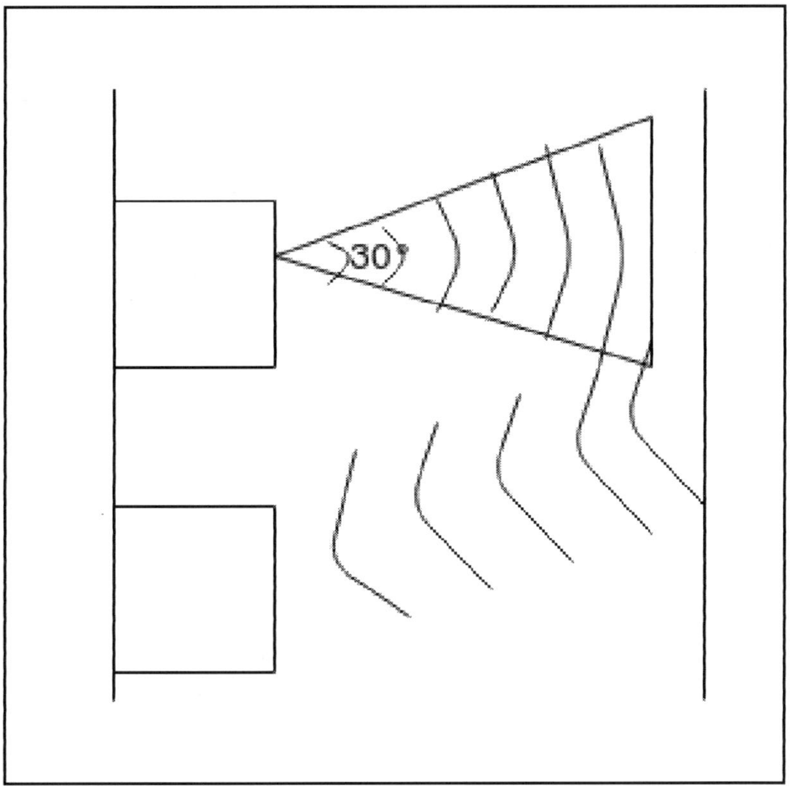

Figure 1. Measuring angle of ultrasonic sensor.

The distance data is used to compute the map of the environment. Since, the angle of measurement is 15 degree the width of the beam increases with the increase in distance(figure 1). For example, at 4 meters the beam should be as wide as the meters i.e. the obstacle's probability increases with distance. Thus, the map generated using ultrasonic range finding device is considerably inaccurate.

4. Proposed methodology

A laser range finder works similar to the ultrasonic sensor. But instead of ultrasonic sound waves it used finely focussed pulse of light to the target and detects the reflection. LIDAR (light detection and ranging) is one of the laser range finding sensors.

The ultrasonic sensor is replaced by a 2D LIDAR (single channel LIDAR) which has a single laser range finder rotated all through 360°taking samples at equal interval. The list of distances will be retrieved after a full rotation which will have the range data of the obstacles around. Thus, the map can be generated using the LIDAR range data. In order to localize the robot within the map, encoder is used. The encoder works on the principle of hall effect which gives 1000 pulses per revolution (which many vary according to the type of encoder used). Thus, mapping and localisation can be performed.

4.1 Localization theory

The robot is built with two motors with encoders and wheels attached to its shaft. A castor wheel is used to support the robot's base. The robot is of differential drive type. Based on the encoder pulse received by the controller, the distance travelled by the robot could be found as below.

IOP Conf. Series: Materials Science and Engineering **995** (2020) 012037 doi:10.1088/1757-899X/995/1/012037

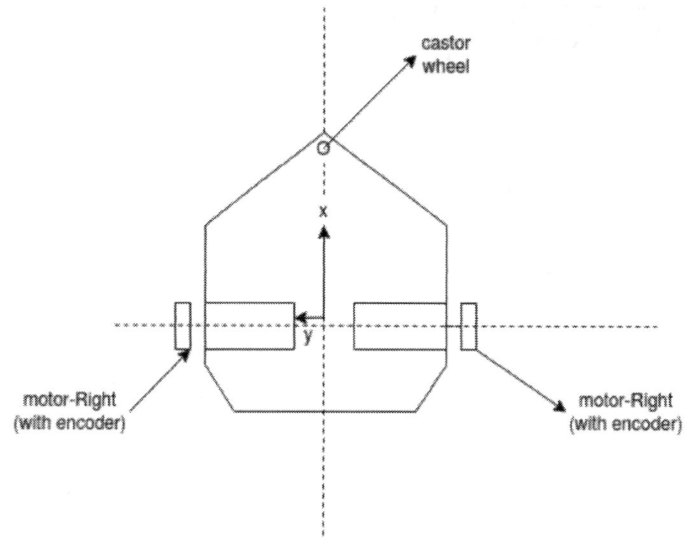

Figure 2. Robot's base.

Let us consider the encoder pulses of left wheel be PL and the right wheel be PR. The pulses received could be related to the distance travelled by the wheels as in equation (1).

$$NR = \frac{\pi * WDia * PR}{PPR}$$

$$NL = \frac{\pi * WDia * PL}{PPR} \tag{1}$$

Where,

N_L = distance travelled by left wheel

N_R = distance travelled by right wheel

PR and PL= Pulse count in right and left encoder

W_{Dia} = wheel diameter

PPR = pulse per revolution of encoder shaft

The robot's base will have a frame attached to the centre of the axes of the wheel(figure 2). Any translational and rotational movement will be performed on that coordinate frame {B}. There are two possibilities that a robot can move.

(i) A pure translation where both the motors being operated in same direction and at the same speed.

(ii) A general transformation in which motors may be operated in different direction and in different speed. In the first case robot will be moving in straight line where as in the second, its trajectory will be a general curve.

4.1.1 $n_L \approx n_R$ (Linear motion)

When the number of pulses in both the encoders is same, the robot would have moved in straight line (figure 3). By the equation (2), both the forward and backward motion will be calculated. To convert the motion with respect to the universal frame{U}, equation(2) could be applied,

$$^{U}T_{B2} = {}^{U}T_{B2} * {}^{B1}T_{B2}$$

$$^{B1}T_{B2} = \begin{bmatrix} 1 & 0 & 0 & n \\ 0 & 1 & 0 & 0 \\ 0 & 0 & 1 & 0 \\ 0 & 0 & 0 & 1 \end{bmatrix} \tag{2}$$

(where n=nL=nR)

$^{U}T_{B1}$ = Robot's old position in the map (Initially an identity matrix)

$^{U}T_{B2}$ = Robot's current position in the map

$^{B1}T_{B2}$ = Robot's displacement

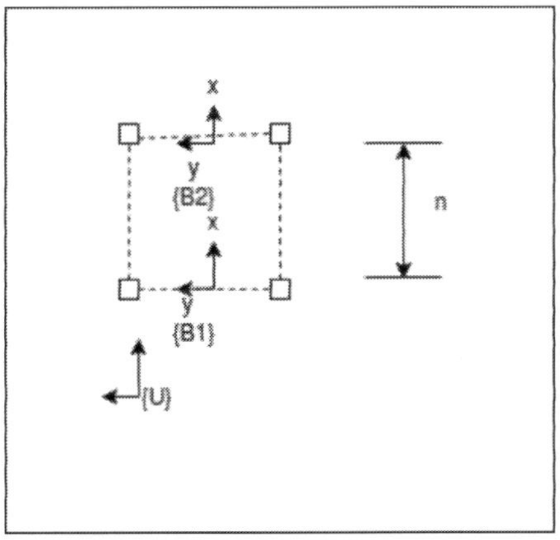

Figure 3. Linear motion of robot.

4.1.2 $nL \neq nR$

If the wheel's direction and speed are not same, the robot will be moving in a general curved path. It could be considered that the robot rotates with respect to an imaginary frame{c} located at instantaneous central point of the rotational motion Translation of robot frame could be found by finding instantaneous centre{c}, angle of rotation(θ) and radius of trajectory of the robot (r) as in equation (3).

$$r = \frac{nL * W}{nR * nL}$$

$$\theta = \frac{nR * nL}{W} \quad (3)$$

$$^{C}T_{B1} = \begin{bmatrix} 1 & 0 & 0 & n \\ 0 & 1 & 0 & -r-(w/2) \\ 0 & 0 & 1 & 0 \\ 0 & 0 & 0 & 1 \end{bmatrix} \quad (4)$$

The transformation of robot's current position$\{B_1\}$ with respect to the centre {c}, $^{C}T_{B1}$, can be calculated using equation(4). From that, the robot's current position {B2} with respect to the centre {c} can be calculated by rotating the {B1} with respect to {C} about θ (figure 4).

$$^{C}T_{B2} = \text{Rot Z}(\theta) * {}^{C}T_{B1} \quad (5)$$

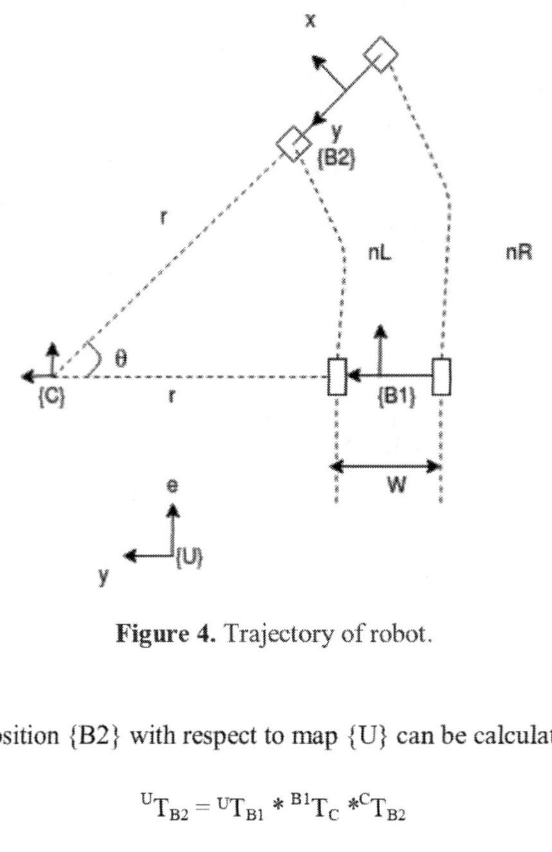

Figure 4. Trajectory of robot.

The robot's current position {B2} with respect to map {U} can be calculated as shown below

$$^{U}T_{B2} = {}^{U}T_{B1} * {}^{B1}T_{C} * {}^{C}T_{B2} \quad (6)$$

Where,

$^{U}T_{B1}$ = Robot's previous position (Initially set to identity matrix)

IOP Conf. Series: Materials Science and Engineering **995** (2020) 012037 doi:10.1088/1757-899X/995/1/012037

$$^{B1}T_C = (^{C}T_{B1})^{-1}$$

Thus, using equation (6), the robot can be localised within the map and $^{U}T_{B2}$ will be the current transformation which will define the translation and rotation of the robot with respect to the frame {m}.

4.2 *Mapping*

Mapping is the process of generating a map of the environment which can be used to navigate within the same. In this work, grid mapping is used i.e. the obstacles in the work environment are mapped in matrix of dimension 1000X1000. The range data provided by the LIDAR will be based on the frame {L} attached to the LIDAR(figure 5).

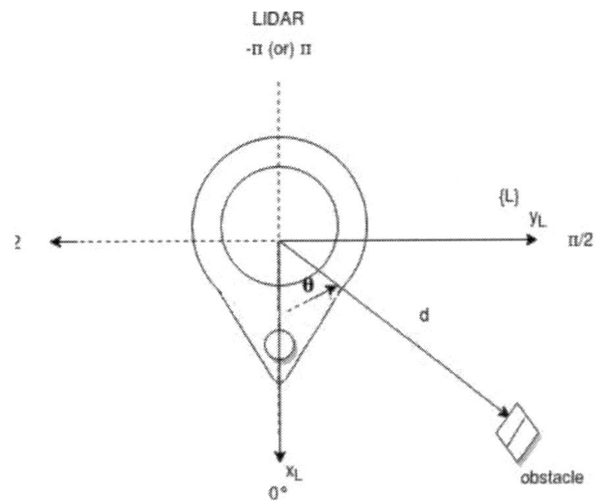

Figure 5. Illustration of frame{L}.

According to the frequency of the LIDAR, the n samples (spaced equally per revolutions) are retrieved. Angle and distance d defines the exact position of the obstacles in the environment.The map will be stored as a 2-D array which will have a height h and width w. In order to generate the map some assumptions are made and those are the following,

- Initially the robot is placed at the centre of the map and mapping starts from there.
- Each cell will represent the distance of resolution(m) per cell of the array.
- Array index starts from 0.
- Number of columns defines width and number of rows defines the height of the map.
- A frame {G} is attached to start of the map (i.e. row = 0 and column = 0).
- A frame {m} is attached to the centre of the map (row = height/2 and column = width/2).
- Robot's transformation is recorded with respect to this frame.

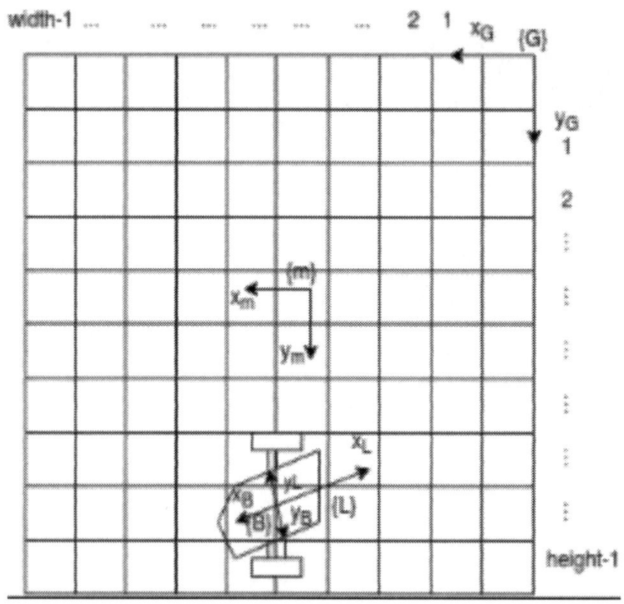

Figure 6. Representing bot's position in map.

The map can be imagined as shown in the above figure 6. Thus resolution affects the accuracy of the map considerably.

$$\text{Resolution} \ \alpha \ (accuracy)^{-1} \tag{7}$$

The resolution, height and width of the map determines the size and of the map. For each revolution the distance points are retrieved along with the angle. The corresponding x, y coordinate of the obstacle can be retrieved using the equation (8),

$$X = \frac{d}{resolution} * \cos\theta$$

$$Y = \frac{d}{resolution} * \sin\theta \tag{8}$$

These points are based on the frame {L} and in order to update into the map, the points need to be represented in the frame {B}. $^{U}T_{B}$ will also need to be converted from distance to cell conversion by dividing the xo position by revolution before proceeding.

Let us assume a obstacle (x) located at coordinate at m and n with respect to {L}

$$Lx = \begin{bmatrix} m \\ n \\ 0 \end{bmatrix}$$

For each revolution 'N' number of points which will be retrieved and we can represent it in frame {L} as,

$$LP = \begin{bmatrix} m1 & m2 & m3 & \dots\dots\dots & mn \\ n & n2 & n3 & \dots\dots\dots & nn \\ 0 & 0 & 0 & \dots\dots\dots & 0 \end{bmatrix}$$

Since the LIDAR is mounted on the top of the robot rigidly, {L} frame's transformation with respect to {B} frame will be a constant.

$$BTL = \begin{bmatrix} -1 & 0 & 0 \\ 0 & -1 & 0 \\ 0 & 0 & 1 \end{bmatrix}$$

Both frames origin coincide since both the LIDAR and the robot is assembled in such a way as shown in the map visualization frame {M} is fixed rigidly on the frame {G} and hence the transformation will be

$$^{G}T_{M} = \begin{bmatrix} 1 & 0 & width/2 \\ 0 & 1 & height/2 \\ 0 & 0 & 1 \end{bmatrix}$$

To convert the distance from the {C} frame to the {G} frame, equation (9) is used.

$$G_P = {}^{G}T_M * {}^{M}T_B * {}^{B}T_L * L_P \quad (9)$$

The first row of the G_P matrix will represent the columns of the array and the second row of the G_P matrix will represent the rows of the array where the distance points from the LIDAR is converted with respect to the grid on applying the above transformations. Thus, the data from the array can be updated after this conversion.

To generate the map, assume the maximum range for the LIDAR (i.e. above which the range data will be discarded).In the map, the obstacle will be marked as 0, floor area will be marked as 1 and unexplored will be marked as -1.Initially the whole map will be set as unexplored since the exploration is not yet started. As the robot moves in the area, the LIDAR data is used to find out the obstacles at that particular time. Since the obstacles around the robot may be dynamic, the presence of object in the map must be checked every time when the obstacles come into the vicinity of LIDAR range.

Each and every time the map has to be updated by the latest LIDAR distance data of obstacles. For this a circular region of radius LIDAR maximum range will be replaced by newly computed data based on the following equation (10).

$$\text{New Update} = (\text{old obstacle} > \text{New floor}) \,||\, \text{New obstacle} \quad (10)$$

4.3 Calculating old obstacle

Obstacle is calculated by taking the values from the map for only the circular region of radius of current maximum LIDAR range and robot position (figure 7).

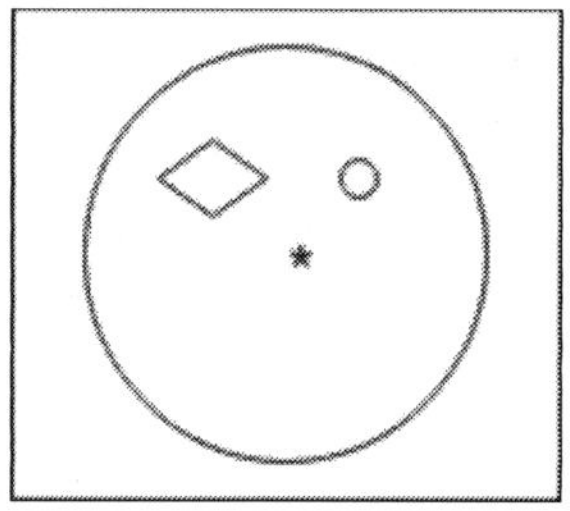

Figure 7. Calculating old obstacle.

4.4 Calculating new floor

New floor area can be calculated by altering the screen range dta and the altered data can be used to find the boundary of the floor. The LIDAR range value having greater than the LIDAR maximum range is replaced with the LIDAR maximum range (figure 8).

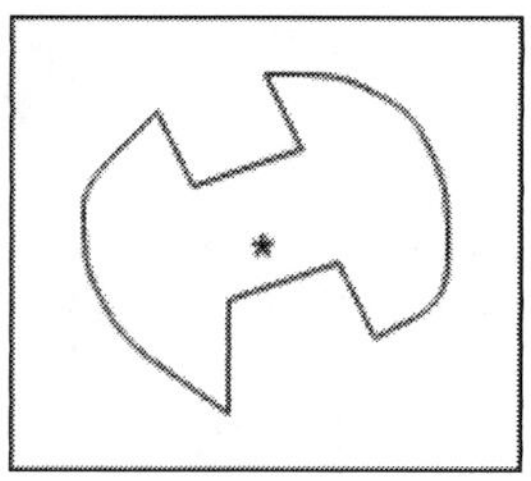

Figure 8. Calculating new floor.

By connecting all the data points, a polygon is made within which all the values are 1 and outside the polygon all the values are 0. This is the new floor data we needed.

4.5 Calculating new obstacles

New obstacles will be calculated by discarding the LIDAR range scan distance data greater than the LIDAR maximum range and setting the data points with the value '1' and the other points as '0' (figure 9).

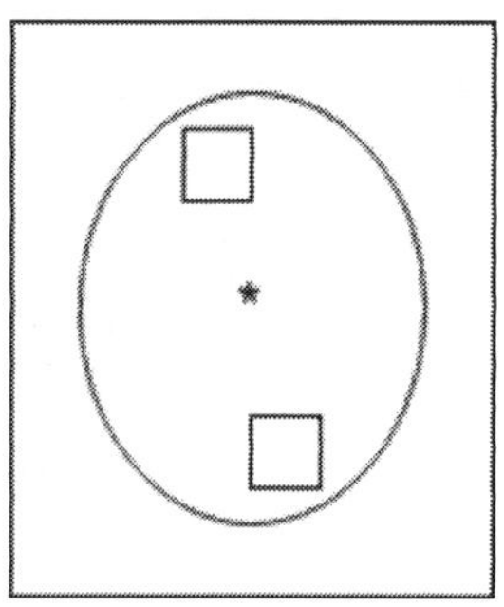

Figure 9. Calculating new obstacle.

4.6 Updating map with new obstacle

Thus, the new value to be updated can be calculated. For uncovered area, the older uncovered data and the new uncovered data are taken and binary AND operation is performed to compute the newer uncovered area. Thus, the value is updated with the newer circular region (figure 10)

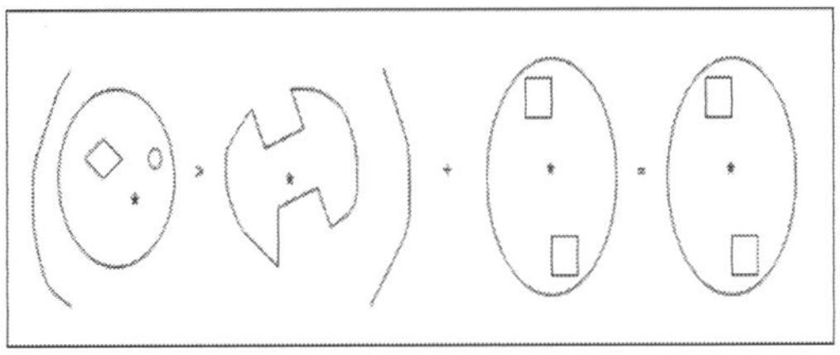

Figure 10. Combining all together.

4.7 Path finding

In the generated map, the robot has to navigate in order to explore the unexplored area in the map and to reach another point in the map. In order to achieve this task we are in need of a path from source(robot's current location) to the destination point.

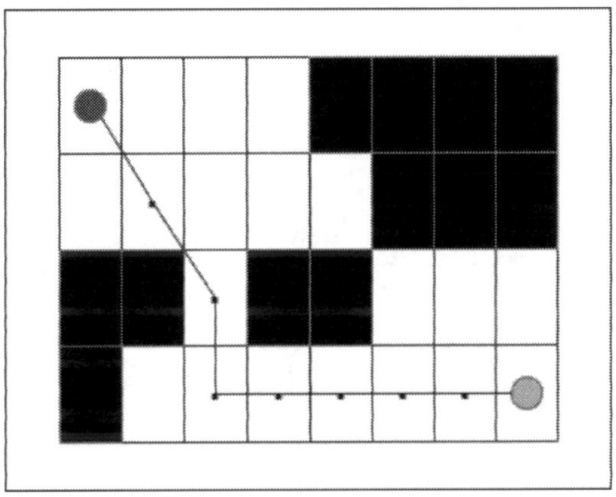

Figure 11. Path finding.

The path can be traced using A* algorithm which is build according to our map representation (figure 11). The algorithm iterates through the possible oaths and finds the appropriate and shortest path among them. A* algorithm is an extension of edsger wybe dijkstra's algorithm which uses heuristic algorithm to estimate the cheapest or shortest path to the destination point from the current position by avoiding the other possible points.Using the path the robot can navigate within the map by calculatingthe required location.

5. Result and analysis

The end results demonstrated that the autonomous mobile robot built using 2D LIDAR generated map with comparatively higher accuracy when compared to the map generated using ultrasonic sensor's data. It is found that the angular divergence of ultrasonic sensor is 15 deg where as when LIDAR is used the divergence come to below 1deg.

The position of the obstacle is well defined in the currently generated map, whereas in the ultrasonic sensor it depicts the possibility of obstacle's presence within its entire angle of divergence range(figure 12).

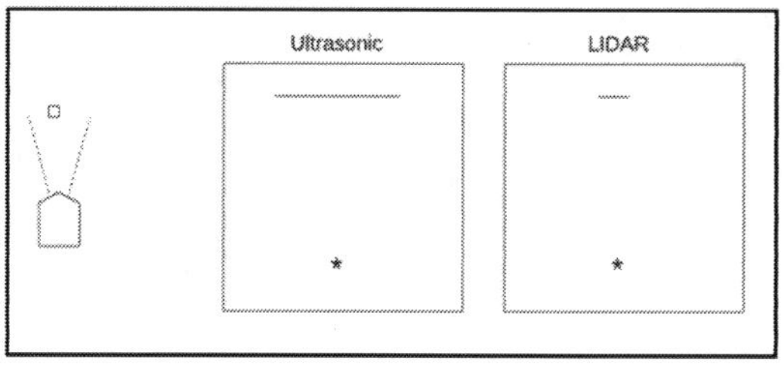

Figure 12. Map detecting the obstacle probability of LIDAR and ultrasonic.

Table 1 Encoder data analysis.

Trials	BOT (X axis in m)	BOT (Y axis in m)	Encoder output (X axis in m)	Encoder output (Y axis in m)
1	1.77	0.17	1.78	0.16
2	1.86	0.27	1.89	0.20
3	1.91	0.40	1.93	0.38
4	1.57	0.24	1.57	0.25
5	1.54	0.23	1.54	0.24
6	1.66	0.30	1.65	0.29
7	1.60	0.29	1.60	0.26
8	1.54	0.24	1.54	0.23
9	1.56	0.24	1.56	0.23
10	1.54	0.30	1.55	0.24

From the experiments conducted to calibrate the encoders (table 1), it is found that the positional coordinated obtained from using encoders are sufficiently accurate.

6. Conclusion

This project has described the localisation of autonomous mobile robot using 2D LIDAR. The autonomous mobile robot can be used to perform indoor applications like drug delivery, taking pictures in a hazardous environment and updating the map of that environment. This project can be further developed and can be made to explore the vast area. For achieving better accuracy IMU can be incorporated with encoder in further developments.

Due to the financial constraints, this work is limited to 2D LIDAR. Using costlier 3D LIDAR enable us to go for SLAM based localization and mapping. The localization and mapping relies on the encoder data obtained from the wheel encoders. Slippage of wheels causes error accumulation during long run. This could be reduced by using SLAM along with IMU and encoder.

Acknowledgement
We would like to express our heart filled gratitude to our Professors, fellow faculty members, Graduate engineering trainees and students, for their constructive ideas, inspirations, encouragement, excellent guidance and much needed technical support extended to complete our project.

References

[1] Genghang Zhaung, Shengjie Chen, Jianfeung and Kai Haung, , (2017), A Real-Time Embedded Localization in Indoor Environment using LiDAR Odometry". National Conference on Embedded System Technology, pp. 210-222. *Springer*, Singapore

[2] Ilya Afanasyev,Artur Sagitov, and Evgeni Magid: "ROS-Based SLAM for a Gazebo- Simulated Mobile Robot in Image-Based 3D Mode of Indoor Environment". International Conference on Advanced Concepts for Intelligent Vision Systems, pp. 273-283. *Springer*, Cham, (2015).

[3] John Kerr and Kevin Nickels: "Robot Operating Systems:Bridging the Gap between Human and Robot" 44th IEEE South eastern Symposium on System Theory in the proceedings of (2012).

[4] Hugh Durrant-Whyte, and Tim Bailey: "Simultaneous Localisation and Mapping (SLAM):Part I The Essential Algorithms" IEEE International Conference, *Robotics and Automation magazine*, (2015).

[5] Jie Hu, Wang gen Wan and Xiaoqing Yu: "A Pathfinding Algorithm in Real-time Strategy Game based on Unity3D" audio, language and image processing of international conference in the year (2012).

[6] Biswas J and Veloso M.: "localization and navigation for autonomous indoor mobile robots". International journal of Computing Communications & Instrumentation Engineering (2015).

[7] Wang C C, Thorpe C and Thrun S: "Online simultaneous localization and mapping with detection and tracking of moving objects", IEEE Conference on Robotics & Automation (2013), *IEEE*.

[8] Deans M and Hebert M: "Experimental comparison of techniques for localization and mapping using a bearing-only sensor". In International Conference on IEEE on Robotics and Automation (Cat. No. 03CH37422), vol. 1, pp. 842- 849. *IEEE*, (2015).

[9] Kim J and Sukkarieh S: "Airborne simultaneous localisation and map building". In IEEE International Conference on Robotics and Automation (Cat. No. 03CH37422), vol. 1, pp. 406-411. *IEEE*, (2013).

Particle swarm optimization of tensile strength on treated Bermuda grass Fiber as reinforcement in polyester composites

M.Raghunah[1], K.Nijandhan[1 *], R.Dhanasekar[2], Sukesh P, Nishanth A, Megantha P[3]

[1]Department of Mechatronics, [2]Department of Electronics and Instrumentation Engineering,

[3]Department of Electrical and Electronics Engineering

Bannari Amman Institute of Technology, Sathyamangalam, Tamilnadu, India

nijandhan@bitsathy.ac.in, raghunath@bitsathy.ac.in, dhanasekar@bitsathy.ac.in,

sukesh.mc18@bitsathy.ac.in, nishanth.ec18@bitsathy.ac.in, megantha.ee18@bitsathy.ac.in

Abstract. The tensile strength of chemical treated Bermuda grass fiber reinforced polymer composites were assessed in this work. This investigation enhanced the binding nature of the Bermuda grass fiber through NaOH treatment, which is used as reinforcement in composite material. The difficult of reduced compatibility among fiber and the resin matrix has been diluted by creating rough surface of Bermuda grass fiber through sodium hydroxide (NaOH) solution. The result of surface modification parameters on the tensile strength of Bermuda grass-polyester composite was analyzed and found that it influence majorly in the tensile strength of composites. The value of 19 MPa, was found as maximum tensile strength through NaOH alkali treatment. The Response Surface Methodology (RSM) approach was used to model the mechanical strength of treated Bermuda Grass-polyester composites. The mathematical equation formed using RSM and the Particle Swarm Optimization (PSO) technique is used to find the better mechanical properties.

Keywords: Bermuda grass fiber, polyester, Compression moulding Tensile strength, Particle Swarm Optimization

Introduction

Many of the researchers found the natural fibers are potential alternatives to artificial fibers which available in the market. The natural fiber-polymer composites have the advantages of light weight, high stiffness, less density, ecofriendly, low cost, less tool wearing etc. These unique characteristics are kindle the designer to select the natural fiber composites for their design. The natural fiber composites are exploited for viable applications such as Automobile, Construction, Aerospace, Sports etc. [1] The primary advantages of natural fibers are low density, plentiful in nature, and low cost, easy processing etc., In addition to that they have less energy consumption, and easy handling [2]. Many researchers have been carried outwork on the altered kinds of natural fibers such as jute, flax, coir, bamboo, and sisal and the possibilities of utilization as lightweight materials are studied [3-6]. Unfortunately, the performance of natural fiber as a reinforcement in polymer composites is insufficient and not similar even with other synthetic fibers due to lack of interfacial bonding strength between fiber and polymer. The Problem of poor interfacial strength between fiber and matrix is attributable to existence of superfluous non cellulosic content in the fiber exterior part. The removal of this extra lignin through surface modification process enhances the surface roughness and increase the bonding strength between fiber and polymer [7]. The use of surface modified fiber as reinforcement which gives significantly improves the mechanical strength of composites [8-9]. This investigation is objectives to increase the mechanical strengths of Bermuda Grass-polyester composites by means of NaOH treatment process parameter namely NaOH alkali solution concentration and fiber soaking time. The optimal process parameters were selected for maximum tensile strength using Swarm intelligence.

1. Experimental Procedure

1.1 Materials and Methods

In the present study, the Bermuda grass fiber used as reinforcement which is available in the local places of sathyamangalam. The used fibers was extracted by means of simple water retting process and it was treated with NaOH aqueous solution with the fiber treatment parameters namely; soaking time and solution concentration as shown in the Table.1.

Table 1. Full factorial design levels

Parameters	Solution Concentration %	Soaking Time (Hrs)
Level1	2	24
Level2	4	48
Level3	6	72
Level4	8	96
Level5	10	120

During the fabrication, the fibers keep in the random orientation in the resin system containing of isopathalic polyester resin, cobalt actuate as acted accelerator and MEKP as acted catalyst in the ratio of 1:0·015:0·015. The compression molding machine was used manufacture the composite plate in the dimension of 300 × 300 × 3 mm^3 and pressed at a force of 0.5 kN. After 1 hour, released the pressure and fabricated composite plate was recover from mold and the plates were permitted to cure at room temperature for 24 hours. The composite specimens were cut to ASTM -D638 standards and were tested for its tensile properties. The five tensile samples were tested for every plate and stated the average value for the statically significant result.

2. Mathematical modeling and Optimization.

Using RSM, the experimental data obtained from tensile testing of Bermuda Grass-polyester composite, the nonlinear regression equation was formed. The mathematical equation formed by RSM, was optimized using non heuristic Particle Swarm Optimization (PSO) Technique. The parameters assigned during optimization of tensile strength using PSO parameters are given (Table.2).

Table 2 The parameters used in PSO

No of particles	100
Penalty	300
Multiplication parameter	1.2
No. of cycles	5
No of iterations	100

3. Result and Discussion.

3.1 Effect of NaOH alkali treatment parameters on the tensile strength of Composites

Figure.1 shows the association between the alkali treatment parameters and tensile properties in MPa. The maximum tensile property was obtained in all the levels of the 72 hours soaking time. After 72 hours the strength of composites were dwindled, because of higher elimination of lignin, pectin, wax and ash etc., material from the fiber may weaken the fiber. The same kind of situation was happened in the solution concentration after 6 %. More solution concentration may damages the fiber by excess removal of non-cellulosic content and less concentration may remove the less amount of non-cellulosic contents available in the fiber which could not sufficient to develop the enough interfacial strength. A less value of the tensile property of 5 MPa was attained in 24 hours and 2% of solution concentration, and a superior range of tensile property of 19 MPa was obtained in 72 hours and 6 % of solution concentration.

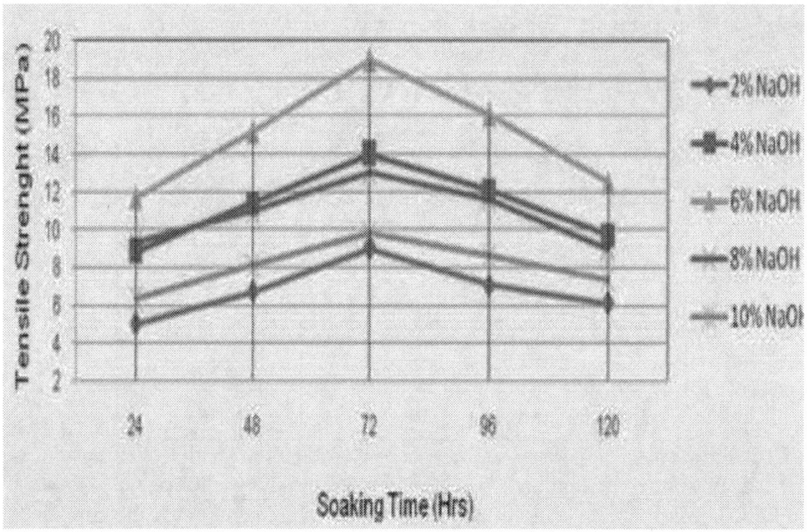

Figure.1 Tensile strength of Bermuda grass fiber reinforced composites

3.2 Mathematical Modeling Using RSM.

The regression equation of alkali treated bermuda grass fiber composites after tensile tested was identified. The response *(t_s)* and the input parameters (alkali process parameters) were obtained from the coefficients resulting from the Design experiment 8.0.4 software. Table 3 shows the mathematical equation for tensile strength model.

The equation was identified mathematically significant or not using co efficient of correlation (R^2). The coefficient of determination (R^2) value shows high range (0.88) which indicate the correctness and accuracy of predicting of the regression equation.

Table 3 .The nonlinear regression of tensile properties

Response	Mathematical Equation	R^2 Value
ts	$-8.77500+4.88379\ f_1+0.24682f_2-4.79167E-004\ f_1f_2-0.39536\ f_1^2-1.63442E-003\ f_2^2$	*0.88*

3.3 Optimization using PSO algorithms

The mathematical equation formed by Response Surface methodology technique, was optimized using PSO Technique. PSO is used to find the solution in the n-dimensional solution space. The each iteration observes the fitness value them and near particles also. The PSO is suitable for the where the solution in global maxima and local maxima. It is a simple method to find the mimic of flocks bird and it find the food searching process. They involved to find thir food in a simple and easiest method. It has been observed and converts as coding to find the better alkali treatment process for the better mechanical properties. The algorithm of PSO has been given in Figure 2.

```
For each particle
   Initialize particle
END

Do
   For each particle
      Calculate fitness value
      If the fitness value is better than the best fitness value (pBest) in history
         set current value as the new pBest
   End

   Choose the particle with the best fitness value of all the particles as the gBest
   For each particle
      Calculate particle velocity according equation (a)
      Update particle position according equation (b)
   End
While maximum iterations or minimum error criteria is not attained
```

Figure 2. The working of PSO algorithm

PSO was established by James Kennedy and Russell Eberhart in the year of 1995.. It is related to evolution-inspired problem solving methods such as genetic algorithms. The PSO algorithm accurately predicted the maximum value of tensile strength. This could be inferred from the Table. 4 shown below.

Table 4.Optimul values

Parameters	Value
Solution Concentration (%)	6.04
Soaking Time (Hrs)	73
Tensile Strength (MPa)	19.005 (PSO-Prediction)
	19 (Experimental value)

The PSO plot obtained for the tensile property model of treated Bermuda grass fiber - polymer composites is shown in the Figure.3

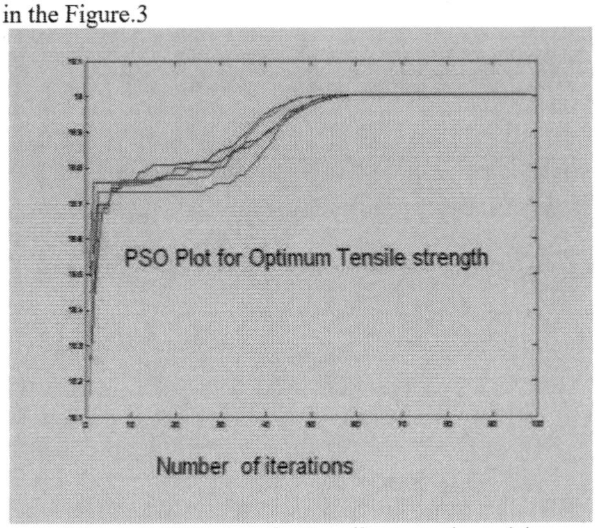

Figure.3 PSO plots for Tensile strength model.

4. Conclusion

The tensile properties of treated Bermuda grass fiber - polyester composites were evaluated as per ASTM D638 standard in this investigation. The surface modification process of Bermuda grass fiber in 6% aqueous solution for 72 h gave in 26.3 % improved in tensile strength. Hence it can be proved that NaOH surface modification process can successfully improve the mechanical strengths of Bermuda grass fiber - polymer composites The mathematical equation for predicting tensile strength over the different environments was formulated using RSM technique. The extreme value of tensile strength for arrangement of process parameters were obtained using PSO algorithm. The PSO algorithm accurately anticipated the maximum value of tensile strength.

References

1. Wallenberger, F. T., Weston, N.: Natural Fibers, Plastics and Composites Natural, Materials Source Book from C.H.I.P.S. Texas, (2004).
2. U.U. Modibbo, B.A. Aliyu, I. I. Nkafamiya and A. J. Manji, The effect of moisture imbibition on cellulosic bast fibres as industrial raw materials, In J Phy Sci 2007; 2 (7):163-168.
3. Satyanarayana, K. G., Sukumaran, K., Mukherjee, P. S., Pavithran, C., Pillai, S. G. K.: Natural Fiber–Polymer Composites, J Ceme. and Conc. Compos. 12 (2), 136 (1990).
4. Mansur, M. A., Aziz, M. A.: Study of Bamboo-Mesh Reinforced Cement Composites, Int.Ceme. Compos. and Lightweight Conc, 5(3), 171 (1983).
5. Gowda, T. M., Naidu, A. C. B., Chhaya, R.: Some Mechanical Properties of Untreated JuteFabric-Reinforced Polyester Composites, J. Compos. Part A: App. Sci. and Manuf. 30(3),284 (1999).
6. Mishra S, Mohanty AK, Drzal LT, Misra M, Parija S, Nayak SK, Tripathy, SS (2003) Studies on mechanical performance of biofiber/glass reinforced polyester hybrid composites Compos. Sci. Technol. 63:1377–1385.
7. Harish S, Peter Michael D, Bensely A, Mohan Lal D , Rajadurai A(2009) Mechanical property evaluation of natural fiber coir composite. Materials Characterization 60:44-49.
8. Monteiro SN , Terrones LAH , D'almeida JRM(2008) Mechanical performance of coir fiber/polyester composites Polymer Testing 27:591– 595.
9. Mishra S, Mohanty AK, Drzal LT, Misra M, Parija S, Nayak SK, Tripathy, SS (2003) Studies on mechanical performance of biofiber/glass reinforced polyester hybrid composites Compos. Sci. Technol. 63:1377–1385.

Monitoring And Control Of Cooler Tank Level Measurement In Gas Cleaning Plant (GCP)

Ms.P Rajalakshmi[1]& Dr.M Ponni Bala [2], Mr.C Aravind [3], Dr. S J Suji Prasad[4]

1P.G Scholar, Kongu Enginnering College, Perundurai, Erode, Tamilnadu, India.
2,3,4 Department of Electronics and Instrumentation Engineering, Kongu Engineering College, Perundurai, Erode, Tamilnadu, India.

ponnibala@kongu.acin, rajikavi5398@gmail.com, sjsujiprasad@kongu.ac.in, aravind@kongu.ac.in

Abstract. The Cooler Tank Level plays a major role in Gas Cleaning Plants (GCP) and it is a very important parameter to be measured and controlled. If the tank level is not maintained in nominal range, the dust particles will accumulate on the Cooler Tank and affect the process. The production of copper metal at Vedanta, Thoothukudi include Blending of various grades of copper concentrates, Smelting and Refining. Sulphur dioxide is generated from during the smelting of copper and is converted to Sulphuric acid as a byproduct. This process involves three steps such as cleaning of sulphur dioxide gas from Furnace, Catalytic conversion of S02 to SO3 and amalgamation of SO3 with sulphuric acid. The conversion process is exothermic, reversible and adiabatic reaction and hence if the temperature increases it destroys the formation of sulphur trioxide. Hence in order to achieve high conversion efficiency, the converter gases need to be cooled. The SO2 gas from smelter contains metallic dust, fume, acid mist, water vapour and other impurities like halides which will contaminate the product. Therefore the gases are to be cooled and cleaned in Gas Cleaning Plant (GCP). Cooler tank used to remove water vapour from the gas and dust settle at the bottom of cooler. Cooler tank level needs to be maintained in order to avoid overflow of water into scrubber which will block the inlet gas. The level of cooler tank is to be continuously monitored and controlled to achieve the optimum results. The limitations of using Differential Pressure Transmitter for level measurement are overcome by ultrasonic based level sensor and automated using Distributed Control System and Human Machine Interface (HMI).

Key Words: Gas Cleaning Plant, Cooler Tank Level control, HMI, Ultrasonic level sensor, Distributed Control System.

1. Introduction

In Industries, the storage tank holds various liquids such as oil, byproducts and other chemicals. These processes needs a systematic monitoring of liquid level in tank and detects the exact quantity of level in storage tanks [1]. In an inner part of coolant tank, the level it occupies and the flow rates they cross are independent variables. Therefore it is indispensable to control its level in the tank. There is a prerequisite of controller for the adjustment of level and maintains constant level in the tank.[2, 3]. Normally, the estimation of liquid level is done by various sensors but there needs a contact with the liquid. An existing methodology uses a level sensor in an interior part of a coolant tank in Gas Cleaning Process, mounted on it. However this method has poor efficiency and it comprises only a single sensor. The storage tanks are usually spread among huge surfaces and use manual identification and are more ridiculous. Therefore time consumption to monitor the liquid level in the tank is too high[4].Therefore remote monitoring are necessary to collect information from the coolant tank. As a result , it is essential to attain higher accuracy [5]. The most frequently Ultrasonic sensors does this. The data acquisition senses the difference in the level and its stored [6]. A server gather information from the DCS system and output will exhibit in a LCD module. It will display the position of liquid in the tank and its volume. [7-10] .

The endeavour for the design of gas cleaning plant are persistent on the difficulty of dust particles catching[11]. In gas cleaning plant it contains a broad range of dust particles that cannot be easily detached. Therefore the gas cleaning plant uses devices like vibration generators i.e., ultrasonic Sensors and programs for monitoring and predicting resonant conditions of different surface area in the tank[12].Due Installation of transducer would provides a better results[13]

Hence there is a necessary to automate the plant thereby to avoid those drawbacks. The proposed methodology is to implement DCS to control the level of cooler tank. For that the stimulation is done using Yokogawa CENTUM Vp.

This paper is structured as follows as. Section 2 introduces the existing system, about level measurement. In section 3 the DCS Simulation and control algorithm is outlined. Experimental setups and Results are projected in section 4. Finally, conclusions and future scope are presented in section 5.

2. Existing Method

The gas coming from smelter contains many impurities like moisture, solid particles, dust, etc. During the cleaning of gas passes through mixing chamber, quencher, non-retention vessel, venturi scrubber, cooler dehumidifier, primary ESP, secondary ESP and finally Sulphuric Acid Plant. The block diagram of the cleaning process in Sulphuric acid plant is shown in the Figure 1.

IOP Conf. Series: Materials Science and Engineering **995** (2020) 012039 doi:10.1088/1757-899X/995/1/012039

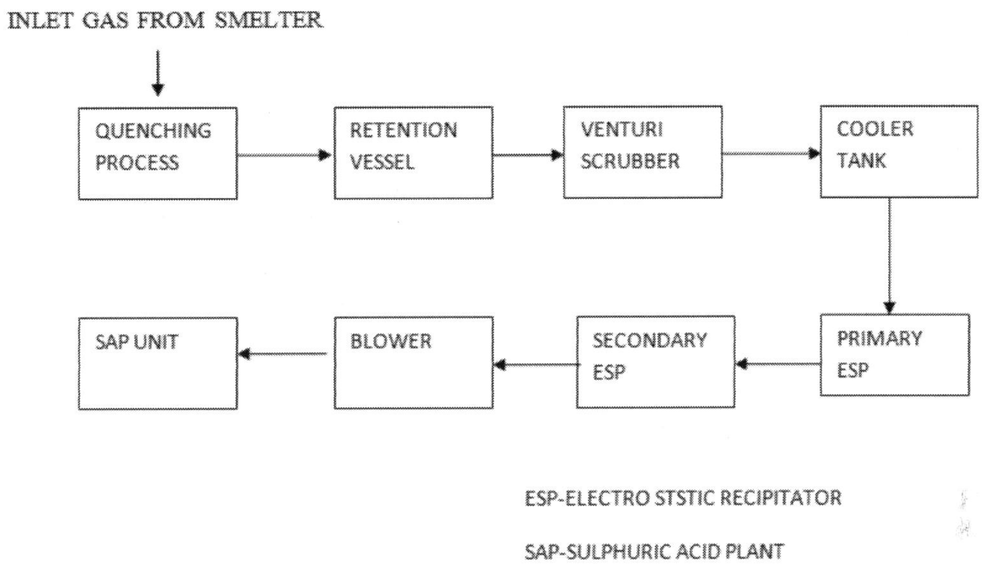

Figure 1 Block Diagram for Gas Cleaning Plant

2.1 Cooler Tank Level Measurement Using Differential Pressure Transmitter (DPT)

In Gas Cleaning Plant, cooler is the heart of the process because it only removes the water vapour from the gas. During suction by blower- negative pressure impact inside cooler-water sprayed on top of cooler. The sprayed water contact with Inlet gas (65DegC) -remove the water vapor and dust present in the gas reduces the temperature to 35Deg C. Dust water collected at the bottom of cooler. Level of water inside the cooler should be maintained at a height of 1meter to avoid water overflow into the venturi scrubber for non blocking the inlet gas.The schematic diagram of DPT level measurement shown in Figure 2.

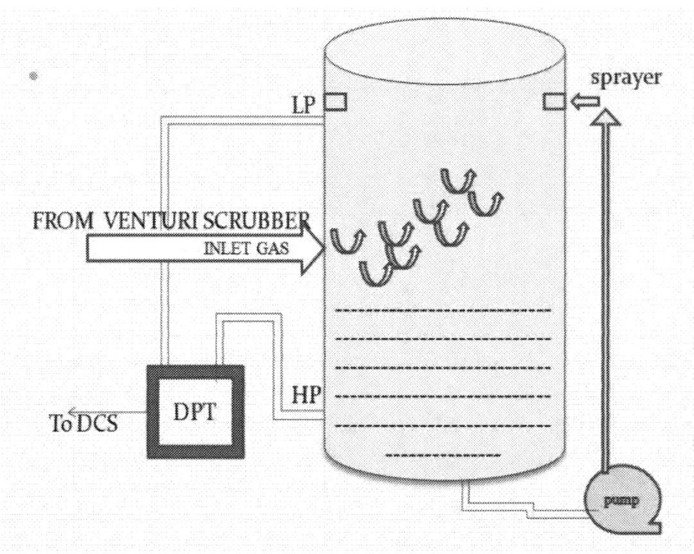

Figure 2 Schematic Diagram of DPT based Level measurement

2.2 Disadvantages of Existing Method

The disadvantages of existing method is Inaccurate level measurement in DPT due to spray water flow into LP side in cooler, Calibration range varies often due to this problem, In order to avoid the above said limitation, Water filled manually inside the LP side of DPT. There is an inaccuracy due to suction pressure by blower.

3. Ultrasonic Based Level Measurement

The drawbacks in using conventional level measurement techniques are eliminated with the help of ultrasonic based level sensor is used for level measurement in cooler tank. This sensor mainly reduces the inaccuracy and produces higher efficiencies. Ultrasound can be used to measure the wind direction, the speed, the tank or channel fluid level, and the speed through air or water. The speed and direction can be measured with the help of multiple detectors and by measuring the relative distances speed will be calculated. To determine tank level, the sensor detects the distance to the surface of fluid. The schematic diagram of proposed method of Ultrasonic sensor based Level measurement is given in Figure 3.

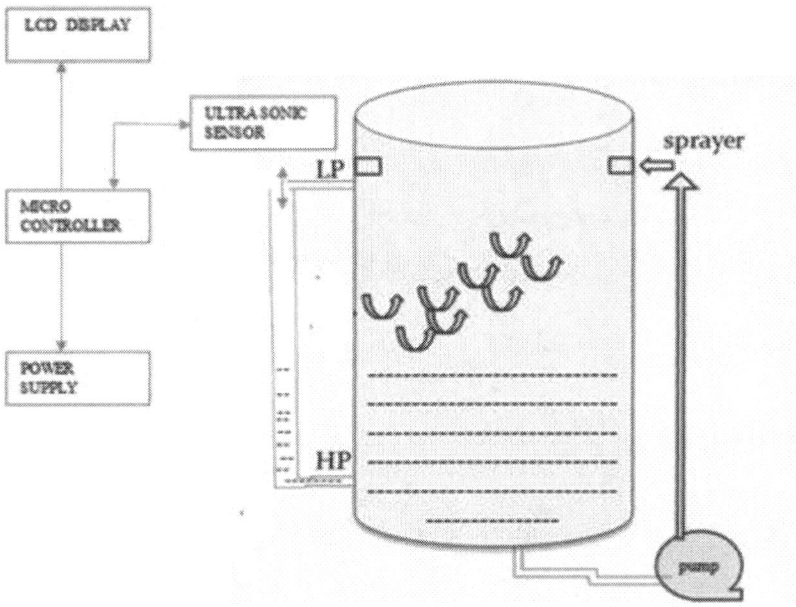

Figure 3 Schematic Diagram of Ultrasonic sensor based Level measurement

Ultrasonic sensor is used in side column of the tank for measuring the level of the tank. The ultrasonic pulse transmitted to side column of the tank and reflected. The reflected pulse distance is measured and displayed by using micro controller. The level was continuously measured by measuring the speed and time travel of ultrasonic pulse. The output of level was indicated by LCD DISPLAY.

3.1. Ultrasonic Sensor

The HC-SR04 type of ultrasonic sensor utilizes a sonar to start distance to object as such of bats do. It provides tremendous exposure without physical contact with high level of accuracy. It is not affected by any rays i.e mainly ultraviolet rays. The ultrasonic pulse is accurately reflected by water and the level is measured easily.

3.2 HC-SR04 Ultrasonic Module Working

The basic principle of the sensor uses an IO trigger for about 10us signal. Therefore, the Module automatically detects whether the signal is back or not. The working diagram of the module is shown in Figure 4.

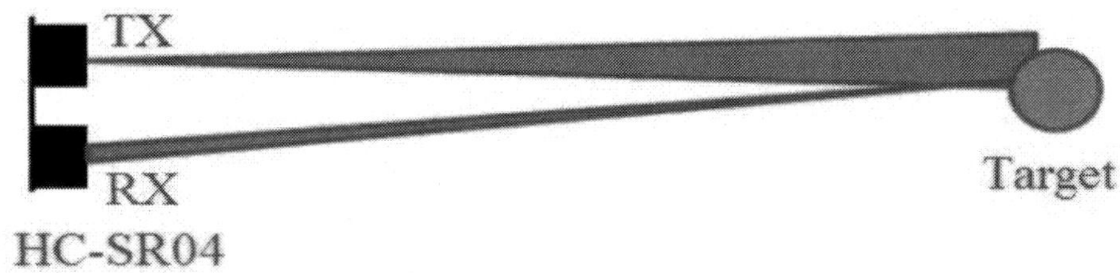

Figure 4 Ultrasonic Module Operations

3.3 Magnetic Float Switch

The level of water in the tank is continuously measured ultrasonic sensor and display unit. For controlling the level of water in tank, two float type magnetic switches connected at top and bottom of the tank. The schematic diagram of float type magnetic switch is shown in Figure 5.

Figure 5 Magnetic Float Switch

3.4 Distributed Control System

A Distributed Control System (DCS) is a digital control architecture which monitor and control the entire process and operations of various plants and series of process in Industry. It has autonomous controllers and supervisory data acquisition system in a central control room which can be able to control, optimise and minitor the complete operation in a plant. Human machine Interface and controllers in DCS system used to increase reliability of the process and minimise installation costs. Monitoring from distant places and supervisory control are achieved with the help of HMI and DCS.

- Here level was controlled by using Distributed Control System. When the level goes high, magnetic switch turned ON the solenoid valve opened to drain water in the tank by using DCS.
- If the level falls down, another solenoid valve is activated to fill the tank.

4. Results and Discussion

The Function Block Diagram (FBD) is a programming language for DCS used to design sequential and logical programme of various input and output parameters. With the help of the set of predefined blocks the sequential operation is programmed in CentumVp DCS system. The input and output parameters used for the process are defined by the basic blocks and connected through connection links both by software and hardware. The simulation of program was created using CENTUM VP software. And the hardware are interfaced with DCS input and output ports. The function diagram with the sequence table of measurement and monitoring of level using controller is shown in Figure 6.

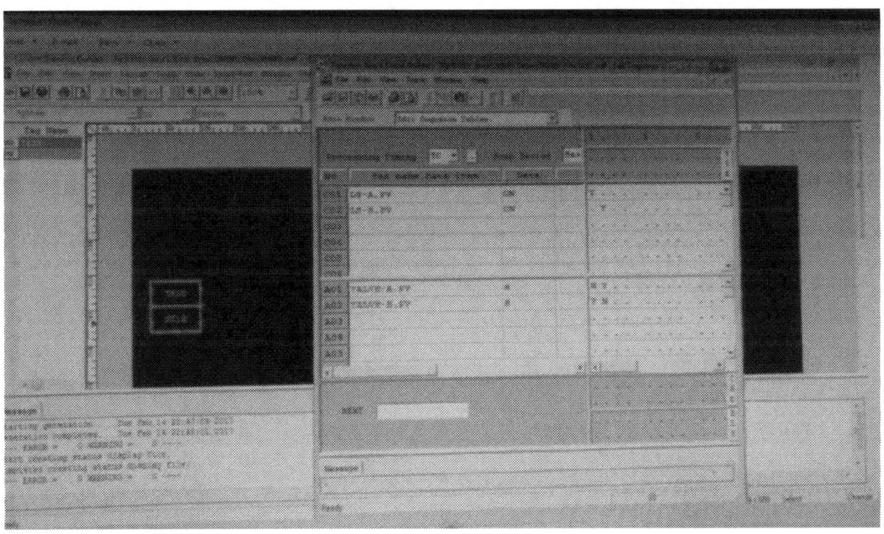

Figure 6 Function Block Diagram and Sequence Table

The accurate values of the variable Gas cooler tank level is measured using ultrasonic based level sensor and the sequential function of float switches 1 and 2 with various level is given in the following Table 1. The results of tank level control by Sequential operation using DCS is shown in Figure 7.

Table 1: Results of various Tank level and valve action

Gas Cooler Tank Level (cm)	Condition	Action
> 45 cm	Float switch 1 energise	Solenoid valve 2 (SV2) open
> 2cm to < 45 cm	Float switches 1 and 2 de-energise	SV1 and SV2 close
< 2 cm	Float switch 2 energise	Solenoid valve 1 (SV1) open

Figure 7 Face plate View of Level Control

If the level goes high (45cm), Magnetic switch 1 is ON and drain the level of tank using solenoid valve. If the level falls down (2cm), Magnetic switch 2 is ON and fills water in the tank. This sequential process is implemented using sensors, solenoid valves and interfaced with digital I/O module of Field control System. The sequential program is developed using ST-16 in CentumVp DCS system. The Graphical system view developed using CentumVp DCS and HMI for Level monitoring and control of Cooler Tank in Gas cleaning Plant is shown in Figure 8.

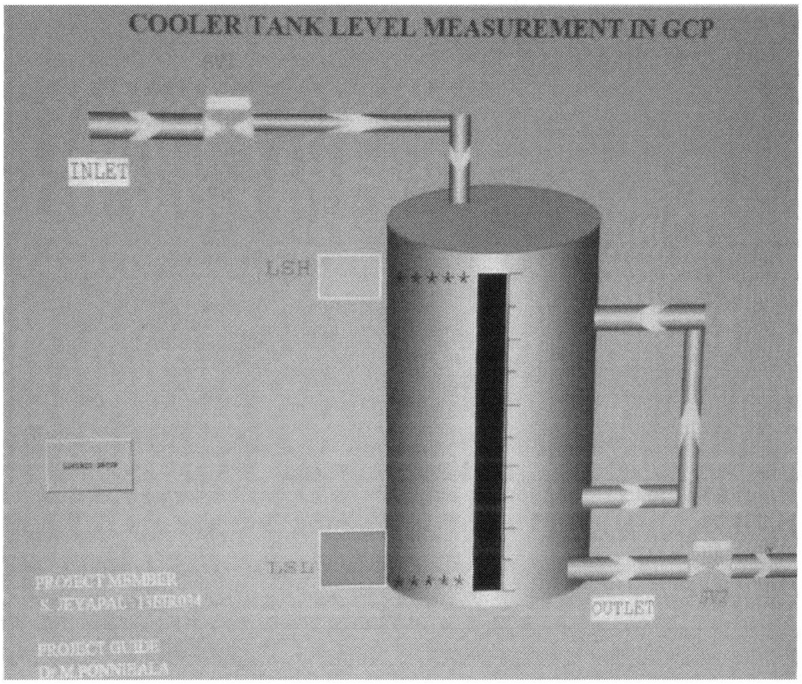

Figure 8 Graphic View of Level Control in Cooler Tank

The liquid level of the cooler tank is measured and controlled by using ultrasonic based level measurement and Distributed Control System respectively. Most of the real time chemical process system exhibits multiple steady states due to non-linear phenomena occuring in the systems. The ultrasonic based level measurement is the basic evalutionary computational method which provides a better satisfactory result. The DCS being an effective tool in automation can automate any process logically along with the usage of I/O.

The main advantages of the proposed method using ultrasonic sensor based Cooler Tank Level measurement in Gas cleaning plant are Independent to the process and vessel pressure, Reduced manpower and Most economic.

5. Conclusion And Future Scope

The level of water in a cooler tank is measured accurately using ultrasonic based level measurement. And the level is controlled and monitored by Distributed Control System and HMI in an effective manner. The proposed method is cost efficient and minimises manual work for continuous monitoring and recording. The ultrasonic based level measurement and control using distributed control system improves plant efficiency, reliability and maximum optimization of the measured variables. It also provides continuous monitoring of level in storage tank and maintains the variable at a predetermined value in response to various load and fluctuations during the process. It also totally records the data in HMI which will be useful for further design and analysis of the system during different loading conditions. DCS based sequential programming reduces complexity, improves speed of response and present accurate output within the stipulated time.

In future, this work can be extended by using the interface tank with magnetic level gauge. If a level gauge is used the wiring can be reduced and its efficiency is high. So that accurate settling time can be achieved this improves economy of the process. Advanced control algorithms such as Fuzzy based controller and Model predictive controllers can also be used with various parameters for efficient operation of the plant.

References

1. Patrascioiu, C. and G. Stamatescu, *Petroleum Fractions Liquid–Vapor Equilibrium Simulation using Unisim Design*. Rev. Chim.(Bucharest), 2015. **66**: p. 1867-1876.
2. Raduca, M., et al., *Fuzzy controller for adjustment of liquid level in the tank*. Annals of the University of Craiova-Mathematics and Computer Science Series, 2011. **38**(4): p. 33-43.
3. Elhosseini, M.A., et al., *Modeling and control of an interconnected combined cycle gas turbine using fuzzy and ANFIS controllers*. Electrical Engineering, 2018. **100**(2): p. 763-785.
4. Siddula, S.S., P. Jain, and M.D. Upadhayay. *Real Time Monitoring and Controlling of Water Level in Dams using IoT*. in *2018 IEEE 8th International Advance Computing Conference (IACC)*. 2018. IEEE.
5. Reza, S.K., S.A.M. Tariq, and S.M. Reza. *Microcontroller based automated water level sensing and controlling: design and implementation issue*. in *Proceedings of the world congress on engineering and computer science*. 2010.
6. Viswanath, S., et al., *Low-power wireless liquid monitoring system using ultrasonic sensors*. International Journal on Smart Sensing & Intelligent Systems, 2015. **8**(1).
7. Husni, M., et al. *Liquid volume monitoring based on ultrasonic sensor and Arduino microcontroller*. in *International Conference on Innovation in Engineering and Vocational Education*. 2016.
8. Kadhim, R.A., A.K.K.A. Raheem, and S.A.H. Gitaffa, *Implementing of liquid tank level control using Arduino-LabView interfaceing with ultrasonic sensor*. Kufa Journal of Engineering, 2017. **8**(2): p. 29-41.
9. Bucur, G., A.-G. Moise, and C. Popescu, *Using Ultrasonic Sensors to Develop a Wireless Level Control in Reservoirs with Liquid Chemical Products*. REVISTA DE CHIMIE, 2019. **70**(2): p. 365-372.

10. Annuar, K.M., et al., *Design and construction of liquid level measurement system.* Journal of Advanced Research in Applied Mechanics, 2015. **12**(1): p. 8-15.
11. KHMELEV, V.N., et al., *Efficiency increase of wet gas cleaning from dispersed admixtures by the application of ultrasonic fields.* Archives of Acoustics, 2016. **41**(4): p. 757–771.
12. Diaz, C.A., et al., *Liquid level measurement based on FBG-embedded diaphragms with temperature compensation.* IEEE Sensors Journal, 2017. **18**(1): p. 193-200.
13. Patrascioiu, C., *The modeling and simulation of the convection section of the atmospheric distillation plant heaters.* Rev. Chim.(Bucharest), 2016. **67**: p. 1599-1606.

A Review on Mechanical Behaviour of Aluminium Reinforced Composites

P Raghuvaran[1], M Suresh[2], S Aakash[3], M Balaji[3], K Dinesh Kumar[3], S Hari Prasath[3]

1 Assistant Professor, Department of Mechanical Engineering, Sri Krishna College of Engineering and Technology, Coimbatore, Tamilnadu, India.
2 Associate Professor, Department of Robotics and Automation Engineering, PSG College of Technology, Coimbatore, Tamilnadu, India.
3 UG Scholars, Department of Mechanical Engineering, Sri Krishna College of Engineering and Technology, Coimbatore, Tamilnadu, India.

ragumechanical@gmail.com

Abstract. Aluminium Matrix Composites (AMC) finds applications in automobiles and aerospace fields as they bear better mechanical and physical properties. Compared to available traditional materials these composites are found to have improved wear, tensile and other similar properties due to the addition of reinforcements to the matrix. The property of base alloy gets improved when it gets mixed up with different reinforcements having various properties. The effects of addition of various reinforcements on aluminium alloy were given overview in this paper. It has been noted after review that fabricated composites yield better results if they are free from agglomeration, have uniform distribution of matrix and reinforcement and ensure better bonding of fiber with matrix. It is obvious from investigations that the life and behaviour of composites are superior for Aluminium matrix composites than that of the traditionally available materials. The studies show that because of reinforcement addition the mechanical properties of the AMC will have an average improvement of at least around 10% approximately while comparing with base material alone.

Keywords: Aluminium, Composites, Stir casting, Mechanical Properties, Wear

1. Introduction

Various methods are available for fabrication of composite materials and a researcher can select one among them depending on type of matrix and reinforcement chosen. Production methods are categorized into three namely solid, liquid and semi-solid phases of fabrication [1]. Stir casting, compo casting, rheocasting and liquid infiltration are some of the liquid processing techniques available [2]. Out of various fabrication methods available, one of the less expensive methods is the stir casting method. Hence, it is used for bulk production of composites. Many researchers earlier have used stir casting technique and found to have improved wear and mechanical properties in the composites fabricated [3]. Melt stirring process is successful in achieving better bonding of matrix and particle, easy in controlling the structure of matrix, applicable for mass production, flexible, simple and inexpensive. On the other hand, non uniform distribution of reinforcement with matrix material and poor wettability are the quite common problems faced in stir casted Aluminium Matrix Composites (AMC).

In order to overcome problems related with stir casting, few techniques were followed by the researchers. Wettability of reinforcement can be improved by pre heating of reinforcement particles before introducing into the molten metal melt. This method helps in adsorption of gases from the reinforcement surface [4]. Reinforcement particles, if injected with the help of inert carrier into metal melt will result in homogeneous distribution of particles into the melt [5]. Wettability of reinforcement particle is found better with the three step stir casting method followed in the newly designed equipment by Sajjadi et al [6].

Combining harder ceramic reinforcements with metallic matrix results in metal matrix composite (MMC). Less production cost and improved properties of Aluminium composites make them best suited for marine and aerospace applications [7]. They are also used in pistons, cylinder liners and brake rotors [8]. Properties like low density, ductility and formability of aluminium and its alloys make them to choose as matrix material for MMC. The performance of composite material depends mainly on its microstructure. Distribution of reinforcement in the alloy, fibre or particulate size and shape are some other essential parameters to be noted on which physical property of the resulting composite depends [9].

For application of the fabricated composites in various areas, proper choice of matrix and reinforcement will play a prominent role. Reinforcements such as SiC, Al_2O_3, B_4C and TiB_2 are widely used in Aluminium composites as reinforcements [10].

Most of the studies on casting analyse the mechanical properties like hardness, tensile, wear and similar other properties and evaluate the microstructure of the resulting specimens. Whereas Finite Element Analysis (FEA) [11], Response Surface Methodology (RSM) and Artificial Neural Networking (ANN) [12] are the methods used for analysing the performance of welded specimens. Behaviour of AMC on using different reinforcements were discussed and reviewed.

2. Effect of SiC on AMC

Rajesh Kumar Bhushan [13] used stir casting method and provided 5, 10 and 15 wt% SiC with size of the particle from 10 to 40 µm to AA 7075 to produce AMC. Scanning electron microscopy (SEM) and energy – dispersive X-ray analysis (EDAX) tests were conducted for examining the samples and studied the interaction and distribution of SiC particles with of matrix. Hardness of samples increased with increasing SiC reinforcement from 5% to 15%. Composite sample with 10% SiC exhibited maximum tensile properties.

Abhijit et al [14] evaluated wear behaviour of Al7075 composite reinforced with SiC particles. 20 µm grain sized SiC particles at 0, 3, 6 and 9% are used for fabricating components by stir casting route. Uniform particulate mixing was observed using X-Ray Diffraction method. Wear tests were performed on all samples with a constant load of 30 N, stable sliding distance of 2000 m and varying sliding velocity as 1, 2, 3 and 4 m/s. Friction between disc and pin resulted in decrease in COF and increase in wear rate. Minimum material removal was achieved in worn out surface Al7075 composite pin having 9% of SiC.

Anshuman et al [15] used ultrasonic stir casting method for fabricating Al-SiC composites with 0, 3, 5 , 8 and 10 weight % of SiC particles. Probe was used for 5 minutes in order to disperse the reinforcement properly in the melt. Tensile, compressive, hardness and density tests were conducted. All the properties were found to be increasing with increasing wt% of SiC.

Soundararajan et al [16] mixed 20wt% SiC to A356 alloy to make AMC. They used three different methods of casting namely gravity die cat, stir cast and stir cum squeeze casting techniques to make composites. Hardness, tensile and wear tests were conducted for samples produced by three different methods and the surface morphology were analysed for the casted and worn out samples. Compared to the other two casted samples squeeze cum stir casted samples exhibited better properties.

Sonagiri Suresh et al [17] manufactured nano composite using Aluminium 7075with SiC particle measuring 50 nm. 1, 2, 3 and 4 wt.% of SiC particles are used for preparation of composite specimens. Wear test with 20, 30 and 40 N loads using round pin on a steel disc and Micro hardness test were conducted. Micro structures were examined on the worn out surfaces. In comparison to the base metal the nano composites exhibited better wear and hardness properties.

Juwen Zhu et al [18] achieved uniform distribution in SiCnp/Al6082 AMC fabricated through combined squeeze and stir casting method. Yield strength, ultimate tensile strength and elongation of SiCnp / Al6082 AMC were examined and found to be increasing than that of matrix material alone. The compactness and grain refinement improved for the stir casted SiCnp / Al6082 composites.

Jaya Prasad Vanam et al [19] studied the effect of mechanical, microstructure and wear properties of Al5083-SiC composite. Stir casting process is used for preparing AMC with SiC wt% as 3, 5 and 7. With normal loads as 10, 20 and 30 N, pin on disc wear test was conducted. Having 3.14 m/s as sliding velocity and 942 m as sliding distance tribological behaviour was evaluated. Wear mechanisms of the worn out surfaces were analysed on the worn out specimens using optical microscope. Micro hardness number and tensile strength of the samples was measured using micro hardness and tensile tests. SEM and XRD analysis were done and the microstructures were observed. Superior tribological properties were noticed with increased addition of particles of SiC.

Karthikeyan et al [20] discussed about the behaviour of wear on Al6063/SiC composite having 5, 7 and 9 vol.% of SiC manufactured through stir casting technique. 10 N load and 500 m sliding distance were set as parameters for conducting wear test. Wear loss increased on increasing sliding velocity and it decreased on increasing reinforcement amount.

Pawar et al [21] studied the outcome of addition of 2.5%, 5%, 7.5% and 10% mass ratio of SiC composite into Aluminium matrix using stir casting technique. Microstructure and hardness tests were conducted and suggested that the composites are used as replacements for gears made of conventional materials. Table 1 given below shows the properties of composites made with SiC reinforcements.

Table 1. Properties of composites made with SiC reinforcements

Matrix material	Reinforcement	Wt.% or Vol.% of reinforcement	Reinforcement particle size	Manufacturing method	Properties Obtained		References
					Maximum Tensile strength	Maximum Hardness	
AA7075	SiC	5, 10 and 15 wt.%	10-20 μm, 20-40 μm	Stir casting	272 MPa	118 BHN	[13]
Al5083	SiC	0, 3, 5, 8 and 10 wt.%	-	Stir casting (Ultrasonic assisted)	254.9 MPa	88.8 BHN	[15]
A356	SiC	20 wt.%	45 μm	Stir casting	331 ± 4.50 Mpa	950 ± 3.80 MPa	[16]
Al7075	SiC_{np}	1, 2, 3 and 4 wt.%	50 nm	Stir casting	-	90 to 110 HV	[17]
Al6082	SiC_{np}	1 wt.%	50 nm	Stir casting	344.21 MPa	-	[18]
Al5083	SiC	3, 5and 7 wt.%	-	Stir casting	226 MPa	112 HV	[19]
Al	SiC	2.5, 5, 7.5 and 10 wt.%	-	Stir casting	-	60.3 BHN	[21]

3. Effect of B_4C on AMC

Suresh Gudipudi et al [22] used ultrasonic supported stir casting technique to overcome problems related to distribution, wetting and incorporation on increase of reinforcement in AA6061-B_4C composite. 2, 4, 5, 6 and 8 wt.% of reinforcements were used to make composites. Among the various wt.% of reinforcements used AA6061-4wt.%B_4C exhibited better microstructural, compressive and hardness behaviour.

Chandrasheker et al [23] made an attempt through electromagnetic stir casting method in synthesising Al7050/B_4C AMC with 3, 6 and 9 wt.% of B_4C particles. Hardness of the composites was found to be increasing on rising wt% of reinforcement. Tensile strength of composites increased until adding of reinforcement up to 6%. Further addition of reinforcement resulted in formation of clusters and decreased the tensile strength. Microstructure analysis helped to check uniformity on reinforcement distribution.

Radhika et al [24] made a comparative study to analyse the friction and wear behaviour of as-cast, A359/10 wt% B_4C composite (heat treated (HT) Functionally Graded (FG)) and A359/6 wt% B_4C (homogeneous) composite made of gravity casting and horizontal centrifugal technique. FG composites were found to have reinforcement particles in a decreasing grade and homogeneous composites observed uniform reinforcement throughout, when analysed using metallographic test. Mechanical properties like tensile and micro hardness are superior for FG composites. Wear resistance is superior for HT FG composite. Wear mechanism were studied using morphological analysis.

Yu Li et al [25] studied the effect of AA6061−31%B_4C composite made through stir casting technique. Scanning electronic microscope (SEM) and X-ray diffraction (XRD) tests were done and the uniform distribution of reinforcements was conformed. Tensile strength of AA6061−31%B_4C composites was found to be superior to that of AA1100−31%B_4C composite.

Siddhartha Prabhakar et al [26] investigated the tribological behavior of Al/5 wt.% B4C composite made of stir casting route. Microstructure tests confirmed even distribution of reinforcement particles with matrix. Wear tests by Taguchi technique were done with wear load of 10, 20 and 30 N, velocity of 1, 2 and 3 m/s and sliding distance of 1000, 1500 and 2000 m, optimum parameters were found. Co-efficient of friction and wear rate decreased with distance and velocity increased with load. Wear mechanism was analysed using Scanning Electron Microscope. Table 2 given below shows the properties of composites made with B_4C reinforcements.

Table 2. Properties of composites made with B_4C reinforcements

Matrix material	Reinforcement	Wt.% or Vol.% of reinforcement	Reinforcement particle size	Manufacturing method	Properties Obtained		References
					Maximum Tensile strength	Maximum Hardness	
AA6061	B_4C	2, 4, 5, 6 and 8 wt.%	30 μm	Stir casting (Ultrasonic assisted)	~220 MPa	36.5 BHN 56.01HV	[22]
Al7050	B_4C	3, 6 and 9 wt.%	-	Electromagnetic stir casting	153 MPa	84.27 HV	[23]
A359	B_4C	10 wt.%	18 μm	Centrifugal and gravity casting	239 Mpa	161.8 HV	[24]
AA6061	B_4C	31 wt.%	23 μm	Stir casting	340 MPa	-	[25]

4. Effect of Al_2O_3 on AMC

Mohanavel et al [27] produced AA6082 matrix composite reinforced with alumina particles of size 50 nm with the help of stir casting route. The volume fractions of reinforcement added vary from 0 to 3% in steps of 1 to produce AMC. SEM analysis was conducted to check microscopic behaviour and dispersion of reinforcement particles with matrix and is found to be homogeneous. Mechanical properties like hardness and tensile strength of composites were compared with base metal and were found to be increasing with increased addition of alumina into the matrix because of better binding between alloy and reinforcement.

Abhishek Kumar et al [28] discussed about the mechanical properties and microstructural behaviour of A359/Al_2O_3 composites fabricated using electromagnetic stir casting method. 30μm sized 2, 4, 6 and 8 wt.% of Al_2O_3 were used as reinforcements in making AMC. Hardness, tensile and microstructure analysis were done on the composite samples. Increase in weight fraction of reinforcement increased the hardness and tensile strength of composites. Casted MMC results in lesser grain size and better matrix reinforcement bonding because of electromagnetic stirring action.

Abdullah Dhayea Assi [29] investigated the effect of Al_2O_3 nanoparticle addition into AA6082-T6 matrix. Experiments were conducted on AMC made with nano Al_2O_3 reinforcements having varying weight fractions ranging from 5 to 30 in steps of 5 fabricated through Stir Casting Technique (SCT). Tensile strength, density and hardness of samples were found increasing on increasing the amount of reinforcement whereas the Young's modulus and yield strength decreased on increasing reinforcement amount.

Prakash et al [30] investigated the performance of various feeding techniques of stir casted Al70705-Al_2O_3 nano composites. SEM micrographs revealed the sample's porosity, particle agglomeration and bonding between ceramic interface and matrix. Mechanical properties like yield strength, micro hardness and ultimate tensile strength were measured and homogeneous reinforcement particle distribution yielded increasing results for composites than that of alloy alone.

Hanuna Haritsa et al [31] investigated the behaviour of A356 composite reinforced with Al_2O_3. Reinforcement added varied from 2 to 15 vol.% and composite specimens were made through stir casting method. Magnesium has been added to improve the wettability of reinforcement. Die was preheated to 800°C before introducing the molten metal into it. Agglomeration and porosity made the composite specimens to result in less tensile strength than that of alloy alone. Even though, hardness of the composite was better than that of base material.

Islam et al [32] studied the result of nano- Al_2O_3 addition with A356 matrix composite made of stir casting method. 0.1 to 1.2 volume fraction of reinforcement is used as reinforcement. 10% of magnesium is added as wetting agent during fabrication. Agglomeration of reinforcement reduced the tensile strength of the resulting samples. Hardness of the composite was found to have an increasing value. Table 3 given below shows the properties of composites made with Al_2O_3 reinforcements.

Table 3. Properties of composites made with Al_2O_3 reinforcements

Matrix material	Reinforcement	Wt.% or Vol.% of reinforcement	Reinforcement particle size	Manufacturing method	Properties Obtained		References
					Maximum Tensile strength	Maximum Hardness	
A359	Al_2O_3	2, 4, 6 and 8 wt.%	30 µm	Electromagnetic stir casting	148.7 MPa	72.8 HRC	[28]
AA6082	Al_2O_{3np}	5, 10, 15, 20, 25 and 30 wt.%	20 - 40 nm	Stir casting	340 MPa	92 BHN	[29]
Al7075	Al_2O_{3np}	2.5 wt.%	50 nm	Stir casting	387 MPa	878.6 MPa	[30]
A356	Al_2O_3	2, 5, 8, 10 and 15 vol.%	-	Stir casting	131.55 MPa	37.43 HRB	[31]
A356	$Al_2O_{3\,np}$	0.1, 0.2, 0.5, 1, 1.2 vol.%	80 nm	Stir casting	164 MPa	47 BHN	[32]

5. Effect of TiB_2 on AMC

Pazhouhanfar et al [33] reported the effect of addition of 3 6 and 9 wt.% of TiB_2 reinforcement addition into Al matrix during manufacture of composite employing stir casting method and studied mechanical properties, microstructure and fracture surface of tensile specimens. Stirring speed, duration and preheating temperatures were optimized initially before manufacture of composites. Reinforcement pre heating and K2TiF6 addition during casting process improved the wettability and matrix reinforcement bonding of the composite, which was revealed during SEM analysis. Uniform reinforcement distribution was found during SEM analysis. Tensile test results were found to be increasing with increasing addition of volume fraction of reinforcement.

Semegn Cheneke et al [34] characterized and investigated the mechanical properties and microstructure of AA2024-TiB_2 composite made of stir rheocast method. 2, 4 and 6 wt.% of reinforcements were used for making composite with working temperature kept as 640°C. Field Emission SEM analysis confirmed even distribution of reinforcements in matrix. Agglomeration of reinforcements was found on composites having 6wt.% of TiB_2. Hardness and tensile strength of composite specimens were compared with liquid cast samples and found better. Dimples and voids present in fractography analysis confirmed the ductile fracture of test specimens.

Fei Chen et al [35] discussed the effect of mechanical stirring of Al-TiB_2 composites using stir casting technique. The study is quite different from others. Instead of mixing the reinforcement directly into the metal melt, K_2TiF_6 and KBF_4 salts were added resulting in formation and uniform distribution of TiB_2 by chemical reaction. Optimized processing parameters were used and the microsturcture and mechanical properties were evaluated. The novel method exhibited improved ductility in addition to increased yield and tensile strengths.

Loganathan et al [36] employed two step stir casting process to produce Al7075-TiB_2 composite with 5, 10 and 15 wt.% of reinforcement. Hardness, tensile, density and impact tests were performed to examine the mechanical property of the composite. SEM and XRD tests were employed to observe the factography of the composite specimens. Grain refinements were seen from those observations on addition TiB_2 which lead to improved mechanical properties of composite. Load bearing capacity of the composite increased because of better matrix reinforcement bonding. Table 4 given below shows the properties of composites made with TiB_2 reinforcements.

Table 4. Properties of composites made with TiB_2 reinforcements

Matrix material	Reinforcement	Wt.% or Vol.% of reinforcement	Reinforcement particle size	Manufacturing method	Properties Obtained		References
					Maximum Tensile strength	Maximum Hardness	
Al6061	TiB_2	3, 6 and 9 wt.%	2 to 10 μm	Stir casting	~260 MPa	~150 (HV 0.25)	[33]
Al	TiB_2	1, 4 and 7 wt.%	-	Stir casting	104.4 ± 2.4 MPa	24.4 ± 0.3 BHN	[35]
AA7075	TiB_2	5, 10 and 15 wt.%	25 μm	Stir casting	~220 MPa	152 HV	[36]

6. Conclusion

It is evident from the reviewed researches that:

- Aluminium composites are finding applications in aerospace and automobile fields and are capable of replacing the conventional materials.
- Adding 10-20μm SiC of 10wt.% with AA7075 increase tensile strength upto 9.67%. 10.48% of increment in hardness can be seen because of addition of 15% of SiC.
- Heat Treated (HT) Functionally Graded (FG) composites exhibit better mechanical and wear behaviour than as-cast samples.
- Wettability between matrix and reinforcement can be improved by adding magnesium and K_2TiF_6 into the metal melt in the casting process.
- Preheating of reinforcements before introducing it into molten metal improves the matrix reinforcement bonding and load carrying capacity of the composites increases if better matrix reinforcement bonding is achieved.

References

[1] Naher S, Brabazon D and Looney L 2004 Development and assessment of a new quick quench stir caster design for the production of metal matrix composites 166 430–9
[2] Reddy P S and Vijaya R K B 2017 Investigation of Mechanical Properties of Aluminium 6061-Silicon Carbide , Boron Carbide Metal Matrix Composite Silicon
[3] Coyal A, Yuvaraj N, Butola R and Tyagi L 2020 An experimental analysis of tensile, hardness and wear properties of aluminium metal matrix composite through stir casting process SN Appl. Sci. 2 1–10
[4] Hashim J, Looney L and Hashmi M S J 2001 The enhancement of wettability of SiC particles in cast aluminium matrix composites 119 329–35
[5] Hashim J, Looney L and Hashmi M S J 1999 Metal matrix composites : production by the stir casting method 93 1–7
[6] Sajjadi S A, Ezatpour H R and Beygi H 2011 Microstructure and mechanical properties of Al-Al2O3 micro and nano composites fabricated by stir casting *Mater. Sci. Eng. A* 528 8765–71
[7] Yuvaraj N, Aravindan S and Vipin 2017 Wear Characteristics of Al5083 Surface Hybrid Nano-composites by Friction Stir Processing *Trans. Indian Inst. Met.* 70 1111–29
[8] Gurusamy P, Prabu S B and Paskaramoorthy R 2015 Influence of processing temperatures on mechanical properties and microstructure of squeeze cast aluminum alloy composites *Mater. Manuf. Process.* 30 367–73
[9] Benal M M and Shivanand H K 2006 Influence of heat treatment on the coefficient of thermal expansion of Al (6061) based hybrid composites *Mater. Sci. Eng. A* 435–436 745–9
[10] Kumaran S T and Uthayakumar M 2015 Application of Grey relational analysis in high speed machining of AA (6351) -SiC-B 4 C hybrid composite Adam Slota and Jerzy Zajac *Int. J. Mater. Prod. Technol.* 51 17–31

[11] Vigneshkumar M, Varthanan P A and Raj Y M A 2019 Finite Element-Based Parametric Studies of Nugget Diameter and Temperature Distribution in the Resistance Spot Welding of AISI 304 and AISI 316L Sheets *Trans. Indian Inst. Met.* 72 429–38
[12] Vigneshkumar M and Varthanan P A 2019 Comparison of RSM and ANN model in the prediction of the tensile shear failure load of spot welded AISI 304/316 L dissimilar sheets *Int. J. Comput. Mater. Sci. Surf. Eng.* 8 114–30
[13] Bhushan R K 2020 Effect of SiC particle size and weight % on mechanical properties of AA7075 SiC composite *Adv. Compos. Hybrid Mater.*
[14] Bhowmik A, Chakraborty D, Dey D and Biswas A 2020 Materials Today: Proceedings Investigation on wear behaviour of Al7075-SiC metal matrix composites prepared by stir casting *Mater. TODAY Proc.* 2–5
[15] Chaubey A, Dwivedi R, Rana R S and Choudhary K 2020 Materials Today: Proceedings Experimental inspection of aluminium matrix composites reinforced with SiC particles fabricated through ultrasonic assisted stir casting process *Mater. Today Proc.* 8–11
[16] Soundararajan R, Sivasankaran S, Babu N and Adhithya G P R 2020 Appraisal of tribological properties of A356 with 20% SiC composites under dry sliding condition *J. Brazilian Soc. Mech. Sci. Eng.* 42 1–12
[17] Suresh S, Gowd G H and Deva Kumar M L S 2019 Tribological Behavior of Al 7075/SiC Metal Matrix Nano-composite by Stir Casting Method *J. Inst. Eng. Ser. D* 100 97–103
[18] Zhu J, Jiang W, Li G, Guan F, Yu Y and Fan Z 2020 Microstructure and mechanical properties of SiCnp/Al6082 aluminum matrix composites prepared by squeeze casting combined with stir casting *J. Mater. Process. Technol.* 283 116699
[19] Vanam J P, Chiranjeevi R, Kumar R S, Ramana V V and Kumar A S 2018 Effect of SiC on Mechanical, Microstructure and Tribological properties of Aluminum MMC processed by Stir Casting *IOP Conf. Ser. Mater. Sci. Eng.* 455
[20] Karthikeyan A and Nallusamy S 2017 Experimental analysis on sliding wear behaviour of aluminium-6063 with SiC particulate composites *Int. J. Eng. Res. Africa* 31 36–43
[21] Pawar P B and Utpat A A 2014 Development of Aluminium Based Silicon Carbide Particulate Metal Matrix Composite for Spur Gear *Procedia Mater. Sci.* 6 1150–6
[22] Gudipudi S, Nagamuthu S, Subbu K, Prakasa S and Chilakalapalli R 2020 Engineering Science and Technology , an International Journal Enhanced mechanical properties of AA6061-B 4 C composites developed by a novel ultra-sonic assisted stir casting *Eng. Sci. Technol. an Int. J.*
[23] Chandrasheker J and Raju N V S 2020 Materials Today : Proceedings Mechanical characterization of al7050 metal matrix composite reinforced with b 4 c by electromagnetic stir casting method *Mater. Today Proc.*
[24] Radhika N, Sasikumar J, Jl S and Kishore R 2019 Dry reciprocating wear and frictional behaviour of B 4 C reinforced functionally graded and homogenous aluminium matrix composites *Integr. Med. Res.* 1–15
[25] LI Y, LI Q lin, LI D, LIU W and SHU G gang 2016 Fabrication and characterization of stir casting AA6061—31%B4C composite *Trans. Nonferrous Met. Soc. China (English Ed.* 26 2304–12
[26] N S P, Radhika N and Raghu R 2014 Analysis of tribological behavior of aluminium / B 4 C composite under dry sliding motion *Procedia Eng.* 97 994–1003
[27] Mohanavel V, Rajan K, Senthil P V and Arul S 2017 ScienceDirect Mechanical behaviour of hybrid composite (AA6351 + Al 2 O 3 + Gr) fabricated by stir casting method *Mater. Today Proc.* 4 3093–101
[28] Kumar A, Lal S and Kumar S 2013 Original article Fabrication and characterization of A359 / Al 2 O 3 metal matrix composite using electromagnetic stir casting method *Integr. Med. Res.* 2 250–4
[29] Assi A D 2020 Influence of Al 2 O 3 Nanoparticles Addition to AA6082-T6 on Mechanical Properties by Stir Casting Technique Influence of Al 2 O 3 Nanoparticles Addition to AA6082-T6 on Mechanical Properties by Stir Casting Technique
[30] Sasikumar S P R and Natarajan E 2020 Influence of Feeding Techniques in Bottom Tapping Stir Casting Process for Fabrication of Alumina Nano-filler-reinforced Aluminium Composites *Trans. Indian Inst. Met.*
[31] Haritsa H, Akira M W and Zulfia A 2020 The Role of Al 2 O 3 Particles on Mechanical Properties and Microstructure of Aluminum A356 Composites Produced by Stir Casting Method 1000 185–92
[32] Nabila Akmalita Khairul Islam , Nayona Ega Wicaksana and Anne Zulfia 2020 Effect of Nano-Al 2 O 3 on Characteristics of Aluminium A356 Matrix Composite Produced by Stir Casting Route * 1000 160–6

[33] Pazhouhanfar Y and Eghbali B 2018 Microstructural characterization and mechanical properties of TiB2 reinforced Al6061 matrix composites produced using stir casting process *Mater. Sci. Eng. A* 710 172–80
[34] Cheneke S and Karunakar D B 2019 Microstructure characterization and evaluation of mechanical properties of stir rheocast AA2024 / TiB 2 composite
[35] Chen F, Chen Z, Mao F, Wang T and Cao Z 2015 Materials Science & Engineering A TiB 2 reinforced aluminum based in situ composites fabricated by stir casting *Mater. Sci. Eng. A* 625 357–68
[36] Loganathan P, 2020 Microstructural Characteristics and Mechanical Behaviour of AA7075/TiB 2 Composite *Mater. Sci. Forum* 979 40–6

A REVIEW ON INFLUENCING WELDING PARAMETERS OF SUPER ALLOYS

S Raja[1], R Saranya[2], V SaiYaswanth[2], A SachinAdithya[2], M B Rohith Kumar[2]

1 Assistant Professor, Department of Mechanical Engineering, Sri Krishna College of Engineering and Technology, Coimbatore, India.
2 UG students, Department of Mechanical Engineering, Sri Krishna College of Engineering and Technology, Coimbatore, India.

raja@skcet.ac.in, 18eumc133@skcet.ac.in, 18eumc127@skcet.ac.in, 18eumc126@skcet.ac.in, 18eumc124@skcet.ac.in

Abstract Manufacturing industries play a vibrant role in economic development and significant growth of a country. Among all different industries, sheet metal industry is irreplaceable in our routine life. Sheet metal applications include in automobiles, aircrafts, computer panels, building roofs etc. Various manufacturing process of sheet metal industry consists of shearing, bending, welding and deep drawing. Welding is one of the vital process in engineering business. Welding is the best operative method of joining similar and dissimilar components together. Choice of influencing process parameters is very much critical in welding for the purpose of obtaining good quality weldments with preferred mechanical properties and enhanced serviceability. This is impossible with single welding method that yields the optimum weld in all conditions. The choice of welding process depends upon the service environment, time, cost, and deposition rate. Superalloy is an alloy having greater mechanical strength and high corrosion resistance in high temperature applications. Superalloys are of, Nickel, Cobalt and Iron built possessing the crystal structure of Face Centered Cubic (FCC) austenitic. Typical examples of super alloys include Inconel, Hastelloy, Waspaloy, Rene alloys, Incoloy. Among these, Hastelloy is a nickel based superalloy containing the major proportion as Nickel 57%, Chromium 15.5% and Molybdenum 16% with the remaining in smaller proportions. Hastelloys are employed by the chemical industries for their Corrosion-Resistant property. The characteristics of Hastelloy include high resistance to localized corrosion in welding and fabrication. Hastelloy are of different grades optimised for specific purpose. Hastelloy alloys of "C-type" alloys are the most versatile. Hastelloy C276 alloy is a nickel alloy engaged in petro chemical and related industries possessing corrosive chemicals. Hastelloy C276 is ductile in nature, suitable for welding, forming and has high resistance to corrosion. Hastelloy C276 alloy is welded by the fusion welding process. The low quantity of carbon existence prevents carbide precipitation to keep resistance to inter granular attack in thermally affected zones. Every industry requires to attain a good quality of weld with great productivity and low processing cost with commanding weld joint.

Keywords: Hastelloy C276, Tensile strength, Microstructure

1. Introduction

Superalloys are alloys that can service in high temperature about 1000°C in rocket engines. They are also termed as thermal-resistant alloys. They possess superior corrosion resistance, mechanical and thermal fatigue (Sims et al. 1987). Superalloys are mentioned to as nickel-base, cobalt-base or iron-base. Major alloying elements are nickel, chromium, cobalt and molybdenum. Nickel-base superalloys contain nickel 38% to 76%; chromium 27% and cobalt 20%. Hastelloy, Inconel, Nimonic, Rene, Udimet, Astroloy and Waspaloy series. Cobalt-base superalloys contain cobalt from 35% to 65%, chromium from 19% to 30% and nickel up to 35%. Iron base superalloys contain iron from 32% to 67%, chromium from 15% to 22% and nickel from 9% to 38%. In general, nickel-chromium alloys have excellent oxidation resistance owing to the oxide development of chromium and aluminum.

1.1 DEVELOPMENT OF MATHEMATICAL MODEL FOR WELD BEAD PREDICTION

Yang et al. (2019) proposed an algorithm for weld seam path identification for robot welding process. Improvement of the welding automation level is depends on the path extraction and identification of weld seam. In industries it majorly affect by the following characteristics of the workpiece metallic surfaces, textures and imperfection. To overcome the difficulty welding robots was taught by the new method of path identification. Feature vectors will be constructed by shape of weld seam. The results shows that ANFIS algorithm will improve optimal performance and accurately identify the weld seam.

Bal et al. (2018) made an attempt to investigate the influencing parameters of the laser beam welding in penetrated welding zone area. Welding zone has a geometry of throat, crown, neck and root width. From full factorial design the regression equation was formed for all the geometrical features. Parameters optimization was conducted to minimize the welding zone area. Minimization of cross sectional geometry of weld bead was confirmed by conducting experiment. It shows that melt zone shape obtained by formula is same as that of experiment. Output response is mainly affected by scanning speed and laser power.

Bal et al. (2018) made an attempt to investigate Ytterbium fiber laser for the failure of Hastelloy C276 weld zone. Failure occur in Hastelloy weld joint was identified by uni-axial tensile test which is equal to 50% reduction of length of Hastelloy specimen. The major cause for the failure is due to micro segregation in solidification and cause brittleness behaviour of specimen. Micro segregation was reduced by limiting the required linear hat input to welding. Ytterbium fiber have higher melting efficiency compared to other sources. 3 millimeter thick C276 Hastelloy sheet is welded with Ytterbium fiber and the tensile strength of welded specimen were studied. It was noted that weld specimen prepared with Ytterbium fiber fails at base metal simultaneously percentage of elongation was improved twice while compared to the previous C276 Hastelloy specimen.

Moosavyet al. (2014) made an attempt to investigate various welding parameters influence of strengthened nickel-base superalloy welded by fiber laser beam welding with constant heat input. Vacuum induction melting furnace is used to produce 5 nickel base superalloy with various Nb and Ti content. Six axis anthropomorphic robot used to perform the welding operations at. Experimental results shows that porosity formation is directly proportional to the velocity and welding power. In 400 W and 1000 W welding power there is no hot solidification and liquation crack but in all heat-affected zones the liquation phenomenon was observed at the same time it does not alter the hardness of the specimen. Results shows that welding power 400 W and 1000 W with velocity of 40 mm/s and 100 mm/s yields defect free welding.

Manikandanet al. (2014) investigated the various mechanical properties and microstructure in laser welding process of alloy C276 by continuous neodymium: yttrium aluminium garnet. Welding trials was conducted with various laser beam scan speed ranging from 550 to 1200 mm/min. EDS analysis is used to predict the extent of segregation in weld joint. Specimen strength will be measured by tensile test. Fine dendritic structure results high hardness in weld zone compare to coarse structure of base metal. Studies indicate that fusion zone micro segregation was acceptable range with the evidence for macro segregation. Defect free weld joint with good tensile strength and ductility produced. Due to narrow heat input there is no significant heat affected zone.The author reported on improvement of microstructure and mechanical properties in the year 2014 and 2015.

Kocharet al. (2019) utilized ANN to predict the weld bead shape in Laser welding with the aid of wire feed, joining speed, voltage, current, and laser power as inputs. The bead profiles of the experiments are expressed in polar coordinates (r,□). With the various simulation of results an optimized parameters are find out which captures the physics of the welding process. Among the various parameters, the most influencing parameters, which directly affect the bead shape are welding travel speed and wire feed shape. The results obtained are a high degree of accuracy. The developed algorithms can be utilized for the other welding process too.

Akkaset al. (2013) employed ANN and ANFIS to forecast the relationship between weld bead geometry and the corresponding influencing parameters in submerged Arc welding process. Optimum parameters are find out and the corresponding experiments are performed to check the predictability of both the system. Both the experiment and the developed system imparts that weld penetration are influenced by arc voltage, current and welding travel speed. Increase of welding thickness depends on arc voltage and arc current. With the minimum of arc current the welding thickness is minimum with increased arc voltage and Vice versa. With the both systems, ANFIS results are more reliable than ANN.

Chen et al. (2019) developed XGBoost algorithm to forecast the depth of penetration and reinforcement in metal active gas welding. Quality of the joint was assessed by two criteria namely penetration depth to thickness of the plate and reinforcement. On performing various trials, Penetration coefficient provides 11.06% relative error and reinforcement penetration provides 20.5% relative error. From the above results, it is concluded that XGBoost algorithm can be used for better prediction for accessing the quality of weld.

Kumar (2011) made an attempt to develop a mathematical model for Submerged Arc Welding which has many advantages in welding of pipes. Mathematical model developed establishes a relation among the influencing process parameters and geometry of the bead. The developed model not only predicts the geometry but also the quality of the weld. Response surface methodology has been developed. The developed model controls the geometry of the bead along with the choice of choosing the range of process parameter for the best quality of the weld. The Developed model directly assist the robotic welding in accessing the quality of the Weld. Similar works are reported by Srivastava and Garg (2017).

Devakumaranet al. (2010) made an attempt to predict the geometrical features of the weld bead in Pulsed GMAW on high strength low alloy steel. The various influencing pulse parameters along with the thermal characteristics influencing the bead geometry has been correlated as mathematical model. The developed model has good agreement with min deviation of 8 to 10% while comparing with the experimental result.

Muruganet al. (1999) made an attempt to predict the Bead Geometry in submerged arc welding by the development of mathematical model that predicts the shape relationships in weld bead geometry. The utilization of response surface methodology with four factor five level design matrix with the regression analysis helps in choosing of optimum parameters for acquiring the expected quality of weld.

Kannan et al. (2018) employed a hot wire gas tungsten arc welding (HW-GTAW) process in joining of T92 steel and S304HCu stainless steel with different filler wires namely. Weld bead depositions are found at two welding currents (130A, 110A) with constant wire feed rate and welding speed. Presence of active surface elements and temperature distribution are main sources for the different shape of weld bead.

Sripriyan and Ramu (2016) proposed a new attempt of defining the weld bead geometry in conventional Gas Metal Arc Welding (GMAW). Varying of the geometry of the filler wire technique is employed in improving the quality of the weld bead. A flat wire has been proposed for investigating the characteristics of weld Bead.

Adak et al. (2015) developed a multiple linear regression model for Gas Metal Arc Welding process to predict output of various input variable to forecast the bead geometry and grain structure on the plate. The responses expected are in decent agreement with the experimental results which shows the acceptability of the developed model.

Muruganet al.(1999b) performed calculations for thermal input and the thermally affected zone for the established model in SAW. Comparative study between the actual joint and bead on trials reveals that the developed model needs to be modified further in accessing the quality of the weld bead. The calculated area for bead on trials is greater that of the joint and the effect of heat input is same for both the joint and bead on plate.

Hedayatiet al. (2017) conducted a heat treatment process of Hastelloy and stainless steel bi metal for microstructural evaluation and interfacial diffusion. The samples were heat treated under different temperature of 800 C, 900 C, 1000 C and 1100 C for a soaking period of 3hrs. Short time exposing to high temperatures develop coarse carbide on the interface, making the weld to failure.

Dhobale and Mishra (2015) made an attempt to study the strength of the weldment which depends upon the thermal input. The weldment hardness increases with increase in thermal input where as it decreases in the thermally affected zone. The Micro constituents are also affected by heat input. The microstructure reveals that the small dendrite structure with less spacing in HAZ zone and vice versa in Weld Zone possessing high thermal input.

Cornacchiaet al. (2018) made an attempt to execute a comparative study evaluate the properties of conventional MIG with Cold Metal transfer and hybrid Laser MIG welding process. Mechanical properties like Hardness, tensile strength, and yield strength are found to be superior than the former process. The material used for this investigation was 6005A-T6. Fractographic observations were made using SEM/EDS to analyze microstructural deviations after joining.

Yu et al. (2017) employed an infrared sensing system to monitor the changes during welding. Different process parameter like welding speed, current and nose to tip distance are varied to capture the changes in the weld pool. Defects like undercut, lack of penetration and humping are been used for validation of the proposed system. Experimental results showed that the proposed monitoring system is very much useful in detecting the defects in welding.

Sumeshet al. (2018) proposed an algorithm to monitor the quality of weld bead in automated environment. The process parameter's like welding current and welding voltage are chosen to define the quality of the weld. Gas metal Arc Welding is employed in welding of carbon plates without the defects like porosity and burn through. Decision tree algorithm are used to classify the defects. The proposed algorithm signatures in defect classification during the welding.

Son et al. (2017) employedneural network to predict the quality of weld with intelligent model Back propagation and Levenberg-Marquardt. The developed model predicts the effects on bead height to access the quality of the Weld in gas metal arc welding process.

Chakiet al. (2015) developed an integrated soft computing model by combining of three different techniques for weld zone's strength prediction in hybrid CO2 laser–MIG welding of aluminium alloy with the input parameters of travel speed , feed and power of laser. A well trained ANN is employed by the soft computing model to perform the task. The most influencing factor for welding strength is welding speed.

Ibrahim et al. (2012) investigated the effect of welding current, Arc voltage, and welding speed to predict the depth of mild steel plate of thickness 6mm in Metal inert gas welding. Depth of penetration increases with the increase in current and vice versa in welding speed and arc voltage. Changes in the boundaries of grains were observed for different level of parameters.

Kumar et al. (2013) investigated the process parameter influenced in welding of two different metals by using Taguchi's technique. GMAW process is employed to join the low carbon and AISI 304 steel plates. L9 orthogonal array is used to prepare the no of weldment required for the investigation. Welding current, voltage and gas flow rate are considered as input parameters. Design of experiment is used to find the controllable factor to calculate the signal to noise ratio. Also he identify and conclude the significant input parameter for tensile strength and hardness of metal by ANOVA technique.

1.2 OPTIMIZATION IN WELDING PROCESS

Narwadkar and Bhosle (2016) made an attempt to address the major problem of angular distortion in butt welded joints. It was overcome by optimizing the input parameter by taguchi technique. Gas flow rate and welding current, voltage are considered as controllable input parameter. Three levels and factors are consider for DOE. L9 orthogonal array is formed and the different butt weld sample was produced by MIG welding process. After ANOVA technique is used to calculate the distortion angle of the samples. Results were validated by confirmation test.

Malviya and Pratihar (2011) made an attempt to discover four approaches for tuning of neural networks for MIG process. PSO for both forward and reverse mapping. Performance of four approaches were identified and compared with each other to solve the problem caused during MIG welding. First two approaches is used to tuning of multilayer forward mapping and the rest two approaches used to decide the network structure. Third and fourth approaches are provide better results compared to first two approaches. Similar work has been carried out by Nageshet al. (2002)

Meenaet al. (2017) made an attempt to study the influence of input process parameters in preparation of weld bead geometry to determine the load carrying capacity of weldments. Weld bead geometry was examined with various thickness of shielding gas. Strength of weldments is purely depends on the weld bead geometry and metal composition and bead shape relationship. Also the metallurgical investigation carried out to find hardness of weldment in base metal, heat affect zone and weld metal zone. Results was correlated with process variables. Similar works were reported by Kumar et al. (2019) Response-Surface Methodology (RSM)

1.3 MECHANICAL AND METALLURGICAL PROPERTIES OF WELDMENTS

Ganjigattiet al. (2007) attempted to study the regression analysis of input – output relationship of MIG welding in cluster wise and global approaches. The study was carried out by data obtaining from full factorial design of experiments. Comparison results of above two approaches indicates that regression analysis of cluster wise approach perform better than other one.

Patel et al. (2016) made an attempt to investigate the importance of hard facing operation in metal to metal seal creation by MIG welding process. Input process parameter and bead geometry plays major role in weld quality. Influence of process parameter in weld height was analysed in this study. Different modelling techniques like ANN, Fuzzy and Regression are employed to determine the optimal weld height. Results from different techniques was compared with experimental data to predict the best modelling technique.

Chaudhariet al. (2018) made an attempt to investigate the influence of oxide flux in MIG welding process. SiO_2 and Cr_2O_3 are the different fluxes used in weld bead. Specimens were prepared with different process parameters. L9 orthogonal array is used to prepare the samples with SiO_2, with Cr_2O_3 and without flux. ASRS and AHP-MOORA methods used to optimize the process parameter and results illustrate that superior penetration obtained with flux coating. Metallurgical characteristics of weldment was evaluated.

Khanna and Maheshwari (2018) made an attempt to promote the mathematical model for predicting the shape relationship of weld bead geometry in MIG welding. Design matrix for experiments were carried out by using statistical technique. Significance and capability of the model was analysed by t-test and F-test. Newly developed mathematical model is more helpful to predict the optimal process parameter and weld bead dimensions.

IOP Conf. Series: Materials Science and Engineering **995** (2020) 012041 doi:10.1088/1757-899X/995/1/012041

Bal et al. (2019) made an attempt to investigate the fracture mechanism of Hastelloy sheet welded by LBW process at weld zone in different heat input condition (60 and 120 J/mm). After PHWT of AW-Hastelloy it was witnessed that the sample prepared at 60 Joules per mm failed at the fusion zone, while the other sample prepared at 120 Jules per mm fractures at the parent metal. AW and PWSHT samples, XRD and SEM results indicate the occurrence of chromium carbide in base metal of sample welded at 120 Jules per mm. compared to AW samples hardness is significantly increasing in PWSHT samples.

Ramkumaret al. (2014) made an attempt to investigate the behavior of nickel alloys (Monel 400, Hastelloy C276) joint by PCGTA welding. ERNiCrMo-3 filler is used to join the dissimilar metals of nickel alloy. Study indicates that Monel 400 subjected to tensile failure due to the absence of unmixed zone in both sides of HAZ. Better ductility was observed in dissimilar metal in bend test. Metallurgical and mechanical properties of dissimilar metals are associated with the use of current pulse.

Panditet al. (2014) made an attempt to study the need of bimetallic (Monel 400, Hastelloy C276) combinations in high temperature applications were studied. GTAW technique used to join these metals by employing ERNiCrMo-3 filler. Weldments characterization was carried out to observe the metallurgical and mechanical properties of weldments. Results obtained from tensile test proves that the hardness of weld zone in bimetallic combination metals is greater than that of parent metals. EDAX/SEM and optical microscope techniques are used to predict the structural property.

Wu et al. (2014) made an attempt to investigate the requirement of Hatselloy C276 in nuclear industry. Studies indicates that weldment less than 1 mm width produce the defect free joint. Fracture mechanism of Hatselloy C276 was observed from SEM analysis at different temperatures (200, 300, 400o C). Results shows that few sample were broken in weld zone and some samples were broken in base metal. AS welded samples have higher strength compared to the base metal.

Saravananet al. (2019) made an attempt to correlate the laser butt weld microstructure of Hastelloy C276 joint by experimental and numerical method. Joints are made at different process parameters viz., pulse energy and duration, welding speed. Experiment was done in 600 W Nd: YAG LWM, FEA is used to find the numerical microstructure of Hastelloy. Results indicates the temperature variation, distribution of heat in weld zone and hardness of weld zone.

Bal et al. (2019) made an attempt to investigate the welding zone, adjacent HAZ and unpretentious base metal correction rate and consequently measure the weakness to correction of weldment without preparing a separate samples by implementing the laser displacement sensor scanning technique in EBW process. Corrosion test indicates that weld zone corrosion rate is greater than corrosion rate of base metal and HAZ.

Bal et al. (2019) made an attempt to optimize the influencing process parameter to increase the quality of weld bead area in Hastelloy C276. Optimisation technique is used to prepare the full factorial design with the various parameters to prepare the required number of samples and the output responses are the various bead geometry. The regression equation was established to optimise the process parameter after validating the experimental result with conformation test. Corrosion occurred in the bead geometry, weld zone and base metal are studied by conducting the potentiodynamic polarization test.

Guoet al. (2014) made an attempt to develop the 3D model by ANSYS code to disclose the behaviour of Hastelloy C276 butt joint made by pulsed laser welding. Experiment were conducted and compared with the simulation result to check the residual distortion and welding temperature of weldments. Finite element method used to investigate the influence of cooling distance in residual distortions and stresses in weldment. Cooling distance plays the vital role in residual distortion and stress distribution compared to traditional PLW and better results were obtained during reducing the cooling distance.

2. Conclusion

From the literature review, many researchers have focus on the welding of superalloys. Since superalloys find it's working in very high temperature. Different welding techniques have been applied on the superalloys. Large number of works have focused on the side of automating the welding. In order to automate the welding process, the most influencing process parameter which affects the quality of the weld has to be determined. Generally the weld quality depends upon the geometry of the bead and the composition of the weld. Each process parameter has influence on the various geometry of the bead. Geometry of the Bead include width of the bead, depth of penetration and height of reinforcement. From the bead's geometry, the weld quality without any defects can be assed manually. Defects like lack of penetration, lack of fusion, spatter, large reinforcement, more depth of penetration, undercut, dilution etc. can be visualized by naked eye. Therefore it is necessary to choose the combination of the better parameter to obtain good quality of the weld. For to obtained the best combination, large of trials are been made on the plate to quantify the range of the parameter. Lot of work described on establishing the relationship among the process parameter to improve the quality of weld.

IOP Conf. Series: Materials Science and Engineering **995** (2020) 012041 doi:10.1088/1757-899X/995/1/012041

References

[1] Adak DK, Mukherjee M, Pal TK. Development of a direct correlation of bead geometry, grain size and HAZ width with the GMAW process parameters on bead-on-plate welds of mild steel. Transactions of the Indian Institute of Metals. 2015 Oct 1;68(5):839-49.

[2] Ahmad M, Akhter JI, Akhtar M, Iqbal M, Ahmed E, Choudhry MA. Microstructure and hardness studies of the electron beam welded zone of Hastelloy C-276. Journal of Alloys and Compounds. 2005 Mar 22;390(1-2):88-93.

[3] Akkas N, Karayel D, Ozkan SS, Oğur A, Topal B. Modeling and analysis of the weld bead geometry in submerged arc welding by using adaptive neurofuzzy inference system. Mathematical Problems in Engineering. 2013 Jan 1;2013.

[4] Bal KS, Majumdar JD, Choudhury AR. Minimization of bead geometry by optimization of regression equations for laser-beam bead-on-plate welded Hastelloy C-276 sheet. Journal of the Brazilian Society of Mechanical Sciences and Engineering. 2018 Sep 1;40(9):451.

[5] Bal KS, Majumdar JD, Choudhury AR. Study on uni-axial tensile strength properties of Ytterbium fiber laser welded Hastelloy C-276 sheet. Optics & Laser Technology. 2018 Dec 1;108:392-403.

[6] Bal KS, Majumdar JD, Choudhury AR. Effect of post-weld heat treatment on the tensile strength of laser beam welded Hastelloy C-276 sheets at different heat inputs. Journal of Manufacturing Processes. 2019 Jan 1;37:578-94.

[7] Bal KS, Majumdar JD, Choudhury AR. Optimization of melt zone area for electron beam welded Hastelloy C-276 sheet and study of corrosion resistance of the optimized melt zone in 3.5 wt% NaCl Aqueous Solution. Arabian Journal for Science and Engineering. 2019 Feb 12;44(2):1617-30.

[8] Bal KS, Majumdar JD, Choudhury AR. Investigation into the intergranular corrosion behaviour of electron beam welded Hastelloy C-276 sheet using laser displacement sensor. Measurement. 2019 Oct 1;144:345-65.

[9] Chaki S, Shanmugarajan B, Ghosal S, Padmanabham G. Application of integrated soft computing techniques for optimisation of hybrid CO2 laser–MIG welding process. Applied Soft Computing. 2015 May 1;30:365-74.

[10] Chaudhari PG, Patel PB, Patel JD. Evaluation of MIG welding process parameter using Activated Flux on SS316L by AHP-MOORA method. Materials Today: Proceedings. 2018 Jan 1;5(2):5208-20.

[11] Chen H, Xue JX, Heng GC. A novel sine wave modulation method of pulse current during double wire MIG welding. The International Journal of Advanced Manufacturing Technology. 2017 Jun 1;90(9-12):3025-36.

[12] Chen K, Chen H, Liu L, Chen S. Prediction of weld bead geometry of MAG welding based on XGBoost algorithm. The International Journal of Advanced Manufacturing Technology. 2019 Apr 19;101(9-12):2283-95.

[13] Cornacchia G, Cecchel S, Panvini A. A comparative study of mechanical properties of metal inert gas (MIG)-cold metal transfer (CMT) and fiber laser-MIG hybrid welds for 6005A T6 extruded sheet. The International Journal of Advanced Manufacturing Technology. 2018 Feb 1;94(5-8):2017-30.

[14] Dhobale AL, Mishra HK. Review on effect of heat input on tensile strength of butt weld joint using mig welding. International Journal of Innovations in Engineering Research and Technology. 2015 Sep;2(9):1-3.

[15] Ganjigatti JP, Pratihar DK, Choudhury AR. Global versus cluster-wise regression analyses for prediction of bead geometry in MIG welding process. Journal of materials processing technology. 2007 Jul 6;189(1-3):352-66.

[16] Goward GW. Current research on the surface protection of superalloys for gas turbine engines. JOM. 1970 Oct 1;22(10):31-9.

[17] Gunaraj V, Murugan N. Application of response surface methodology for predicting weld bead quality in submerged arc welding of pipes. Journal of materials processing technology. 1999 Apr 15;88(1-3):266-75.

[18] Gunaraj V, Murugan N. Prediction and comparison of the area of the heat-affected zone for the bead-on-plate and bead-on-joint in submerged arc welding of pipes. Journal of Materials Processing Technology. 1999 Oct 15;95(1-3):246-61.

[19] Hedayati O, Korei N, Adeli M, Etminanbakhsh M. Microstructural evolution and interfacial diffusion during heat treatment of Hastelloy/stainless steel bimetals. Journal of Alloys and Compounds. 2017 Jul 25;712:172-8.
[20] Ibrahim, I.A, Mohamat, S.A, Amir, A. &Ghalib, A, 2012. 'The Effect of Gas Metal Arc Welding (GMAW) processes on different welding parameters'. Procedia Engineering, vol. 41, pp. 1502-1506.
[21] Kannan PR, Muthupandi V, Devakumaran K. On the effect of temperature coefficient of surface tension on shape and geometry of weld beads in hot wire gas tungsten arc welding process. Materials Today: Proceedings. 2018 Jan 1;5(2):7845-52.
[22] Khalid MN, Naranje V, Gaidhane VH. Prediction of Best Weld Quality Using Artificial Neural Network. In2019 Amity International Conference on Artificial Intelligence (AICAI) 2019 Feb 4 (pp. 213-217). IEEE.
[23] Khanna P, Maheshwari S. Development of mathematical models for prediction and control of weld bead dimensions in MIG welding of stainless steel 409M. Materials Today: Proceedings. 2018 Jan 1;5(2):4475-88.
[24] Kumar GS, Raghukandan K, Saravanan S, Sivagurumanikandan N. Optimization of parameters to attain higher tensile strength in pulsed Nd: YAG laser welded Hastelloy C-276–Monel 400 sheets. Infrared Physics & Technology. 2019 Aug 1;100:1-0.
[25] Kumar P, Roy BK. Parameters Optimization for Gas Metal Arc Welding of Austenitic Stainless Steel (AISI 304) & Low Carbon Steel using Taguchi's Technique. International Journal of Engineering and Management Research (IJEMR). 2013;3(4):18-22.
[26] Malviya R, Pratihar DK. Tuning of neural networks using particle swarm optimization to model MIG welding process. Swarm and Evolutionary Computation. 2011 Dec 1;1(4):223-35..
[27] Manikandan M, Arivazhagan N, Rao MN, Reddy GM. Microstructure and mechanical properties of alloy C-276 weldments fabricated by continuous and pulsed current gas tungsten arc welding techniques. Journal of Manufacturing processes. 2014 Oct 1;16(4):563-72.
[28] Manikandan M, Arivazhagan N, Rao MN, Reddy GM. Improvement of microstructure and mechanical behavior of gas tungsten arc weldments of alloy C-276 by current pulsing. Acta Metallurgica Sinica (English Letters). 2015 Feb 1;28(2):208-15.
[29] Manikandan M, Hari PR, Vishnu G, Arivarasu M, Ramkumar KD, Arivazhagan N, Rao MN, Reddy GM. Investigation of microstructure and mechanical properties of super alloy C-276 by continuous Nd: YAG laser welding. Procedia Materials Science. 2014 Jan 1;5:2233-41.
[30] Meena SL, Butola R, Murtaza Q, Jayantilal H, Niranjan MS. Metallurgical Investigations of Microstructure and Micro hardness across the various zones in Synergic MIG Welding of Stainless steel. Materials Today: Proceedings. 2017 Jan 1;4(8):8240-9.
[31] Moosavy HN, Aboutalebi MR, Seyedein SH, Goodarzi M, Khodabakhshi M, Mapelli C, Barella S. Modern fiber laser beam welding of the newly-designed precipitation-strengthened nickel-base superalloys. Optics & Laser Technology. 2014 Apr 1;57:12-20.
[32] Nagesh DS, Datta GL. Prediction of weld bead geometry and penetration in shielded metal-arc welding using artificial neural networks. Journal of Materials Processing Technology. 2002 Apr 30;123(2):303-12.
[33] Narwadkar A, Bhosle S. Optimization of MIG welding parameters to control the angular distortion in Fe410WA steel. Materials and Manufacturing Processes. 2016 Dec 9;31(16):2158-64.
[34] Pandit S, Joshi V, Agrawal M, Manikandan M, Ramkumar KD, Arivazhagan N, Narayanan S. Investigations on mechanical and metallurgical properties of dissimilar continuous GTA welds of Monel 400 and C-276. Procedia Engineering. 2014 Jan 1;75:61-5.
[35] Planckaert JP, Djermoune EH, Brie D, Briand F, Richard F. Modeling of MIG/MAG welding with experimental validation using an active contour algorithm applied on high speed movies. Applied Mathematical Modelling. 2010 Apr 1;34(4):1004-20.
[36] Ramkumar KD, Joshi V, Pandit S, Agrawal M, Kumar OS, Periwal S, Manikandan M, Arivazhagan N. Investigations on the microstructure and mechanical properties of multi-pass pulsed current gas tungsten arc weldments of Monel 400 and Hastelloy C276. Materials & Design. 2014 Dec 1;64:775-82.
[37] Saravanan S, Raghukandan K, Kumar GS. Comparison of numerical and experimental macrostructure in Nd: YAG laser welding of Hastelloy C-276. Optik. 2019 Feb 1;180:562-8.
[38] Sims CT, Stoloff NS and Hagel WC., 1987.Superalloys II, John Wiley and Sons, NewYork,
[39] Son JS, Park MH, Jin BJ, Yun TJ, Kim IS. A Study on Modeling and Controlling Bead Height in The Robotic GMA Welding. International Journal of Applied Engineering Research. 2017;12(13):3631-8.

[40] Sripriyan K, Ramu M. Analysis of weld bead characteristics on GMAW by changing wire electrode geometry. Journal of Achievements in Materials and Manufacturing Engineering. 2016;78(2):49-58.
[41] Sumesh A, Nair BB, Rameshkumar K, Santhakumari A, Raja A, Mohandas K. Decision tree based weld defect classification using current and voltage signatures in GMAW process. Materials Today: Proceedings. 2018 Jan 1;5(2):8354-63.
[42] Wu DJ, Ma GY, Niu FY, Guo DM. Pulsed laser welding of Hastelloy C-276: high-temperature mechanical properties and microstructure. Materials and manufacturing processes. 2013 May 1;28(5):524-8.
[43] Yang L, Li E, Fan J, Long T, Liang Z. Automatic extraction and identification of narrow butt joint based on ANFIS before GMAW. The International Journal of Advanced Manufacturing Technology. 2019 Jan 16;100(1-4):609-22.
[44] Yu P, Xu G, Gu X, Zhou G, Tian Y. A low-cost infrared sensing system for monitoring the MIG welding process. The International Journal of Advanced Manufacturing Technology. 2017 Oct 1;92(9-12):4031-8.
[45] Zhou S, Ma G, Dongjiang W, Chai D, Lei M. Ultrasonic vibration assisted laser welding of nickel-based alloy and Austenite stainless steel. Journal of Manufacturing Processes. 2018 Jan 1;31:759-67.

Predict the Tensile Strength of Friction Welded Steel/ASS304L Dissimilar Joints

C Rajendran[1], E Suryaprakash[2], V Sivasankar[3], P Suryaprakash[4], and K Rohith[5]

1 Associate Professor, Sri Krishna College of Engineering and Technology, Department of Mechanical Engineering, Coimbatore, Tamil Nadu, India.
2-5, UG Scholar, , Sri Krishna College of Engineering and Technology, Department of Mechanical Engineering, Coimbatore, Tamil Nadu, India.

rajendranc@skcet.ac.in, 18eumc158@skcet.ac.in, 18eumc144@skcet.ac.in, 18eumc159@skcet.ac.in, 18eumc123@skcet.ac.in

Abstract. Due to their availability and unique features, mild steel and ASS 304 L were used for various applications such as shipbuilding, boiler, aircraft, and automotive industries. This alloy is welded by using method of fusion welding. In addition, the strength of welded joints has touched 60 percent of the strength of base materials. However, it is difficult to maintain quality weld when dissimilar steel is joined. Therefore, the choice of electrode, filler wire and other process parameters like current, voltage, shielding gas, etc. requires proper attention. Solid state joining process, friction welding (FW) process is an ideal process to get sound joint. Three major FW parameters such as rotational speed, friction force, and forging force have been used in this work to optimize FW parameters to maximize strength using statistical methodology such as variance analysis and experimental design methodology. From the results, the joint is welded at a rotational speed of 1000 rpm, forging pressure of 30 MPa and friction pressure of 30 MPa

Keywords: Friction welding, Tensile strength, Mild steel, ASS304 L, ANOVA, RSM

1. Introduction

The corrosion resistant steel, particularly austenitic stainless steel (ASS), typically have good resistance to hydrogen embrittlement, weld ability and formability. Although exhibits high strength and ductility. However, it conceived significantly lower strength in hot condition [1]. The 304 grade is the most widely used in the 300series austenitic stainless steels, due to its good toughness at low temperature and corrosion resistance [2, 3]. Low-carbon steel is steel in which carbon in the range of 0.12 2.0 percent is the predominant interstitial alloying constituent. There are a variety of issues that need to be discussed in the arc welding of two different materials, in addition to those associated with welding of identical materials. Second, from a practical point of view, if the melting points of the two materials are too different, it might not be feasible to make a fusion weld, since it is important to have regulated melting simultaneously on both sides of the joint [4-10]. Second, even though this condition is met, if the two materials are metallurgically incompatible, it might not be possible to create an acceptable joint [11-15]. Metallurgical inconsistency can result in uncontrollable welding of metal / HAZ cracking or a welded metal microstructure which cannot provide adequate mechanical or corrosion efficiency, such as containing inappropriate martensite or intermetallic phases[16-22]. Even though it is possible to prevent cracking and deposit sufficient weld metal, there are other potential issues. There will be a band adjacent to the fusion boundary, usually very narrow, within which a steep composition the gradient of the melting point may be present. It is known from previous work that welding of dissimilar materials by friction welding is scarce.

2. Experimental Work

In this experiment, mild steel and ASS304 L were used as parent metal (PM)for friction welding, the chemical properties and characterization of mechanical data of which are shown separately in Table 1 and Table 2 respectively.

Table 1 Chemical composition (% wt.) of parent metals

Materials	C	Si	Mn	Ni	Cr	Cu	Fe
MS	0.13	0.11	0.39	0.01	0.02	0.02	99.1
ASS304 L	0.09	0.29	8.80	0.41	14.0	1.2	74.53

Table 2 Characterization mechanical data of parent metal

Material	0.2% yield strength (MPa)	Micro hardness 0.5kg, 15 s (Hv)	% Elongation 50 mm gauge length	Ultimate tensile strength (MPa)
MS	177	139	14	223
ASS304L	649	157	57	383

A 75 mm length and 12 mm diameter rod used for this investigation. The joints were created using an FW machine with numerical control (RV-FW, capacity-30kN, spindle speed-3000 rpm,). In order to find the feasible range of each process parameter, trail runs were performed (Fig.1). Each process parameter was modified from minimum to maximum, while the others were kept constant. The main influencing factors and their levels are shown in Table 3.

F (MPa/Ssec.)
F < 20 MPa/sec
F > 40 MPa/sec
P (MPa/sec.)
P < 20 MPa/sec
D > 40 MPa/sec
N (rpm)
(N) < 800 rev/sec
(N) > 1200 rev/sec

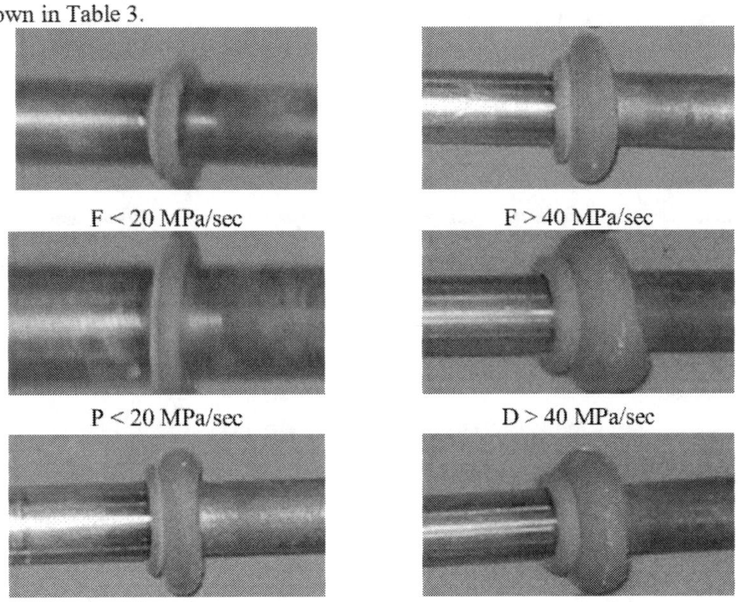

Fig.1 Working range of friction welding parameters

Table 3 Working window of FW parameters

Factors	Levels				
Rotation speed "rpm"	-1.682	-1	0	+1	1.682
	800	900	1000	1100	1200
Friction pressure " MPa"	20	25	30	35	40
Forging Pressure "MPa"	20	25	30	35	40

Design expert software was used to limit the number of trails. Three variables were selected to reduce no experimental conditions, five level composite rotatable concept matrixes. The welded joints photograph, as shown in Fig.2.

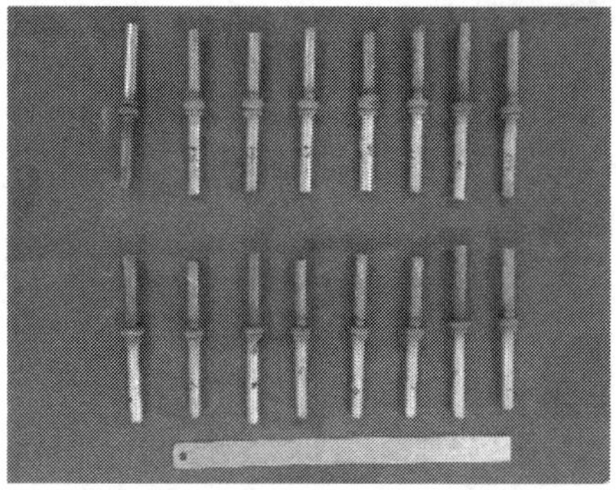

Fig.2 Photograph of Fabricate FW joints

Where, X is the coded value needed for a variable from X min to X max; X min is the variable's lowest level; and X max is the variable highest.

$Xi = 2 [2X - (X_{max}+X_{min})] / (X_{max}-X_{min})$ [1]

Tensile specimens were manufactured using the ASTM E8M04 standard to test the tensile strength of the joints. UTM was used to measure the strength of each joint, and the average results of the three specimens are shown in Table.4.

3. Formulating an empirical relationship

Joining parameters such as friction pressure (F), rotation speed (N), forging pressure (P) are calculated by the tensile strength of the FW 304L with AISI 1020 bimetallic rod.

TS=f (P, F, N) [2]

The chosen polynomial equation is expressed as

$TS= b_0 + b_1 (N) + b_2 (F) + b_3 (P) + b_{11} (N^2) + b_{22} (F^2) + b_{33} (P^2) + b_{12} (NF) + b_{13} (NP) + b_{23} (FP)$ [3]

To measure the coefficient of each factor, ANOVA was used. The student t and P test values were determined using ANOVA and shown in Table 5. Values of less than 0.05 for "Prob > F" mean that the terms of the model are important. Important model words are in this case N, F, P, NF, NP, FP, N^2, F^2, and P^2. Values more than 0.1 are negligible in terms of model. By uses of this coefficient, the final equation has been established and is given below.

$TS = \{ 491.15 + 6.42 (N) + 9.62 (F)- 4.33 (P) + 7.72 (NF) - 1.24 (NP) + 2.0(FP)-9.86(N^2)-29.15(F^2)-17.20(P^2)\}$ MPa [4]

Table 4 Experimental runs and its response

	Coded value			UTS	Average UTS
Exp.no	Rotation speed (N) "rpm"	"MPa"	Forging pressure(P) "MPa"	"MPa"	"MPa"
1	-1	-1	-1	426,430,432	429
2	1	-1	-1	428,431,437	433
3	-1	1	-1	428,432,436	432
4	1	1	-1	440,460,480	460
5	-1	-1	1	418,422,426	422
6	1	-1	1	410,418,414	414
7	-1	1	1	432,420,426	426
8	1	1	1	450,462,456	456
9	-1.682	0	0	442,462,452	452
10	1.682	0	0	480, 474,462	472

11	0	-1.682	0	390,392, 391	391
12	0	1.682	0	415,425,414	424
13	0	0	-1.682	438,442,432	448
14	0	0	0	430,440,432	434
15	0	0	1.683	490,492,492	492
16	0	0	0	480,488,481	487
17	0	0	0	485,497,491	491
18	0	0	0	490,496, 490	492
19	0	0	0	480,498, 489	489
20	0	0	0	480,485,505	490

4. Optimization of FW parameters

To optimize the FW parameters considered in this research, the response surface methodology (RSM) wasused.RSM is a series of numerical methods that are important for designing a number of tests, Constr ucting a computational formula, determining the optimal combination of various parameters and presentin g the value graphically.[15-16]. Figure 3 shows the perturbation plot for FW joint answer TS. This graph provides the shape of the response and depicts the TS transition as each FW variable shifts from the point of reference.

Table 5 ANOVA test results

Source	SS	Dof	Mean Square	F Value	p-value	
Model	18345.4	9	2038.3	436.5	< 0.0001	significant
N	562.2	1	562.3	120.4	< 0.0001	
F	1266.1	1	1266.1	271.1	< 0.0001	
P	259.5	1	259.6	55.6	< 0.0001	
NF	480.4	1	480.5	102.9	< 0.0001	
NP	12.4	1	12.5	2.6	0.132	
FP	32.0	1	32.0	6.8	0.025	
N^2	1404.8	1	1404.93	300.9	< 0.0001	
F^2	12239.0	1	12239.11	2621.3	< 0.0001	
P^2	4312.1	1	4312.28	923.6	< 0.0001	
Residual	46.7	10	4.67			
Lack of Fit	27.7	5	5.57	1.47		Not significant
Pure Error	18.7	5	3.77			
Cor Total	392.2	19				

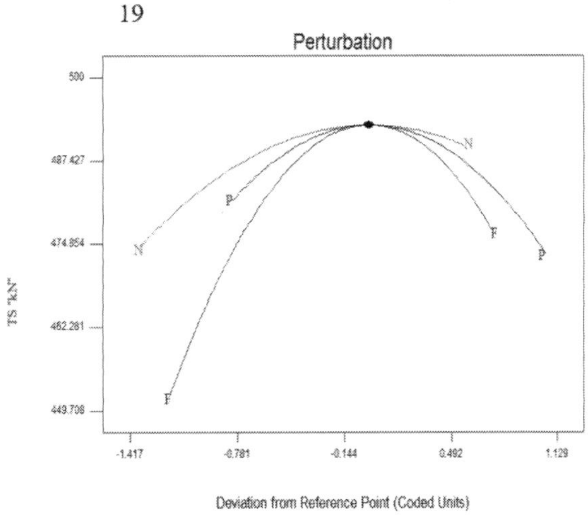

Fig. 3 Pertubation graph

Fig. 3 shows the halfway tier, and deviation from reference and its response. These contours of response will assistance to predict TS of any region. Peak of the response plot indicates the overall TS achieved. RSM plays a very key role in studying the surface of a result.

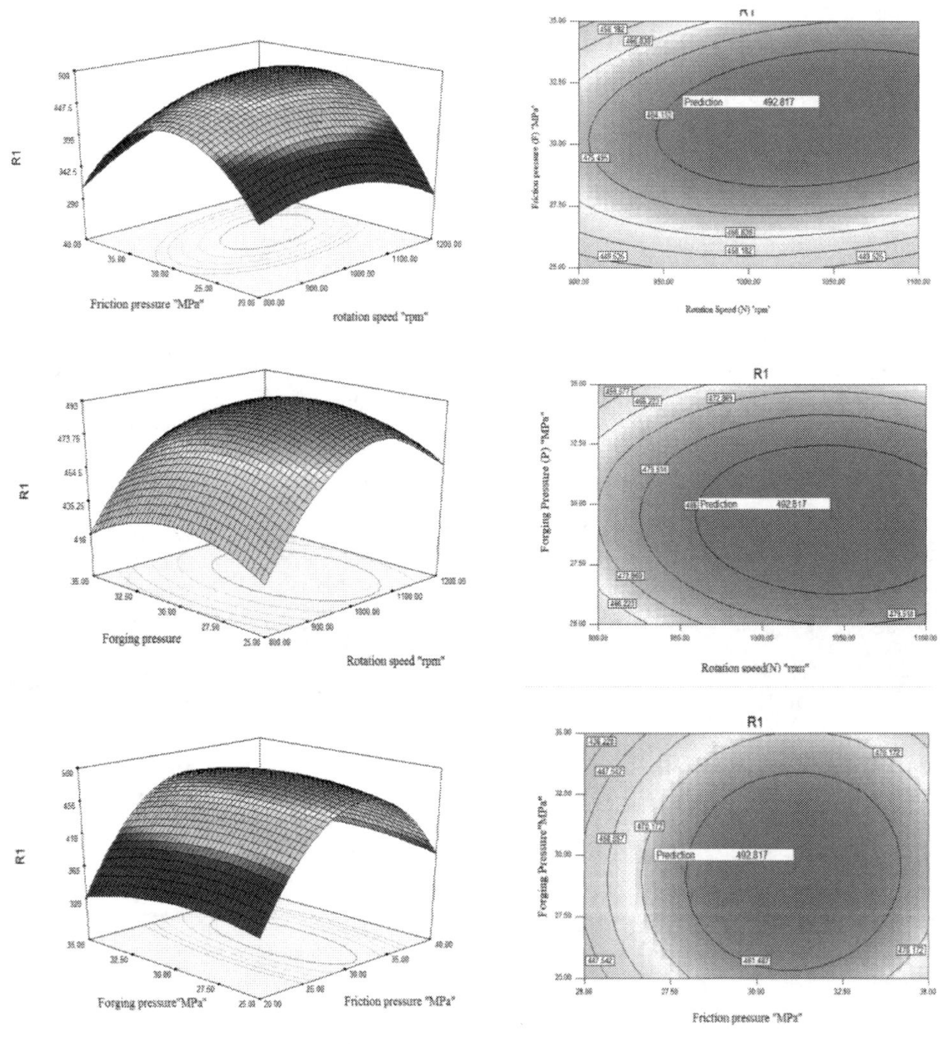

Fig. 4 Response surface graphs and contour plots

It is obvious from this that when the TS increase to a certain value by increasing the friction strength, rotational speed and forging strength and then lowered. It is also well-known that the initial rise in friction pressure raises the TS to the certain value, and additional friction pressure keeps a TS stable. The maximum achievable TS value is found to be 492.817 MPa. From the RSM plots (Fig.4). A rotational speed of 1041.92 rpm, a friction pressure of 31.08 MPa, and a forging pressure of 29.36 MPa are the corresponding parameters that yield this optimum value. The higher value of the F ratio means the respective levels are more important. It can be inferred from the F ratio value that the key factor leading to the exploitation of TS is the friction pressure, accompanied by the rotational speed and the forging pressure for the range considered in this investigation. Three confirmation experiments with the FW parameters selected independently from the viable working range were performed to validate the established relationship [21]. The average of three measured outcomes was determined as the actual response. The experimental values, expected values and the variance of FW joints are summarized in Table 6.

Table 6 Validation test results of optimization procedure

Sl. No	Rotation speed "rpm"	Friction pressure "MPa"	Forging pressure "MPa"	Actual strength "MPa"	Predicted strength "MPa"	% Variation
1	900	35	25	431,430,431	430.93	+0.07
2	1200	30	30	493,492,490	491.4	+0.6
3	1100	25	35	414,410,418	414.2	-0.20

Fig.5 shows the fractograph of tensile tested sample at rotational speed of 1000 rpm, forging pressure of 30 MPa, and frictional pressure of 30 MPa. It reveals shear ridges and populated dimples with varying in size and shape. It implies that large stretch zone was present at the tip of crack. This indicates the mode of fracture was ductile. It may be attributed due to optimal heat input during friction welding of dissimilar steel.

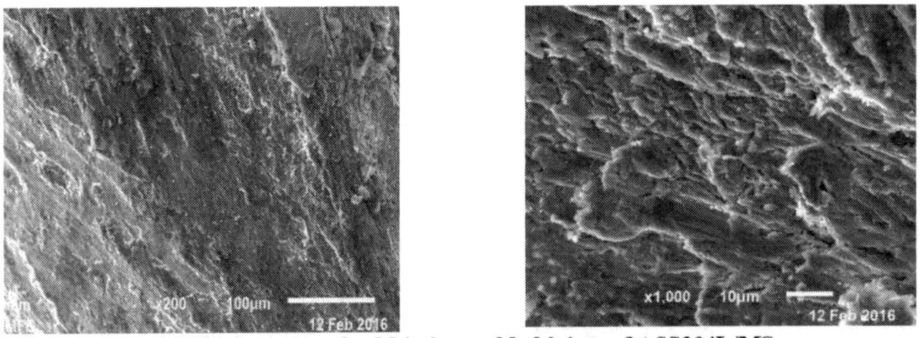

Fig. 5 SEM fractograph of friction welded joints of ASS304L/MS

Conclusions

i) Parameters for the friction welding process have been optimized using surface response methodology to achieve optimum joints tensile strength. It is observed that these optimized values are quite similar to the experimental values.

ii) Under rotational speeds of 1000 rpm, frictional pressure of 30 MPa and forging pressure of 30 MPa, a maximum ultimate tensile strength of 492 MPa could be achieved.

iii) The combination of friction welding parameters, i.e., friction pressure of 31.08 MPa / s, forging pressure of 29.36 MPa / s and rotational speed of 1041,92 rev / s yielded maximum tensile strength (492.81 MPa) of dissimilar steel joints.

References

[1] Radha, R.M., Visnu, K.T., and Rajesha, S,2014, A Study of Tensile Strength of Mig and Tig Welded Dissimilar Joints Of Mild Steel And Stainless Steel, International Journal of Advances in Materials Science and Engineering, Vol.3(2),pp 23-32.

[2] Pouraliakbara, B.H., Hamedia, M., Kokabia, A.H., and Nazarib, A, 2014, Designing of CK45 Carbon Steel and AISI 304 Stainless Steel Dissimilar Welds, Materials Research, Vol. 17(1), pp 106-114.

[3] Tayyab, I: Analysis of Dissimilar Metal Welding of 1020 Mild Steel and 304 Stainless Steel, 2014, Master Thesis, Department of Mechanical Engineering, National Institute of Technology, Rourkela.

[4] Oladele, I.O.; Omotoyinbo, J.A.; and Akinwekomi, A.D, 2010, The Effect of Weld Goemetry and Post Weld Heat Treatment on the Corrosion Behaviour of Austenitic Stainless Steel Immersed in 1.0 M NaCl Solution, Material Research, vol.13(1), pp 405-418.

[5] Giridharan, S. Kannan, T.M., Balamurugan, K.; Kumar, A.., and Vignesh: Experimental Investigation and Analysis of Dissimilar welding of AISI 316L and IS 2062 using GTAW, ,2016,International Journal of Innovative Research in Science, Engineering and Technology, vol.5(6), pp 11052-11058.

[6] Oladele, I.O., Betiku, O.T., and Fakoya, M.B,2017, Effect of Post Weld Heat Treatment on the Mechanical and Corrosion Behaviour of Welded Al-Fe-Si Alloy Joints, Leonardo Electronic Journal of Practice and Technologies,vol.30, pp. 75-86.

[7] Tanaka, T.; Morishige, T.; and Hirata, T, 2009, Comprehensive Analysis of Joint Strength for Dissimilar Friction Stir Welds of Mild Steel to Aluminum Alloys, Scripta Materialia, vol. 61, pp 756–759.

[8] Shahid, F., Khan, A.A., and Hameed, M.S, 2015, Mechanical and Microstructural Analysis of Dissimilar Metal Welds, International Journal of Research and Reviews in Applied Sciences, vol.25(1),pp 6-14.

[9] Jamaludin, S.B.; Noor, M.M.; Kadir, S.K.; and Ahmad, K.R, 2013, Mechanical Properties of Dissimilar Welds between Stainless Steel and Mild Steel, Advanced Materials Research, vol.795,pp 74-77.

[10] ASTM A370-08A: Standard Test Methods and Definitions for Mechanical Testing of Steel Products, 2008, ASTM international.

[11] ASTM E190-92: Standard Test Method for Guided Bend Test for Ductility of Welds, , 2008, ASTM International, Re-approved.

[12] Muralimohan, C.H.; Haribabu, S.; Hariprasada, R., Muthupandi, V and Siva Prasad, K, 2014, Evaluation of Microstructures and Mechanical properties of Dissimilar Materials by Friction Welding, Procedia. Materials Science, vol.5, pp 1107-1113.

[13] Andrés, G.F., Rafael, S.A., Leiry, C.B., and Alberto, V.R, Crack growth study of dissimilar steels (Stainless-Structural) butt welded unions under cyclic loads,2011, Procedia. Engineering, vol.10,pp.1917–1923.

[14] Hsieh C.C., Lin D.Y., Chen M.C.., and Wu W, 2007, Microstructure, Recrystallization, and Mechanical Property Evolutions in the Heat-Affected and Fusion Zones of the Dissimilar Stainless Steels, Materials Transactions, vol. 48(11),pp 2898-2902.

[15] Rajendran, C., K. Srinivasan, V. Balasubramanian, H. Balaji, and P. Selvaraj. 2018,Identifying the combination of friction stir welding parameters to attain maximum strength of AA2014-T6 aluminum alloy joints. Advances in Materials and Processing Technologies vol.4, no. 1, pp. 100-119.

[16] Rajendran, C., K. Srinivasan, V. Balasubramanian, H. Balaji, and P. Selvaraj. 2019, Identifying combination of friction stir welding parameters to maximize strength of lap joints of AA2014-T6 aluminium alloy. Australian Journal of Mechanical Engineering vol.17, no. 2, pp 64-75.

[17] Fewell M. P. and J. M. Priest, 2008, High-order diffractometry of expanded austenite using synchrotron radiation, Surf. Coatings Technol., vol. 202,no 9, pp 1802–1815).

[18] Christiansen T and M. A. J. Somers,2004, On the crystallographic structure of S-phase, Scr. Mater., vol.50, no.1,pp 35–37.

[19] Farrell K, E. D. Specht, J. Pang, L. R. Walker, A. Rar, and J. R.2005, Mayotte, Characterization of a carburized surface layer on an austenitic stainless steel, J. Nucl. Mater., vol.343,no 1–3, pp 123–133.

[20] Michal G M, F. Ernst, and A. H. Heuer,2006, Carbon para equilibrium in austenitic stainless steel, Metall. Mater. Trans. A, vol. 37, no. pp 1819–1824.

[21] Oddershede J, T. L. Christiansen, K. Ståhl, and M. A. J. Somers, Extended X-ray absorption fine structure investigation of nitrogen stabilized expanded austenite, Scr. Mater., vol.62, no 5, pp 290–293.

[22] Oddershede, J, T. L. Christiansen, K. Ståhl, and M. A. J. Somers,2011, Extended X-Ray absorption fine structure investigation of carbon stabilized expanded austenite and carbides in stainless steel AISI 316, Steel Res. Int., vol.82, no 10,pp 1248–1254.

Flow conditions analysis of Gas Turbine Combustor

R Abinaya[1], D Rohini[2], B Soundarya[3], S Balakrishnan[4], B Dakshinamurthy[5] and P Rajeshwari[6]

1, 2 Assistant Professor, Bannari Amman Institute of Technology, Department of Aeronautical Engineering, Erode, Tamilnadu, India.
3,4,5,6 Student, Bannari Amman Institute of Technology, Department of Aeronautical Engineering, Erode, Tamilnadu, India.

abirahul5@gmail.com, rohinidharamaraj90@gmail.com, soundarya.ae19@bitsathy.ac.in, balakrishnan.ae19@bitsathy.ac.in,dakshinamurthy.ae19@bitsathy.ac.in, rajeswari.ae19@bitsathy.ac.in

Abstract. Gas turbines are equipped for working effectively on a wide assortment of energizes. Combustor is directed to a great extent by the requirement for its length and frontal territory to stay inside the cut-off points set by other motor segments by the need for a diffuser to limit pressure problems, another issue of expanding significance is that acoustic reverberation. The decrease of combustor size and weight is another significant necessity for air motors. The aim of the present work is to analyse the combustion chamber geometry using FINE/OPEN of NUMECA software and to predict the hot and cold total pressure loss, (circumferential and radial) pattern factors and combustion efficiency. This paper mainly work towards understanding the effect of cold and hot flow in a practical combustor systems using CFD and to estimate the various parameters like pressure loss, pattern factor (circumferential and radial), velocity distribution through diffuser, dome, swirler, three zones and combustor holes. The flow analysis is further extended to investigate the effects of temperature distribution of the flow that reaches the turbine inlet. The combustor model is created using CATIA and parts are then cleaned individual using CADfix suitable for meshing. The performance of the turbulence model is analysed and compared. Present work deals with the implementation of Computational Fluid Dynamics (CFD) for three-dimensional analysis of flow in an annular combustor of an aero gas turbine engine. FINE/OPEN of NUMECA commercial code will be used to analyse the Combustion chamber geometry. The analysis is carried out both in FLUENT and NUMECA and the results are compared.

Keywords: Combustor, pressure loss, pattern factor, CADfix model, Fluent and NUMECA analysis

1. Introduction

Air-breathing plane impetus began toward the start of the twentieth century. The working liquid enters as natural air which is ducted through a channel diffuser into the motor, the motor fumes gas comprises somewhat of burning gas and halfway of air. The fumes gas is extended through a push spout or spouts to surrounding pressure. Thrust can be characterized as the demonstration of pushing or the condition impelled. It states that the study of propulsion includes the study of the propelling force, the motion caused, and the bodies involved. Propulsion involves an object to be propelled and additional bodies, called propellant. Combustor is directed generally by the requirement for its length and frontal region to stay inside the cutoff points set by other motor segments by the need for a diffuser to limit pressure misfortune and by the prerequisite of a liner (fire tube) to give stable activity over a wide scope of air/fuel proportions. During the past 50 years, burning weights have ascended from 5 to 50 climates, delta air temperatures from 450 to 900K and outlet temperatures from 1100 to 1850K. The present combustors display near 100% ignition productivity over their typical working reach. Moreover the future of air motors liners has ascended from simply a couple hundred hours to a huge number of thousands of hours. Gas turbines are fit for working productively on a wide assortment of fills: strong, fluid and gas. Another issue of expanding significance is that acoustic reverberation. It happens when ignition insecurities become combined with the acoustics of the combustor. This issue is significant for the future advancement of lean premixed combustors. The decrease of combustor size and weight is another significant prerequisite for air motors. This prompts the exploration on new materials and new techniques for creation. This rearranges essential combustor plan and lessens cost. It has brought about the improvement of cutting edge divider cooling strategies. The metal suitable temperatures can be upgraded by the utilization of warm boundary hard-headed coatings inside the ignition liner. The ignition cycle is significant in a gas turbine cycle as the cycle of compound energy of the fuel is changed over to warm energy. It is changed over into work by the turbine. Burning is maybe portrayed most basically as an exothermic response of a fuel and an oxidant. In gas turbine applications, the fuel might be vaporous or fluid, yet the oxidant is consistently air. Ignition happens in numerous structures, not which are all joined by fire or iridescence.

- Deflagration
- Detonation
- Flame-tube cooling

Requirements of Combustion Chamber,

- Combustion efficiency should be higher (i.e., the fuel should be completely.)
- Ignition should be reliable and smooth
- An outlet temperature dispersion (design factor) that is custom fitted to augment the daily routines of the lives of the turbine cutting edges and spout direct vanes.

Factors affecting combustor design,

- The temperature of the gases after burning must be relatively low to suit the exceptionally focused on turbine materials.
- Toward the finish of the ignition space the temperature dispersion must be of known structure if the turbine cutting edges are not to experience the ill effects of neighborhood overheating.
- Ignition must be kept up in a surge of air moving with high speed in the area of 30-60 m/s and stable activity is required over a wide scope of air/fuel proportion from full burden to sitting conditions.

2. Methodology

The stream in the combustor is profoundly perplexing. The significant plan necessities rely upon the interior optimal design. The combustor design comprises of packaging, liner and fuel injector. The fast stream in the combustor lessens the living time accessible for fuel planning, blending and consuming of fuel with vitiated air.

The blending and burning cycle in a combustor are predominantly affected by stream design, distribution made by fire stabilizer, pressure misfortune and temperature impacts. To accomplish these prerequisites, the significant boundaries to be considered are:

- velocity distribution
- hot and cold pressure distribution
- pattern factor (circumferential and radial)

It is dominatingly guided by test techniques and past experience. Anyway trial techniques are naturally moderate and furthermore exorbitant particularly at motor working conditions. These disadvantages and developing need to comprehend the unpredictable stream field marvel included, has prompted the improvement of mathematical strategies for foreseeing streams in combustor.

These models are utilized to advance the plan of the combustor and lessen the volume of the trial work and cost included.

IOP Conf. Series: Materials Science and Engineering **995** (2020) 012043 doi:10.1088/1757-899X/995/1/012043

So as to reenact the genuine stream circumstance in a combustor, a far reaching model joining every one of these components are communicated through overseeing differential conditions. The arrangement of these conditions is tedious. Notwithstanding, with quick advancement in PC memory and speed, more sensible stream reenactments are progressively endeavored now. In the most recent decade's critical endeavors have been made to create mathematical techniques for anticipating stream in the combustor. Precise expectations of such streams are needed to plan a superior and dependable annular combustor.

3. Design of Model

The computational method dependent on the arrangement of preservation conditions of mass, force, energy and species are utilized. The disturbance impacts are recreated utilizing SpallaTalmalrs condition. The principle steps advanced in the examination of combustor are,

3.1. Software used for working procedure

- **Geometry modeling:** Modelling the Annular Combustor using CAD software (CATIA V5).
- **Geometry cleaning:** Geometry is cleaned using CADfix software that helps meshing easily.
- **Grid generation:** Grid generation for the sector model using HEXPRESS of NUMECA software.
- **Flow analysis**: Flow will be analysed using Fine/Open of NUMECA software.
- **Post Processing**: The results will be obtained using CFView of NUMECA software.

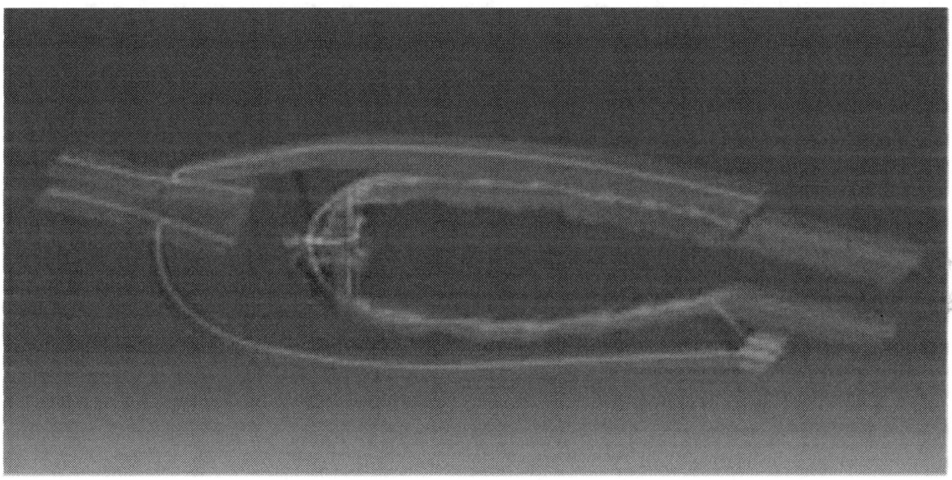

Figure 1: Gas turbine annular combustor

The model consists of a diffuser. The diffuser is attached with a manifold solid and the combustor design consists of an outer casing that is preceded with the diffuser. The combustor design consist of outer casings where the casings are seen as (2 layers at top and bottom) i.e. 2 inner liners which is covered by 2 outer liners. It covers the inner liner of the combustor. The upstream hemisphere end of the liner is called dome. At the center of the dome is swriler is located and within the swriler igniter is present. The swriler is surrounded by the atomizer and are enclosed by the cooling rings and cowl.

IOP Conf. Series: Materials Science and Engineering **995** (2020) 012043 doi:10.1088/1757-899X/995/1/012043

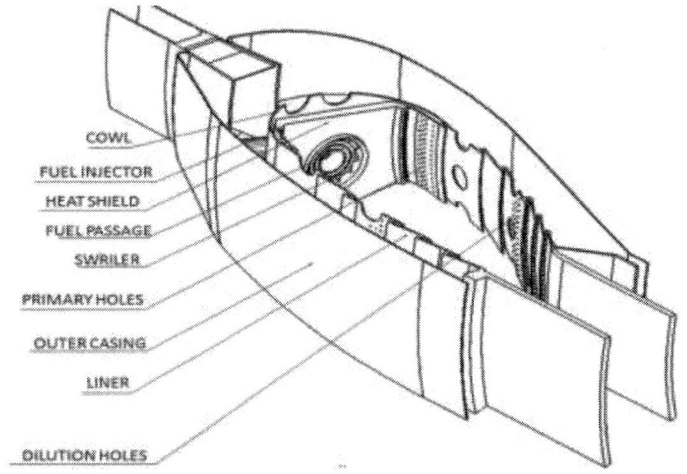

Figure 2: Annular combustor with parts

There is large number of small holes called flare holes to cool the interior surface. The liner and cowl will have heat shield with large number of holes to prevent it from burning. There are two big holes in the primary zone for allowing air for combustion and cooling process. At downstream of the liner there are large number small holes called muff holes that are located to reduce the temperature of the air that reaches the turbine. At appropriate points the liner will have slots and louvers to have film cooling air to control the temperature of the inner surface of the liner.

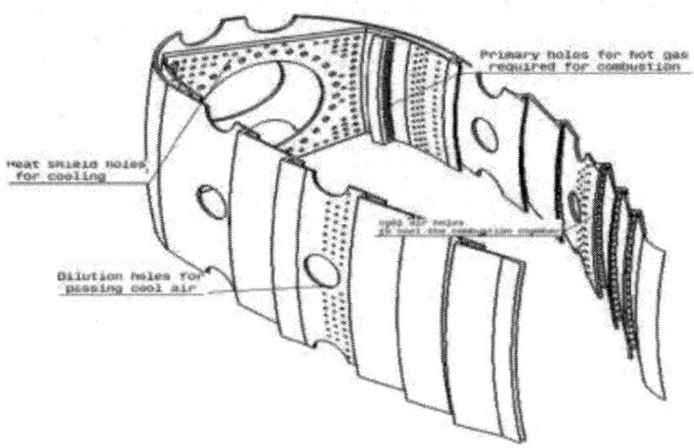

Figure 3: Annular combustor holes

To have an easy meshing the annular combustor in CATIA can be divided into outer, inner and cowl part separately.The inner liner part in CATIA is shown below where the holes are removed, closed and meshed separately. The combustor is cleaned separately using CADfix.

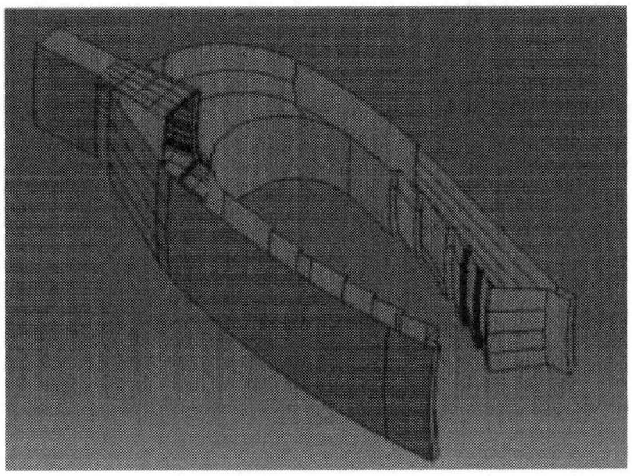

Figure 4: Outer casing

The CATIA parts are then cleaned singular utilizing CADfix appropriate for cross section. CADfix is driving CAD interoperability instrument which handles the ever present issues of 3D model information trade and re-use between various designing applications. CADfix permits the client to import CAD information, productively fix and adjust it, and fare in the most reasonable structure for reuse in the downstream framework, disposing of the requirement for costly CAD model revise.

3.2. CADfix application in the Model

The combustor model is then imported in CADFIX which helps to visualize and rectifies the surfaces, faces and unwanted double lines of CAD model. When the errors are rectified in CADFIX it is then suitable for importing in HEXPRESS of NUMECA software which requires a CAD clean, water tight model and where grids are generated and meshing is done.

CADfix is a software application that translates, heals and repairs (gaps, silver faces, reversed normal's, duplicate entities).Minimizes rework associated with CAD data exchange and obtaining usable geometry downstream.

Figure 5: CADfix cleaned holes

CADfix isn't a CAD system in which some data is so badly corrupted that it cannot be automatically repaired and its missing data is too complex to re-create in CADfix, e.g., fillets and blends.
For meshing the domain (combustor), HEXPRESS of NUMECA software is used. HEXPRESS can only understand closed models and requires cad clean and air tight model.

3.3. Domain discretization

When blunders are eliminated in CADFIX the combustor is then gets appropriate for bringing in HEXPRESS. Utilizing CAD control instrument bar the combustor model is checked. The combustor is imported in HEXPRESS as parasolid space record. For persuade the external packaging, liner, gaps and cowl part are imported independently in CADFIX and HEXPRESS by shutting the model in CATIA and afterward it is additionally amassed together.

The combustor is divided into 7 domains and grids are generated individually which are then connected using FNMBs. The vertices of the domain are connected to more than 3 edges.

Figure 6: Outer domain

3.4. Grid Generation

For meshing the domain (combustor), HEXPRESS of NUMECA software is used. HEXPRESS can only understand closed models and requires cad clean and air tight model. Hence the model is imported in CADFIX (software that create cad clean model) as separate parts and cleaned separately.

Figure 7: Meshed part of combustor

The values for the initial mesh and refinements for different domains in combustor is given in the table 1 and 2.

Table 1: Mesh details

Domains	Initial mesh X direction	Initial mesh Y direction	Initial mesh Z direction	No. of cells
Outer casing	16	7	4	448
Inner +heat shield	19	8	6	912
Cowl	6	9	8	432
Swriler	4	14	14	784
Swriler path	5	9	8	360
Fuel ring	10	10	10	1000
Fuel path	8	11	11	968

Table 2: Mesh requirements

Domains	Max no. of refinements	Curve refinements	Surface refinements	Target cell size
Outer casing	5	10,000	4	0,0,0
Inner + heat shield	7	10,000	4	0,0,0
Cowl	4	10,000	3	0,0,0
Swriler	3	10,000	3	0,0,0
Swriler path	4	10,000	4	0,0,0
Fuel ring	3	10,000	3	0,0,0
Fuel ring	4	10,000	3	0,0,0

4. Optimization

In general, combustion problems are characterized by a strong variation of density. Since combustion in most technical applications takes place at low speeds, the density variations are not caused by the variation in the pressure, but by the heat released from the exothermic reaction between fuel and oxidizer. Most technical combustion problems can therefore be simulated with models developed for low Mach number flow. From a mathematical point of view this means, that the pressure variations in the flow have to be small compared to the thermodynamic background pressure of the system. Based on this definition, even systems with a transient background pressure (e.g., internal combustion engines) can be modelled as low Mach number problems. In the most general approach, combustion is described as a convection-diffusion-reaction problem involving n species and m reactions between these species as described by a reaction mechanism.

The combustion problem in two sub problems,

- Mixture Fraction
- Radiation model

4.1. Validation case study

Validation is the process of deciding how much a model is a precise portrayal of this present reality from the viewpoint of the proposed employments of the model. Fuel Methane is provided from the focal aspect of the combustor. The air is provided from the annular surface which is called essential air stream. The essential air is additionally provided from the half of the length of the combustor.

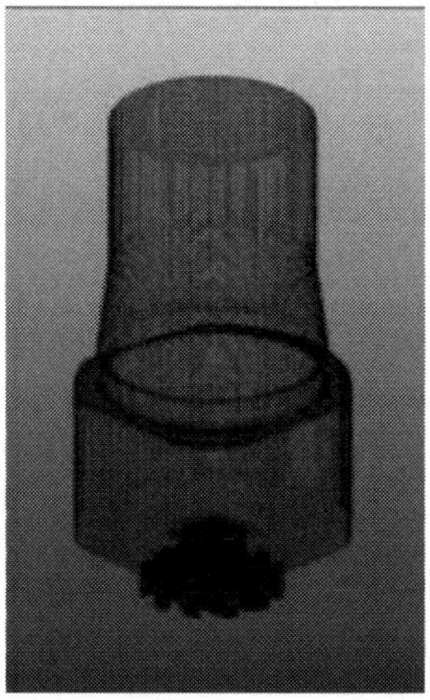

Figure 8: Grid of the combustor model

4.2. Boundary conditions

The boundary conditions are used to define the flow and thermal variables on the boundaries of the model. This section gives the details of the boundaries given for the flow simulation. Boundary conditions are employed so as to bring the computational domain in accordance with the real time conditions.

Inlet: Mass flow rate is specified at the core inlet and bypass inlet along with total temperature and turbulent intensity and hydraulic diameter.

Outlet: The boundary condition has been applied at the exit of the domain in the form of pressure outlet. Static pressure and total temperature are defined.

Periodic: The side faces are made periodic in order to map the flow solutions on each other. Periodic are defined on radial sectors to obtain identical solutions on either sides of the sector considered.

Physical Modelling

The time-averaged governing equations for fluid flow in the Cartesian tensor form,

- Continuity Equation
- Momentum Equation
- Energy Equation
- Navier-Stokes Equation

Table 3: Boundary conditions

Parameters	Inlet
Total Pressure	17 bar
Total Temperature	812K
Outlet Static Pressure	15,28000 bar
Mass flow	2.933 kg/s

IOP Conf. Series: Materials Science and Engineering **995** (2020) 012043 doi:10.1088/1757-899X/995/1/012043

Table 4: Outlet boundary conditions for the annular combustor

Regions	**Values**
Core outlet	2.2 kg/s
Bleed outer	½ 0.1965 kg/s
Bleed Inner 1	½ 0.117 kg/s
Bleed Inner 2	0.108 kg/s

The Boundary conditions applied in the combustor model is listed in the table 1 and the outlet boundary condition for the combustor model is provided in the table 2.

5. Results and Discussion

The comparison of models can also be made by observing the change in pressure from the diffuser to the dilution zone, temperature pattern, Mach no. variations and velocity variations along the annular combustor section.

Figure 9: Primary Zone

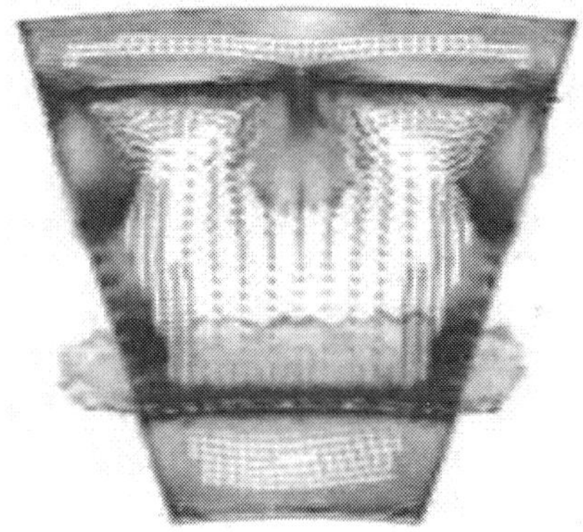

Figure 10: Primary Holes

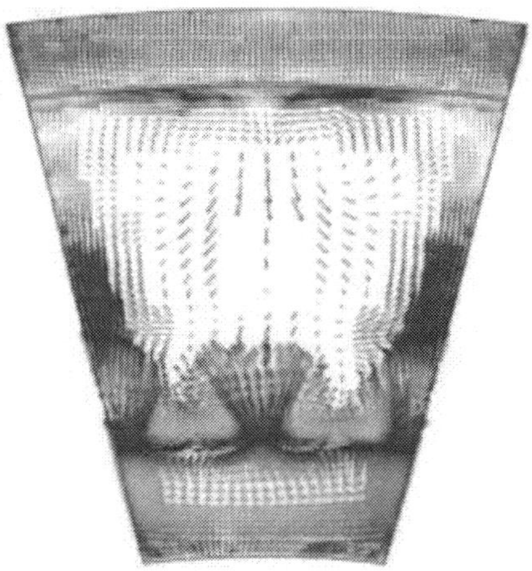

Figure 11: Intermediate Zone

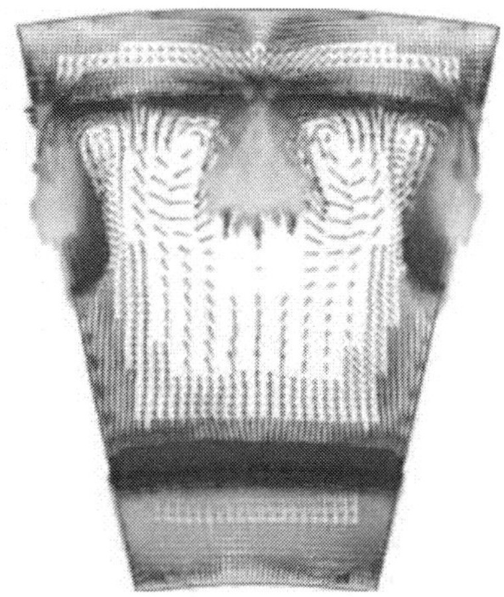

Figure 12: Exit of Intermediate

The velocity distribution near the primary zone is obtained as shown above. The velocity at the primary holes at the upper and lower part of the liner is high compared to the flow in the inner part of the liner. The velocity distribution at the cut planes i:e near the primary zones exist and near the holes of the intermediate zone are shown above

The velocity distribution is about 150m/s at the holes and quite low at the other sides. The mass flow is more near the lower part of the swirler exist hence the velocity distribution is more at that point. The velocity vectors and its distribution is low compared to the primary zone.

5.1. Flow Analysis

The combustor works in different modes i.e. the combustion can be premixed non premixed or partially premixed combustion. The analysis done in this report is for non-premixed combustion. The flow inside the combustor is hot and cold flow resulting in hot and cold pressure loss. This is due to the combustion process leading to flow called hot flow and dilution process (flow without combustion can also be included) leading to flow called cold flow.

The inlet total temperature at the primary zone is very high and the total temperature in the region where combustion takes place has a very high total temperature. As the flow advances, the temperature near the wall regions reduces gradually due to the cooling holes and boundary layer interactions and hence, the high temperature is concentrated at the center of flow.

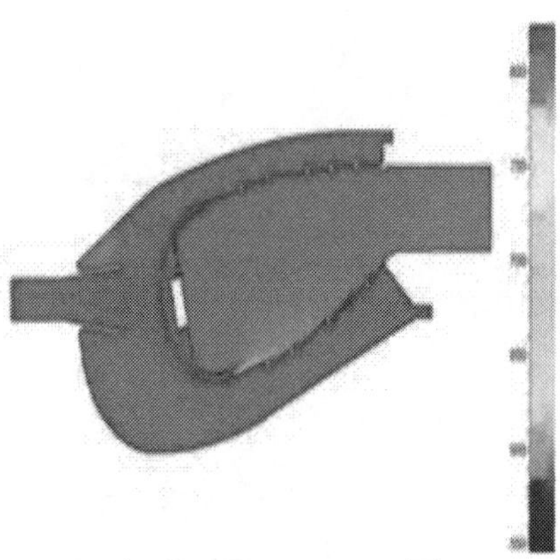

Figure 13: Absolute Temperature (K) cut section

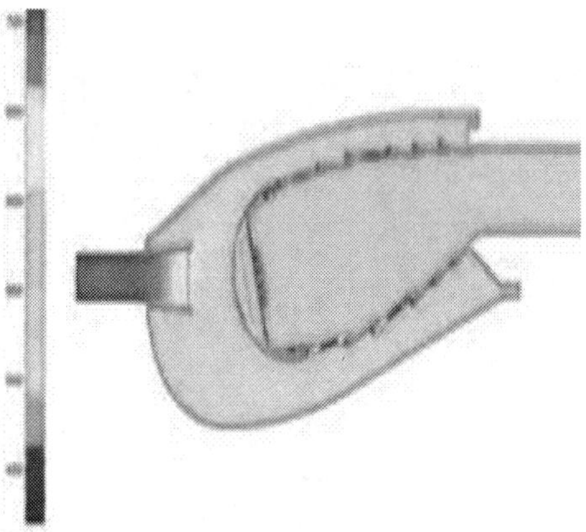

Figure 14: Absolute Temperature (K)

From the above analysis we can see the difference in temperature before and after combustion. The Absolute total temperature is increases drastically after combustion and reduces by cooling effects of slots and holes of the liner.

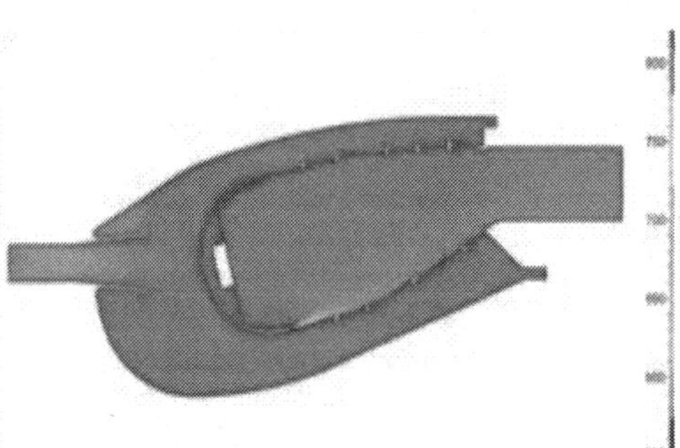

Figure 15: Static temperature (K) Cut section

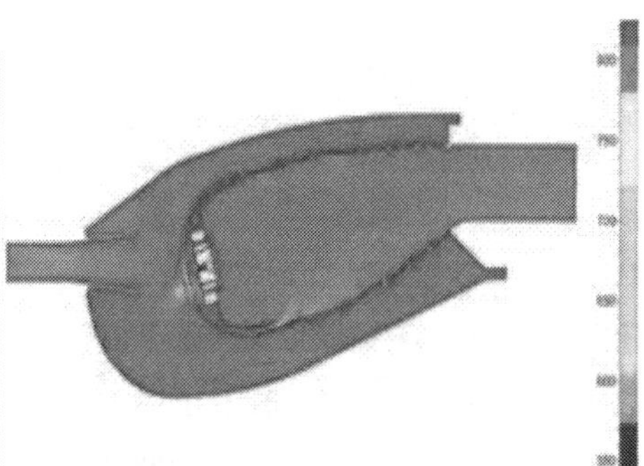

Figure 16: Static temperature (K)

The pressure at the diffuser increases by decreasing the velocity of the flow. The inlet total pressure gradually increases near the dilution zone. The total pressure loss is obtained by the summation of inlet, annulus and outlet pressure loss

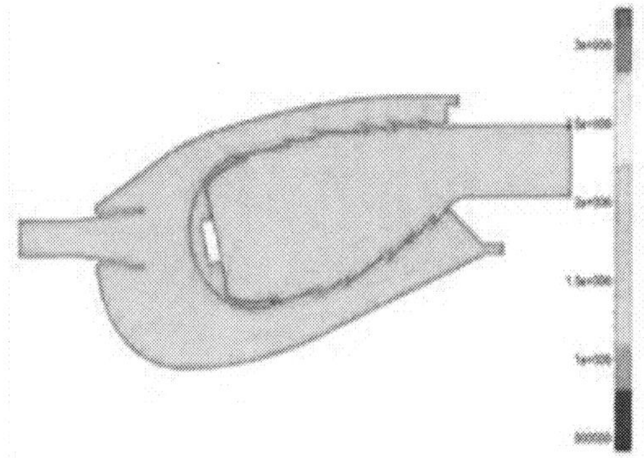

Figure 17: Absolute Total Pressure (Pa) Cut Section

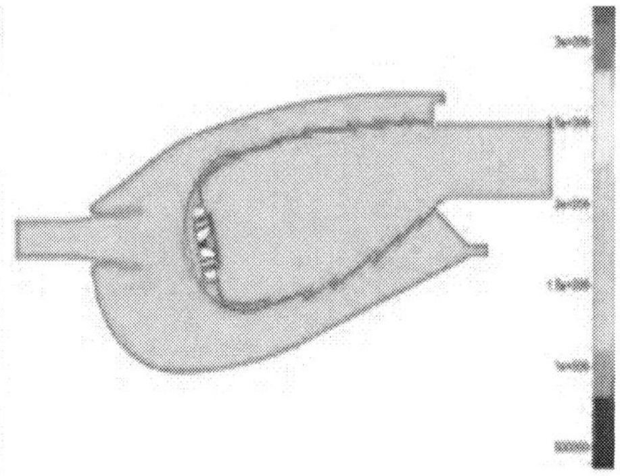

Figure 18: Absolute Total Pressure (Pa)

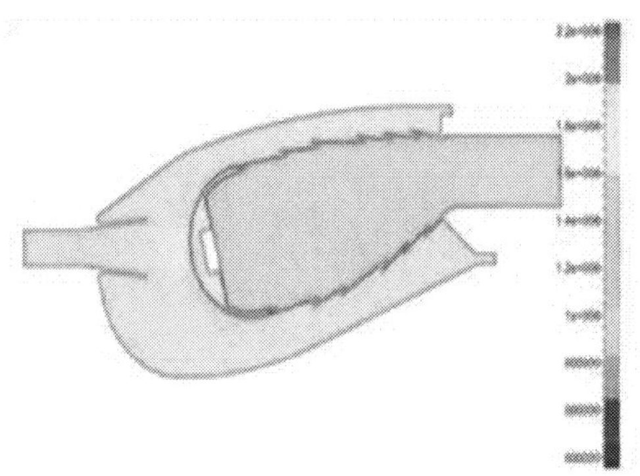

Figure 19: Static Pressure (Pa) Cut Section

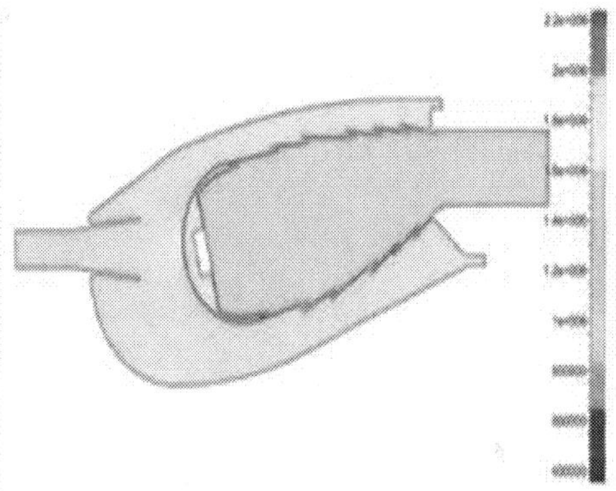

Figure 20: Absolute Static Pressure (Pa)

Table 5: Pressure Distribution

Pressure Distribution		
Region	**Static Pressure (Pa)**	**Absolute Temperature (K)**
Outer Annulus	200000	3e^6
Dome	1.6 e^6 -1900000	3e^6
Swriler	1800000	2e^6
Cooling ring	1600000	3e^6
Primary Holes	1.4e^6	3e^6 - 2e^6
Flare Holes	1.2e^6 -1.6 ^6	3e^6 - 2e^6
Dilution Holes	1800000	3e^6 - 2e^6
Muff Holes	180000	3e^6 - 2e^6

The pressure distribution at various parts of the combustor is given in the table3. The Mach number of the flow is obtained. It can be seen that the recirculation zone formed near the end of diffuser section and near the cowl is well defined.

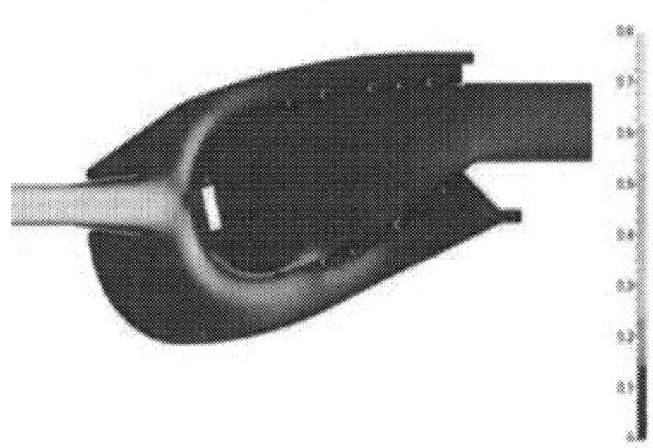

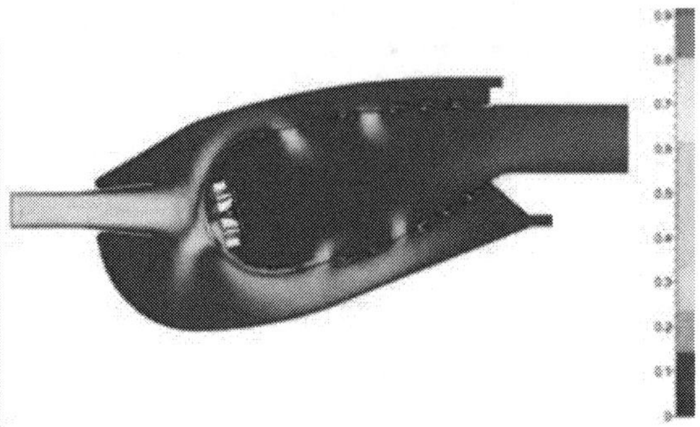

Figure 21: Absolute Mach number distribution in Cut Section **Figure 22:** Absolute Mach number

5.2. Comparison between fluent and NUMECA results

The annular combustor is also analysed using fluent software and the pressure loss results obtained from NUMECA is compared with the pressure loss results obtained from Fluent. The pressure loss at inner annulus, outer loss and the overall pressure loss are compared by plotting graph between Mach no. and Pressure loss.

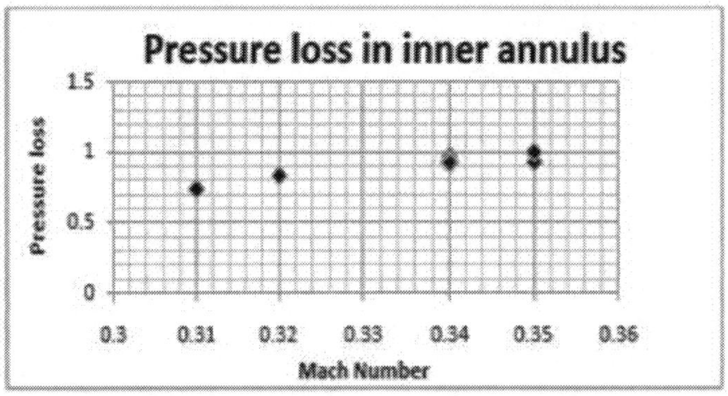

Figure 23: Graph of Pressure Loss at Inner Annulus (Fluent)

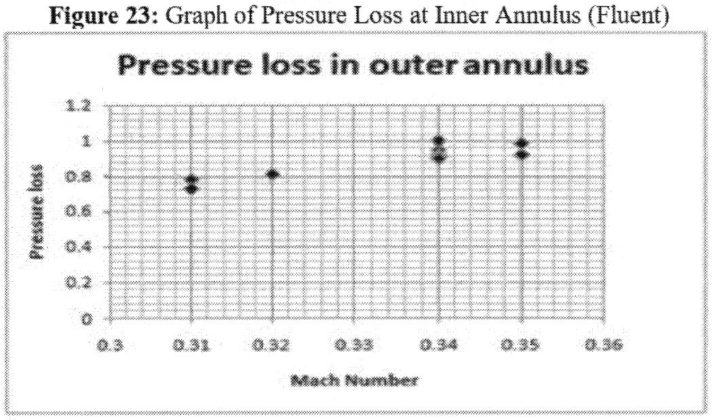

Figure 24: Graph of Pressure Loss at Outer Annulus (Fluent)

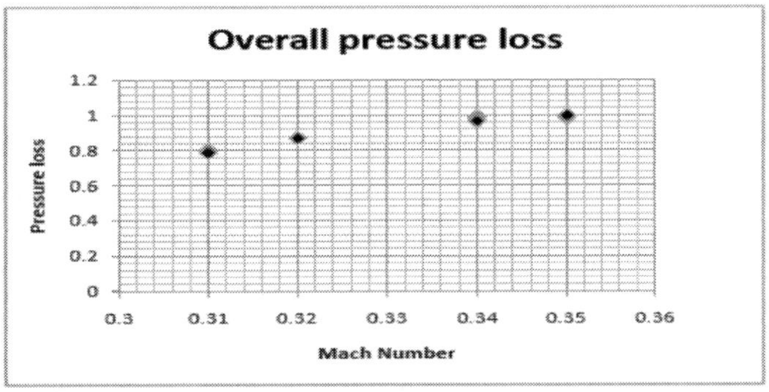

Figure 25: Graph of Pressure Loss at Overall Combustor (Fluent)

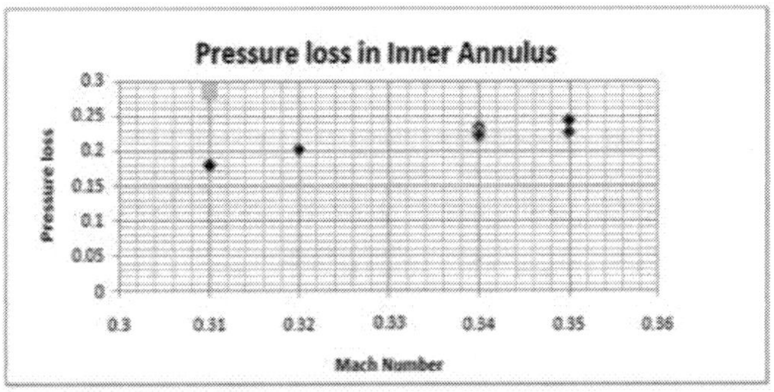

Figure 26: Pressure variation in inner annulus (NUMECA)

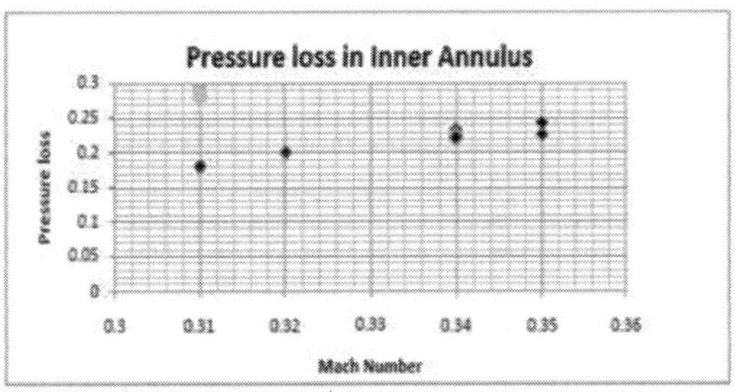

Figure 27: Pressure variation in outer annulus (NUMECA)

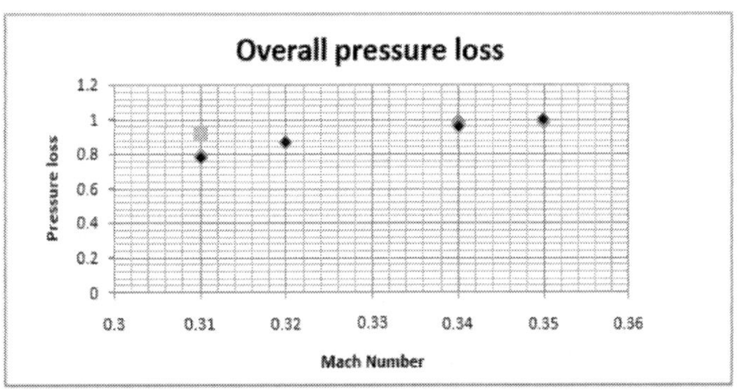

Figure 28: Pressure variation in overall Combustor (NUMECA)

6. Conclusion

Numerical prediction of flow, combustion and combustion processes are carried out inside the gas turbine combustor under cold flow and hot flow conditions. Emphasis is put on the evaluation of temperature distribution in the dilution zone and pressure loss at the diffuser. The following conclusions are drawn from this study.

- ✓ The flow decelerates smoothly along the diffuser passage. There is flow separation at the end of diffuser leading to a small recirculation zone created by the swirler increasing turbulence.
- ✓ The flow field analysis reveals that the air from the compressor enters into the diffuser and then flows through the swriler, heat shield and primary holes, cooling rings and through the dilution slots and holes and finally to the turbine.
- ✓ The diffuser reduces the velocity. The dome and the swriler creates recirculation zone upstream of the combustor, primary zone.
- ✓ The flow decelerates in the diffuser from a Mach number of about 0.3to 0.25
- ✓ The pressure loss in cold flow condition is around 7% when measured from inlet to core outlet of the combustor. This pressure loss can also occur due to extra turbulence created by the dome and the swriler.
- ✓ The pressure loss in outer annulus is 3% and at inner annulus is of 2.8%
- ✓ The hot flow analysis reveals that the combustion increases the pressure loss from 3% at the primary zone to 9% at the inlet to the core outlet during mass flow condition (case2).
- ✓ The maximum temperature on liner surface for both cases is found to be within the service temperature of material selected as the temperature must also be suitable for turbine blades and is of about 812K.
- ✓ Therefore, an additional volume of air can be used to cool the combustion mixture at the dilution zone to keep the metal temperature within the limit.

References

[1] F.N.Egolfopolous and P.Dagut "Combustion and Flames" is published to have high quality work from experimental, theoretical and computational investigation on the phenomena and closely allied matters.
[2] Marc R.J. Charest, "Design methodology for lean premixed pre-vaporized Can combustor", Ottawa-Carleton Institute, Ottawa, Ontario.pp 103-121 & 162-178, April 2005.
[3] Eric Baudoin, "Large Eddy Simulation of Turbulent Premixed and Partially Premixed Combustion", Lund Institute of Technology, Sweden. pp 10-25 December 2010.
[4] K. V. Chaudhari, D. B. Kulshreshtha, S.A. Channiwala/ International Journal of Engineering Research and Applications (IJERA) ISSN: 2248-9622 www.ijera.com Vol. 2, Issue 6, November-December 2012, pp.1641-1645 1641.
[5] K. Sreenivasarao, " CFD Modeling Of An Aero Gas Turbine Combustor For A Small Gas Turbine Engine", International Journal of EngineeringResearch & Technology Vol. 2 Issue 1, January-2013.
[6] Mohd. Ikhwan Saifulism Bin Roslan, "Computational fluid dynamics (CFD) simulation on gas turbine combustion chamber", University Malaysia Pahang, 28- 43 November 2007.
[7] Arthur H. lefebvre, "Gas turbine combustion", Aerodynamics of combustion chamber, 1999.
[8] Sachin Bhalerao, Dr. A.N.Pawar. "Thermal mapping of a can type gas turbine combustion chamber using CFD" Jawaharlal Darda Institute of Engineering and Technology, Yavatmal India, Government Polytechnic, Amaravati.
[9] F.X. Nicoul, " A study of the influence of a swirled injection on micro scale non- premixed combustion". Fundamental and Applied Energetics Department, ONERA, Palaiseau, France.2009.
[10] S N Singh, V Seshadri, K Saha, K K Vempati, S Bharani "Effect of inlet swirl on the performance of annular diffusers having the same equivalent cone angle" Proceedings of The Institution of Mechanical Engineers Part G-journal of Aerospace Engineering.

[11] P.S. Jeyalaxmi Dr.G.Kalivarathan Research Scholar, CMJ University, Meghalaya, Shillong. IAEME 39 "CFD analysis of flow characteristics in a gas turbine-a viable approach to predict the turbulence". Principal/ PSN Institute of Technology and Science, Tirunelveli, Tamilnadu, Supervisor,CMJ University, Shillong. International Journal of Mechanical Engineering and Technology (IJMET), ISSN 0976 –6340(Print), ISSN 0976 – 6359(Online) Volume 4, Issue 2, March - April (2013).

[12] P.Sravan Kumar1, P.Punna Rao2 " Design and Analysis of Gas Turbine Combustion Chamber " PG Student, Department of Mechanical Engineering, Nimra College of Engineering & Technology. Assistant Professor, Department of Mechanical Engineering, Nimra College of Engineering & Technology, Vijayawada, AP, INDIA.

[13] Syed Alay Hashim "Design and Fabrication of an Annular Combustion Chamber for the Micro Gas Combustion and its control are essential for the existence of an engine" Published in: International Journal of Engineering Research & Technology Website: www.ijert.org Volume/Issue: Vol.2 - Issue 8 (August - 2013) e-ISSN: 2278-0181.

[14] Wang Cheng-jun, "Tuburlence combusion modeling in the Gas Turbine combustor" Third International Conference on Measuring Technology and Mechatronics Automation, 2011.

[15] Jack D.Mattingly "Elements of Gas Turbine Propulsion " Published by Tata McGraw-Hill Publishing Company Limited , 7 West Patel Nagar , New Delhi 110 008 , and printed at Sai Printo Pack Pvt. Ltd., Delhi 110 020

Impact of Baffle on Liquid Structure Connection in Fuel Tank Sloshing

D Rohini[1], R Abinaya[2], D Lokesharun[3], K Karthik[4], V Sovishnuchringth[5], P Sivaprakash[6]

1, 2 Assistant Professor, Bannari Amman Institute of Technology, Erode, Tamilnadu, India
3 Research Associate, PARK Research Centre, PARK Institution, Coimbatore, India
5, 4&6 Student, Bannari Amman Institute of Technology, Erode, Tamilnadu, India

rohinidharmaraj90@gmail.com, abiragul5@gmail.com, lokesharun90@gmail.com, Karthik.ae19@bitsathy.ac.in, sovishnuchringth.ae19@bitsathy.ac.in, sivaprakash.ae19@bitsathy.ac.in

Abstract.

The current age has a desire to do explore in sloshing because of its intriguing applications with regards to different fields, for example, estimation of surface profile, DIC [Digital Image Correlation] and so on. The specialists have accomplished effective statures in sloshing strategy, yet answer for sloshing is a major test. This is a direct result of its exceptionally nonlinear component. This undertaking is managed the examination of sloshing elements in airplane fuel tank and the impact presenting confounds. This undertaking is managed the examination of sloshing elements in airplane fuel tank and the impact presenting confounds. This additionally provides data for assurance in the powerful area of confounds through the extent of the tank to refine execution also limit disposition of the vehicle. This correlation is finished using outcomes from computational examination of both 2D and 3D.

Keywords: Effect of sloshing, Baffle, CFD Analysis 2D & 3D

1. Introduction

Sloshing is the liquid movement within the tank. Slosh dynamic direct on the top of liquid will be based on the liquid properties, condition of the compartment [3]. The limits can impact the quality of the tank which moves and decrease its effect [4], Hydrodynamic load calculation in compartments which moves contains two extraordinary portions which is achieved by the fluid moving with the speed and increases the tank speed. The resulting one is surface and liquid association which is required weight [5]

There are numerous kinds of capacity tanks relying upon the structure, development material, substance, volume, and capacity condition [5]. Fluid stockpiling tanks can be worked by steel or cement. Because of extraordinary harms on steel tank, the solid stockpiling tanks are commonly utilized these days. Fortified Concrete (RC) has been utilized in natural designing structures, for example, water repositories and sewage treatment tanks. Water tanks are these days utilized massively for different applications, for example, stockpiling of drinking water, horticultural cultivating and animals, fire concealment, and numerous different applications.

Fluid structure affiliation happens on account of the coupling of helper curving and tricky fluid stream [4]. It is a two-way coupling of weight and redirection. Its application consolidates airbag showing, fuel tank sloshing, heart valve illustrating, helicopter crash appearances, etc Motivation behind inspecting FSI is that fluid mechanics may impact and be affected by the fundamental mechanics, and the opposite way around [6]. From now on for this circumstance the coupling of the fluids pressure and the development of the structure is essential.

1.1. *Surface representation*
Various methods exist are available for following immiscible interfaces, which can be grouped under thr principle of classifications as per physical and numerical methodology:

- Lagrangian approach or Moving Grid approach (Capturing).
- Eulerian Approach or Fixed grid method (Tracking).
- Combined Approach - Lagrangian and Eulerian.

1.2. Suppression devices
As the size of fluid compartments builds the hydrodynamic powers and minutes become huge especially in the area of reverberation. While trying to keep away from auxiliary disappointment or bothersome powerful conduct, one must acquaint a few methods with smother or to slice the dynamic burdens [4]. We can lessen dynamic burden by utilizing bewilders [6]. Puzzles change the liquid normal recurrence relying upon the shape, size and position. Against slosh perplexes are relied upon to keep the all-out fluid power from surpassing a specific recommended most extreme incentive under all conceivable filling conditions, tank direction and outside excitation[6].

The factor impacting the plan of slosh-concealment are as per the following
- The vehicle's main goal profile and direction.
- The damping prerequisites for a given holder
- The physical qualities of the tank, for example, its calculation, flexible distortion and protection.
- The fluid filling and depleting necessities.
- Properties of the fluid.
- The taking care of, slosh and sway stacks that must be continued by the gadgets.

2. Physical Model

2.1. Effective parameters
The main important parameter used in the simulation that plays viral role has been listed below:
- Tank geometry
- Propellant properties
- Acceleration field
- Effective damping
- Level of propellant in the tank
- Perturbing motion of tank

2.2. Physical model description
The description of the physical model shown in figure 1,3D rectangular tank is given in the table 1.Tank is set up with vertical astound with thickness 0.005m of distinction in stature to decrease the effect of sloshing on tank divider, it depends on the water level inside the tank. Physical model with different measurements are given in figure 1.Variety of tank shape is available like cubic, spherical, rectangular, prismatic and circular , in this project rectangular tank is preferred because a rectangular tank with baffles mounted at the bottom is the common engineering remedy applied to loaded liquid containers.

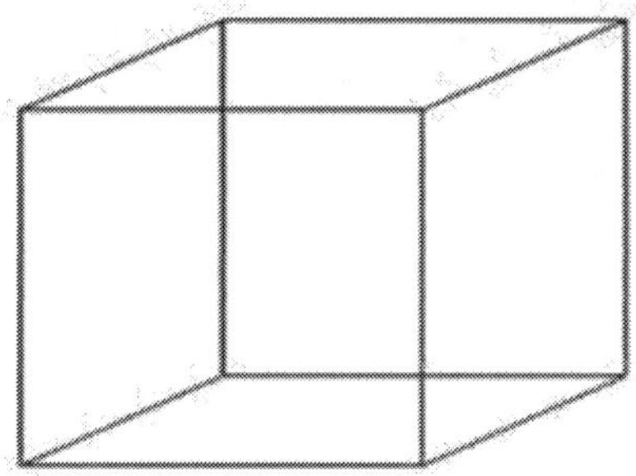

Figure 1. 3 Dimensional Rectangular Tank Model

Table 1. Physical model description – Material properties

Variables	Values
Density – kerosene	780Kg/m^3
Density – Air	1.223 Kg/m^3
Dynamic viscosity	0.0024Kg/m-s
Tank Area(2D)	0.5*0.5 in m^2
Tank area (3D)	0.5*0.5*.0.5 in m^3

2.3. Motion of tank

The simulation has been done for uniform acceleration of 1.3 m/s2. In the negativity x direction and negativity g in y direction for 1 seconds initially and then undisturbed. Simulation works are carried for a minimum of 2 seconds.

2.4. Fuel tank with baffle

Sloshing happens because of movement of liquid inside mostly filled tank [1]. Thus, to decrease the impact of sloshing puzzles are presented in the tank at fitting areas. In the current examination, re-enactment is finished with and without perplex and contemplated the adequacy of various tallness of confound on sloshing powers and weights applied on tank divider. The confound thickness of .005m is remembered for the centre of tank divider vertically corresponding to YZ plane is shown in figure 2. The math for the current examination is demonstrated as follows

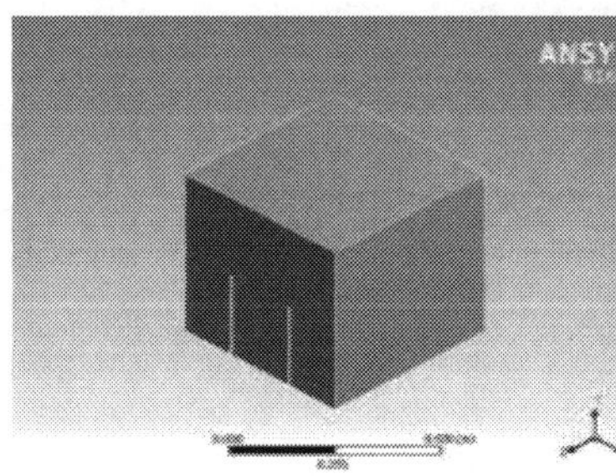

Figure 2. 3-D Model with baffle

2.5. Geometry modelling

2-Dimensional rectangular tank of dimension 0.5m×0.5m without baffle and with both horizontal and vertical baffle is drawn. Thickness and height of the vertical baffle is 0.005 m and 0.25 m. Thickness and position of the horizontal baffle is 0.005m and 1) 0.25m, 2) 0.125m from the bottom of the tank. Rectangular tank model created for 2-D Geometry without baffle, 2-D Geometry with single baffle, 2-D Geometry with double baffle, 2-D Geometry with holed baffle, 2-D Geometry with baffle at 50%, 2-D Geometry with baffle at 25%. A 3-Dimensional rectangular tank with dimension 0.5m×0.5m×0.5m without baffle and with baffle are drawn. Thickness and height of the baffle is 0.005m and 0.25m. Model created for 3-D Geometry without baffle, 3-D Geometry with single baffle, 3-D Geometry with double baffle. (Reference diagram is shown in figure 2)

3. Computational Analysis

Computational liquid elements is mostly used to foreseeing liquid stream, compound response, heat move, and related wonders. Numerically solved by the set of governing mathematical equation.

- Mass conservation
- Momentum conservation
- Energy conservation
- Body force effect

Liquid stream and warmth move are administered by essential thing conditions. Thus, expectation of liquid stream should be possible in the accompanying methodologies

- Experimental Approach
- Analytical Approach
- Numerical Approach

The advantages of CFD are it can handle LPDE, jumbled physical marvels can be managed, time appraisal of stream can be procured and we can recreate in the ideal condition. There are no such disadvantages in CFD approach aside from there is a couple of botches in this cycle, for instance, truncation botch, alter, and machine botch, etc.

3.1. Mesh generation

After creation of model, meshing is done in the ANSYS pre-processor. In present cases uniform quadrilateral mesh is generated. Mesh is generated for 2-D Mesh without baffle, 2-D Mesh with single baffle, 2-D Mesh with double baffle, 2-D Mesh with holed baffle, 2-D Mesh with baffle at 50%, 2-D Mesh with baffle at 25%, 3-D Mesh without baffle, 3-D Mesh with single baffle, 3-D Mesh with double baffle. (A reference mesh diagram is shown in figure 3).

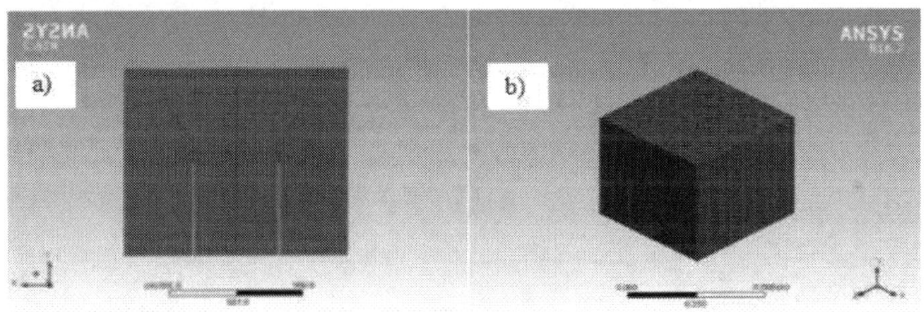

Figure 3. a) 2-D Mesh with double baffle, b) 3-D Model without baffle

3.2. Boundary condition – Fluent analysis

When cross section is being done, work document is next sent out to FLUENT. In present examination 2-D and 3-D two fold exactness familiar solver with equal handling is utilized. The operating conditions are mentioned in the table 2.

- Total time for simulation 2sec and time step size is 0.1 and number of time step is 20.

Table 2. Operating Condition

Boundary conditions	Values
Pressure	1 bar
Acceleration for first one second	X= -6 m/s^2
Acceleration for next second	Y= -9.81 m/s^2
Density	1.225Kg/m^3

4. Results and Discussion

In this part addresses a point by point discussion about the eventual outcomes of the current examination. In this examination a computational multiplication is finished for a fuel tank when is presented to uniform speeding up field. In view of this animating development, sloshing powers is made in the tank. Amusement is finished as tank without baffle and with befuddle at three particular circumstances along the tank.

Results of the analysis is discussed from Case 1 – Case 12, and for each and every cases effect of sloshing, impact of sloshing, sloshing height is showed for different time seconds like 0.2,0.4,0.6,0.8 and 1 .All the effects are shown for 2-D and 3-D with and without baffle. Case 1 represent slosh dynamics in the fuel tank without baffles, case2 represent slosh elements of fuel tank when the baffles are embedded at half of the length of the tank, case 3 represent slosh dynamics of the tank if the baffles are introduced at 30% & 70% with respect to the length of the fuel tank, case 4 represent slosh dynamics of fuel tank if the hole baffle is introduced at 50% with respect to the length of the tank, case 5 represent slosh dynamics of the tank when the horizontal baffle is introduced at 50% of the height of the tank, case 6 represent slosh dynamics of the tank when the horizontal baffle is inserted at 25% of the height of the tank, case 7 represent slosh dynamic of the tank when the comparison of horizontal baffle at 25% and 50% at 1,case 8 represent slosh dynamic of the tank when the comparison of horizontal baffle at 25% and 50% at 1,case 9 represent 3-D slosh dynamics of fuel tank without baffles, case 10 represent 3-D slosh dynamics of the tank when the baffles are inserted at 50% of the length of the tank, case 11 represent 3-D slosh dynamics of fuel tank if the baffles are introduced at 30% & 70% with respect to the length, case 12 represent slosh dynamic of the tank when the comparison of the theoretical, 2-D and 3-D value of sloshing height.

4.1 CASE 1: At the point when the tank without baffle is exposed to speeding up for 1 second at 2-D:
Figure 4 represents the slosh dynamics in the fuel tank without baffles.

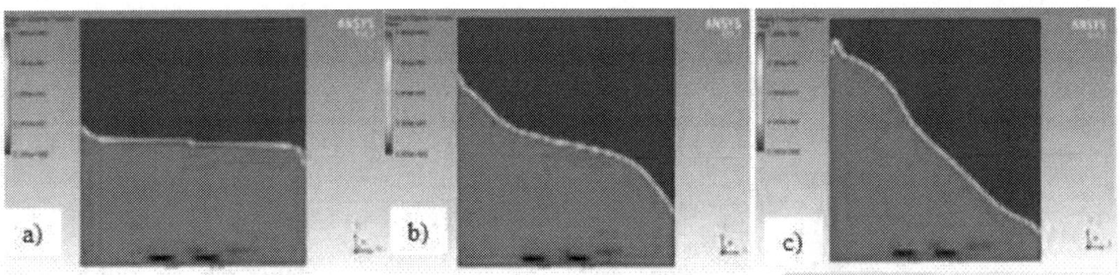

Figure 4. a) At 0.2 seconds, b) At 0.6 seconds, c) At 1.0 second

4.2 CASE 2: At the point when the tank with baffle at half is exposed to speeding up for 1 second at 2-D:
The following figure 5 shows the slosh elements of fuel tank when the baffles are embedded at half of the length of the tank

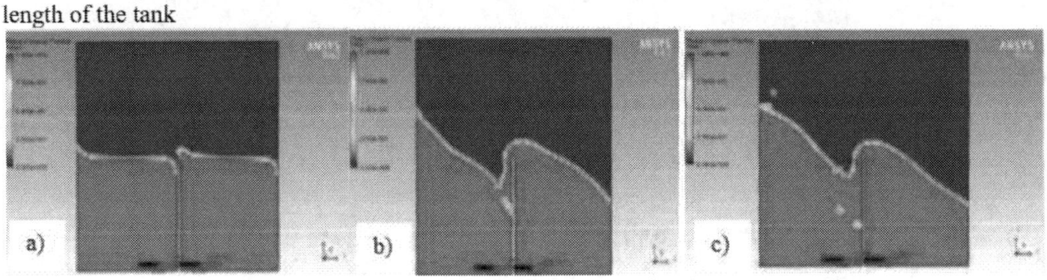

Figure 5. a) At 0.2 seconds, b) At 0.6 seconds, c) At 1.0 second

4.3 CASE 3: If the tank with 2 baffles at 30% & 70% is subjected to acceleration for 1 second at 2-D:
The following figure 6 shows the slosh dynamics of the tank if the baffles are introduced at 30% & 70% with respect to the length of the fuel tank.

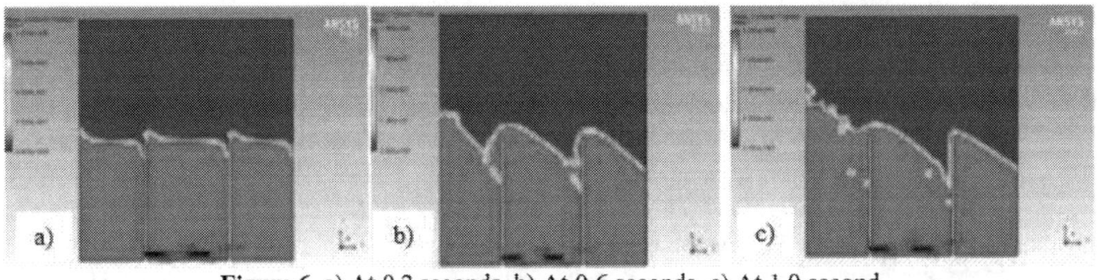

Figure 6. a) At 0.2 seconds, b) At 0.6 seconds, c) At 1.0 second

4.4 CASE 4: If the tank with hole baffles at 50% is subjected to acceleration for 1 second at 2-D:
The following figure7 shows the slosh dynamics of fuel tank if the hole baffle is introduced at 50% with respect to the length of the tank.

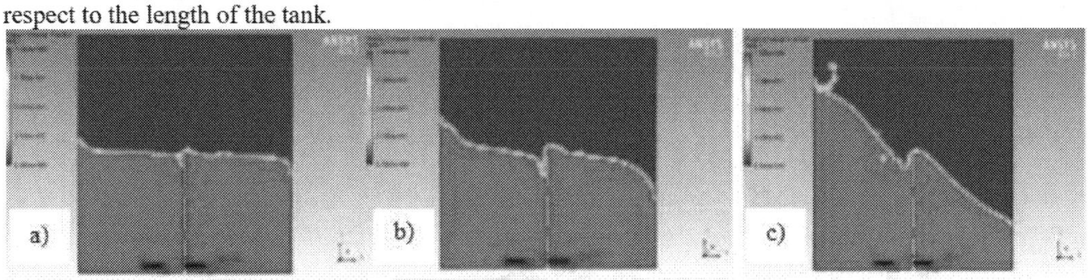

Figure 7. a) At 0.2 seconds, b) At 0.6 seconds, c) At 1.0 second

4.5 CASE 5: If the tank with horizontal baffle at 50% is subjected to acceleration for 1 second at 2-D:
The following figure 8 shows the slosh dynamics of the tank when the horizontal baffle is introduced at 50% of the height of the tank.

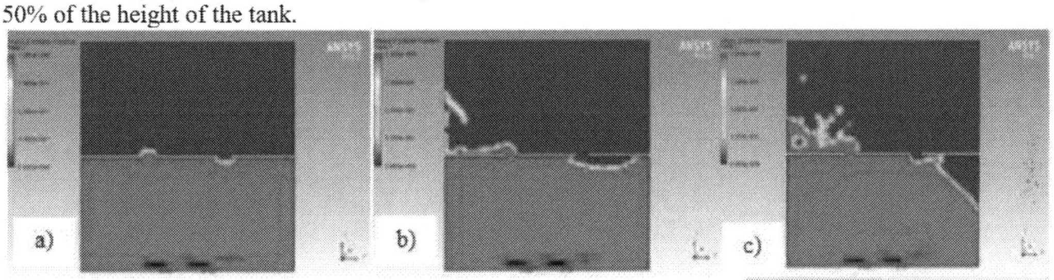

Figure 8. a) At 0.2 seconds, b) At 0.6 Seconds, c) At 1.0 Second

4.6 CASE 6: At the point when the tank with even baffle at 25% is exposed to quickening for 1 second at 2-D:
The following figure 9 shows the slosh dynamics of the tank when the horizontal baffle is inserted at 25% of the height of the tank.

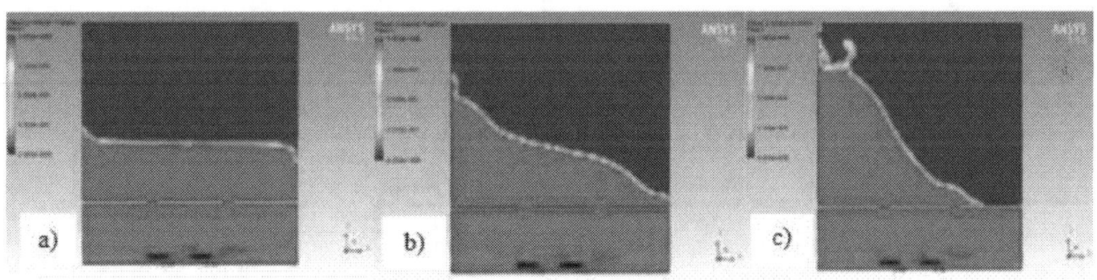

Figure 9. a) At 0.2 seconds, b) At 0.6 Seconds, c) At 1.0 second

4.7 CASE 7: Comparison of tank's sloshing for baffle with hole and without hole for 1 second at 2-D:
The following figure 10 shows the slosh dynamic of the tank when the comparison of baffle with hole and without hole at 1 second.

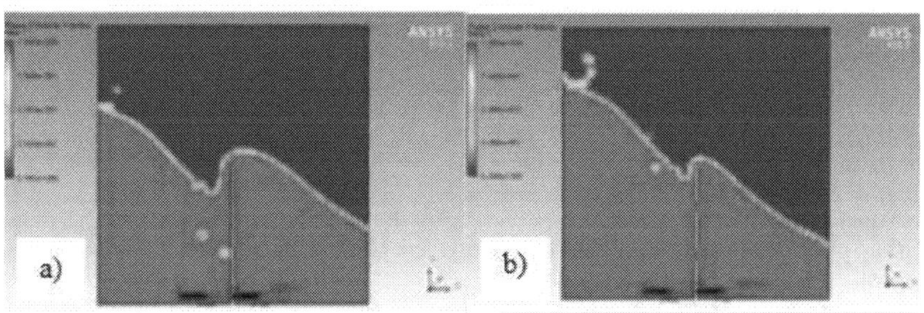

Figure 10. a) Without hole at 1 Second, b) with hole at 1 second

4.8 CASE 8: Comparison of tank's sloshing horizontal baffle at 25% and 50% for 1 second at 2-D:
The following figure 11 shows the slosh dynamic of the tank when the comparison of horizontal baffle at 25% and 50% at 1

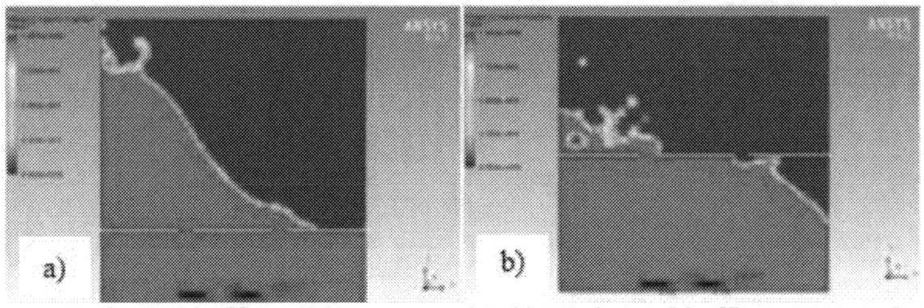

Figure 11. Horizontal baffle at 25% at second 1, b) Horizontal baffle at 50% At 1 Second

4.9 CASE 9: Fuel tank without baffle subjected to acceleration for one second at 3-D:
The following figure 12 diagram shows the slosh dynamics of fuel tank without baffles.

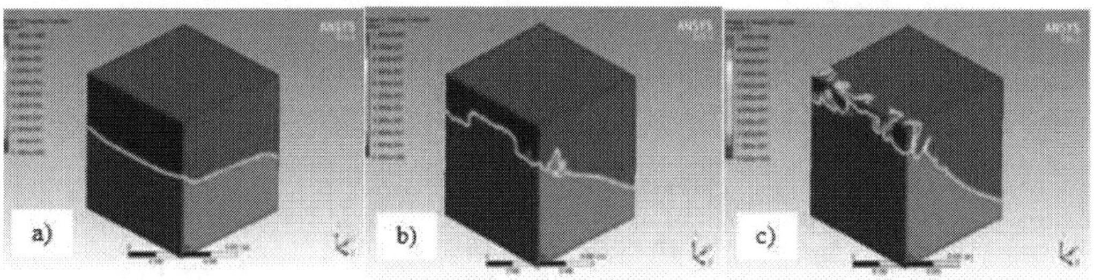

Figure 12. a) At 0.2 seconds, b) At 0.6 seconds, c) At 1.0 second

4.10 CASE 10: Fuel tank with baffle at 50% is subjected to acceleration for 1 second at 3-D:
The following figure 13 shows the slosh dynamics of the tank when the baffles are inserted at 50% of the length of the tank

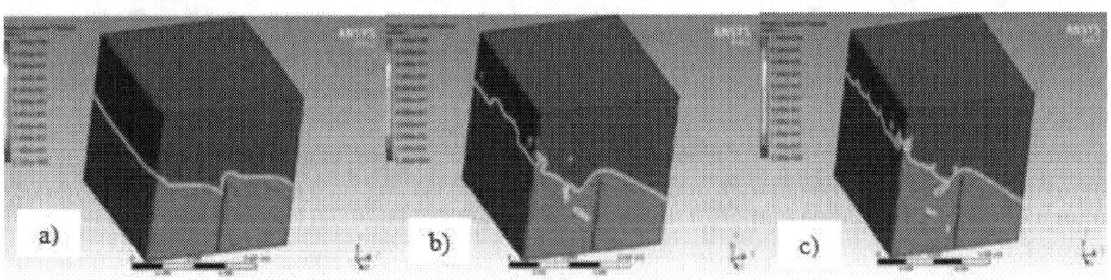

Figure 13. a) At 0.2 seconds, b) At 0.6 seconds, c) At 1.0 second

4.11 CASE 11: Fuel tank with 2 baffles at 30% & 70% is subjected to acceleration for 1 seconds at 3-D:
The following figure 14 shows the slosh dynamics of fuel tank if the baffles are introduced at 30% & 70% with respect to the length.

Figure 14. a) At 0.2 seconds b) At 0.6 seconds, c) At 1.0 second

4.12 CASE 12: Comparison of theoretical, 2-D and 3-D value of sloshing height:
The following figure 15 shows the slosh dynamic of the tank when the comparison of the theoretical, 2-D and 3-D value of sloshing height.

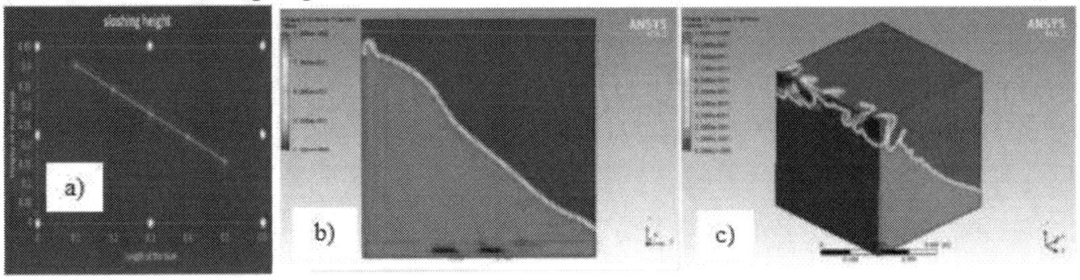

Figure 15. a) Theoretical sloshing height- 0.4021 m, b) Sloshing height- 0.431 m, c) Sloshing height- 0.41229 m

5. Conclusion

Unsteady flow simulation is carried out, the results are validated for different time seconds and compared the theoretical results with the numerical results. From the outcome, it is noted that, the inclusion of puzzles diminishes the impact of sloshing. The vertical confuse hoses the sloshing impact. As the quantity of astounds builds the slosh tallness can be diminished yet the heaviness of tank increments. The slosh tallness can be successfully diminished by putting even confounds precisely on the interface of fluid and air. The indented capacity of the flat confounds lessens when it set inside the interface of fluid and air. When contrasting perplexes and gap and without opening, the weight lessens and fuel limit increments in the latter case. Moreover the variety in slosh tallness is unimportant when looking at those two cases. From the liquid and structure perspective, the permeable perplex kept at level position precisely on the interface is proclaimed as the best model. The investigative 2D and 3D arrangement is corresponded with the hypothetical arrangement and is discovered that the rate mistake is 6.7 %.

References

[1] Mi-An Xue,1, 2 Jinhai Zheng,1, 2 and Pengzhi Lin3, 2012, Numerical Simulation of Sloshing Phenomena in Cubic Tank with Multiple Baffles, Journal of Applied Mathematics, Article ID 245702, 21 pages, doi:10.1155/2012/245702BhattacharyyaB J and Munda 2003Experimental investigation into electrochemical micromachining (EMM) processJournal of Materials Processing Technology 140 pp 287291.

[2] De-Zhi Ning, Wei-Hua Song, Yu-Long Liu, and Bin Teng, 2012, A Boundary Element Investigation of Liquid Sloshing in Coupled Horizontal and Vertical Excitation, Journal of Applied Mathematics, Article ID 340640, 20 pages, doi:10.1155/2012/340640.

[3] M. J. Ketabdari and H. Saghi ,2013, Numerical study on behavior of the trapezoidal storage tank due to liquid sloshing impact, International Journal of Computational Methods, Vol. 10, No. 06, 1350046

[4] J. HallT.C.S. RendallC.B. AllenH. Pee, 2015, A multi-physics computational model of fuel sloshing effectson aeroelastic behaviour, Journal of Fluids and Structures, page no:11-32, http://dx.doi.org/10.1016/j.jfluidstructs.2015.04.003

[5] Hyunjong Kim, Nanjundan Parthasarathy, Yoon-Hwan Choi and Yeon-Won Lee, 2018, Reduction of sloshing effects in a rectangular tank through an air-trapping mechanism – a numerical study, Journal of Mechanical Science and Technology, page no: 1049~1056, DOI 10.1007/s12206-018-0206-9

[6] Zhan Liu, Yuyang Feng, Jie Cui, Gang Lei and Yanzhong Li, 2019, Effect of Excitation Types on Sloshing Dynamic Characteristics in a Cryogenic Liquid Oxygen Tank, American Society of Civil Engineers, Page no : 04019096-1 - 04019096-12, DOI: 10.1061/(ASCE)AS.1943-5525.0001094

IOP Conf. Series: Materials Science and Engineering **995** (2020) 012045 doi:10.1088/1757-899X/995/1/012045

Experimental Analysis on Water Absorption Behaviour of Natural Composite Materials

D Lakshmanan[1], R Naveen[2],P Saravanan[3], D Nivitha[4], R R Mathi Vathana[5]

1, 2Assistant Professor, Bannari Amman Institute of College of Technology, Department of Aeronautical Engineering, Sathyamangalam, Tamilnadu, India.
3 Professor, Mahendra College of Engineering, Department of Mechanical Engineering, Namakkal, Tamilnadu, India.
4, 5Scholar, Bannari Amman Institute of College of Technology, Department of Aeronautical Engineering, Sathyamangalam, Tamilnadu, India

lakshmanand@bitsathy.ac.in, naveenr@bitsathy.ac.in,saro.aero@gmail.com, nivitha.ae17@bitsathy.ac.in

Abstract. This research work deals the fabrication process of naturally available composite fibers with appropriate proportions. Though Natural fiber based composite materials are occupies the suitable position in the Aircraft Manufacturing process. Compositions of cellulose and lignin in the fibres are the low influencing alternative for the other form of composites and that comes under the category of Natural Fibre Composites. In order to strengthen the Natural Composite Materials the fibre reinforced composite materials are basically reduce the percentage of volume comparatively to the natural Composites very high volume fraction especially to provide strength and stiffness. To enhance the materials quality of Natural Composite Materials the water absorption as well as the tensile analysis has been experimented with the combinations of coir and hemp natural composite. The surface treatment process said to be Alkaline was applied to the combinations of coir and hemp. Hand layup method has been used to fabricate the expected laminates with proper thickness and dimensions. Finally the fabricated samples were subjected to water absorption test through which the mechanical property has been observed. During this process coir and hemp were proportionally mixed in the ratio of 0:100, 100:0, 25:75, 50:50 and 75:25. During the heat treatment process, the simplest equipment Hot Air oven has been used for maintain the desired temperature to dry the samples.

Keywords: Natural Composites, Coir, Hemp, Water Absorption

1. Introduction

Animals and Plants are the major sources for making natural composite fibres. These sources are readily available as the renewable one. Recent days the Automobile and Aerospace Industries have been effectively utilizing the composite materials for making the structural components. According to the materials manufacturing process, numbers of industries are being involved in producing different formats of structural components for variety of applications. Considering the factor of safety and light weight, Aluminum is always one of the best for making Aerospace related structural parts. Recent days the uses of composite materials are increased among the Aerospace Industries. Because of their low density as well as the very good thermal insulating properties, the Natural composite Materials are low weight and high strength to carry more loads. So that the Aerospace Industries are developing their structural components with Natural Fibre based materials especially aims to cost reduction purposes. It has been observed that the combinations of Hemp and Coir compositions are good enough to withstand more mechanical strength. Coir is a kind of coconut based natural fiber which is taken out from the outer layer of the coconut husk. The coir fibers are always more resistance to water, corrosion resistance property is more unless it is processed with sea water. Most Probably the Fresh water is used for making the Coir in brown colour, then the mixer of sea water and fresh water are used for making the coir in white colour. While the Hemp is predominantly used for producing the daily life products such as textiles, shoes, paper, bio fuel, plastics and insulating materials.

With respect to the phenomenon such as biodegradable, recyclable and sustainable, the hemp based fiber materials has been captured the market with attractive profits. And the usages also increased among the public in recent years. Some Plant which has stem and are used for making Hemp with very high strength and stiffness which is essential requirements for the reinforced composite materials. However the mechanical and structural properties are in variability for Hemp based natural composite materials.

If the Hemp composite materials are mixed with materials such as thermoset and thermoplastics, then then it can gain more or good mechanical properties with efficient structural stability. There are many surface treatment process were applied to the Hemp fiber matrix to make the internal structure of hemp with very strong bonding medium. It has been observed that the basic physical properties of coir consists of 30% of breaking elongation with the diameter of 0.1 to 1.5 mm as well as the Modules of rigidity about 1.8924 dyne/cm2 and are swelling the water with the diameter of 5%.

The basic treatment commonly called as Alkaline Process a kind of chemical treatment has been generally applied to the natural fibres. During this treatment process, the Sodium Hydroxide (NaOH) in a concentrated form might be used approximately at given temperature over the period of time. Naturally the Cellulose materials has the effective component of stiffness. So that the Alkaline processed Natural fibers exhibits very high mechanical properties and improves the stiffness of matrix components. Sometimes Alkaline process has been applied prior to water absorption test for special case of Natural fibers such as hemp and coir.

2. Experimental Details

During the preliminary investigation process, it has been decided to choose coir and hemp as the reinforcement materials in order to make the smooth fabrication of laminates. The company called Covai Seenu, Pvt Ltd supplied the special hardener HY951 and Epoxy resin (LY556) to make the perfect laminates. Also the high quality natural composites such as Coir and Hemp were taken from the Sathyamangalam Tiger Reserve Forest. The mixing ratio of resin and hardener were given to 20:1. While 20 ml of epoxy resin and 1 ml of Hardener for mixing the selected natural composites. The test specimens are prepared as per the ATSM D570 standards.

3. Chemical Treatment

According to the investigations, the chemically treated natural fiber composite materials were utilized for further industrial applications. While conducting the chemical treatment process, the Hemp and coir were mixed with proper proportions so that the expected level of laminates may encounter the effective mechanical properties. But the Major disadvantage is the poor compatibility between natural fiber elements and matrix element which are low rate of moisture absorption properties. In order to make the effective mechanical surface properties of natural fibers, Alkaline based chemical treatments were performed with experimental set-up.

3.1. Hand Layup Method

One of the low cost investment based composite manufacturing method has used in this paper to produce coir-hemp laminated sections. Here it has been produced with Hand layup method for the fabrication of Natural composites. During the beginning of Fabrication process, the fibers are prepared a woven type, and are physically inserted in the midland also the special brush has been used in order to apply the resin matrix on the reinforcing matrix component.

Thereafter the Hand roller has been utilized for rolling process to wet the proportion of composites and also ensured the uniformity of natural fiber elements as well as resins are in the expected level of thickness with proper distributions. Then the final product of laminates has been subjected to curing process with Hot Air Oven at the temperature of 90 Degree Celsius for 120 Minutes. Overall this experimental process were divide into as such as Preparation of Molding, Applying the gel coat, Hand layup and Curing action. Hand layup method has been effectively used for making marine and aerospace components based on the natural fiber reinforcement process.

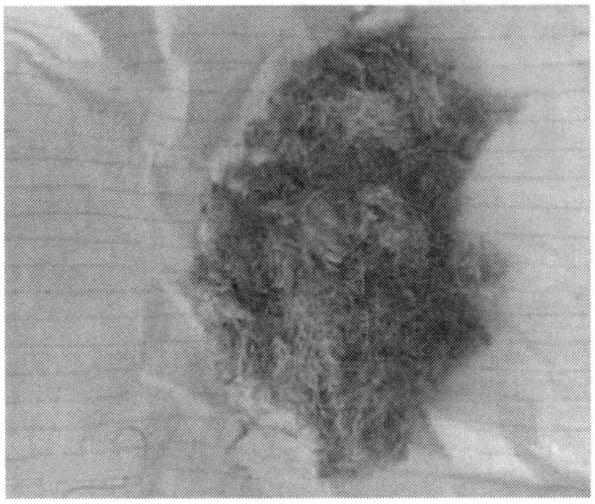

Figure 1: After alkaline treatment of hemp and Coir

4. Water Absorption Test

After the chemical treatment process, the Fabricated Natural fiber based composites were taken out and immersed in the water setup basically at room temperature. It has been shown that the more water observed then the chances for increasing the mechanical property.

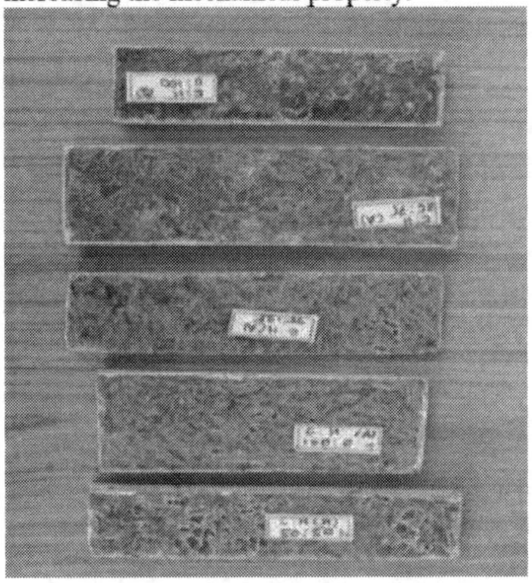

Figure 2: Fabricated Laminates – Hemp and Coir

For the betterment of calculating the level of water observed the laminates, the weight difference between the dry and wet samples were found out with digital meter and also the values are taken for the further process.

Figure 3: Laminates under Water Absorption Test

Due to the chemical structural arrangements of laminated composite materials, micro molecules are attracted by the water elements that creates the strong polymer chain bonding structure. The water absorption test were conducted for the estimated time periods. 30min, 60 Min, 120 Min, 240 min and 600 min.

5. Result And Discussion

The combination of coir and hemp shown their natural characteristics in which the combinations are test with different intervals and are projected in the following Table.5.A.

Table 1 Water absorption of composite after alkaline (%)

Time (Min)	H:C(100:0)	H:C(0:100)	H:C (50:50)	H:C(25:75)	H:C (75:25)
10	.265	.652	.589	.698	.364
30	.479	.85	.796	.895	.558
90	.898	1.23	.982	1.23	.789
120	1.63	1.96	1.256	1.56	1.64
24	1.098	2.25	1.36	1.89	1.54
360	1	2	1.28	2.04	1
600	1.054	1.78	1.08	1.16	1.34

It has been observed from the experimental data, as time increases the water absorptions of the laminates were also increased .The laminates are fabricated as per the proportions mention in the table 5.2.A.

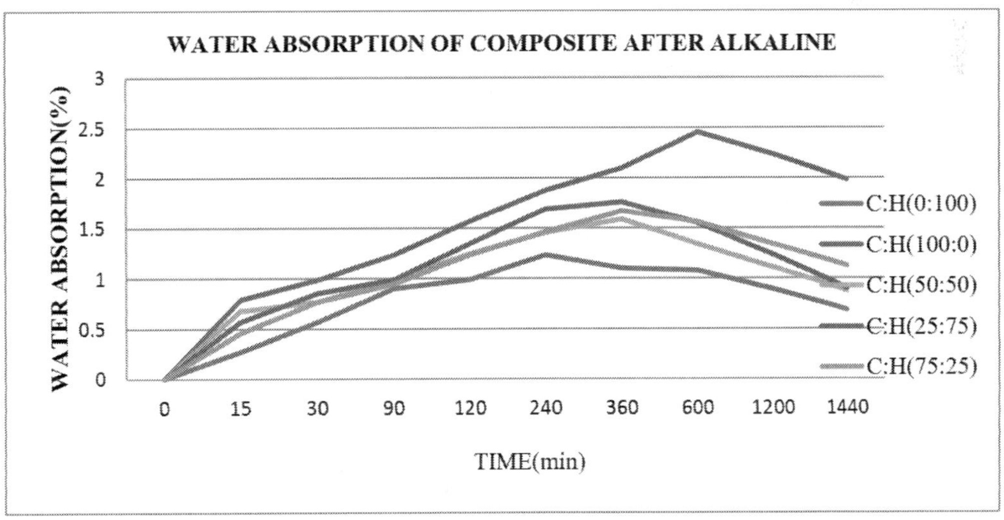

Figure 4: Graphical Representation of Water Absorption Natural Fiber: Hemp and Coir

The filler element Basalt could be added during the surface treatment process according to the dimensions of the laminates structure so that the resultant mechanical behaviors have been enhanced to the maximum levels. The improved water absorption properties are observed as 69.7%, 75.1%, 77.3%, 98.5% and 85.8% respectively for the combinations such as C:H (0:100, 100:0, 50:50, 25:75, 75:25)

6. Conclusion

It has been observed that the Coir and Hemp based Natural fibers are practically very low cost based materials and the fabrication method used here does not emits CO2. According to the water absorption characteristics of Natural Composites such Coir and Hemp combinations are good enough to observe the water molecules. Overall the fabrication process gives the smooth surface based Coir-Hemp laminates that can be used for making the different components. It has been observed from the mechanical test, the tensile load carrying capacity of wet samples are greater than the Dry laminates. Finally it was concluded that the water absorption test supports the natural composites to increase the mechanical properties. The water intake properties results in the experimentation that exactly improved the qualities of Hemp – Coir combinations of Natural fibers. The hybrid composite 25% Coir and 75% of Hemp have more water absorption properties than the remaining proportions of Coir and Hemp. The percentage of water impact characteristics varies from 69.7% to 98.5% and it is sufficient for making durable mechanical structures. Hence it has observed that the expected fabricated laminates are very close to the experimental results though effectively used for different applications. Improvement of water absorption was more for Hemp and Coir based materials with high mechanical strength compared to the carbon and banana based natural fibers.

Acknowledgment

This research was carried out at the Aircraft structures and Repair Laboratory of Aeronautical Engineering Department Bannari Amman Institute of Technology Sathyamangalam.

References

[1] G Thilagavathi, N Muthukumar S Neela Krishnanan & T.Senthilram Development and Characterization of Pineapple Fibre Nonwovens for Thermal and Sound Insulation Applications Journal of Natural Fibers Vol.17 Pp. 1391-1400 Year 2020
[2] V M Manickavasagam B Vijay Ramnath1 C Elanchezhian1 V Aravinthan1 and A Vignesh, Natural Fibre Composites A Review Published IoP Series International Conference on Materials and Manufacturing Engineering Vol.390 March 2018
[3] C Elanchezhian B Vijaya Ramnath G Ramakrishnan Review onMechanical properties of natural fiber composites Materials today Proceedings Volume 5 Issue 1 Part 1 2018 Pp1785-1790
[4] Pradeep Ranga Dr Sandeep Singha A Review Paper on Natural Fiber Reinforced Composite International Journal of Engineering Research & Technology (IJERT) ISSN 2278-0181 Vol. 3 Issue 2 February 2014
[5] S Kandwal, S Singh Bhupendra kumarProcessing and Characterization of Natural Fiber Reinforced Polymer Composite International Journal of Engineering and Advanced Technology (IJEAT) ISSN 2249 8958 Volume-9 Issue-2 December 2019
[6] S J Eichhom Review Current International Research into Cellulosic Fibers and Composites J. of Mat. Sci. 36 (2001) pp. 2107–2131
[7] JGeorge M S Sreekala and S Thomas A Review on Interface Modification and Characterization of Natural Fiber Reinforced Plastic Composites Poly. Eng. and Science pp. 1471–1485
[8] Sathyanarayana K G Wypych, F Guimares J L Studies on natural fibers of Brazil and green composites Metals Materials and Processes Vol.17 pp. 183–194
[9] Sanjay M R Madhu, P Jawaid, M Characterization and properties of natural fiber polymer composites A comprehensive review Journal of Cleaner Production Vol.172 pp. 566–581 Jan. 2018

Comparative Analysis of the Conventional Heat Treatment and Electro-Pulsing Treatment for the Microstructural Analysis

Shubhayan Mukherjee[1], Rakibul Biswas[2] and Sannik Acharya[3]

1Department of Electronics, West Bengal State University, Barasat, India- 700126
2Department of Physics, BarrackporeRastraguruSurendranath College, Barrackpore, India- 700120
3Department of Physics, Bangabasi College, Kolkata, India- 700009

shubhayan.edu.29@gmail.com

Abstract. Heat Treatment is the most common application in the metallurgical analysis. For alternating some of the properties of the metal like physical and chemical, heat treatment plays a crucial role. Conventional heat treatment is a very common and traditional treatment in the study of material science. To make a sample as hot and cold, conventional treatment is very common but for the microstructural analysis like recrystallization, change in mechanical properties conventional heat treatment is not sufficient to make the precious analysis. Electro-pulsing treatment (EPT) creates the difference from the conventional heat treatment for the microstructural evaluation. In this work, we have reviewed a few of the samples (Eg. Cu strip, Si steel strip, Titanium alloy, etc.) from the experimental data and made a comparison between conventional heat treatment and EPT by analyzing the experimental data. Mainly we have focused on the Tensile strength, Elongation and Yield strength for both of the mentioned heat treatment. This work will be helpful to get a clear comparative view of the heat treatment of the two different methods.

Keywords: Electro-pulsing treatment, Conventional Heat Treatment, Microstructural analysis

1. Introduction

Physical and Chemical properties of the material is the study of the metallurgical analysis. Heat treatment plays the key role of this material analysis study. Conventional heat treatment is the widely used processes in which a piece of metal has been showered through heat from an external heat chamber[1-5]. It has to go several repetitive heating processes [7] that results the enhancing of its mechanical properties [8] like increasing of tensile strength [2-3], yield strength, elongation [6], increasing of ductility, removing of internal stress etc. It is worth to be noted that, it is a continuous flow of heat unlike the Electropulsing treatment (EPT) [4] in which some discrete electric pulses have been carried out through the metal and causes Joule heating effect [11-13].

2. Conventional Heat Treatment

When the temperature difference is created, the heat flows continuously from higher temperature end to lower end through the metal by conduction until the equilibrium arises. The thermal energy flowing per unit area per unit time has been characterized by:

$$Q = -K\, dT/dy \qquad (1)$$

where K= thermal conductivity and dT/dy = temperature gradient. Steady- state condition has been satisfied when dQ/dt=0[22]

Recrystallization is another property that is to be considered. Heat treatment causes deformation in the internal structure of metals, which leads concentration gradient in the atoms. This arises the phenomenon of diffusion and is characterized by the coefficient of diffusion which further leads to the recrystallization in the internal structure. The number of atoms crossing the given surface per unit area per unit time is given by:

$$P = -D\frac{dn}{dy}; \tag{2}$$

where D= coefficient of diffusion and dn/dy = concentration gradient[26]

Several heat treatment methods [18] are used for the improvement of the metals. Some methods include:

Annealing is the process in which the metal has to be heated up to a certain temperature (critical point) and then it goes down to a slow cooling rate. By that, a well- refined microstructure reproduces which improves machinability of the metal. In ferrous alloy, annealing results the formation of pearlite. Annealing often soften the metal and through recrystallization it repairs the defects caused by plastic deformation.

In Normalizing technique, the grain size gains uniformity throughout the metals. It also makes the metals harder and stronger but less ductile than annealing. In this technique, the steel is heated is about 40 degree above its upper critical temperature.

Stress relieving method is used to reduce the internal stresses of metals, in which the metals are heated below the lower critical temperature and then it is cooled indiscriminately. Tempering is used to improve the ductility of metals and thus to reduce the brittleness. In tempering, the steel is heated under the lower critical temperature (often from 205 to 595 degree centigrade) to impart some toughness, though some yield strength is lost.

3. Electropulsing Treatment

Electropulsing treatment (EPT) [9][10], is also known as high density electric current pulse [12] (ECP), is a process of heating of different material. Like conventional heating, Electropulsing not only heats of the material but also it improve the microstructural property [1][19] of the material like, recrystallization[15]grain size of the material ,different strain ,stress ,and elongation[14][16] .Since the pioneering work of Trotsky in 1960's this technique used in different material like cupper ,steel ,and aluminium alloys [20].The term electro pulse means the pulse of high density current with different time periods, which control by charging and discharging of the capacitor[17] using the circuit of electro-pulse system. The temperature of the material is increase due to increasing heat, produced by the electropulsing treatment of the material. Using the given equation, we can measure the temperature of the material,

$$\Delta T = \rho j^2 t\, (C_p d)^{-1} \tag{3}$$

P	= Resistivity of the material
j	= Current density
t	= Time
Cp	=Specific heat of the material
D	= Density of the material
T	=Temperature

4. Comparative Analysis

4.1. For Deformed ZrTiAlV Alloy

We have also studied about another alloy to investigate its electropulsing effect on mechanical property [4]. The chemical formula of that alloy is Zr40Ti5Al4V.The electropulsing heat treatment is first-heating method. After this treatment on the material we can see the improvement of both microstructure of the material and also mechanical property. In case of conventional heat treatment of deformed ZrTiAlV alloy, the phase transition and recrystallization occur. But the improvement of microstructure like grain size is better in case of electropulsing treatment than the conventional heat treatment.

The sample was prepared with 40.2 wt. % Ti, 4.5 wt. %Al, 4.2 wt. % V and remain Zr was melting in a vacuum consumable electrode arc furnace and forged in to around bar with a size of 50 mm. The length of the specimen is 63 mm, height is 15 mm, and width is 2.86 mm. The applying current density of the electro-pulse on the sample with charging volt of 2kV,4 kV,6 kV and 8 kV [1].

The time interval between two pulses of the electropulsing effect is 400 μs. The electropulsing treatment is applied one time on each sample with current density 0.8, 0.93, 1.44, and 1.92 kA/mm^2 respectively.

IOP Conf. Series: Materials Science and Engineering **995** (2020) 012046 doi:10.1088/1757-899X/995/1/012046

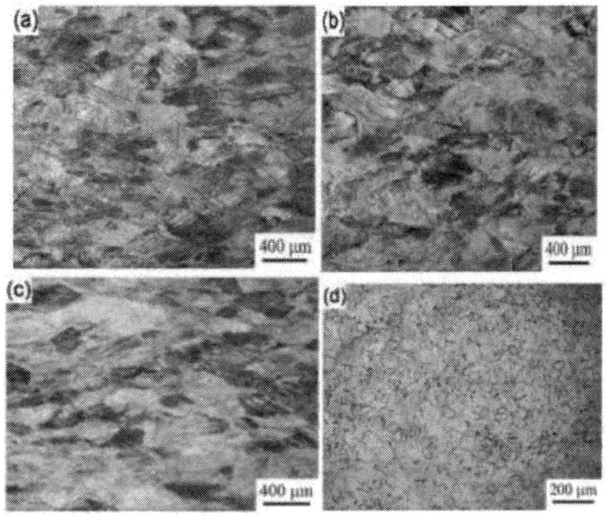

Figure 1: Here, specimen (a) is treated with 2kV,specimen (b) is 4kV,and specimen (C) is 4kV and specimen (d) is 8kV charging.From the above picture, wecan see thatwith increasing the charging voltage to 8k the material is fully recrystallizethe average grain size is 26±5 μm,shown in figure (d).

It is again proved here that the grain size of the specimen during the conventional heat treatment is grater then the EP treatment, which is about 61 μm [23].
We also say that the microstructural change is neglectable when the charging voltage is not above 4kV.The temperature increment due to electro pulsing treatment is calculated by the equation,

$$\Delta T = \rho(C_p d S^2)^{-1} \int_0^{t_p} I^2 dt \qquad (4)$$

Cp = 478 J/kg ˚C is the specific heat
D = 5.20×10^3 kg/m3 the density of sample
P = 1.78×10^{-6} V Ωm the resistivity
t_p =400 μs
I = Js, J= current density, s = surface area.
The calculated temperature of the specimen is 86 ˚C, 284 ˚C, 614 ˚C, 1076 ˚C at 2kV, 4kV, 6kV and 8 kV respectively.

Table 1: Mechanical property [1]

Specimens	σ0.2 (Yield strength) (MPa)	σUTS(Tensile strength) (MPa)	∂ (%)	Φ (%)
ST	739	922	21.0	22.1
Cold-rolle	1110	1280	7.2	10.7
Charging volt 2kV	989	1231	7.9	12.5
Charging volt 4kV	1050	1263	5.3	9.6
Charging volt 6kV	1164	1337	2.0	1.9
Charging volt 8kV	817	959	22.0	35.2

4.2. For TC4 titanium alloy sheet

There is another useful material Titanium alloy which also annualized here. Titanium alloy (TC4) is mainly used in aeronautics and astronautics application for its good mechanical properties [21]. It has also few limitations like lower thermal conductivity, causing higher- internal stress and forming cracks [4]. To improve the mechanical property and its microstructure we use high density electro pulsing treatment. The sample was prepared along the rolling direction from the TC4 sheet. The length of the sample was 17 mm, width 4 mm and thickness were 1.4 mm [4]. The samples were divided in three parts sample A, sample B and sample C. Sample A was untreated by electro pulsing and it use to compare the influence of

IOP Conf. Series: Materials Science and Engineering **995** (2020) 012046 doi:10.1088/1757-899X/995/1/012046

electro pulsing. Sample A and B were treated by electro pulsing. The schematic diagram of sample is shown in Figure 2.

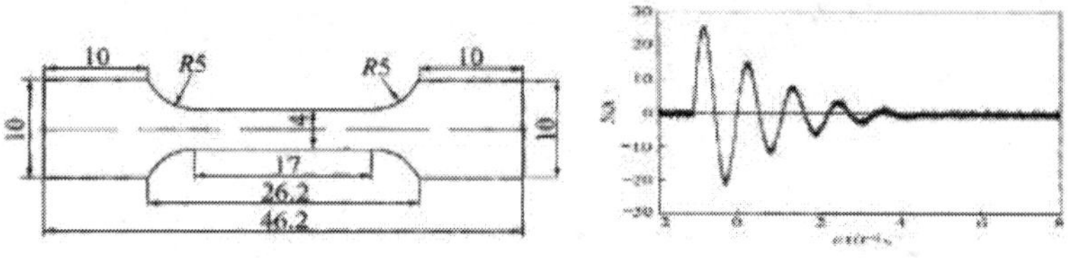

Figure 2: Specimen and Waveform

The wave form of the electro pulsing treatment isshown right side.The value of the period of the pulse is same (110 μs)for both the samples but the value of the currentdensity is different. The value of current density for sample B was J = 5.26 KA/mm^{2}and for sample C J = 5.09 KA/mm^2.The experimental result is shown blow Table 2 [4]. Figure 3 represents the stress-strain curves.

Table 2: Experimental Result

Specimen	State	Tensile strength σb / MPa	Yield strength, σ0.2/MPa	Total elongation %	Uniform elongation %	Yield to tensile ratio	σ0.2/E
A	Annealed	1033	936	15.6	10.0	0.91	0.0086
B	Pre-EPT	1026	769	17.90	13.0	0.75	0.0071
C	Inter-EPT	947	750	23.18	18.01	0.79	0.0069

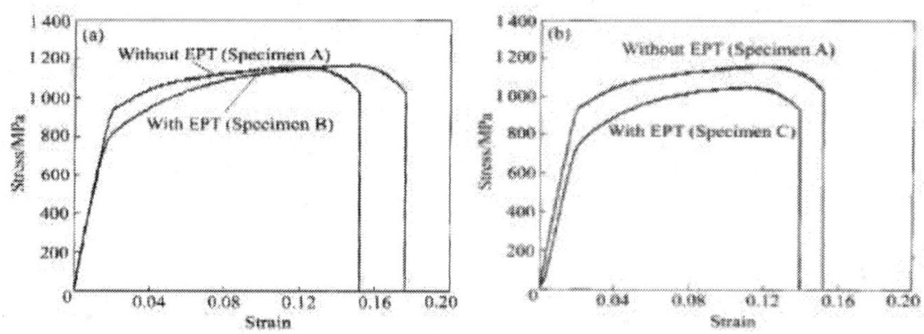

Figure 3: The distribution curve between true-stress vs Strain

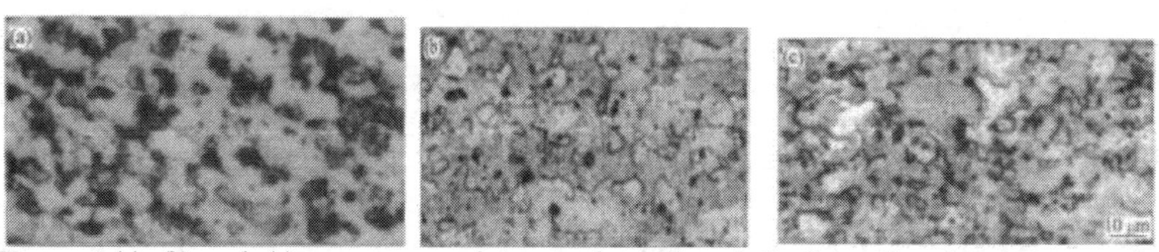

Figure 4:(a) (b) and (c) are correspond to sample A B and C. From the picture we see that afterElectropulsing treatment the microstructure of the sample have change significantly. The recrystallization occurs in the sheet. And the grain growth was not found.

The possible reasons of electro pulsing effect are the effect of electron wind, joule heating effect and the mosaic thermal action, and the thermal compressive stress [4]. The temperature raising formula of the sample due to joule heating effect is given by

$$\Delta T = (C_p \rho S^2)^{-1} \int_0^{t_p} \gamma I^2 dt \qquad (5)$$

I	= Amplitude of currant
T	= Time duration of pulsing
S	= Cross-sectional area of the sample
γ	= Electrical resistivity of the material (1.8 x 10 ^-6 Ωm)
ρ	= Density of the material (4.44 x 10^3 Kg /m^3)
Cp	= Specific heat capacity of the material (750 J/kg. C)

From the above equation the temperature raise of the sample A and C are -728 and 629 ˚c respectively.

4.3. Cu Strip

The specimen is a cold drawn Copper (99.9%) wire whose thickness is 0.60 mm and the width is 3.12 mm has been kept in room temperature [2].Following table represents the tensile strength, elongation and the electrical resistance of the cu strip for the conventional and after the EPT processing (Table 3). In tradition case, the tensile strength is 395 MPa and the elongation is only 3.4% [2].

Table 3: Mechanical and electrical properties of the EPT CU with different frequencies

Sample No.	Electropulsing frequency(Hz)	Tensile strength(MPa)	Elongation (%)	Electrical resistance (* 10^-8 ohm)
0	(conventional treatment)	395	3.4	1.73
1	200	321	5.1	1.69
2	300	261	25.0	1.68
3	400	254	32.0	1.67
4	500	238	42.6	1.65

The optical microstructure of Cu strips before and after EPT is presented in the following figure. It shows the evident of recrystallization microstructure in which grains grew to 20 μm (Figure 4).

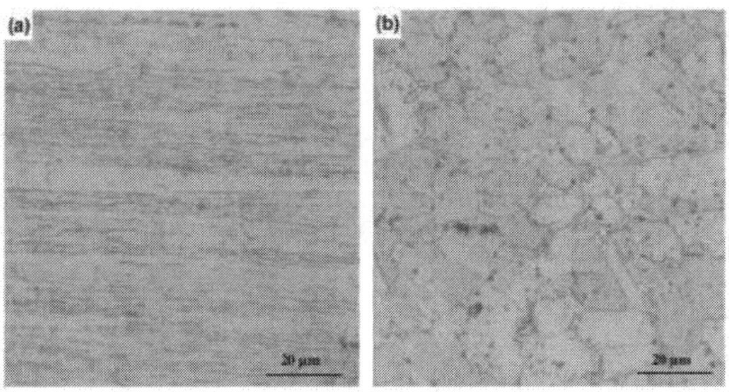

Figure 5: The microstructure of the Cu strip (a) cold rolling (b) Treated by EP with the frequency of 500 Hz [2]

EBSD orientation map of Cu strips after EPT is shown in the following figure (a)-(d). It seems the average grain sizes increases from 5 to 18 μm as frequency increases [2] (Figure 5).

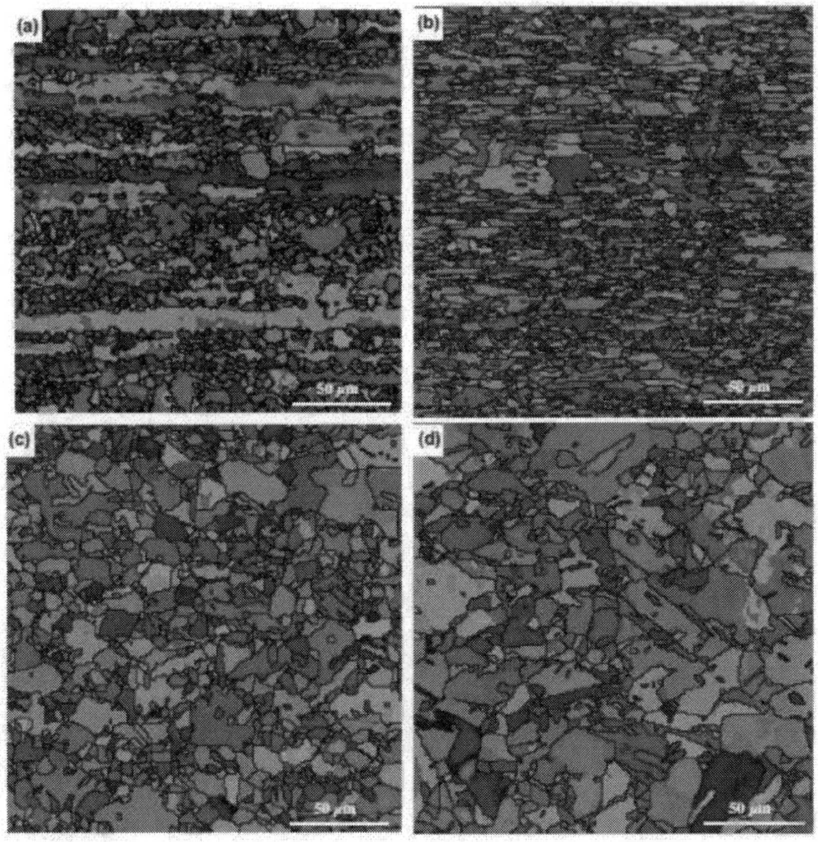

Figure 6:EBSD orientation map of Cu strip annealed by electropulsing: a)200 Hz, b)300 Hz, c)400Hz, d)500 Hz

4.4. Silicon Steel Strips

The material composes with Al(0.024 wt.%), N(0.026 wt.%),Cu(0.10 wt.%), Si(3.0 wt.%),Fe in balance. The width of the alloy strip is 5 mm and the thickness is 0.3 mm which has been produced after a final reduction of 23%of cold rolling [3] (Table 4).

Table 4: Variation in the tensile stress & elongation with the temperature of both EPT & THT

Temperature (°C)	UTS by THT/MPa	UTS by EPT/MPa	Elongation by THT (%)	Elongation by EPT (%)
420	672	668	7	6.5
480	613	618	13	16
520	548	562	17	20
560	519.5	532	22	26
630	512	520	24	32.5
720	508	512	27.5	26
800	498	487	23	22

Following figure revealed the relation between temperature and mechanical properties of both the THT and EPT specimens. After EPT, the dislocation accumulation developed, especially at the low angle grain boundaries, so dislocation density has increased which causes further development in the elongation and tensile strength [3].

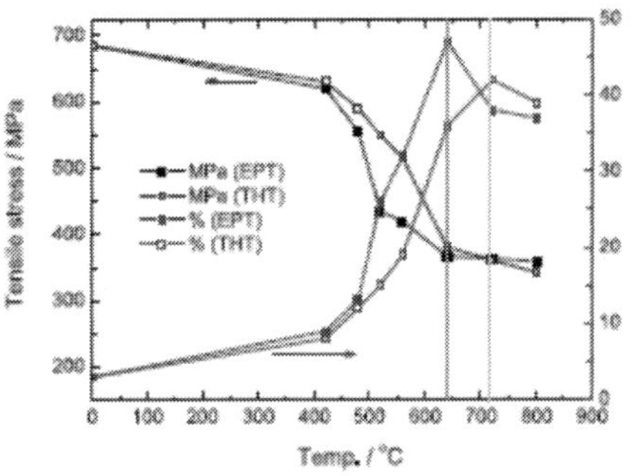

Figure 7: Relationship of elongation and tensile strength of both THT and EPT with temperature

5. Conclusion

After this comparative study of the various samples from the various paper, we have mainly focused to the heat treat process and tried to find out the physical parameters. The collective physical parameters from the different experimental works we listed out few of the parameters to make a constructive analysis to study the defect mechanism of the materials. In this review work we are mainly focused on the ZrTiAlV Alloy, TC4 titanium alloy, Cu Strip and Silicon Steel Strip. From the above review it may concluded that, for the Electro pulsing treatment the tensile strain for the Cu strip &Si strip has decreased and the elongation increased. In terms of grain size, for the first two samples the grain size has decreased. Grain mainly signify the defect mechanism of the material for the micro structural analysis.

References

[1] Yang ZN Jiang F Wang XB 2019 Effect of Electropulsing Treatment on Microstructure and Mechanical Properties of a Deformed ZrTiAlV Alloy Materials (Basel) DOI:10.3390/ma12213560
[2] RF Zhu JN Liu GY Tang SQ Shi MW Fu 2012 Properties, microstructure and texture evolution of cold rolled Cu strips under Electropulsing treatment Journal of Alloys and Compounds
[3] Guoliang Hu Yaohua Zhu Guoyi Tang ChanhungShekJianan Liu 2011 Effect of Electropulsing on Recrystallization and Mechanical Properties of Silicon Steel Strips Journal of Material Science & Technology
[4] SONG Hui WANG Zhong-jin GAO Tie-jun 2004 Effect of high-density Electropulsing treatment on formability of TC4 titanium alloy sheet
[5] Yizhou Zhou SuhongXioaJingdongGuo 2004 Recrystallized microstructure in cold working brass produced by electropulsing treatment Materials Letter
[6] M Gao G H He F Yang J D Guo Z X Yuan B L Zhou 2002 Effect of electric current pulse on tensile strength and elongation of casting ZA27 alloy Material Science & Engineering: A
[7] M H Park, A Shibata, N Tsuji "Grain refinement of 2Mn-0.1C steel by repetitive heat treatment and recrystallization". August 2015. IOP Conference series Materials Science and Engineering 89(1). DOI: 10.1088/1757-899X/89/1/012041
[8] Yanbin Jiang, Lei Guan, Guoyi Tang, Zhihao Zhang "Improved mechanical properties of Mg-9Al-1Zn alloy by the combination of aging, cold-rolling and electropulsing treatment". Journal of Alloys and Compounds. 25 March 2015
[9] Lei Guan, Guoyi Tang, Yanbin Jiang, Paul K Chu "Texture evolution in cold-rolled AZ31 magnesium alloy during electropusing treatment". Journal of Alloys and Compounds. 13 November 2009
[10] Yanbin Jiang, Guoyi Tang, Lei Guan, Shaonan Wang, et al "Effect of electropulsing treatment on solid solution behavior of an aged Mg alloy AZ61 strip". October 2008. Journal of Materials Research 23(10):2685-2691. DOI:10.1557/jmr.2008.0328
[11] ReiyuChein, Yeong Chin Yang, Yushan Lin "Estimation of Joule heating effect on temperature and pressure distribution in electrokinetic-driven microchannel flows". February 2006. Electrophoresis 27(3):640-9. DOI:10.1002/elps.200500314
[12] W J Lu, R S Qin "Stability of martensite with pulsed electric current in dual phase steels". Material Science and Engineering: A. 20 November 2016
[13] Xin ZHANG, Hongwei LI, Mei ZHAN "Mechanism for the macro and micro behaviors of the Ni-

based superalloy during electrically-assisted tension: local Joule heating effect". Journal of Alloys and Compounds. 25 April 2018.

[14] Yinying Sheng, Youlu Hua, Xiaojian Wang, Xueyang Zhao, et al "Application of High-Density Electropulsing to improve the performance of Metallic Materials: Mechanisms, Microstructure and Properties". January 2018. Materials 11(2):185. DOI: 10.3390/ma11020185

[15] Yang Liu, Jianfeng Fan, Hua Zhang, Wei Jin, Hongbiao Dong, BingsheXu "Recrystallization and microstructure evolution of the rolled Mg-3Al-1Zn alloy strips under electropulsing treatment". Journal of Alloys and Compounds. 15 February 2015

[16] Zhu R F Tang G Y Shi S Q 2013 Effect of electropulsing treatment on the microstructure and super elasticity of TiNi alloy Applied Physics A 111(4). DOI: 10.1007/s00339-012-7342-3

[17] JianyiZHENG Wen HE Runjie SHEN 2010 Development of Electropulsing Treatment Device Based on Capacitor Energy Storage and Discharge Advanced Materials Research 139-149:163-166. DOI: 10.4028/www.scientific.net/AMR.139-141.163

[18] Hasan MF 2016 Analysis of Mechanical Behavior and Microstructural Characteristics Change of ASTM A-36 Steel Applying Various Heat Treatment Journal of Material Science & Engineering 05(02). DOI: 10.4172/2169-0022.1000227

[19] W Zhang M L Sui Y Z Zhou D X Li 2003 Evolution of microstructures in materials induced by electropulsing Micron

[20] Phillip J Noell Jeffrey M Rodelas Zahra N Ghanbari Chris M Laursen 2020 Microstructural modification of additively manufactured metals by electropulsing Additive Manufacturing

[21] Yizhou Zhou JingdongGuo Wei Zhang G H He 2002 Influence of electropulsing on nucleation during phase transformation December Journal of Material Research pp3012-3014. DOI: 10.1557/JMR.2002.0438

[22] Garg S C Bansal R M Ghosh C K (2012). Thermal Physics: Kinetic Theory, Thermodynamics and Statistical Mechanics (2nd edition) McGraw Hill Education (India) Private Limitedpp 2.23,2.26

AUTHOR INDEX